Rudiments of

Ordinary Level

MATHEMATICS

Nji Emmanuel Ndi
GBHS MANKON
Tel: (+237) 676 684 050
Email: manuelndike@gmail.com

Revised and updated Edition

Printed by CreateSpace, an Amazon.com Company

EStore address: www.CreateSpace.com/6354942

Available from Amazon.com, CreateSpace.com, and other retail outlets

Available on Kindle and other retail outlets

Also by the same author:

Complete Ordinary Level Mathematics Passport

Advanced Level Mathematics Key Facts

ISBN-10: 1534755004
ISBN-13: 978-1534755000

DEDICATION

Dedicated to all lovers, learners and users of Mathematics.

TABLE OF CONTENTS

ACKNOWLEDGEMENT

Many thanks go to Mr. Akoko Godfred Amanda of G.B.H.S. Kedjom-Keku, the North West Regional Pedagogic Inspector for Mathematics Mr. Nfor Samuel Ndi who edited the manuscript and gave ample advice, which went a long way to reshape the document. I heartily thank the Former North West Regional Pedagogic Inspector for Mathematics Mr. Nji Samuel Tatah who made a very commendable effort to edit the Mathematics content of the book. I cannot forget the encouragements and advice, which the National inspector of Mathematics Mme Babila Emilia inspired and gave me.

I equally pay much tribute to my students on which this material was tested.

I cannot end here without thanking my wife Nji Irene Nfih and my children who encouraged and supported me in one way or the other during the course of the work.

Many thanks go to the North West Mathematics Pedagogic Office, the Mathematics Teacher's Association (MTA), the Teacher's Resource Centre (TRC), the WAEC and the CGCE Board for allowing me to use their past examination questions directly or indirectly.

Nji Emmanuel Ndi
G.B.H.S.Mankon, Bamenda
North West Region
Cameroon
TEL: (+237) 676 684 050
E-mail: manuelndike@gmail.com

FORWARD

Countries and Examination boards often review their syllabuses to meet up with new challenges in Education. The Cameroon General Certificate of Education Board recently reviewed her Examination syllabuses and will henceforth base their Examinations on these new syllabuses. There was therefore need for a corresponding review of textbooks used in teaching subjects examined at these Examinations. The "Rudiments of Ordinary Level Mathematics" is one of those books written for this purpose. The book is very snappy and straight to the point and has avoided the use of many words which turn to obscure the necessary objectives. This book is equally useful for examinations examined by other Examination bodies. The book has included new topics such as Logic, Networks, and Flow diagrams etc, introduced into the new syllabus for Ordinary level Mathematics. Also one of the potentials which give the book extra strength is the fact that most concepts and skills have been illustrated with examples or diagrams where necessary.

The book is highly recommended as a companion to Mathematics teachers and students offering Ordinary Level Mathematics especially those in the evening schools who might not have enough time to go through very bulky documents.

Nfor Samuel Ndi
Regional Pedagogic Inspector
(RPI) for Mathematics
North West Region
Cameroon

PREFACE

Following the recent trends in life and the corresponding changes in the syllabuses and the evaluation method of many examination Boards, there has been dear need for material which can assist both students and teachers. It is for this reason that this book has been written. The author is aware that many students dread very voluminous books. That is why though he has written a more elaborate book "Complete Ordinary Level Mathematics Passport" which explains concepts, skills and their applications very detailly, he has still sympathized with these students to write this smaller book, though ensuring that almost every objective is treated. It is hoped that the work will go a long way to simplify the task of both teachers and students of this subject especially as the work has been intentionally presented as short notes. Readers who require more detailed treatment and explanations of the material may find "Complete Ordinary Level Mathematics Passport" more valuable.

Most of the questions are selected from past questions of the West African Examination Council and the Cameroon General Certificate of Education. The author has rephraimed a good number of questions sometimes by adapting and restructuring past examination questions from different examination passed questions.

Most concepts and skills are illustrated with examples and diagrams. Each section begins with short notes and examples and end with a good number of multiple choice questions. By carefully going through these notes, many of the questions will be tackled without much ado.

The answers have voluntarily not been given so as to ensure that students really work through the exercises and not memorize answers.

Nji Emmanuel Ndi
G.B.H.S.Mankon, Bamenda
North West Region
Cameroon
TEL: (+237) 676 684 050
E-mail: manuelndike@gmail.com

CHAPTER 1

NUMBERS AND NUMERATION

1.1 NUMBERS AND MANIPULATION

Numbers and Numerals

A **number** is an idea that expresses a quantity.
A **numeral** is a symbol which represents a number.

Example
State whether each of the following is a number or a numeral.
(a) Five oranges (b) 𝍸 (c) 5

Solution
(a) number (b) numeral (c) numeral

The Four Basic Operation

Addition, Sum or Plus (+)

$$\overbrace{\underbrace{5}_{\text{term}}}^{\text{addend}} + \overbrace{\underbrace{7}_{\text{term}}}^{\text{addend}} = \underbrace{12}_{\text{sum}}$$

Subtraction or Difference (–)

$$\overbrace{\underbrace{12}_{\text{term}}}^{\text{minuend}} - \overbrace{\underbrace{7}_{\text{term}}}^{\text{subtrahend}} = \underbrace{5}_{\text{difference}}$$

Multiplication, Times or Product (×)

$$\overbrace{\underbrace{4}_{\text{factor}}}^{\text{multiplicand}} \times \overbrace{\underbrace{6}_{\text{factor}}}^{\text{multiplier}} = \overbrace{\underbrace{24}_{\text{multiple}}}^{\text{product}}$$

Division or Quotient (÷)

$$\overbrace{\underbrace{24}_{\text{dividend}}}^{\text{multiple}} \div \overbrace{\underbrace{4}_{\text{divisor}}}^{\text{factor}} = \overbrace{\underbrace{6}_{\text{quotient}}}^{\text{factor}}$$

PROPERTIES OF OPERATIONS

1. **The identity element Property**

   An **identity element** is a number which leaves another number unchanged when the two numbers are combined.
   The identity element for addition is 0. Example: $0 + 4 = 4 + 0 = 4$.
   The identity element for multiplication is 1. Example: $1 \times 4 = 4 \times 1 = 4$.

2. **The Inverse element Property**

   Two numbers are said to be the inverse of each other if on combining them, the result is the identity element for the operation.

   Under the operation addition, 4 and – 4 are inverses of each other, since $4 + (-4) = (-4) + 4 = 0$ and 0 is the identity element for addition.
   Under the operation multiplication, $\frac{3}{5}$ and $\frac{5}{3}$ are inverses of each other, since $\frac{5}{3} \times \frac{3}{5} = \frac{3}{5} \times \frac{5}{3} = 1$ and 1 is the identity element for multiplication.

   The identity element is self inverse since under the operation of addition, $0 + 0 = 0$ and under the operation of multiplication, $1 \times 1 = 1$.

3. **The Commutative Property**

   (i) **Commutative property of addition:** interchanging the addends does not change the sum. For instance $3 + 8 = 8 + 3 = 11$.

   (ii) **Commutative property of multiplication:** interchanging the multiplicand and the multiplier does not change the product.

   For instance $3 \times 5 = 5 \times 3 = 15$.

4. **The Associative Property**

   (i) **Associative property of addition:** Changing the grouping of the addends does not change the sum.

   For instance $3 + 4+7 = (3 + 4) +7 = 3 + (4+7) = 14$.

   (ii) **Associative property of multiplication:** Changing the grouping of the numbers does not change the product.

   For instance $3 \times 4 \times 2 = 3 \times (4 \times 2) = 3 \times (4 \times 2) = 24$.

   Note that only addition and multiplication exhibit the commutative and associative properties. Subtraction and division do not.

5. The Distributive Property

The distributive property of multiplication over addition and subtraction is the property that *the product of a number and the sum or difference of terms is the same as the sum or difference of the product of the number and the individual terms*. Thus,

(i) $3(4+2) = 3(4) + 3(2) = 12 + 6 = 18$ and $3(4+2) = 3(6) = 18$

(ii) $3(4 - 2) = 3(4) - 3(2) = 12 - 6 = 6$ and $3(4 - 2) = 3(2) = 6$.

6. The Multiplicative Property of Zero

The result of multiplication by 0 is 0.

Applications of the Properties of Numbers

The properties of numbers can be used to simplifying calculations as follows.

(a) Grouping the numbers to form **compatible numbers** (numbers which are easy to manipulate).
(b) Grouping inverse elements to simplify (numerical) expressions.
(c) Breaking numbers into units that can easily be simplified.
(d) Factoring common factors out to simplify (numerical) expressions.

Sequence of Operations

When evaluating expressions the sequence of operation is; **Brackets, Exponents, Of , Division, Multiplication, Addition, Subtraction.** The acronym **BEODMAS** may be useful as a mental aid.

Example

Evaluate $16 + 8 \div 4 - 3 \times 2 + (9 - 3)$

Solution

$16 + 8 \div 4 - 3\times2 + (9 - 3) = 16 + 2 - 6 + 6 = 18$

MULTIPLE CHOICE EXERCISE 1:1

1. The statement that refers to numeral is:
   [A] When counting 7 comes before 8 [B] The sum of 2 and 6 is 8
   [C] In 78, 7 comes before 8 [D] 78 is the sum of 50 and 28
2. The statement that refers to numbers is:
   [A] In 23, 2 comes before 3 [B] When counting, 2 comes before 3
   [C] 2 combined with 3 is either 23 or 32 [D] 23 consist of 2 and 3

3. In $5+0=5$, the property that applies is:
   [A] Commutative property of zero [B] Associative property of zero
   [C] Distributive property of zero [D] Additive identity property of zero
4. In $2\times0+7=7$, the property of 0 that applies is:
   [A] The multiplicative property [B] The additive property
   [C] The distributive property [D] The identity property
5. In $3\times1+2=5$, the property of 1 that applies is:
   [A] The distributive property [B] The additive property
   [C] The multiplicative property [D] The identity property
6. $(6\times2)\times50=6\times(2\times50)$. The property used is:
   [A] The commutative law of multiplication
   [B] The associative law of multiplication
   [C] The distributive law of multiplication
   [D] The multiplicative property of numbers
7. $(73+25)+75=73+(25+75)$. The property applied is:
   [A] The associative law of addition [B] The commutative law of addition
   [C] The distributive law of addition [D] The addition property of numbers
8. In $3\times7=21$, 3 is called:
   [A] the multiplier [B] the multiplicand [C] the minuend [D] the dividend
9. In $8-5=3$, 5 is called:
   [A] the difference [B] the minuend [C] the dividend [D] the subtrahend
10. $8+5(4-2)$ is equal to:
    [A] 12 [B] 50 [C] 18 [D] 22

1.2 NATURAL NUMBERS

The set of natural numbers is denoted by $\mathbb{N}$ where $\mathbb{N}=\{0,1,2,3\ldots\}$.

The Place Value System in Whole Numbers

Digits are the symbols used in a number system, to form numerals. The digits in our number system are 0, 1, 2, 3, 4, 5, 6, 7, 8 and 9. Our number system is a place value system. In a place value system, the value of each digit is equal to the product of the digit and its place value in the numeral. For instance, the value of 3 in 846302 is three hundred and in 8463020 is three thousand.

Place Value of Whole Numbers

Billions			Millions			Thousands			Ones		
Hundred billions	Ten billions	Billions	Hundred millions	Ten millions	Millions	Hundred thousands	Ten thousands	Thousands	Hundreds	Tens	Units
8	6	3	4	0	1	7	4	9	3	0	2

We can use the above table to read and write number names up to hundreds of billions.

Example
Write out 897,543 in words.

Solution
Eight hundred and ninety seven thousand five hundred and forty three.

Example
Write in figures; ninety million three hundred and one thousand and five.
Solution

$$\begin{array}{r} 90{,}000{,}000 \\ 300{,}000 \\ 1{,}000 \\ +\quad 5 \\ \hline \underline{\underline{90{,}301{,}005}} \end{array}$$

Example
What is the value of 6 in the number 356,789,743?

Solution
$6 \times 1{,}000{,}000 = 6{,}000{,}000$ i.e. Six million

Index Notation or *Exponential notation* (*Index Form*)

5^{6} ← exponent or index or power; 5 ← base

5^6 means 'multiply 5 by itself seven times'. i. e. $5^6 = 5 \times 5 \times 5 \times 5 \times 5 \times 5$.

Any real number other than 0 raised to the power 1 is the number.
Any real number other than 0 raised to the power 0 is 1. 0^0 *is meaningless.*

Powers of Ten

To write any power of ten in index form, raise 10 *to a power equivalent to the number of zeros in the power of ten.*

Example
Write in exponential form.
(a) 100,000 (b) 10,000,000 (c) 1,000,000,000

Solution
(a) $100{,}000 = 10^5$ (b) $10{,}000{,}000 = 10^7$ (c) $1{,}000{,}000{,}000 = 10^9$

Multiplication and Division Laws of Indices

To multiply together quantities written to the same base in index form, raise the base to the sum of the exponents.
To divide quantities written to the same base in index form, raise the base to the difference of the exponents.

Example
Evaluate leaving your answer in index form.
(a) $3^4 \times 3^2$ (b) $7^8 \div 7^5$

Solution
(a) $3^4 \times 3^2 = 3^{4+2} = 3^6$ (b) $7^8 \div 7^5 = 7^{8-5} = 7^3$

FACTORS AND MULTIPLES

A multiple of a whole number is the product of the number and any non-zero whole number.
A factor of a whole number is the quotient of the number and any non-zero whole number less than or equal to the number.

Example
Write down the set of all the factors of 36.

Solution
Factors of 36 = {1, 2, 3, 4, 6, 9, 12, 18, 36}

Example
Write down the first ten multiples of 3.

Solution
Multiples of 3 = {3, 6, 9, 12, 18, 9, 21, 24, 27,10}

Divisibility Rules

We can easily find the factors and multiples of a number if we master the following rules. A number is divisible by
2 if its last digit is 0, 2, 4, 6 or 8.
4 if the number formed by the last two digits is divisible by 4.
8 if the number formed by the last three digits is divisible by 8.
3 if the sum of its digits is divisible by 3.
9 if the sum of its digits is divisible by 9.
6 if its last digit is 0, 2, 4, 6, 8 and the sum of its digits is divisible by 3.
12 if the number is divisible by 3 and 4.
5 if its last digit is 5 or 0.
50 if its last two digits are 50 or 00.
25 if its last two digits are 25, 50, 75 or 00.
10 or a power of ten if the number of zeros in the number is at least equal to the number of zeros in the power of 10.
7 if the difference between twice the last digit and the rest of the number is 0 or is divisible by 7.
11 if the difference between the sum of the odd digits and the sum of the even digits is 0 or is divisible by 11.

Odd and Even Numbers

Even numbers are numbers that, when divided by two, leave no remainder. The first six even numbers are 2, 4, 6, 8, 10 and 12.
Odd numbers are numbers that, when divided by two, leave a remainder of 1. The first six odd numbers are 1, 3, 5, 7, 9, and 11.

Prime and Composite Numbers

A **prime number** is a natural number that has exactly two factors, the number itself and one.
A **composite number** is a number that has more than two factors.

Note that the number 1 is neither a prime number nor a composite number because it has only one factor, only 1 itself.

Prime Factorization

Prime factorization is the process of expressing a composite number as a product of its prime factors.

Example

Express each of the following as a product of prime factors.

(a) 30 (b) 24

Solution

(a) $30 = 2 \times 15 = 2 \times 3 \times 5$ (b) $24 = 2\times12 = 2\times2\times6 = 2\times2\times2\times3 = 2^3 \times 3$

Techniques of Prime Factorization

Example

Write 7290 as a product of prime factors, using

(a) the peeling (or repeated division) method (b) the factor tree method.

Solution

(a)

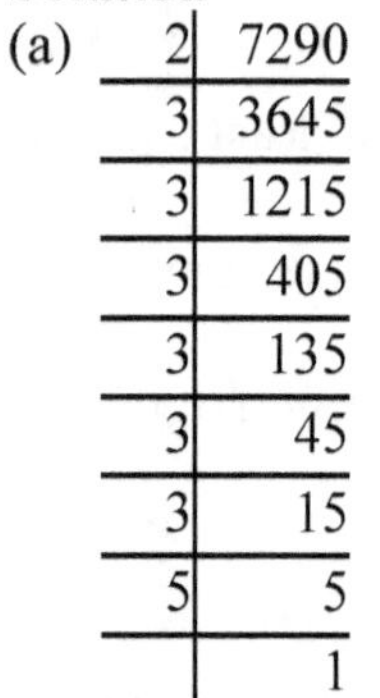

(b)

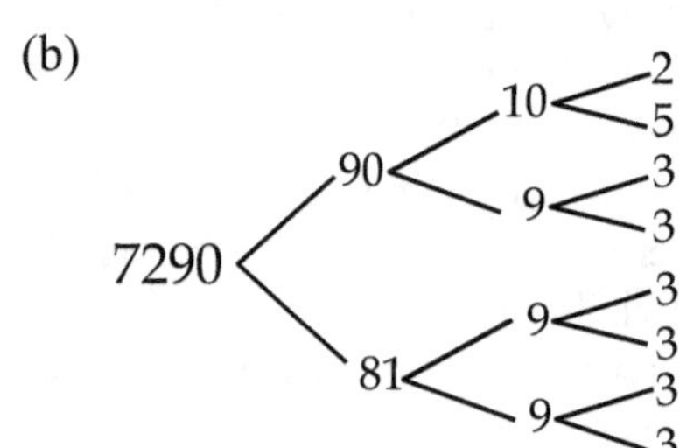

Using either method, $7290 = 2 \times 3^6 \times 5$

Common Factors

The common factors of two or more whole numbers are the factors of these numbers that belong to both or all the numbers.

Example

List the common factors of 18 and 24.

Solution

Factors of 18 = {1, 2, 3, 6, 9, 18}

Factors of 24 = {1, 2, 3, 4, 6, 8, 12, 24}

Therefore, the common factors of 18 and 24 = {1, 2, 3, 6}

Highest Common Factor

The highest common factor (HCF) also called the greatest common divisor (GCD) of two or more given numbers is the largest of the common factors of the numbers.

Example

Find the HCF of 42 and 48.

Solution

Factors of 42 = {1, 2, 3, 6, 7, 14, 21, 42}
Factors of 48 = {1, 2, 3, 4, 6, 8, 12, 16, 24, 48}
Common factors of 42 and 48 = {1, 2, 3, 6}
Therefore, the HCF of 42 and 48 = 6

Common Multiples

The common multiples of two or more whole numbers are the multiples that belong to both or all the numbers.

Example

List the first 4 common multiples of 2 and 3.

Solution

Let M_2 = multiples of 2 and M_3 = multiples of 3
Then,
M_2 ={2,4,6,8,10,12,14,16,18,20,22,24,26...}
M_3 ={3, 6, 9,12,15,18,21,24,27...}
∴ First 4 common multiples of 2 and 3 = {6, 12, 18, 24}

Least Common Multiple (LCM)

The Least common multiple (LCM) of two or more members is the smallest of their common multiples. Practically, the LCM is the product of the highest powers of the prime factors of given numbers.

Example

Find the LCM of 3 and 4.

Solution

Multiples of 3 = {3, 6, 9, 12, 15, 18, 21, 24, 27...}
Multiples of 4 = {4, 8, 12, 16, 20, 24, 28...}

Common multiples of 3 and 4 = {12, 24…}
∴ LCM of 3 and 4 = 12

HCF and LCM by Prime Factorization

To find the LCM, by the peeling method, peel the numbers until no two of them have common factors.

HCF = Product of highest powers of common prime factors.
LCM = product of the highest powers of the prime factors.

Example
Find the HCF and LCM of 16, 20 and 24.

Solution

Using the peeling method

2	16	20	24
2	8	10	12
2	4	5	6
	2	5	3

Using the method of product of primes

$16 = 2 \times 2 \times 2 \times 2 = 2^4$
$20 = 2 \times 2 \times 5 = 2^2 \times 5$
$24 = 2 \times 2 \times 2 \times 3 = 2^3 \times 3$

HCF = product of common prime factors $= 2 \times 2 = 4$
LCM = product of the highest powers of the prime factors
$= 4 \times 2 \times 2 \times 3 \times 5 = 240$

Squares of Natural Numbers

The square of a number is the product of the number and itself.
A whole number, which is the square of another whole number, is called a **perfect square**. For instance 49 is a perfect square because $49 = 7^2$.

Square Roots of Natural Numbers

The square root of a given number is a number that when multiplied by itself gives the given number.
To find the square root of a number, peel the number, group the common factors in twos and find the product of the factors selected each from each group of two.

Example
Find (a) $\sqrt{64}$ (b) $\sqrt{900}$

Solution

(a)

$$\begin{array}{l} 2\left\{\begin{array}{l|r} 2 & 64 \\ \hline 2 & 32 \\ \hline \end{array}\right. \\ 2\left\{\begin{array}{l|r} 2 & 16 \\ \hline 2 & 8 \\ \hline \end{array}\right. \\ 2\left\{\begin{array}{l|r} 2 & 4 \\ \hline 2 & 2 \\ \hline \end{array}\right. \\ \quad\;\;\begin{array}{l|r} & 1 \end{array} \end{array}$$

$$\therefore \sqrt{64} = 2 \times 2 \times 2 = 8$$

(b)

$$\begin{array}{l} 2\left\{\begin{array}{l|r} 2 & 900 \\ \hline 2 & 450 \\ \hline \end{array}\right. \\ 3\left\{\begin{array}{l|r} 3 & 225 \\ \hline 3 & 75 \\ \hline \end{array}\right. \\ 5\left\{\begin{array}{l|r} 5 & 25 \\ \hline 5 & 5 \\ \hline \end{array}\right. \\ \quad\;\;\begin{array}{l|r} & 1 \end{array} \end{array}$$

$$\therefore \sqrt{900} = 2 \times 3 \times 5 = 30$$

Cubic Numbers or Cubes

A cube of a number is the product of the numbers number by itself twice.

Example

Evaluate (i) 7^3 (ii) 13^3

Solution

(i) $7^3 = 7 \times 7 \times 7 = 49 \times 7 = 343$ (ii) $13^3 = 13 \times 13 \times 13 = 169 \times 13 = 2197$

Cube Roots of Whole Numbers

The cube root of a given number is a number that, when multiplied by itself twice, gives the given number.

To find the cube root of a number, peel the number, group the common factors in threes and find the product of the factors selected each from each group of three.

Example

Find the cube root of (a) 729 (b) 1000

Solution

(a)

$$\begin{array}{r}
3\left\{\begin{array}{c|r} 3 & 729 \\ \hline 3 & 243 \\ \hline 3 & 81 \\ \hline \end{array}\right. \\
3\left\{\begin{array}{c|r} 3 & 27 \\ \hline 3 & 9 \\ \hline 3 & 3 \\ \hline & 1 \end{array}\right.
\end{array}$$

(b)

$$\begin{array}{r}
2\left\{\begin{array}{c|r} 2 & 1000 \\ \hline 2 & 500 \\ \hline 2 & 250 \\ \hline \end{array}\right. \\
5\left\{\begin{array}{c|r} 5 & 125 \\ \hline 5 & 25 \\ \hline 5 & 5 \\ \hline & 1 \end{array}\right.
\end{array}$$

$\sqrt[3]{729} = 3^{6\div 3} = 3^2 = 3 \times 3 = 9$, $\sqrt[3]{1000} = 2^{3\div 3} \times 5^{3\div 3} = 2 \times 5 = 10$

NUMBER BASES

Number System and their Digits

Our number system is called a **base ten system**, a **denary system** or a **decimal system** because counting is done in groups of ten. Our number system is a place-value system. A place-value system can use any base, one of the symbols must be zero and the number of digits must be equal to the base used.

The following table shows other place value systems and their digits.

Base	Name	Digits Used
Base two	Binary system	0,1
Base three	Ternary or tertiary system	0,1,2
Base four	Quadrinal system	0,1,2,3
Base five	Quinary system	0,1,2,3,4
Base six	Hexal or senary system	0,1,2,3,4,5
Base seven	Heptademal system	0,1,2,3,4,5,6
Base eight	Octal system	0,1,2,3,4,5,6,7
Base nine	Nonal system	0,1,2,3,4,5,6,7,8
Base ten	Denary or a decimal system	0,1,2,3,4,5,6,7,8,9
Base eleven	Unidecimal system	0,1,2,3,4,5,6,7,8,9, t
Base twelve	Duodecimal or duodenary system	0,1,2,3,4,5,6,7,8,9, t,e

Conversion of Number Bases

The following show how numbers can be converted from one base to another.

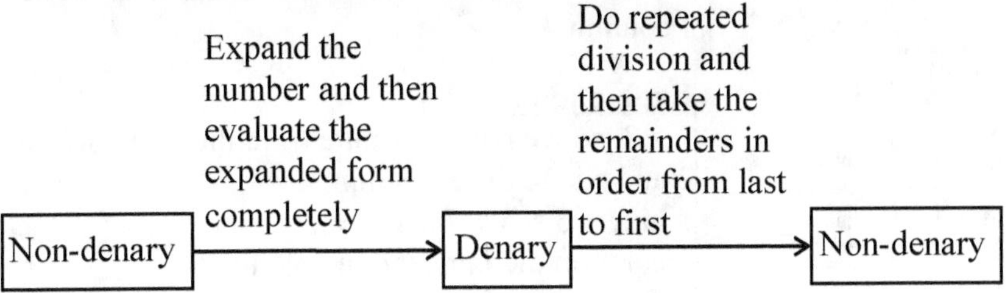

Converting Non-Denary to Non-Denary

To convert a number from a non-denary base to another non-denary base, we first convert the number to base ten then, convert to the destination base.

Example

Convert 1346_{seven} to base five.

Solution

$1346_{seven} = 1 \times 7^3 + 3 \times 7^2 + 4 \times 7^1 + 6 = 524_{ten}$

524_{ten} is then changed to base five using repeated division by 5.

5	524
5	104R 4
5	20R 4
5	4R 0
	0R 4

$\therefore 1346_{seven} = 4044_{five}$

Arithmetic in Other Bases

Arithmetic in other bases is done in a way similar to base ten, remembering that the bundle is the base indicated.

Example 18:10

Evaluate (a) $103_{four} + 213_{four}$ (b) 853_{nine} and 237_{nine}.

(c) $42_{five} \times 34_{five}$ (d) $4346_{eight} \div 42_{eight}$

Solution

(a)

$$\begin{array}{r} 1\ 0\ 3_{four} \\ +\ 2\ 1\ 3_{four} \\ \hline 3\ 2\ 2_{four} \\ \hline\hline \end{array}$$

Explanation
In column 3: $3_{four} + 3_{four} = 12_{four}$
Write the 2 under column 3 and add 1 to the sum in column 2.

(b)

$$\begin{array}{r} 8\ 5\ 3_{nine} \\ -\ 2\ 3\ 7_{nine} \\ \hline 6\ 1\ 5_{nine} \\ \hline\hline \end{array}$$

Explanation
$7 > 3$, so we borrow 1 nine from the 5 in column 2 and add to the 3 column 3.
1 nine + 3 = 5 + 7 and 5 + 7−7 = 5.
4 is left in the minuend in column 2.
4−3 = 1 and 8−2 = 6

(c)

$$\begin{array}{r} 4\ 2 \\ 3\ 4 \\ \times\qquad \\ \hline 3\ 2\ 3 \\ +\ 2\ 3\ 1\ \ \ \\ \hline 3\ 1\ 3\ 3 \\ \hline\hline \end{array}$$

Explanation
$4 \times 2 \div 5 = 1$ R 3. Write 3 and take over 1.
$4 \times 4 \div 5 = 3$ R 1. Add the 1 taken over to the remainder and take over 3.
$3 \times 2 \div 5 = 1$ R 1. Write 1 and take over 1.
$3 \times 4 \div 5 = 2$ R 2. Add the 1 taken over to the remainder and take over 2.
Add the two rows of products.

(d) Keep in mind that the base is eight!

$$\begin{array}{r} 103 \\ 42_{eight}\ \overline{)\ 4346_{eight}} \\ 42_{eight}\quad \\ \hline 146_{eight} \\ 146_{eight} \\ \hline \end{array}$$

Explanation
$43 \div 42 = 1$, $1 \times 42_{eight} = 42_{eight}$,
$43 - 42 = 1$
Bring down 4. Now 42 cannot divide 14 so we write 0 in the answer and bring down 6.
$146 \div 42 = 3$, $3 \times 42_{eight} = 146_{eight}$
$146 - 146 = 0$

$\therefore 4346_{eight} \div 42_{eight} = 103_{eight}$

The Binary System

The binary system is very important in electronics and computers which all use the 16-8-4-2-1 format. In fact switches and all two state devices are binary system devices.

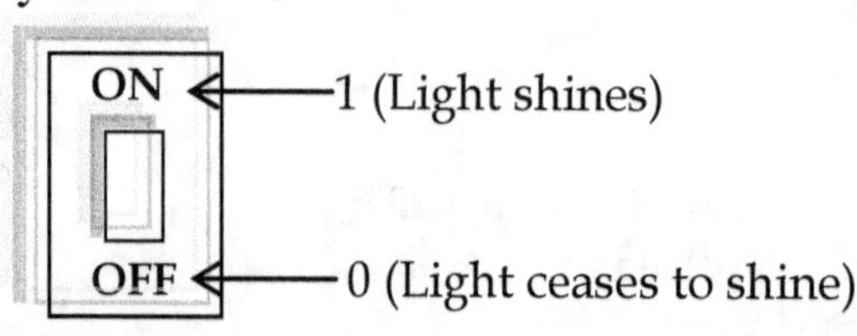

1	0
On	Off
Up	Down
Odd	Even
Left	Right
Good	Bad

Example

Evaluate the following

(a) $100101_{two} + 110101_{two}$ (b) $10110101_{two} - 110101_{two}$

(c) $1001_{two} \times 101_{two}$ (d) $1000001_{two} \div 1101_{two}$

Solution

(a)
$$\begin{array}{r} 100101_{two} \\ +\ 110101_{two} \\ \hline 1011010_{two} \\ \hline\hline \end{array}$$

(b)
$$\begin{array}{r} 1011010_{two} \\ -\ 110101_{two} \\ \hline 100101_{two} \\ \hline\hline \end{array}$$

(c)
$$\begin{array}{r} 1001_{two} \\ \times\ 101_{two} \\ \hline 100100_{two} \\ 1001_{two} \\ \hline 101101_{two} \\ \hline\hline \end{array}$$

(d)
$$\begin{array}{r} 101_{two} \\ 1101_{two}\overline{)1000001_{two}} \\ 1101_{two} \\ \hline 1101_{two} \\ 1101_{two} \\ \hline \end{array}$$

MULTIPLE CHOICE EXERCISE 1:2

1. The Hindu Arabic numeral representing "Two hundred and four thousand and four" is:
   [A] 20404 [B] 240004 [C] 24400 [D] 204004
2. One million three hundred and fifty four is written as:
   [A] 1000354 [B] 1030054 [C] 1300054 [D] 1354000
3. Ninety nine thousand and ninety nine written in figures is:
   [A] 990099 [B] 9999 [C] 99099 [D] 90999
4. The number 605, 080 is read:
   [A] Sixty thousand and five thousand and eighty
   [B] Six hundred and five hundred and eighty
   [C] Six thousand and five hundred and eighty
   [D] Six hundred and five thousand and eighty
5. The amount 2,300,240 francs is read:
   [A] Twenty three million four hundred and twenty francs
   [B] Two million, three hundred thousand four hundred and twenty francs
   [C] Two million three hundred thousand two hundred and forty francs.
   [D] Twenty three million two hundred and forty francs
6. The value of the digit 6 in the number 726251 is:
   [A] six hundred [B] six hundredth
   [C] six thousandth [D] six thousand
7. The value of 5 in 2753 is:

[A] Tenth [B] tens [C] Hundredth [D] hundredth

8. When the value of 6 in 5624 is divided by the value of 3 in 2639, the result is:
   [A] 2 [B] 5 [C] 20 [D] 16
9. The product of the value of 7 in 2721 and the value of 3 in 5837 is:
   [A] 21000 [B] 26677 [C] 21 [D] 2100
10. The sum of eleven thousand and one thousand hundred is:
    [A] 11100 [B] 12100 [C] 11110 [D] 111000
11. In 6,36<u>7</u>,804, the value of the underlined digit is:
    [A] 7 [B] 700 [C] 7,000 [D] 70,000
12. Four million and six is represented by:
    [A] 4,000,600 [B] 4,000,006 [C] 4,600 [D] 4,006
13. The Hindu Arabic numeral representing "Two hundred and four thousand and four" is:
    [A] 20404 [B] 240004 [C] 24400 [D] 204004
14. The number of prime numbers between 1 and 20 is:
    [A] 9 [B] 8 [C] 7 [D] 6
15. The first four prime numbers are:
    [A] 1, 2, 3, 4 [B] 1, 3, 5, 7 [C] 2, 3, 5, 7 [D] 2, 4, 6, 8
16. 84 as a product of prime factors is:
    [A] $2^2 \times 3^2 \times 7^2$ [B] $2^2 \times 3^3 \times 7$ [C] $2^2 \times 3 \times 7$ [D] $2^2 \times 3^2 \times 7$
17. As a product of prime factors 200 is equal to:
    [A] 2×5 [B] $2^3 \times 5^2$ [C] $2^2 \times 5^3$ [D] $2^2 \times 5^2$
18. Leaving your answer as a product of prime factors in index form, 72 equals:
    [A] $2^2 \times 3^2$ [B] $2^2 \times 3^3$ [C] $2^3 \times 3^3$ [D] $2^3 \times 3^2$
19. The number that is not a prime number is:
    [A] 9 [B] 7 [C] 3 [D] 2
20. The number of factors a prime number has is:
    [A] 0 [B] 1 [C] 2 [D] 3
21. The prime number among the following is:
    [A] 15 [B] 13 [C] 9 [D] 1
22. 17 is a prime number because:
    [A] it is not divisible by 2 [B] it is a sieve of Erasthodenes
    [C] it has no factor other than itself [D] it has only two factors
23. The prime number among the following is:
    [A] 57 [B] 61 [C] 63 [D] 69
24. Two of the numbers 11, 21, 31, 77, 112 are prime numbers. The number lying exactly half way between these two prime numbers are:
    [A] 54 [B] 26 [C] 21 [D] 16
25. Three of the numbers 11, 21, 31, 77, 112 have a common factor. The common factor is:
    [A] 14 [B] 11 [C] 7 [D] 2
26. The number that is the product of two consecutive prime numbers is:

[A] 8 [B] 15 [C] 18 [D] 21

27. The LCM of 6 and 14 is:
[A] 42 [B] 24 [C] 14 [D] 84
28. The LCM of 8,9, and 12 is:
[A] 29 [B] 72 [C] 96 [D] 108
29. The result of dividing the LCM of 8 and 12 by 3 is:
[A] 12 [B] $10\frac{2}{3}$ [C] $9\frac{1}{3}$ [D] 8
30. The HCF of the numbers 30,120 and 125 is:
[A] 3 [B] 5 [C] 10 [D] 15
31. The HCF of 18,24, and 36 expressed as a product of prime factors is:
[A] $2^2 \times 3$ [B] 2×3^2 [C] 2×3 [D] $2^2 \times 3^2$
32. Dividing the LCM of 24 and 30 by their HCF gives:
[A] 2 [B] 20 [C] 24 [D] 25
33. The result of dividing the LCM of 12,16 and 24 by their HCF is:
[A] 12 [B] 11 [C] 10 [D] 9
34. The result of squaring the number 6 is:
[A] 12 [B] 26 [C] 36 [D] 62
35. $\sqrt{7744}$ in index form is:
[A] 26 [B] $2^3 \times 13$ [C] $2^3 \times 9$ [D] $2^3 \times 11$
36. $\sqrt{3136}$ in index form is:
[A] 2×7^2 [B] $2^3 \times 7$ [C] $2^2 \times 7$ [D] 2×7
37. The smallest number by which $3^2 \times 5$ can be multiplied to give a perfect square is:
[A] 5 [B] 6 [C] 15 [D] 25
38. The least number, which multiplies 54 to make a perfect square, is:
[A] 3 [B] 4 [C] 6 [D] 8
39. 75_{ten} is the same as:
[A] 300_{five} [B] 400_{five} [C] 500_{five} [D] 600_{five}
40. When expressed as a binary number 27_{ten} is equal to:
[A] 1110_{two} [B] 1111_{two} [C] 11011_{two} [D] 1001_{two}
41. The denary (base ten) number 37, written in binary (base two) is:
[A] 100011_{two} [B] 100111_{two} [C] 100001_{two} [D] 100101_{two}
42. The denary number 39 written in binary is:
[A] 100111_{two} [B] 100101_{two} [C] 1001_{two} [D] 10001_{two}
43. The value of the decimal number 89 as a binary number is:
[A] 101101_{two} [B] 1011001_{two} [C] 1001001_{two} [D] 1001101_{two}
44. 35 in base two is:
[A] 1000_{two} [B] 10011_{two} [C] 100011_{two} [D] 110010_{two}
45. The equivalence of 11111_{two} in base ten is:
[A] 9 [B] 17 [C] 19 [D] 31

46. 101101_{two} in expanded form is:
    [A] $1\times 2^{-6} + 1\times 2^{-4} + 1\times 2^{-3} + 1\times 2^{-1}$
    [B] $1\times 2^{-5} + 1\times 2^{-3} + 1\times 2^{-2} + 1\times 2^{0}$
    [C] $1\times 2^{5} + 1\times 2^{3} + 1\times 2^{2} + 1\times 2^{0}$
    [D] $1\times 2^{6} + 1\times 2^{4} + 1\times 2^{3} + 1\times 2^{1}$
47. As a number in base ten 321_{five} is equal to:
    [A] 85_{ten} [B] 86_{ten} [C] 32_{ten} [D] 43_{ten}
48. The value of $3310_{five} - 1442_{five}$ is:
    [A] 2131_{five} [B] 1313_{five} [C] 1103_{five} [D] 4302_{five}
49. If $540_{seven} - 253_{seven} = x_{seven}$, x must be:
    [A] 457 [B] 475 [C] 254 [D] 284
50. The base of the addition $324 + 135 = 503$ is:
    [A] 3 [B] 4 [C] 5 [D] 6
51. Evaluating $2002_{three} - 202_{three}$ gives:
    [A] 100_{three} [B] 101_{three} [C] 1010_{three} [D] 1100_{three}
52. 42_{five} is equivalent to:
    [A] 21_{ten} [B] 10101_{two} [C] 112_{four} [D] 212_{three}
53. 24_{five} is equivalent to:
    [A] 40_{three} [B] 112_{three} [C] 11000_{two} [D] 120_{ten}
54. The possible binary number in the following is:
    [A] 112 [B] 101 [C] 102 [D] 211
55. The sum of 1111_{two} and 111_{two} is:
    [A] 101100_{two} [B] 101101_{two} [C] 10110_{two} [D] 10011_{two}
56. Given that 1101, 11011 and 11 are binary numbers, their sum will be:
    [A] 11001_{two} [B] 111001_{two} [C] 10110_{two} [D] 101011_{two}
57. The difference between 1110_{two} and 101_{two} is:
    [A] 1010_{two} [B] 1001_{two} [C] 101_{two} [D] 1011_{two}
58. When 1101_{two} is subtracted from 110011_{two}, the result is:
    [A] 10110_{two} [B] 111000_{two} [C] 100110_{two} [D] 110011_{two}
59. The square of 101_{two} is:
    [A] 1010_{two} [B] 1111_{two} [C] 1011_{two} [D] 11001_{two}
60. 111_{four}, 22_{eight}, 11011_{two} in ascending order of size is:
    [A] 11_{four}, 22_{eight}, 11011_{two} [B] 22_{eight}, 111_{four}, 11011_{two}
    [C] 11011_{two}, 22_{eight}, 111_{four} [D] 11011_{two}, 111_{four}, 22_{eight}
61. When evaluated, $101_{two} \text{x}\ 11_{two}$ equals:
    [A] 1111_{two} [B] 1110_{two} [C] 1101_{two} [D] 101_{two}
62. The average of 1011_{two} and 111_{two} is:
    [A] 1000_{two} [B] 10010_{two} [C] 1001_{two} [D] 10001_{two}
63. When 101_{two} and 10_{two} are multiplied the answer is:
    [A] 1001_{two} [B] 1010_{two} [C] 1011_{two} [D] 1110_{two}
64. $1001_{two} \times 101_{two}$ is equal to:
    [A] 101011_{two} [B] 101101_{two} [C] 110110_{two} [D] 110101_{two}
65. On dividing 1010_{two} by 101_{two}, the result is:

[A] 101_{two} [B] 11_{two} [C] 10_{two} [D] 100_{two}

66. Two numbers 24_x and 36_y are equal in value when converted to base ten. The true equation under this condition is:
    [A] $2x-3y=2$ [B] $3y-2x=2$ [C] $3y=x+2$ [D] $3y=2-x$
67. If $104_x = 68$. The value of x is:
    [A] 5 [B] 7 [C] 8 [D] 9
68. Given that $4P4_{five} = 119_{ten}$, the value of P is:
    [A] 0 [B] 1 [C] 2 [D] 3
69. If $M5_{ten} = 1001011_{two}$, the value of M is:
    [A] 5 [B] 6 [C] 7 [D] 8
70. Given that x is a denary number and $x = 111101_{two}$, the value of x in base ten is:
    [A] 29 [B] 61 [C] 62 [D] 63

1.3 INTEGERS AND DIRECTED NUMBERS

Integers, $\mathbb{Z}$

The set of integers is the set of all positive and negative whole numbers.

$$\mathbb{Z} = \{0, \pm1, \pm2, \pm3...\} = \mathbb{Z}^- \cup \{0\} \cup \mathbb{Z}^+.$$

$$\mathbb{Z}^+ \text{ or } \mathbb{Z}^* = \{+1, +2, +3...\} = \{x \in \mathbb{Z}: x > 0\}.$$

$$\mathbb{Z}^- = \{-1, -2, -3...\} = \{x \in \mathbb{Z}: x < 0\}.$$

Meaning of Directed Numbers

A directed number is a number with a positive (+) or negative (–) sign.

The Absolute Value of a Directed Number

The absolute value of a directed number is its value without regard of the sign. For instance, $|-3| = 3$, $|+3| = 3$

Example

Find the absolute value of each of the following.

(a) +8 (b) –17

Solution

(a) $|+8| = 8$ (b) $|-17| = 17$

The Additive Inverse of a number

The additive inverse of a number is its counterpart such that their sum is zero. For instance, – 4 is the additive inverse of +4 since – 4 + (+4) = 0.

Example
State the additive inverse of each of the following.
(a) +17 (b) – 15 (c) 30

Solution
(a) –17 (b) +15 (c) –30

Addition and Subtraction of Directed Numbers

1. To add numbers with the same sign, add their absolute values and maintain their common sign.
2. To add two numbers with unlike sign, subtract the number with the smaller absolute value from the one with the larger absolute value and retain the sign of the number with the larger absolute value.
3. Subtracting a directed number is the same as adding its additive inverse.

Example
Evaluate (a) $(-7)+(-6)+(-4)$ (b) $(-6)+(+8)$
(c) $(-6)-(-8)$ (d) $(-6)-(+8)$

Solution
(a) $(-7)+(-6)+(-4) = -(7+6+4) = -17$
(b) $(-6)+(+8) = +(8-6) = +2$
(c) $(-6)-(-8) = (-6)+(+8) = +2$
(c) $(-6)-(+8) = (-6)+(-8) = -(6+8) = -14$

Multiplication and Division of Directed Numbers

The sign of a product and/or quotient of directed numbers is:
(a) Positive if the number of negative signs is even
(b) Negative if the number of negative signs is odd.

Example
Evaluate (a) $(-3)(-5)(-4)$ (b) $(-6)(+3)$
(c) $\frac{(-6)(-8)(+10)}{(+5)(-12)}$ (d) $\frac{(-6)(+4)(-14)(-3)}{(-21)(-8)(-1)}$

Solution
(a) $(-3)(-5)(-4) = -60$ (b) $(-2)(+3)(+4)(-1) = +24$

(c) $\frac{(-6)(-8)(+10)}{(+5)(-12)} = -8$ (d) $\frac{(-6)(+4)(-14)(-3)}{(-21)(-8)(-1)} = +6$

MULTIPLE CHOICE EXERCISE 1.3

1. The greatest of the following is:
   [A] –2 [B] –4 [C] –6 [D] –8
2. As an integer, 32°F below zero degrees is:
   [A] –31°F [B] 32°F [C] –32°F [D] 31°F
3. The opposite of the integer –3 is:
   [A] 1 [B] 3 [C] –3 [D] $\frac{1}{3}$
4. The absolute value of 5 is:
   [A] 5 [B] –5 [C] $\sqrt{5}$ [D] 5^2
5. The correct order of the integers 8, 15, –1, 6, –6 from least to greatest is:
   [A] −6, −1, 6, 8, 15 [B] −1, −6, 15, 6, 8
   [C] 8, 6, −6, −1, 15 [D] 15, 6, 8, −1, −6
6. The sum 1 and 5 is:
   [A] 6 [B] 4 [C] –4 [D] –6
7. The difference –5 – (–2) is:
   [A] 3 [B] –3 [C] –7 [D] 7
8. When evaluated, –7+ 5 – 2 + 11 – 8 equals:
   [A] +1 [B] –33 [C] +33 [D] –1
9. The value of – 6 – (– 6) is:
   [A] –12 [B] 0 [C] 12 [D] 36
10. The value of (–4)(–3) is:
    [A] 7 [B] –7 [C] 12 [D] –12
11. (–2) × (–3) is equal to:
    [A] –6 [B] 6 [C] –5 [D] 5
12. The value of (+4)(–2) – (–2) is:
    [A] –6 [B] –4 [C] +4 [D] +6
13. –12 + 8.3 equals:
    [A] –4.7 [B] –4.36 [C] –3.7 [D] 3.7
14. A fish was swimming at 180 m below sea level. Then it descended to 315 m below sea level. As an integer in meters, the change in the fish's depth is:
    [A] 135 [B] –135 [C] –495 [D] 495
15. As arctic air moved into a region where the temperature was originally 25°C, the temperature began falling at a steady rate of 4°C per hour. The temperature after 9 hours would be:
    [A] –7°C [B] –11°C [C] –36°C [D] –15°C
16. 8÷ (–4) is:
    [A] –2 [B] $\frac{1}{2}$ [C] 2 [D] $-\frac{1}{2}$

17. Anye owns a small business. There was a loss of 14 FCFA on Thursday and a loss of 12 FCFA on Friday. On Saturday, there was a loss of 11 FCFA, and on Sunday, there was a profit of 16 FCFA. The total profit or loss for the four days is:
    [A] 31 FCFA loss [B] 25 FCFA profit
    [C] 21 FCFA loss [D] 53 FCFA profit
18. Using the following table, Pa Tangwe's profit or loss for the month of January is:
    [A] –4,155 [B] 4,155 [C] 521 [D] –521

Pa Tangwe Income and Expenditure

Month	Income	Expenditure
January	1,817 FCFA	–2,338 FCFA
February	2,271 FCFA	–2,315 FCFA
March	3,243 FCFA	– 1,530 FCFA
April	3,929 FCFA	–1,167 FCFA
May	3,477 FCFA	–1,101 FCFA
June	3,077 FCFA	–834 FCFA

19. $|-1|$ is equal to:
    [A] 0 [B] –1 [C] 1 [D] 2
20. The symbol, which replaces the square to make $|-14| \square |10|$ true is:
    [A] = [B] > [C] < [D] =
21. It is true to say that:
    [A] $|-7|>|5|$ [B] $|-7|<|5|$ [C] $|-7|<|7|$ [D] $|0|>|-5|$

1.4 RATIONAL, IRRATIONAL AND REAL NUMBERS

Rational Numbers, $\mathbb{Q}$

A **rational number** is any number, which can be expressed as the quotient of two integers.

$$\mathbb{Q} = \left\{x: x = \frac{p}{q}; p, q \in \mathbb{Z}, q \neq 0\right\}$$

Some rational numbers are plotted on the number line below.

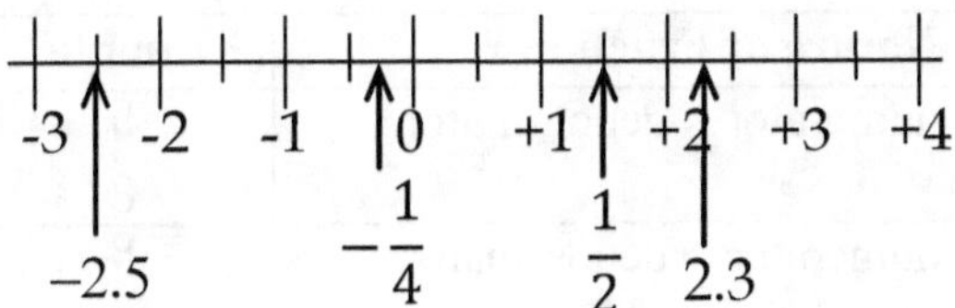

Irrational Numbers, $\mathbb{Q}'$

An irrational number is a number, which cannot be expressed as a ratio of two integers. Examples are 0.1171117111171111117…, $\sqrt{2}$, 0.41586237 etc.

Real Numbers, $\mathbb{R}$

The set $\mathbb{R}$ of real numbers is the set of all rational numbers and all the irrational numbers. $\mathbb{R}^+ = \{x \in \mathbb{R}: x \geq 0\}$.
We can represent any real number by a point on a number line and vice versa.
The following tree and Venn diagram shows the relationship between the sets of numbers.

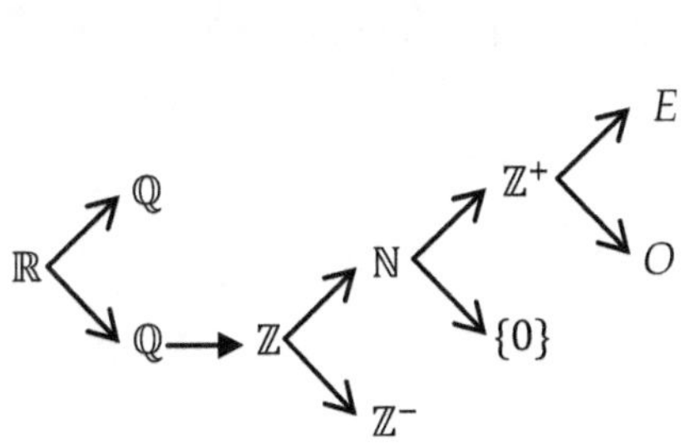

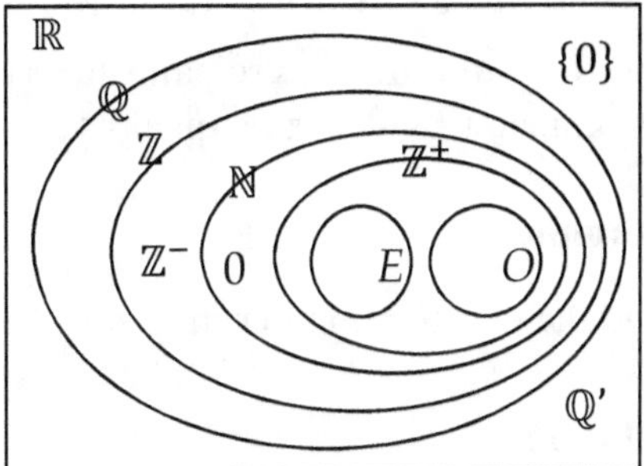

VULGAR FRACTIONS

A **fraction** is a number that expresses part of a whole quantity.

$\frac{4}{3}$ 4 ←numerator or dividend
3 ←denominator or divisor

Types of Fractions

Types of Fraction	Nature of Fraction	Example
Proper fraction	numerator < denominator	$\frac{5}{8}$
Improper fraction	numerator > denominator	$\frac{8}{5}$
mixed number or mixed fraction	whole number part and a proper fractional part	$1\frac{3}{5}$

A fraction with numerator 0 is simply 0. A fraction with denominator 0 is meaningless.

Equivalent Fractions

Equivalent fractions are fractions which represent exactly the same quantity.

For instance $\frac{1}{2} \equiv \frac{2}{4} \equiv \frac{4}{8} \equiv \frac{8}{16} \Rightarrow \frac{1}{2}, \frac{2}{4}, \frac{4}{8}, \frac{8}{16}$ are equivalent fractions.

The value of a fraction remains the same if we multiply both numerator and denominator by the same quantity.

Example

Find the missing number in $\frac{2}{3} \equiv \frac{?}{15}$.

Solution

Since $3 \times 5 = 15$, the missing number is $2 \times 5 = 10$.

A fraction is in its **lowest terms**, if the numerator and denominator have no common factors.

To simplify a fraction to its lowest terms we divide both numerator and denominator by their common factors until there are no common factors left.

Example

Simplify $\frac{48}{120}$.

Solution

$\frac{48}{120} = \frac{2\times 24}{5\times 24} = \frac{2}{5}$

Some Order Symbols

$2 < 5$ is read ‘2 is less than 5’.

$9 > 5$ is read '9 is greater than 5'.

Comparing fractions

To compare fractions having common denominators, compare the numerators.
To compare fractions with different denominators, first convert the fractions to equivalent fractions with the LCM as their denominators then compare the numerators.

Example
Compare the following using >, < or =.

(a) $\frac{5}{8}$ and $\frac{3}{8}$ (b) $\frac{3}{4}$ and $\frac{18}{24}$ (c) $\frac{2}{3}$ and $\frac{4}{5}$

Solution
(a) $\frac{5}{8} > \frac{3}{8}$ (b) $\frac{3}{4} = \frac{18}{24}$ (c) $\frac{2}{3} = \frac{2\times5}{3\times5} = \frac{10}{15}$ and $\frac{4}{5} = \frac{4\times3}{5\times3} = \frac{12}{15} \Rightarrow \frac{2}{3} < \frac{4}{5}$.

An alternative method is to convert all the fractions to decimals before comparing. Watch out!

Converting Mixed Numbers to Improper Fractions

To convert a mixed number to an improper fraction write the whole number part as an equivalent fraction whose denominator is the same as that of the fractional part and add the numerator, then retain the denominator.

Example
Convert $5\frac{7}{8}$ to an improper fraction.

Solution
$5\frac{7}{8} = \frac{40}{8} + \frac{7}{8} = \frac{47}{8}$.

This solution usually done mentally as follows; $5\frac{7}{8} = \frac{8\times5+7}{8} = \frac{47}{8}$.

Addition and subtraction of fractions

1. To add or subtract fractions with equal denominators, keep the common denominator and add or subtract the numerators.
2. To add or subtract fractions with unequal denominators, first convert each of the fractions to equivalent fractions with the LCM of the denominators as their denominators.
3. To add or subtract mixed numbers, convert the mixed numbers to improper fractions before adding or subtracting.

Example

Evaluate the following.

(i) $\frac{3}{5}+\frac{1}{5}$ (ii) $\frac{9}{10}-\frac{3}{10}$ (iii) $\frac{5}{6}+\frac{3}{4}$ (iv) $\frac{7}{8}-\frac{1}{4}$ (v) $\frac{5}{6}-\frac{2}{3}+\frac{1}{10}$

Solutions

(i) $\frac{3}{5}+\frac{1}{5}=\frac{3+1}{5}=\frac{4}{5}$

(ii) $\frac{9}{10}-\frac{3}{10}=\frac{9}{10}-\frac{3}{10}=\frac{6}{10}=\frac{3}{5}$

(iii) $\frac{5}{6}+\frac{3}{4}=\frac{10}{12}+\frac{9}{12}=\frac{19}{12}=1\frac{7}{12}$

(iv) $\frac{7}{8}-\frac{1}{4}=\frac{7}{8}-\frac{2}{8}=\frac{5}{8}$

(v) $\frac{5}{6}-\frac{2}{3}+\frac{1}{10}=\frac{25}{30}-\frac{20}{30}+\frac{3}{30}=\frac{8}{30}=\frac{4}{15}$

Example

Evaluate the following.

(a) $2\frac{2}{5}+3\frac{1}{3}$ (b) $3\frac{2}{5}-2\frac{3}{4}$ (c) $4\frac{3}{5}-2\frac{1}{2}+1\frac{1}{10}$

Solution

(a) $2\frac{2}{5}+3\frac{1}{3}=\frac{12}{5}+\frac{10}{3}=\frac{36}{15}+\frac{50}{15}=\frac{86}{15}=5\frac{11}{15}$

(b) $3\frac{2}{5}-2\frac{3}{4}=\frac{17}{5}-\frac{11}{4}=\frac{68}{20}-\frac{55}{20}=\frac{13}{20}$

(c) $4\frac{3}{5}-2\frac{1}{2}+1\frac{1}{10}=\frac{23}{5}-\frac{5}{2}+\frac{11}{10}=\frac{46}{10}-\frac{25}{10}+\frac{11}{10}=\frac{32}{10}=3\frac{1}{5}$

Reciprocals

Recall that the **reciprocal** or **multiplicative inverse** of $\frac{a}{b}$ is $\frac{b}{a}$.

MULTIPLYING AND DIVIDING FRACTIONS

1. To multiply two fractions, first simplify by dividing by the common factors then multiply the numerators together and the denominators together.
2. To multiply a fraction by a whole number, multiply the numerator of the fraction by the whole number and maintain the denominator of the fraction.
3. We can look at division by a given number as multiplication by the reciprocal of the number.
4. To multiply and or divide mixed numbers, first convert the mixed numbers to improper fractions.

Example

Evaluate (a) $\frac{18}{35} \times \frac{14}{27}$ (b) $\frac{3}{11} \times 2$ (c) $4 \times \frac{5}{9}$

(d) $5\frac{3}{8} \times 4$ (e) $2\frac{1}{2} \times 1\frac{1}{4}$ (f) $\frac{3}{4}$ of 300

Solution

(a) $\frac{18}{35} \times \frac{14}{27} = \frac{2}{5} \times \frac{2}{3} = \frac{4}{15}$

(b) $\frac{3}{11} \times 2 = \frac{6}{11}$

(c) $4 \times \frac{5}{9} = \frac{20}{9}$

(d) $5\frac{3}{8} \times 4 = \frac{43}{8} \times 4 = \frac{43}{2} = 21\frac{1}{2}$

(e) $2\frac{1}{2} \times 1\frac{1}{4} = \frac{5}{2} \times \frac{5}{4} = \frac{25}{8} = 3\frac{1}{8}$

(f) $\frac{3}{4}$ of $300 = \frac{3}{4} \times 300 = 3 \times 75 = 225$

Example

Manfred took $\frac{2}{3}$ of his money to school. At school, he used $\frac{3}{4}$ of what he took. What fraction of his money did he use?

Solution

$$\frac{3}{4} \text{ of } \frac{2}{3} = \frac{\overset{1}{\cancel{3}}}{\underset{2}{\cancel{4}}} \times \frac{\overset{1}{\cancel{2}}}{\underset{1}{\cancel{3}}} = \frac{1}{2}$$

Example

Mundi signed out $\frac{2}{3}$ of the 357,000 Frs. in her credit union account. How much did she sign out?

Solution

Amount signed out $= \frac{2}{3}$ of $357000 = \frac{2}{3} \times 357000 = 238000$ Frs.

Example

Evaluate (i) $\frac{9}{10}\div\frac{3}{5}$. (ii) $1\frac{4}{5}\div 3$

Solution

(i) $\frac{9}{10}\div\frac{3}{5}=\frac{\cancel{9}^{3}}{\cancel{10}_{2}}\times\frac{\cancel{5}^{1}}{\cancel{3}_{1}}=\frac{3}{2}=1\frac{1}{2}$ (ii) $1\frac{4}{5}\div 3=\frac{\cancel{9}^{3}}{5}\times\frac{1}{\cancel{3}_{1}}=\frac{3}{5}$

Expressing One Quantity as a Fraction of Another

$$\text{Quantity } A \text{ as a fraction of quantity } B = \frac{\text{Quantity } A}{\text{Quantity } B}$$

Example
Express 25 as a fraction of 80

Solution

25 as a fraction of 80 $=\frac{25}{80}=\frac{5}{16}$

Example
Out of the 13464 candidates who sat for an examination, 7854 passed. What fraction of the candidates passed?

Solution
Fraction that passed$=\frac{7854}{13464}=\frac{7}{12}$.

DECIMALS

A **decimal fraction** or **decimal** is a fraction whose denominator is a power of ten. A **decimal point** or **decimal marker** is a dot used to separates the whole number part from the fractional part in decimals.

To divide by a power of 10, count the zeros in the power of 10 and move the decimal point the corresponding number of decimal places to the left.

Example 3:21
Rewrite the following decimals using a decimal marker.

(i) $\frac{931}{100}$ (ii) $\frac{5743}{10000}$

Solution

(i) $\frac{931}{100} = 9.31$ (Move the decimal point 2 places to the left because power of ten has two zeros)

(ii) $\frac{5743}{10000} = 0.5743$ (power of ten has four zeros)

Place Value System and Decimals

Each digit to the right of the decimal point represents a number of tenths, hundredths, thousandths etc.

Example

State the value of 3 in each of the following

(a) 0.03 (b) 0.483 (c) 51.25431

Solution

(a) three hundredth (b) three thousandth (c) thirty thousandth

Changing Fractions to Decimals

To change a fraction to a decimal, first write the fraction as an equivalent fraction whose denominator is a power of 10, and then write the decimal using a decimal marker.

Example

Convert the following to decimals. (a) $\frac{4}{25}$ (b) $2\frac{1}{8}$

Solution

(a) $\frac{4}{25} = \frac{4}{25} \times \frac{4}{4} = \frac{16}{100} = 0.16$ (b) $2\frac{1}{8} = 2 + \frac{1}{8} \times \frac{125}{125} = 2 + \frac{125}{1000} = 2.125$

Changing Decimals to Fractions

To change a decimal to a fraction, write the number as a decimal fraction with denominator a power of 10 having the same number of zeros as the number of decimal places in the decimal then simplify the result.

Example

Change the following fractions to decimals

(i) 0.6 (ii) 0. 125

Solution

(i) $0.6 = \frac{6}{10} = \frac{3}{5}$ (ii) $0.125 = \frac{125}{1000} = \frac{1}{8}$

Addition and Subtraction of Decimals

To add or subtract decimals, arrange the numbers vertically aligning the decimal point then add or subtract as with whole numbers.

Example

Compute the following without using a calculator.

(a) $6.4163 + 17.3187 + 5.4128$ (b) $8.217 - 2.831$

Solution

(a)
$$\begin{array}{r} 6.4163 \\ 17.3187 \\ +\ \ 5.4128 \\ \hline 29.1478 \\ \hline\hline \end{array}$$

(b)
$$\begin{array}{r} 8.217 \\ -2.831 \\ \hline 5.396 \\ \hline\hline \end{array}$$

Multiplication of Decimals

1. To multiply decimals, ignore the decimals and multiply as with whole numbers. The number of decimal places in the product is the sum of the number of decimal places in the numbers.
2. To multiply by a power of 10, count the number of zeros in the power of 10, and move the decimal point the corresponding number of places to the right.

Example

Evaluate (i) 136.8 × 47 (ii) 1000 × 2.5813 (iii) 0.236 × 0.3

Solution

(i) 136.8 × 47 = 6429.6 (since 1368 × 47 = 64296)
(ii) 1000 × 2581.3 = 2581.3 (move the decimal point 3 places to the right)
(iii) 0.236 × 0.3 = 0.0708 (236 × 3 = 708 move 4 decimal places to the left)

Dividing Decimals

To divide decimals, multiply both the dividend and divisor by an appropriate power of 10 which eliminates the decimals then simplify or do short or long division as with whole numbers.

Example

Compute $0.4 \div 0.00025$ without using a calculator.

Solution

$$0.4 \div 0.00025 = \frac{0.4 \times 100000}{0.00025 \times 100000} = \frac{40000}{25} = 1600$$

Terminating and Non-terminating Decimals

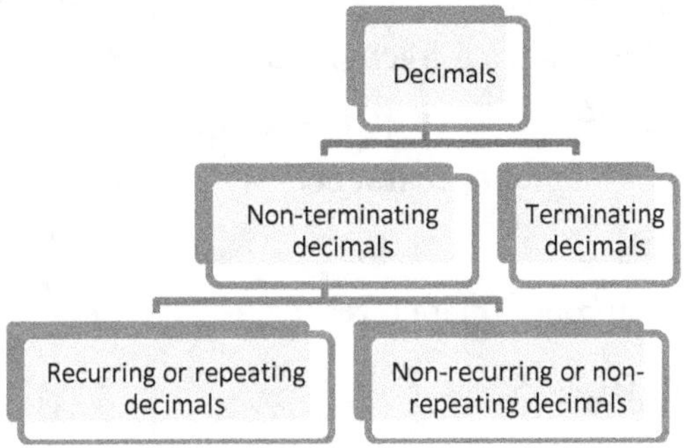

Non-terminating decimals result from the conversion of fractions whose denominators when in their lowest term have prime factors other than 2 or 5. Examples of are $\frac{4}{7}, \frac{7}{12}$ and $\frac{7}{13}$.

Example

Which of the following is likely to result in a non-terminating decimal?

(a) $\frac{6}{42}$ (b) $\frac{28}{70}$ (c) $\frac{39}{104}$ (d) $\frac{21}{77}$

Solution

(a) $\frac{6}{42} = \frac{1}{7}$ (b) $\frac{28}{70} = \frac{2}{5}$ (c) $\frac{39}{104} = \frac{3}{8}$ (d) $\frac{21}{77} = \frac{3}{11}$

(a) and (d) will result in non-terminating decimals.

MULTIPLE CHOICE EXERCISE 1:4

2. The fraction which is equivalent to $\frac{5}{6}$ is:

[A] $\frac{35}{36}$ [B] $\frac{10}{15}$ [C] $\frac{30}{36}$ [D] $\frac{110}{120}$

3. The fraction, which is not equivalent to, $\frac{3}{4}$ is:

[A] $\frac{35}{36}$ [B] $\frac{10}{15}$ [C] $\frac{30}{36}$ [D] $\frac{110}{120}$

4. The fraction which is equivalent to $\frac{5}{4}$ is:

[A] $\frac{14}{12}$ [B] $\frac{14}{8}$ [C] $\frac{20}{18}$ [D] $\frac{30}{24}$

5. The fraction $\frac{12}{16}$ is the same as:

[A] $\frac{2}{6}$ [B] $\frac{1}{4}$ [C] $\frac{3}{4}$ [D] $\frac{1}{3}$

6. In its lowest terms $\frac{60}{108}$ is:

[A] $\frac{7}{9}$ [B] $\frac{5}{9}$ [C] $\frac{15}{27}$ [D] $\frac{20}{36}$

7. The largest among the following fractions is:

[A] $\frac{2}{3}$ [B] $\frac{11}{15}$ [C] $\frac{7}{10}$ [D] $\frac{5}{6}$

8. The fraction, which is an improper fraction, is:

[A] $\frac{3}{4}$ [B] $\frac{7}{8}$ [C] $\frac{4}{3}$ [D] $\frac{1}{5}$

9. $7\frac{4}{5}$ expressed as an improper fraction is:

[A] $\frac{35}{5}$ [B] $\frac{27}{5}$ [C] $\frac{20}{5}$ [D] $\frac{39}{5}$

10. As a mixed number $\frac{40}{3}$ is:

[A] $13\frac{2}{3}$ [B] $13\frac{1}{3}$ [C] $13\frac{1}{4}$ [D] $13\frac{3}{4}$

11. The sum of $\frac{1}{5}$ and $\frac{3}{10}$ is:

[A] $\frac{4}{5}$ [B] $\frac{1}{2}$ [C] $\frac{4}{10}$ [D] $\frac{4}{15}$

12. In its lowest terms $\frac{5}{6} - \frac{5}{8}$ is:

[A] $\frac{5}{12}$ [B] $\frac{10}{48}$ [C] $\frac{35}{24}$ [D] $\frac{5}{24}$

13. On simplification $1 - \left(\frac{1}{4} + \frac{2}{3}\right)$ gives:

[A] 0 [B] $\frac{11}{12}$ [C] $\frac{1}{12}$ [D] $\frac{4}{7}$

14. The value of $6\frac{1}{5} - 2\frac{2}{3} + 1\frac{1}{6}$ is:

[A] $4\frac{3}{10}$ [B] $4\frac{7}{10}$ [C] $5\frac{7}{10}$ [D] $5\frac{9}{10}$

15. A jar is $\frac{4}{5}$ full of water. If Nfor drinks $\frac{5}{9}$ of the water, the fraction of the water left will be:

[A] $\frac{11}{45}$ [B] $\frac{4}{9}$ [C] $\frac{5}{9}$ [D] $\frac{16}{45}$

16. A man gave $\frac{5}{8}$ of his money to his wife and $\frac{1}{4}$ to his son. His fraction of the money left is:

[A] $\frac{7}{8}$ [B] $\frac{1}{4}$ [C] $\frac{5}{8}$ [D] $\frac{1}{8}$

17. $\frac{2}{3}$ and $\frac{1}{4}$ of a floor are covered by tiles and carpet respectively. The fraction of the floor not covered is:

[A] $\frac{1}{12}$ [B] $\frac{5}{12}$ [C] $\frac{11}{12}$ [D] $\frac{3}{4}$

18. The value of $\frac{1}{2} \times \frac{1}{3}$ is:

[A] $\frac{1}{5}$ [B] $\frac{5}{6}$ [C] $\frac{1}{6}$ [D] $\frac{2}{3}$

19. $\frac{7}{9}$ of $6\frac{3}{7}$ is equal to:

[A] 5 [B] 10 [C] 12 [D] 15

20. The result of multiplying $1\frac{2}{11}$ by $1\frac{7}{26}$ is:

[A] $\frac{11}{2}$ [B] $\frac{11}{3}$ [C] $1\frac{1}{2}$ [D] $\frac{2}{3}$

21. Three quarters of 12 is:

[A] 16 [B] 8 [C] 7 [D] 9

22. $2\frac{1}{4} \times 3\frac{1}{2}$ is:

[A] $5\frac{3}{4}$ [B] $6\frac{1}{8}$ [C] $6\frac{7}{8}$ [D] $7\frac{7}{8}$

23. A student ate $\frac{1}{2}$ of $\frac{2}{3}$ of the food he preserved for super. The fraction of the food he preserved left is:

[A] $\frac{1}{2}$ [B] $\frac{2}{3}$ [C] $\frac{1}{3}$ [D] $\frac{3}{5}$

24. The value of the quotient $2\frac{1}{5} \div \frac{1}{5}$ is:

[A] 10 [B] 11 [C] 12 [D] 13

25. $\frac{1}{2} + \frac{1}{2} \times \frac{1}{2}$ is equal to:

[A] $\frac{1}{8}$ [B] $\frac{1}{2}$ [C] $\frac{3}{4}$ [D] $1\frac{1}{2}$

26. The value of $3 \times \left(\frac{1}{2} + \frac{1}{4}\right)$ is:

[A] 1 [B] $\frac{3}{8}$ [C] $3\frac{3}{4}$ [D] $2\frac{1}{4}$

27. When evaluated $\frac{\frac{2}{3} \times 1\frac{1}{2}}{4\frac{4}{5}}$ gives:

[A] $\frac{5}{12}$ [B] $9\frac{3}{5}$ [C] $13\frac{1}{5}$ [D] $\frac{5}{6}$

28. $\left(\frac{2}{3}\times\frac{1}{4}\right)-\frac{1}{12}$ is the same as:

[A] $\frac{1}{12}$ [B] $\frac{1}{6}$ [C] $\frac{1}{4}$ [D] $\frac{1}{3}$

29. $\frac{4}{5}$ of $\left(\frac{1}{2}+\frac{3}{4}\right)$ has the value:

[A] 1 [B] 2 [C] 3 [D] 4

30. When simplified, the result of $\frac{\frac{1}{2}-\frac{1}{3}}{\frac{1}{2}+\frac{1}{3}}$ is:

[A] $\frac{1}{10}$ [B] $\frac{1}{15}$ [C] $\frac{1}{5}$ [D] $\frac{1}{20}$

31. On simplification $2\frac{3}{5}\div 1\frac{1}{5}+\frac{1}{2}$ gives:

[A] $1\frac{1}{3}$ [B] $2\frac{2}{3}$ [C] $2\frac{1}{2}$ [D] $3\frac{1}{2}$

32. The product of $\frac{1}{6}$ and the sum of $\frac{2}{5}$ and $1\frac{1}{3}$ is:

[A] $\frac{13}{45}$ [B] $\frac{14}{45}$ [C] $\frac{15}{45}$ [D] $\frac{17}{45}$

33. When simplified the value of $5+\frac{5}{4}-5\times\frac{5}{4}$ is:

[A] $\frac{25}{16}$ [B] $\frac{5}{16}$ [C] 0 [D] 1

34. The value of $\frac{\frac{3}{4}}{1\frac{1}{4}}\times\left(1\frac{1}{2}-\frac{2}{3}\right)$ is:

[A] $\frac{25}{32}$ [B] $\frac{7}{24}$ [C] $\frac{9}{25}$ [D] $\frac{1}{2}$

35. When simplified, the result of $5\frac{1}{4}\div\left(1\frac{2}{3}-\frac{1}{2}\right)$ is:

[A] $1\frac{3}{4}$ [B] $3\frac{1}{4}$ [C] $4\frac{1}{2}$ [D] $8\frac{1}{2}$

36. The result of $\frac{(-1)(-2)(+3)(-4)}{(+6)(-8)}$ is:

[A] 2 [B] −2 [C] $-\frac{1}{2}$ [D] $\frac{1}{2}$

37. The value of $\left(-\frac{6}{7}\right)\times\left(-\frac{7}{18}\right)\times\left(-\frac{3}{4}\right)$ is:

[A] $-\frac{1}{4}$ [B] $\frac{1}{4}$ [C] −1 [D] 1

38. $1\frac{3}{4}\div\left(-2\frac{1}{4}\right)$ equals:

[A] $\frac{7}{9}$ [B] $\frac{8}{63}$ [C] $-\frac{7}{9}$ [D] $-\frac{16}{63}$

39. Simplifying $\frac{5}{3}$ of $\left(1\frac{1}{5}-2\frac{1}{2}\right)\div 3\frac{1}{3}$ leads to:

[A] $-\frac{2}{20}$ [B] $-\frac{7}{20}$ [C] $-\frac{13}{20}$ [D] $-\frac{39}{20}$

40. The value of $8+30\div 10-2\times 5$ is:

[A] 38 [B] 1 [C] $\frac{95}{4}$ [D] $\frac{19}{20}$

41. The result after evaluating $5\times 4 - 18 \div 6$ is:
[A] $-\frac{17}{3}$ [B] $-\frac{35}{3}$ [C] $-\frac{28}{3}$ [D] 17
42. $8 \times 6 - 10 \div 5 + 12$ after simplification gives:
[A] 58 [B] 34 [C] $\frac{58}{5}$ [D] $-\frac{32}{7}$
43. A labourer's monthly salary is 48000 FRS. If he saves $\frac{1}{5}$ of this amount, in one year he must have saved:
[A] 9,600 FRS [B] 115,20 FRS [C] 180,000 FRS [D] 115,200 FRS
44. A man spends $\frac{3}{4}$ of his monthly salary on food and $\frac{1}{2}$ of the remainder on rent. If he has 15000 FRS left, he surely earns:
[A] 60,000 FRS [B] 90,000 FRS [C] 105,000 FRS [D] 120,000 FRS
45. Given that a pole has $\frac{1}{3}$ of its length in mud, $\frac{2}{5}$ of the remainder in water and the rest 6 m long above the surface of the water, the length of the pole is:
[A] 12 m [B] 10 m [C] 15 m [D] 16 m
46. The fraction which can be subtracted from the sum of $2\frac{1}{6}$ and $2\frac{2}{12}$ to give $3\frac{1}{4}$ is:
[A] $\frac{1}{3}$ [B] $\frac{1}{2}$ [C] $1\frac{1}{2}$ [D] $1\frac{1}{6}$
47. A student spends $\frac{1}{4}$ of her pocket money on books and $\frac{1}{3}$ on a dress. The fraction of her pocket money remaining will be:
[A] $\frac{5}{12}$ [B] $\frac{7}{12}$ [C] $\frac{5}{6}$ [D] $\frac{1}{6}$
48. A man spent $\frac{3}{8}$ of his salary on rent and $\frac{1}{3}$ of the remainder on cloths. The fraction of his salary left is:
[A] $\frac{23}{24}$ [B] $\frac{5}{6}$ [C] $\frac{19}{24}$ [D] $\frac{5}{12}$
49. The reciprocal of 21 is:
[A] 21^2 [B] 0.21 [C] $\sqrt{21}$ [D] $\frac{1}{21}$
50. The reciprocal of 0.02 is:
[A] 500 [B] 50 [C] 0.5 [D] 0.05
51. The reciprocal of 0.0002 is:
[A] 50 [B] 500 [C] 5000 [D] 50,000
52. To one decimal place, the reciprocal of 0.625 is:
[A] 1.6 [B] 0.6 [C] 6.3 [D] 62.5
53. The reciprocal of $\frac{x}{y}$ is:

[A] $-\frac{y}{x}$ [B] $-\frac{x}{y}$ [C] $\frac{y}{x}$ [D] $\frac{1}{xy}$

54. $\frac{5}{16}$ as a decimal is:
[A] 0.3125 [B] 0.5125 [C] 0.4125 [D] 0.2725

55. 25 out of 200 pineapples, are bad. As a decimal, the fraction of pineapples that are bad is:
[A] 0.875 [B] 0.185 [C] 0.225 [D] 0.125

56. 0.375 as a fraction is:
[A] $\frac{1}{8}$ [B] $\frac{3}{8}$ [C] $\frac{5}{16}$ [D] $\frac{11}{16}$

57. 0.72 is equivalent to:
[A] $\frac{18}{25}$ [B] $\frac{7}{10}$ [C] $7\frac{1}{5}$ [D] $\frac{71}{100}$

58. 5 – 0.003 equals:
[A] 0.002 [B] 4.003 [C] 4.007 [D] 4.997

59. The value of 4.7–1.9 + 2.1 is:
[A] 5.9 [B] 8.7 [C] 1.7 [D] 4.9

60. 0.93 + 0.08 is equals to:
[A] 1.01 [B] 1.1 [C] 1.11 [D] 0.101

61. $0.1 \times 0.2 \times 0.3$ is equal to:
[A] 0.06 [B] 0.006 [C] 0.05 [D] 0.005

62. The value of $\frac{2.4}{4}$ is:
[A] 0.6 [B] 6 [C] 60 [D] 2.1

63. The value of 136×47 is 6392. The value of 1.36×4.7 is:
[A] 0.6392 [B] 6.392 [C] 63.92 [D] 639.2

64. The value of 136×47 is 6392. The value of $\frac{63.92}{13.6}$ is:
[A] 47 [B] 0.047 [C] 0.47 [D] 4.7

65. $(0.12)^2$ is equal to:
[A] 1.44 [B] 0.144 [C] 0.0144 [D] 0.24

66. The value of $\frac{1}{0.2} + \frac{1}{0.25}$ is:
[A] 45 [B] 4.5 [C] 2.5 [D] 9

67. $78.75 \div 0.35$ is:
[A] 0.225 [B] 0.25 [C] 22.5 [D] 225

68. In base ten 11.1011_{two} is the same as:
[A] $\frac{59}{4}$ [B] $\frac{59}{8}$ [C] $\frac{59}{16}$ [D] $\frac{59}{32}$

69. When converted to base ten, 10.1001_{two} becomes:
[A] $2\frac{9}{16}$ [B] $2\frac{4}{16}$ [C] $2\frac{1}{16}$ [D] $2\frac{3}{16}$

70. The fraction which will result in a non-terminating decimal is:
[A] $\frac{7}{16}$ [B] $\frac{16}{84}$ [C] $2\frac{55}{125}$ [D] $4\frac{3}{48}$

1.5 ESTIMATIONS AND APPROXIMATIONS

ESTIMATIONS

Estimating sums and differences in whole numbers

Example

Estimate the following (i) $245 + 350 + 570$ (ii) $431 + 53$ (iii) $748 - 394$

Solution

(i) $245 + 350 + 570 \approx 200 + 400 + 600 \approx 1200$
(ii) $431 + 53 \approx 430 + 50 \approx 480$
(iii) $748 - 394 \approx 700 - 400 \approx 300$

Estimating Sums and Differences in Decimals

Example

Estimate to the nearest whole number. (i) $4.68 + 0.71$ (ii) $6.7234 - 3.5138$

Solution

(i) $4.68 + 0.71 \approx 5 + 1 \approx 6$ (ii) $6.7234 - 3.5138 \approx 7 - 4 \approx 3$

Example

Estimate to the nearest tenth. (a) $3.623 + 0.29 + 5.386$ (b) $4.86 - 3.456$

Solution

(a) $3.623 + 0.29 + 5.386 \approx 3.6 + 0.3 + 5.4 = 9.3$
(b) $4.86 - 3.456 \approx 4.9 - 3.5 = 1.4$

Estimating products and quotients in whole numbers

Example

Estimate the following (a) 58×24 (b) 653×56 (c) 48830×750

Solution

(a) $58 \times 24 \approx 60 \times 20 \approx 1200$ (b) $653 \times 56 \approx 700 \times 60 \approx 42000$
(c) $48,830 \times 750 \approx 50,000 \times 800 \approx 4,000,000$

Example

Estimate the following quotients (i) $\frac{562}{18}$ (ii) $\frac{68}{27}$ (iii) $\frac{62}{24}$

Solution

(i) $\frac{562}{18} \approx \frac{600}{20} \approx 30$ (ii) $\frac{68}{27} \approx \frac{70}{30} \approx 2.3$ (iii) $\frac{62}{24} \approx \frac{60}{20} \approx 3$

Estimation of Products and Quotients in Decimals

Example

Give an estimate for the following. (i) 4.755×0.5 (ii) $\frac{89.93}{4.1}$

Solution

(i) $4.755 \times 0.5 \approx 5 \times 0.5 \approx 2.5$ (ii) $\frac{89.93}{4.1} \approx \frac{90}{4} \approx 22.5$

APPROXIMATIONS

Decimal Places

Rounding off to the nearest tenth, hundredth, thousandth etc is the same as rounding off to 1, 2, 3 etc decimal places.

Significant Figures

The significant figures (s.f) of a given number are the figures (or digits) required to express the number to a given degree of accuracy. To count the number of significant figures in a given number, start with the first non-zero digit from the left and, moving to the right, counting all the digits including final zeros if they are to the right of the decimal point or up to the given degree of accuracy in the case of whole numbers.

For example, 3.1047, 3.1040, 0.031 047, 0.003 1040 and 3104.0 all have 5 significant figures but 3104000 may be to 4, 5, 6 or 7 significant figures depending on the given degree of accuracy.

In rounding a number to *n* significant figures, we replace the original number by a number with *n* significant figures.

Example

Round 6526 to the nearest: (a) ten (b) hundred (c) thousand.
In each case, state the number of significant figures.

Solution

(a) 6526 to the nearest ten = 6530 [3 s.f.]
(b) 6526 to the nearest hundred = 6500 [2 s.f.]
(c) 6526 to the nearest thousand = 7000 [1 s.f.]

Example

Round 1.045 to the given number of decimal places, stating the number of significant figures in each case. (a) 2 (b) 1

Solution

(a) 1.045 to 2 d.p.s. = 1.05 [3 s.f.]
(b) 1.045 to 1 d.p. = 1.0 [2 s.f.]

Example

Round 0.01027 to the given number of decimal places, stating the number of significant figures in each case. (a) 4 (b) 3 (c) 2 (d) 1

Solution

(a) 0.01027 to 4 d.p.s. = 0.0103 [3 s.f.]
(b) 0.01027 to 3 d.p.s. = 0.010 [2 s.f.]
(c) 0.01027 to 2 d.p.s. = 0.01 [1 s.f.]
(d) 0.01027 to 1 d.p.s. = 0.0 [0 s.f.]

MULTIPLE CHOICE EXERCISE 1:5

1. 0.015849 expressed correct to three significant figures is:
   [A] 0.0158 [B] 0.0159 [C] 0.0160 [D] 0.020
2. Given that $x = 0.0102$, correct to three significant figures. The value, which cannot be the actual value of x, is:
   [A] 0.01021 [B] 0.01014 [C] 0.01015 [D] 0.01016
3. 6474 correct to three significant figures is:
   [A] 647 [B] 648 [C] 6470 [D] 6480
4. The number 25.973 correct to three significant figures is:
   [A] 25.973 [B] 25.97 [C] 25.9 [D] 26.0
5. The number of people attending a football match is quoted as 27000, correct to 2 significant figures. The greatest possible attendance shown by this figure is:
   [A] 27,000 [B] 27499 [C] 27599 [D] 26999
6. 0.0063 correct to 2 decimal places is:
   [A] 0.006 [B] 0.01 [C] 0.06 [D] 0.10
7. After evaluating 2.35×0.48, the answer to 2 decimal places is:
   [A] 11.28 [B] 1.13 [C] 1.128 [D] 1.10
8. The value of $3.769 \div 0.7$ to the nearest tenth is:
   [A] 5.41 [B] 5.0 [C] 10 [D] 5.4
9. To the nearest whole number, the result of $\frac{6.6\times1.8}{5.4}$ is:
   [A] 2.2 [B] 3 [C] 2 [D] 22
10. $0.000252 \div 0.007$ to two decimal places is:
    [A] 0.04 [B] 0.03 [C] 0.36 [D] 0.40

11. By evaluating $\frac{7+3.32}{9.91-5.11}$, the answer to one decimal place will be:
    [A] 21.5 [B] 2.1 [C] 22.0 [D] 2.2
12. $0.44734 \div 0.01$, evaluated to the nearest hundredth is:
    [A] 44.70 [B] 45 [C] 44.73 [D] 44.00
13. $\frac{6.3\times60\times0.2}{3.6\times1.4}$, when simplified, the answer to the nearest ten is:
    [A] 15 [B] 20 [C] 10 [D] 1.5

1.6 STANDARD FORM

It is better and easier to write very small decimals or very large numbers in **standard form (scientific notation)** rather than in the **normal or decimal form** many initial or final zeros. A number in standard form is of the form

$$N = A \times 10^{n}, \text{where } n \in \mathbb{N} \text{ and } 1 \le |A| < 10$$

If the number is between 0 and 1, n is negative and if the number is greater than 1, n is positive.

Example

Express the following in standard form
(a) 0.000,000,000,000,000,000,000,000,000,911
(b) 300,000,000 (c) 0,00048

Solution

(a) $0.000,000,000,000,000,000,000,000,000,911 = 9.11\times10^{-28}$
(b) $300,000,000 = 3 \times 10^{8}$
(c) $0.00048 = 4.8 \times 10^{-4}$

Calculations in Standard Form

The following laws of indices are very useful when performing calculations in standard form.

$$10^{n} \times 10^{m} = 10^{n+m}$$

$$10^{n} \div 10^{m} = 10^{n-m}$$

$$(10^{m})^{n} = 10^{nm}$$

Example

Evaluate $\sqrt{\frac{(1.8\times10^{9})(4\times10^{-4})}{2\times10^{-2}}}$ leaving your answer in standard form.

Solution

$$\sqrt{\frac{(1.8\times 10^{9})(4\times 10^{-4})}{2\times 10^{-2}}} = \sqrt{0.9\times 4\times 10^{9-4-2}}$$

$$= \sqrt{3.6\times 10^{7}} = 6\times 10^{3}$$

MULTIPLE CHOICE EXERCISE 1:6

1. 930,000,000 in standard form is:
   [A] 93.0×10^{9} [B] 9.3×10^{8} [C] 9.3×10^{7} [D] 9.3×10^{-7}
2. 5238, expressed in standard form is:
   [A] 5.238×10^{3} [B] 5.238×10^{2} [C] 5.238×10^{1} [D] 5.238×10^{0}
3. Expressed in standard form 435600 is:
   [A] 4.536×10^{7} [B] 4.536×10^{6} [C] 4.536×10^{5} [D] 4.536×10^{4}
4. When expressed in standard form 2789 equals:
   [A] 2.789×10^{-3} [B] 2.789×10^{2} [C] 2.789×10 [D] 2.789×10^{3}
5. In standard form, 52006 can be written as:
   [A] 5.2006×10^{3} [B] 5.2006×10^{-4} [C] 5.2006×10^{4} [D] 5.2006×10^{-3}
6. 120,000 written in standard form is:
   [A] 1.2×10^{2} [B] 1.2×10^{3} [C] 1.2×10^{4} [D] 1.2×10^{5}
7. The number 36700 written in standard form is:
   [A] 3.67×10^{3} [B] 3.67×10^{5} [C] 3.67×10^{4} [D] 3.67×10^{2}
8. 325,000 in standard form is:
   [A] 3.25×10^{6} [B] 3.25×10^{-6} [C] 3.25×10^{5} [D] 3.25×10^{-5}
9. 0.00562 in standard form is:
   [A] 5.62×10^{-3} [B] 0.562×10^{-2} [C] 5.62×10^{-2} [D] 5.62×10^{2}
10. Expressed in standard form 0.0462 is:
    [A] 0.462×10^{-1} [B] 0.462×10^{-2} [C] 4.62×10^{-1} [D] 4.62×10^{-2}
11. 0.000834 in standard form is:
    [A] 8.34×10^{-4} [B] 8.34×10^{-5} [C] 8.34×10^{3} [D] 8.34×10^{4}
12. 0.0000027 in standard form is:
    [A] 2.7×10^{6} [B] 2.7×10^{-6} [C] 2.7×10^{5} [D] 2.7×10^{-5}
13. 0.000,000,070,2 in standard form is:
    [A] 7.02×10^{-5} [B] 7.5×10^{-6} [C] 7.02×10^{-7} [D] 7.5×10^{-8}
14. Expressed in standard form 0.000,082,3 becomes:
    [A] 0.823×10^{5} [B] 0.823×10^{-5} [C] 8.23×10^{5} [D] 8.23×10^{-5}
15. Written in standard form 0.000370 is:

[A] 3.7×10^{-1} [B] 7.5×10^{-2} [C] 3.7×10^{-3} [D] 7.5×10^{-4}

16. 46×900 expressed in standard form is:

    [A] 4.14×10^{3} [B] 4.14×10^{5} [C] 4.14×10^{4} [D] 4.14×10^{6}

17. When 4 hours is converted to seconds and expressed in standard form, the result is:

    [A] 1.44×10^{3} [B] 1.44×10^{-3} [C] 1.44×10^{4} [D] 1.44×10^{-4}

18. 258 km when expressed to mm and in standard form becomes:

    [A] 2.58×10^{8} [B] 2.58×10^{7} [C] 2.58×10^{6} [D] 2.58×10^{5}

19. $\frac{8.75}{0.025}$ expressed in standard form is:

    [A] 3.5×10^{-3} [B] 3.5×10^{-2} [C] 3.5×10^{1} [D] 3.5×10^{2}

20. Given that $0.00208 = 2.08 \times 10^{x}$, the value of x is:

    [A] 4 [B] −4 [C] 5 [D] −5

21. The product of 0.06 and 0.09 in standard form is:

    [A] 5.4×10^{2} [B] 5.4×10^{-3} [C] 5.4×10^{1} [D] 5.4×10^{-2}

22. When $0.009 \div 0.012$ is evaluated, the answer in standard form is:

    [A] 7.5×10^{2} [B] 7.5×10^{-3} [C] 7.5×10^{-1} [D] 7.5×10^{-2}

23. 5.7×10^{4} in ordinary form is:

    [A] 5700 [B] 57000 [C] 7500 [D] 75000

24. As a decimal fraction 8.2×10^{-5} is:

    [A] 0.0082 [B] 0.00082 [C] 0.000082 [D] 0.0000082

25. In normal form 3.746×10^{-3} is:

    [A] 0.003746 [B] 0.0003746 [C] 0.03746 [D] 3746

26. 9.258×10^{-3} in normal form correct to 3 significant figures is:

    [A] 926 [B] 0.093 [C] 0.009 [D] 0.00926

27. 7.15×10^{5} in normal form is:

    [A] 71500 [B] 715000 [C] 7150 [D] 7150000

28. 4.5×10^{3} is a number in standard form. The number is:

    [A] 0.0045 [B] 0.045 [C] 4500 [D] 45000

29. The sum of 728.93 and 0.46 expressed in standard form to three significant figures is:

    [A] 72.9×10 [B] 7.29×10^{2} [C] 728×10^{0} [D] 7.28×10^{0}

30. The sum of 2.48×10^{3} and 5.49×10^{4} is:

    [A] 6.148×10^{1} [B] 6.148×10^{4} [C] 6.148×10^{3} [D] 6.148×10^{2}

31. $4 \times 10^{2} \times 2 \times 10^{-2}$ is equal to:

    [A] 8×10^{6} [B] 8×10^{-2} [C] 8×10^{-8} [D] 6×10^{-2}

32. $7.580 \times 10^{9} + 7.677 \times 10^{9}$ is equal to:

[A] 1.5257×10^{10} [B] 1.5257×10^{8} [C] 1.5257×10^{9} [D] 1.5257×10^{7}

33. Given that $x = 5.7 \times 10^6, y = 1.8 \times 10^6, x - y =$:

    [A] 3.9×10^{-6} [B] 3900000×10^{6} [C] 3.9×10^{6} [D] 39×10^{5}

34. 2.52×10^5 is equal to:

    [A] 2.52×10^{5} [B] 2.52×10^{6} [C] 2.52×10^{4} [D] 2.52×10^{-5}

35. The population of two towns *A* and *B* is given as 5.77×10^6 and 3.66×10^6 respectively. The difference in the population of *A* and *B* is:

    [A] 2.11×10^{0} [B] 2.11×10^{-2} [C] 2.11×10^{4} [D] 2.11×10^{6}

36. $7.42 \times 10^{-6} - 4.33 \times 10^{-7}$ is the same as:

    [A] 6.987×10^{-6} [B] 5.987×10^{-6} [C] 4.987×10^{-6} [D] 3.987×10^{-6}

37. $5.72 \times 10^3 - 2.37 \times 10^2$ in the normal form, is equal to:

    [A] 4483 [B] 4473 [C] 5483 [D] 3450

38. The product of 0.012 and 0.0008 in standard form is:

    [A] 9.6×10^{6} [B] 9.6×10^{-6} [C] 9.6×10^{5} [D] 9.6×10^{-5}

39. When simplified, $(2 \times 10^{-5})(4 \times 10^{-2})$ becomes:

    [A] 8.0×10^{-7} [B] 8.0×10^{7} [C] 8.0×10^{3} [D] 8.0×10^{-3}

40. On simplification $8 \times 10^5 + 2 \times 10^3$ gives:

    [A] 4×10^{8} [B] 4×10^{2} [C] $4 \times 10^{\frac{5}{3}}$ [D] $4 \times 10^{-\frac{5}{3}}$

41. 450×70 is:

    [A] 3.15×10^{5} [B] 3.15×10^{4} [C] 3.15×10^{3} [D] 3.15×10^{5}

42. The result of $\frac{0.126}{36}$ is:

    [A] 3.5×10^{4} [B] 3.5×10^{-4} [C] 3.5×10^{3} [D] 3.5×10^{-3}

43. The simplified value of $\frac{(12\times10^{8})(16\times10^{-6})}{1\times10^{-2}}$ in standard form is:

    [A] 1.92×10^{6} [B] 192×10^{4} [C] 1.92×10^{4} [D] 1.92×10^{2}

44. On evaluation $\sqrt{\frac{0.81\times10^{-5}}{2.25\times10^{7}}}$ equals:

    [A] 3.6×10^{-13} [B] 6.0×10^{-7} [C] 3.6×10^{13} [D] 6.0×10^{7}

45. Given that $p = 3.6 \times 10^{-3}$ and $q = 2.25 \times 10^6$ the value of $\sqrt{\frac{p}{q}}$ is:

    [A] 1.6×10^{-9} [B] 4.0×10^{-5} [C] 1.6×10^{9} [D] 4.0×10^{5}

1.7 NUMBERS IN REAL LIFE

WEIGHTS AND MEASURES

Conversion of Metric Units of Length and Mass

			metre			
Kilo—	Hecto—	Deka—	or	deci—	centi—	milli—
			gram			

We can use the above metric unit conversion scale as follows.

(a) To convert to a unit that is to the left, move the decimal point the corresponding number of decimal places to the left.

(b) To convert to a unit that is to the right, move the decimal point the corresponding number of decimal places to the right.

Example

Convert the following

(a) 25 mm to decimetres. (b) 3 kg to decigrams.

(c) 421 hectograms to centigrams. (d) 1200 mm to metres.

Solution

(a) 25 mm = 0.25 dm (move 2 decimal places to the left).

(b) 3 kg = 30000 dg (move 4 decimal places to the right).

(c) 42.1 Hg = 421000 cg (move 4 decimal places to the right).

(d) 1200 mm = 1.2 m (move 3 decimal places to the left).

Conversion of Units of Area

To convert units of area quickly and easily use the conversion scale below.

Km^2 - Ha - a - m^2 - dm^2 - cm^2 - mm^2

To convert to a unit of area that is to the left, count the number of steps and multiply by 2. Next, move the decimal point the corresponding number of decimal places to the left.

To convert to a unit of area that is to the right, count the number of steps and multiply by 2. Next, move the decimal point the corresponding number of decimal places to the right.

Example

Convert (a) 2,000 mm^2 to m^2 (b) 0.42 Ha to cm^2.

Solution

(a) The unit m^2 is 3 steps to the left of mm^2 and $3 \times 2 = 6$.
2,000 mm^2 = 0.002 m^2 (count 6 decimal places to the left).

(b) The unit cm^2, is 4 steps to the right of ha and $4 \times 2 = 8$
$\therefore$ 0.42 Ha = 42,000,000 cm^2 (count 8 decimal places to the right).

Volume Measure (Cubic Measure)

The following table shows the units most commonly used in measuring volume.

Unit of Volume	Relationship to mm^3
Cubic metre m^3	1000,000,000 mm^3
Cubic decimetres dm^3	1000,000 mm^3
Cubic centimetres cm^3	1000 mm^3
Cubic millimetres mm^3	1 mm^3

Capacity

Capacity is the volume of a liquid or gas. Capacity uses all the units of volume. In addition, the following units based on the litre are also used.

litre (l)	decilitre (dl)	centilitre (cl)	millilitres (ml)

$$1 \text{ litre } (l) = 1000 \text{ cm}^3$$

$$1 \text{ millilitre } (ml) = 1 \text{ cm}^3 = 0.001 \text{ litre}$$

$$10 \text{ millilitres } (ml) = 1 \text{ centilitre } (cl)$$

$$10\ cl = 1 \text{ decilitre } (dl)$$

$$10\ dl = 1l$$

$$1000\ cl = 1l$$

Note that sometimes capacity is loosely used to means the space available. For example the capacity of a 72 seater bus is 72 persons.

Time Measure

Historical Time

Larger values of B.C. time stand for earlier	Larger values of A.D. time stand for later

B.C. means before Christ (was born)
A.D. means anno Domini in Latin, translated as in the year of our Lord.

Relationship between Time Units

60 seconds make 1 minute (min)
60 minutes make 1 hour (h)
24 hours make 1 day (d)
7 days make 1 week (wk)
4 weeks make 1 month (mth)
12 months make1 year (yr)
10 years make 1 decade
10 decades make 1 century
10 centuries make 1 millennium

Conversion of Time Units

To convert from a larger unit to a smaller one, multiply.
To convert from a smaller unit to a larger one, divide.

Example
Convert 4 years to seconds.

Solution
4 years = $4 \times 365 \times 24 \times 60 \times 60$ secs
$=126,144,000$ seconds $\approx 1.26144 \times 10^8$ seconds

Example
Convert 18921600 seconds to years.

Solution

$$18921600 \text{ seconds} = \frac{18921600}{60 \times 60 \times 24 \times 7 \times 4 \times 12} \text{ years} = 0.65 \text{ years}$$

Clocks and Watches

For 12 hour time, time before 12 noon carries the suffix a.m. (ante meridian which means before 12 noon), and time after 12 noon carries the suffix p. m. (post meridian which means after 12 noon)

	24 hour time	
12 hour time	before 12 noon (a.m.)	After 12 noon (p.m.)
1	1	13
2	2	14
3	3	15
4	4	16
5	5	17
6	6	18
7	7	19
8	8	20
9	9	21
10	10	22
11	11	23
12	12	24 or oo

Example

(a) Change 2.45 p.m.to 24-hour time. (b) Change 16:22 to 12-hour time.

Solution

(a) 2.45 p.m. = 2.45 + 12 = 14:45 (b) 16:22 =16:22–12:00 = 4:22 p.m.

Inter-conversion of Temperature

$$F = \frac{9}{5}C + 32 \Leftrightarrow C = \frac{(F-32)5}{9}$$

Example

Convert (a) 85 °C from °F (b) 149 °F to °C

Solution

(a) $F = \frac{9}{5}C + 32 = \frac{9}{5}(85) + 32 = 185°\text{C}$

(b) $C = \frac{(F-32)5}{9} = \frac{(149-32)5}{9} = 65°\text{F}$

Conversion of Currency

We normally convert currency using the rate of exchange which usually varies from time to time.

Example

The rate of exchange is such that 1 FF = 120 FCFA and 45 Belgian Franc (BF) = 1 FF. Find in FCFA the cost of a motorcycle, which cost 180,000 BF.

Solution

$45 \text{ BF} = 1 \text{ FF} \Rightarrow 180{,}000 \text{ BF} = \frac{1}{45} \times 180{,}000 \text{ FF} = 4{,}000 \text{ FF}$

$1 \text{ FF} = 120 \text{ FCFA} \Rightarrow 4{,}000 \text{ FF} = 4{,}000 \times 120 \text{ FCFA} = 480{,}000 \text{ FCFA}.$

MULTIPLE CHOICE EXERCISE 1:7

1. Four packets have weights marked: 2 kg, 250 g, 500 g, and 3.5 kg. Their total weight in kilograms is:
   [A] 5.25 [B] 6.25 [C] 6 [D] 13
2. Taken 400 grams out of a bucket containing 25 kg of rice. The number of kilograms left in the bucket is:
   [A] 2.46 [B] 24.6 [C] 246 [D] 2460
3. A metre rule was broken into 4 equal pieces. Each piece is:
   [A] 25 cm [B] 15 cm [C] 10 cm [D] 50 cm
4. 42,600 m expressed in kilometres is:
   [A] 0.426 km [B] 4.26 km [C] 42.6 km [D] 426 km
5. 650 mm written in metres is:
   [A] 6500 m [B] 65 m [C] 0.065 m [D] 0.65 m
6. The sum of 48 km and 20 dm is:
   [A] 48.002 m [B] 480.02 m [C] 4800.02 m [D] 48002 m
7. The number of square centimetres in a square metre is:
   [A] 1,000 cm^2 [B] 10,000 cm^2 [C] 100,000 cm^2 [D] 1,000,000 cm^2
8. $\frac{2}{3}$ of 3 hours is:
   [A] 1200 minutes [B] 120 minutes [C] 1200 seconds [D] 120 seconds
9. The difference between $3x$ minutes and $40x$ seconds is:
   [A] 140 seconds [B] $140x$ [C] 37 seconds [D] $37x$ seconds
10. Bih took 35 minutes to go to school. If she arrived at the school at 7:25 a.m., the time she left her house was:
    [A] 6:50 a.m. [B] 6:55 a.m. [C] 6:45 a.m. [D] 6:50 a.m.

11. The value of 0.75 hours: 20 minutes is:
    [A] 85 minutes [B] 75 minutes [C] 65 minutes [D] 55 minutes
12. The number of seconds in 12 minutes 10 seconds is:
    [A] 720 seconds [B] 730 seconds [C] 490 seconds [D] 630 seconds
13. The number of days in 30 weeks 5 days is
    [A] 305 days [B] 35 days [C] 180 days [D] 215 days
14. $\frac{3}{8}$ of a day is:
    [A] 324 minutes [B] 540 seconds [C] 540 minutes [D] 324 seconds
15. Given that 1000 FCFA = ₦ 240. ₦ 30,000 will be equivalent to:
    [A] 250,000 FCFA [B] 80,000 FCFA [C] 125,000 FCFA [D] 72,000 FCFA
16. Given that \$3 = 720 FCFA. 600,000 FCFA in dollars (\$) is:
    [A] \$2,500 [B] \$14,400,000 [C] \$277 [D] \$200 00
17. At the current exchange rate, 1 dollar (\$) = 500 FCFA and
    1 pound = 1000 FCFA. 200 pounds, exchanged to dollars will be:
    [A] \$400 [B] \$10.00 [C] \$100 [D] \$2.00

FINANCIAL ARITHMETIC

Basic Terms

Term	Symbol	Definition
Principal	P	Amount of money invested.
Interest	I	Charge on an investment.
Time	T	Period of an investment.
Rate	R	Ratio of the interest to the principal percent per annum.
Amount	A	$I + P$
Simple interest	I	Interest charged on the principal only

SIMPLE INTEREST

$$I = \frac{PRT}{100} \Rightarrow P = \frac{100I}{RT}, T = \frac{100I}{PR}, R = \frac{100I}{PT} \text{ and } A = P\left(1 + \frac{PRT}{100}\right)$$

Example

Calculate the simple interest, which Mrs. Fube will get on 640,000 FRS for 2 years 6 months at the rate of $4\frac{1}{2}$% per annum.

Solution

$$P = 640000 \text{ FRS}, \text{T} = 2\frac{1}{2} \text{ years} = 2.5 \text{ years}, \text{R} = 4\frac{1}{2}\% = 4.5\%$$

$$I = \frac{PRT}{100} = \frac{640000 \times 4.5 \times 2.5}{100} = 72000 \text{ FRS.}$$

Example

Danjuma paid a simple interest of 132, 000 FRS., for money borrowed at 8% per annum after 3 years. Calculate the amount of money she borrowed.

Solution

$I = 132{,}000$ Frs., $T = 3$ years, $R = 8\%$ per annum.

$$P = \frac{100I}{RT} = \frac{100 \times 132000}{3 \times 8} = 550{,}000 \text{ Frs.}$$

Example

Mr. Jaiy invested 700,000 Frs. at 4% per annum simple interest. How long will the amount reach 784000 Frs.?

Solution

$A = 784{,}000$ Frs., $P = 784{,}000$ Frs., $R = 4\%$

$$I = A - P = 784{,}000 - 700{,}000 = 84{,}000 \text{ Frs .}$$

$$T = \frac{100I}{PR} = \frac{100 \times 84000}{700000 \times 4} = 3 \text{ years}$$

Example

At what rate must Pa Fru invest 2,500,000 Frs. for 4 years to yield a simple interest of 500,000 Frs.?

Solution

$I = 500{,}000$ Frs., $P = 2{,}500{,}000$ Frs., $T = 4$ years

$$R = \frac{100I}{PT} = \frac{100 \times 500000}{2500000 \times 4} = 5\% \text{ per annum.}$$

COMPOUND INTEREST

Compound interest is the amount of money paid on the principal and the interest accumulated from the past years or months as the case may be. The formula for calculating the compound interest on an investment P at a rate r % per annum for t years is

$$\boldsymbol{A = P\left(1+\frac{r}{100}\right)^t \text{ and } I = A - P}$$

Example

Mr. Mofor invested the sum of 3,000,000 Frs. for 3 years at 10% per annum compound interest. Calculate his interest at the end of the 3 years.

Solution

Let the principals for first, second and third years be $P_1,\ P_2,\ P_3$ and interest for first, second and third years be I_1, I_2 and I_3 respectively.

Then, $R = 10\%, P_1 = 3{,}000{,}000\ Frs, I = \frac{PRT}{100} = \frac{10}{100}P$

$$I_1 = \frac{10 \times 3{,}000{,}000}{100} = 300{,}000 \text{ Frs}$$

$$\Rightarrow \mathrm{P}_2 = \mathrm{P}_1 + \mathrm{I}_1 = 3{,}000{,}000 + 300{,}000 = 3{,}300{,}000 \text{ Frs.}$$

$$I_2 = \frac{10 \times 3{,}300{,}000}{100} = 330{,}000 \text{ Frs}$$

$$\Rightarrow \mathrm{P}_3 = \mathrm{P}_2 + \mathrm{I}_2 = 3{,}300{,}000 + 330{,}000 = 3{,}630{,}000 \text{ Frs.}$$

$$I_3 = \frac{10 \times 3{,}630{,}000}{100} = 363{,}000 \text{ Frs}$$

Compound interest $= I_1 + I_2 + I_3$

$= 300{,}000 + 330{,}000 + 363{,}000 = 993{,}000$ Frs.

Alternatively, we can use the compound interest formula as follows.

Solution

$$A = P\left(1+\frac{r}{100}\right)^t = 3{,}000{,}000\left(1+\frac{10}{100}\right)^3$$

$$= 3000000(1.1)^3 = 3{,}993{,}000 \text{ Frs.}$$

Compound interest I = A – P = 3,993,000 – 3,000,000 = 993,000 Frs.

Compound Interest with Varying Principal

Example

Mr. Nformi borrows 6,000,000 FCFA from his Credit Union at 3% per annum compound interest. He repays 2,000,000 FCFA at the end of each year. How much does he still owe at the end of the third repayment?

Solution

Principal borrowed = 6,000,000 FCFA

$I_1 = \frac{3}{100} \times 6{,}000{,}000 = 180{,}000$ FCFA

Amount owed at end of year 1 = 6,000,000+180,000 = 6,180,000 FCFA

First repayment = 2,000,000 FCFA

Amount owed after first repayment = 6,180,000−2,000,000

= 4,180,000 FCFA

$I_2 = \frac{3}{100} \times 4{,}180{,}000 = 125{,}400$ FCFA

Amount owed at the end of year 2= 4,180,000+ 125,400 =4,305,400 FCFA

Second repayment = 2,000,000 FCFA

Amount owed after second repayment 4,305,400−2,000,000

= 2,305,400 FCFA

$I_3 = \frac{3}{100} \times 2{,}305{,}400 = 69{,}162$ FCFA

Amount owed at end of year 3 = 2,305,400 + 69,162 = 2,374,562 FCFA

Third repayment = 2,000,000 FCFA

Amount owed after third repayment = 2,374,562−2,000,000

= 374,562 FCFA

Therefore, amount owed at the end of the third repayment is 374,562 FCFA.

Depreciation

Depreciation is loss in the market value of assets such as equipment or buildings over a given period.

Suppose an asset with original value *P*, depreciates at the rate of *r*% per annum then its value *V*, after *t* years will be

$$\boldsymbol{V = P\left(1 - \frac{r}{100}\right)^t \text{ and } D = P - V}$$

Example
A car bought at 6,000,000, FCFA depreciates at the rate of 10% per annum. What will be the value of this car after 3 years?

Solution

$$V = P\left(1-\frac{r}{100}\right)^t = 6{,}000{,}000\left(1-\frac{10}{100}\right)^3$$
$$= 6{,}000{,}000(0.9)^3 = 4374000 \text{ Frs.}$$

We can also solve this problem using the progressive method. Try that.

Percentage Change (Difference)

$$\text{Percentage change} = \frac{\text{change in value}}{\text{original value}} \times 100\%$$

Some common types of percentage changes are discount, percentage profit (gain) or loss, percentage error, percentage passed or failed, percentage devaluation, percentage increase or decrease etc.

Example
A student measured the length of a pole as 34.5 m instead of 34.8 m. Calculate the percentage error in the measurement of the pole.

Solution

$$\text{Percenatage increase} = \frac{\text{Difference}}{\text{Real value}} \times 100\%$$
$$= \frac{34.8 - 34.5}{34.8} \times 100\% = 0.86\%$$

Example
In 1985 the number television owners in a certain town was 3000. In 1995 the number increased by 80%. How many people had televisions in 1995?

Solution

$$\text{Percenatage increase} = \frac{\text{increase}}{\text{Original}} \times 100\%$$
$$80\% = \frac{\text{No. in } 1995 - 3000}{3000} \times 100\%$$
$$\Rightarrow \text{No. in } 1995 = 80 \times 30 + 3000 = 5400 \text{ people.}$$

Profit and Loss

Cost price is the price at which a trader buys goods.
Selling price is the price at which a trader sells goods.

Profit = Selling Price – Cost Price

Loss = Cost Price – Selling Price

Example
A trader bought 15 bags of rice at 12,500 Frs. each. If she sold all these bags for 180,000 Frs., what profit or loss, did she make?

Solution
Total cost price for 15 bags = 12,500 ×15 = 187,500 Frs.
Total selling price for 15 bags = 180,000 Frs.

Cost price is greater than the selling price so she made a loss.
⇒Loss = cost price – selling price = 187,500 – 180,000 = 7,500 Frs.

Percentage Profit And Loss

$$\text{Percentage profit} = \frac{\text{profit}}{\text{CP}} \times 100\% = \frac{\text{SP}-\text{CP}}{\text{CP}} \times 100\%$$

$$\text{Percentage loss} = \frac{\text{loss}}{\text{CP}} \times 100\% = \frac{\text{CP}-\text{SP}}{\text{CP}} \times 100\%$$

(1) A profit of p % ⇒ SP = $(100 + p)$ % × CP and $\text{CP} = \frac{100}{(100+p)} \times \text{SP}$.

(2) A loss of l % ⇒ SP = $(100 - l)$ % × CP and $\text{CP} = \frac{100}{(100-l)} \times \text{SP}$.

Example
A Woman bought a car at 4,500,000 Frs. and sold it at a profit of 15%. What was its selling price?

Solution

$$\text{SP} = \frac{100+p}{100} \times \text{CP} = \frac{100+15}{100} \times 4{,}500{,}000 = 5{,}175{,}000 \text{ Frs.}$$

Example
A man sold a house for 3,600,000 Frs., incurring a loss of 40 %. Calculate the amount he paid for the house.

Solution

$$CP = \frac{100}{(100 - l)} \times SP = \frac{100}{(100 - 40)} \times 3{,}600{,}000 = 6{,}000{,}000 \text{ Frs.}$$

Example

A trader bought a television at 250,000 Frs. and sold it for 280,000 Frs. Calculate his percentage profit or loss.

Solution

He made profit because his selling price is higher than the cost price.

$$\text{Percentage profit} = \frac{SP - CP}{CP} \times 100\%$$

$$= \frac{280{,}000 - 250{,}000}{250{,}000} \times 100\% = 12\%$$

DISCOUNT

Discount is a reduction from the standard (or marked) price of a commodity.

Discount = Marked price – SP

$$\text{Percentage discount} = \frac{(\text{Marked price} - \text{SP})}{\text{Marked price}} \times 100\%$$

Example

A trader sells a shirt marked 2800 Frs. at a discount of 5 %. What is its selling price?

Solution

$$SP = \frac{(100 - \text{Percentage discount})}{100} \times \text{marked price}$$

$$= \frac{95}{100} \times 2800 = 2660 \text{ Frs.}$$

Example

A pair of shoes has a marked price of 16000 Frs. A customer buys it at 15000 FRS, what is his percentage discount?

Solution

Discount = 16000 –15000 Frs. =1000 Frs.

$$\text{Percentage discount} = \frac{\text{Discount}}{\text{Marked price}} \times 100\% = \frac{1000}{16000} \times 100\% = 6.25\%$$

Example
A bookseller sells a book at 7200 Frs. Given that, the discount on the book is 25 %. Calculate the marked price.

Solution

$$\text{Marked price} = \frac{100}{\text{Percentage discount}} \times \text{SP} = \frac{100}{75} \times 7200 = 9600 \text{ FRS}$$

TAXES

A **tax** is an amount of money levied by a government or association on its citizens or members for it running.

$$\text{Tax} = \frac{\text{Tax rate}}{100} \times \text{earnings}$$

Example

A man earns 136,000 FRS., per month. If he pays a tax of $3\frac{1}{2}\%$ of his salary, calculate his tax in FRS.

Solution

$$\text{Tax} = \frac{\text{Tax rate}}{100} \times \text{earnings} = \frac{3.5}{100} \times 136{,}000 = 4760 \text{ FRS.}$$

Example
A civil servant has a monthly salary of 189,000 Frs. If his allowances total 42,000 Frs. Calculate his income tax given that, the rate is 21 % and the government does not tax allowances.

Solution
Taxable income = 189,000 −42,000 = 147,000 Frs.

$$\text{Tax} = \frac{\text{Tax rate}}{100} \times \text{earnings} = \frac{21}{100} \times 147{,}000 = 30{,}870 \text{ Frs.}$$

MULTIPLE CHOICE EXERCISE 1:8

1. The cost of 4 articles at 720 FCFA each is:
   [A] 2920 FCFA [B] 2880 FCFA [C] 2820 FCFA [D] 2900 FCFA
2. The cost of 2.5 kg of tomatoes at 360 FCFA per kg is:
   [A] 900 FCFA [B] 860 FCFA [C] 1100 FCFA [D] 720 FCFA

3. The cost of 2 metres of material at 1200 FCFA per metre and 3 metres at 1500 FCFA per metre is:
   [A] 6900 FCFA [B] 13500 FCFA [C] 6600 FCFA [D] 7100 FCFA
4. The cost of 2000 articles at 25 FCFA each is:
   [A] 5,000 FCFA [B] 25,000 FCFA [C] 50,000 FCFA [D] 500,000 FCFA
5. The change from 1000 FCFA after buying 18 buttons at 5 FCFA each is:
   [A] 990 FCFA [B] 890 FCFA [C] 810 FCFA [D] 910 FCFA
6. The cost of 200 articles at 30 FCFA each is:
   [A] 7000 FCFA [B] 6000 FCFA [C] 5000 FCFA [D] 5700 FCFA
7. While doing his Physics practical, Che recorded a reading as 1.12 cm, instead of 1.21 cm. The percentage error he made is:
   [A] 1.17% [B] 6.38% [C] 7.44% [D] 8.5%
8. Miss Yaje sold an article for 7500 FRS instead of 12750 FRS. Her percentage error correct to one decimal place is:
   [A] 41.2% [B] 5.3% [C] 1.7% [D] 1.4%
9. A student measured the length and breadth of a rectangular lawn as 59.6 cm and 40.3 cm respectively instead of 60 cm and 40 cm. The percentage error in his calculation of the perimeter of the lawn is:
   [A] 1.4% [B] 0.1% [C] 0.2% [D] 0.7%
10. A boy estimated his transport fare for a journey as 1900 FRS instead of 2000 FRS. The percentage error in his estimate is:
    [A] 95% [B] 47.5% [C] 5.26% [D] 5%
11. To two significant figures, the percentage error in approximating 0.375 to 0.4 is:
    [A] 6.7% [B] 6.6% [C] 2.5% [D] 2.0%
12. A furniture maker estimated that the cost of making a cupboard would be about 250000 FRS. He bought the materials and found that the cost came to 275000 FRS. The percentage increase in the estimate is:
    [A] 5% [B] 10% [C] 15% [D] 20%
13. The word discount means:
    [A] Money reduced from the price of a hired article.
    [B] Money taken out of your wage.
    [C] Money taken off the price of an article.
    [D] Money owed to a businessperson.
14. During a sale, a shop reduced the price of every article by 10%. The selling price of an article originally priced at 4300 FCFA is:
    [A] 4300 FCFA [B] 3400 FCFA [C] 3870 FCFA [D] 3970 FCFA
15. For his holidays, a man put aside 10% of his 15000 FCFA weekly wage for 40 weeks in the year. The amount saved for his holidays is:

[A] 60000 FCFA [B] 30000 FCFA [C] 15000 FCFA [D] 150000 FCFA

16. A businessperson decided to give 10 % discount on all the purchases from his store. The price a customer pays for a shirt that was marked at 5,400 FRS is:
[A] 5,940 FCFA [B] 5,500 FCFA [C] 5,300 FCFA [D] 4,860 FCFA
17. A girl bought a record for 1500 FCFA and sold it for 1200 FCFA. Her loss as a percentage of the cost price is:
[A] 15% [B] 20% [C] 60% [D] 75%
18. A woman bought a dress for 15000 FCFA. To make a profit of 20% the dress the selling price must be:
[A] 19,000 FRS [B] 18,000 FRS [C] 16,000 FRS [D] 14,400 FRS
19. A trader made a profit of 50% on his cost price by selling a radio at 15,000 FCFA. The record cost him:
[A] 5,000 FRS [B] 7,500 FRS [C] 10,000 FRS [D] 25,000 FRS
20. A development association decreased her budget by 4 percent this year. Given that, the annual budget for last year was 32,000,000 FRS. Its budget for this year in FRS is:
[A] 30,720,000 [B] 25,000 [C] 33,280,000 [D] 2,240,000
21. A man bought a house for 1,000,000 FRS and later auctioned it for 800,000 FRS. The percentage loss was:
[A] 20% [B] 30% [C] 40% [D] 25%
22. 22. By selling some crates of soft drinks for 6000 FRS, a dealer makes a profit of 50 %. The amount, which the dealer pays for the drinks, is:
[A] 12,000 FRS [B] 25,000 FRS [C] 4,500 FRS [D] 4,000 FRS
23. A trader makes a loss of 15 % when selling an article. The ratio, selling price: cost price is:
[A] 3:20 [B] 3:17 [C] 17:20 [D] 20:23
24. A man made a loss of 15 % by selling an article for 59500 FRS. The cost price of the article was:
[A] 60,000 FRS [B] 70,000 FRS [C] 68,425 FRS [D] 89,250 FRS
25. Mr. Anyang bought a piece of land for 2.5 million FCFA and sold it for 3 million FCFA. His percentage profit is:
[A] $18\frac{2}{3}\%$ [B] 20% [C] 16.7% [D] 25%
26. If the simple interest on 200,000 FRS after 9 months is 6000 FRS, the interest rate per annum is:
[A] $2\frac{1}{4}\%$ [B] 6% [C] 5% [D] 4%

27. A cooperative society charges an interest of $5\frac{1}{2}$ % per annum on any amount borrowed by its members. If a member borrows 125000 FRS, the amount he pays back after 1 year will be:
[A] 136,875 FRS [B] 131,875 FRS [C] 128,750 FRS [D] 126,250 FRS
28. The compound interest on 400,000 FRS for 2 years at 8 % per annum is:
[A] 32,000 FRS [B] 34,560 FRS [C] 66,560 FRS [D] 43,200 FRS
29. A car brand under usage depreciates at the rate of 10 % per annum. Given that, a new brand of this car cost 4,000,000 FRS. The cost in FRS of such a car, which is for 3 years old, will be:
[A] 3,440,000 [B] 2,916,000 [C] 3,600,000 [D] 3,240,000
30. The exchange rate for FCFA is 1000 FCFA for 238 Naira. 2,856,000 FCFA changed into Naira would be:
[A] 12 [B] 1200 [C] 679728 [D] 6797280
31. A customer can buy a computer for 189500 FCFA cash or on hire purchase for 9 monthly payments of 23700 FCFA each. The hire purchase method cost is greater by:
[A] 22,100 FCFA [B] 23,800 FCFA [C] 29,500 FCFA [D] 33,200 FCFA
32. A car worth 10,000,000 FCFA depreciates by 10% yearly. The value after 2 years in FCFA will be:
[A] 1,200,000 [B] 8,000,000 [C] 9,000,000 [D] 8,100,000

CHAPTER 2

ALGEBRA

2.1 BASIC ALGEBRAIC EXPRESSIONS

Variables

A variable is a symbol or letter such as n, which stands for any number.
Values are the numbers used to replace or substitute variables.
A **pro-numeral** is any symbol or letter, which stands for a numeral.

Algebraic Sentences (Expressions)

The following table show different key words and the implied operation.

Operation	Key Words
Addition	Add (to), sum , plus, more than, increased by
Subtraction	Subtract (from), difference, minus, less than, decreased by, reduced by, take away, less
Multiplication	Multiply (by), product (of), times, twice, thrice, of (fractions/percentages)
Division	Divide, quotient, share

Variable Substitution

Variable substitution is the process of finding different values of an expression for different values of the variables.

Example

(a) Find $a + b$ when $a = 2$ and $b = 8$

(b) If $t = s + u$, find t when $s = 6$ and $u = -3$.

Solution

(a) When $a = 2$ and $b = 8$, $a + b = 2 + 8 = 10$.

(b) $t = s + u = 6 + (-3) = 6 - 3 = 3$

Unknowns and Constants

An **unknown** is a pro-numeral, which stands particular values which can be found.
A **constant** is a letter or symbol, which has a fixed value. π is a constant.

Terms and Coefficients

The **terms** of an expression are the different parts joined by '+' or '–' signs. The **coefficients** are the numbers which appear in front of each term.

Like and Unlike Terms

Like terms are two or more terms which are constants or which are the same in relation to all variables and their powers.

U**nlike terms** are two or more terms whose variables or powers are different.

Example

State the like terms in $4 + 6xy - 8 - 5xy^2 - 2xy + 3y^2x - x^2y$.

Solution

4 and 8, $6xy$ and $2xy$, $5xy^2$ and $3y^2x$.

Algebraic Rules

The properties of numbers also hold in algebra. We can use the properties of numbers in the following table to simplify expressions.

Properties of addition	For all values of x, y, z
Commutative property	$x + y = y + x$
Associative property	$(x + y) + z = x + (y + z)$
Additive identity property of zero	$x + 0 = x$ and $0 + x = x$

Distributive property	For all values of x, y, z
Distributivity over addition	$x(y + z) = xy + xz$
Distributivity over subtraction	$x(y - z) = xy - xz$

Properties of multiplication	For all values of x, y, z
Commutative property	$xy = yx$
Associative property	$(xy)z = x(yz)$
Multiplicative property of zero	$0(x) = 0$ and $x(0) = 0$
Multiplicative identity property of one	$1(x) = 1$ and $x(1) = 1$

Combining Like-Terms

We can combine like-terms of an algebraic expression by adding or subtracting them. On the other hand, we cannot combine unlike terms.

Example

Compute (a) $a + 2a + 7a$ (b) $5p + \frac{3p}{8} - \frac{5p}{8}$

Solution

(a) $a + 2a + 7a = 10a$ (b) $5p + \frac{3p}{8} - \frac{5p}{8} = \frac{38p}{8} = \frac{19p}{4}$

Example

Evaluate the following.

(a) $19x - 15x - 7x + 4x$ (b) $3x + 2y - 5x + 7y$

Solution

(a) $19x - 15x - 7x + 4x = x$

(b) $3x + 2y - 5x + 7y = (3x - 5x) + (2y + 7y) = -2x + 9y$

Multiplication and Division in Algebra

Example

Evaluate each of the following.

(a) $(4x)(2x)$ (b) $(3x)(5y)$ (c) $(-2ab)(3a)$

(d) $\frac{2x^2y}{xy} + \frac{5xy}{xy}$ (e) $\frac{20pq}{4p^2} - \frac{16q}{4p}$

Solution

(a) $(4x)(2x) = 8x^2$ (b) $(3x)(5y) = 15xy$

(c) $(-2ab)(3a) = -6a^2b$ (d) $\frac{2x^2y}{xy} + \frac{5xy}{xy} = 2x + 5$

(e) $\frac{20pq}{4p^2} - \frac{16q}{4p} = \frac{5q}{p} - \frac{4q}{p} = \frac{q}{p}$

The HCF and LCM of Literal Expressions

To find the HCF and LCM of literal expressions do repeated division until there are no common factors left. Recall that the HCF is the product of the common factors and the LCM is the product of the highest powers of all the factors.

Example

Find the HCF and LCM of $8pq^2r^3, 6p^2q^2, 10pq^2r^2$

Solution

2	$8pq^2r^3$	$6p^2q^2$	$10pq^2r^2$
p	$4pq^2r^3$	$3p^2q^2$	$5pq^2r^2$
q^2	$4q^2r^3$	$3pq^2$	$5q^2r^2$
r^2	$4r^3$	$3p$	$5r^2$
	$4r$	$3p$	5

$\therefore \text{HCF} = 2\times p\times q^2 = 2pq^2$

and $\text{LCM} = 2\times p\times q^2\times r^2\times 4r\times 3p\times 5 = 120p^2q^2r^3$

MULTIPLE CHOICE EXERCISE 2:1

2. The pair of expressions which are like terms are:
   [A] $2xy$ and $3x$ [B] $5xy^2$ and yx^2
   [C] $3x^2y$ and $7yx^2$ [D] $4y$ and $2yx$
2. The pair of expressions which are unlike terms are:
   [A] $2ab^5$ and $3b^5a$ [B] $5xy^2$ and yx^2
   [C] $3x^2y$ and $7yx^2$ [D] $4xy$ and $2yx$
3. The expressions, which are like terms, are:
   [A] $\frac{1}{5}xy$ and $\frac{1}{5}x$ [B] $6xy^2$ and $6yx^2$
   [C] $\frac{1}{5}x^2y$ and $\frac{1}{5}x$ [D] $6y$ and $6yx$
4. The expressions, which are unlike terms, are:
   [A] $\frac{1}{7}ab^3$ and $7b^3a$ [B] $7xy^2$ and $7yx^2$
   [C] $7x^2y$ and $7yx^2$ [D] $4xy$ and $4yx$
5. In the expansion $a(b+c) = ab + ac$, the law used is:
   [A] The associative law [B] The distributive law
   [C] The commutative law [D] The multiplicative law
6. In $31 + (52 + 23) = (31 + 52) + 23$, the law used is:
   [A] Associative law of addition [B] Commutative law of addition
   [C] Distributive law of addition [D] Identity law of addition
7. In $\left(50\times\frac{1}{3}\right)\times 24 = 50\times\left(\frac{1}{3}\times 24\right)$, the law used is:
   [A] Associative law of multiplication

[B] Commutative law of multiplication
[C] Distributive law of multiplication
[D] Multiplicative identity law

8. In $3\frac{2}{5} \times x = x \times 3\frac{2}{5}$, the law used is:
[A] Associative law of multiplication
[B] Commutative law of multiplication
[C] Distributive law of multiplication
[D] Multiplicative identity law

9. In $x(7) = 7x$, the law used is:
[A] Associative law [B] Commutative law
[C] Distributive law [D] Multiplicative identity law

10. In $(p \times q) \times 6 = p \times (q \times 6)$, the law used is:
[A] Associative law of multiplication
[B] Commutative law of multiplication
[C] Distributive law of multiplication
[D] Identity law of multiplication

11. Given that $p = 1, q = -1$ and $r = 0$. The value of $p + q + r$ is:
[A] −2 [B] −1 [C] 0 [D] 1

12. Given that $p = 1, q = -1$ and $r = 0$. The value of pq is:
[A] −2 [B] −1 [C] 0 [D] 1

13. Given that $p = 1, q = -1$ and $r = 0$. The value of $p(q + r)$ is:
[A] −2 [B] −1 [C] 0 [D] 1

14. If $x = 2$, the value of $2x^2 + 3$ is:
[A] 11 [B] 19 [C] 14 [D] 24

15. When b = -1, the value of 5-b-b^2:
[A] 5 [B] 3 [C] 2 [D] 0

16. If $\frac{1}{v} = \frac{1}{f} - \frac{1}{u}$ then the value of v when $f = 2$ and $u = 3$ is:
[A] −1 [B] +5 [C] +6 [D] −6

17. When $x = -1$, the value of $\frac{x^2 + x - 2}{x^2 + x - 3}$ is:
[A] −2 [B] −9 [C] $-\frac{1}{2}$ [D] 1

18. Given that $p = 2, q = 5$ and $r = -4$. When evaluated $3p^2 - q^2 - r^3$ gives:
[A] −77 [B] 77 [C] 51 [D] 101

19. Given that $x = -3$ *and* $y = -7$, $\frac{x^2 - y}{y^2 - x}$ has value:

[A] $-\frac{1}{11}$ [B] $\frac{1}{23}$ [C] $\frac{4}{13}$ [D] $\frac{12}{17}$

20. The expression, which is not equal to, $\frac{1}{2}pq$ is:
    [A] $\frac{pq}{2}$ [B] $\frac{p \times q}{2}$ [C] $\frac{1}{2}qp$ [D] $\frac{1}{2}p \times q$

21. If $\frac{1}{2}p = x + y$, then p equals:
    [A] $2x + y$ [B] $\frac{1}{2}(x+y)$ [C] $x+2y$ [D] $2(x+y)$

22. The LCM of $9p^2q$ and $15pq^2$ is:
    [A] $45p^2q^2$ [B] $45p^2q$ [C] $45pq^2$ [D] $45pq$

23. The LCM of $2a$ and $7a^2$ is:
    [A] $14a$ [B] $14a^2$ [C] $28a$ [D] $28a^2$

24. The LCM of $6a^2b^2, 24ab^2, 40a^2bc^2$ is:
    [A] $120a^2b^2c^2$ [B] $124a^2b^2c^2$ [C] $144a^2b^2c^2$ [D] $96a^2b^2c^2$

25. Given that $\frac{7}{4x} - \frac{5}{x} + \frac{4}{3x}$. The L C M of the denominators is:
    [A] $12x^2$ [B] $12x$ [C] 12 [D] $12x^{-2}$

26. The HCF of $12a^2b$ and $10ab$ is:
    [A] $5ab$ [B] $4ab$ [C] $2ab$ [D] $2a^2b$

27. The HCF of $8ab^2c^3, 6a^2b^2, 10ab^2c^2$ is:
    [A] $8ab^2c^3$ [B] $10ab^2c^2$ [C] $2ab$ [D] $2ab^2$

28. If $b \in \mathbb{R}$, the positive square root of $9b^2$ is:
    [A] $9b$ [B] 9 [C] $3b$ [D] $3b^2$

29. The factor (s) of $4x^2y$ is/are:
    [A] $2x$ [B] xy [C] $4x$ [D] $2x$, xy and $4x$

30. The coefficient of x in $(x - 2)(x + 9)$ is:
    [A] –2 [B] 7 [C] 9 [D] –18

2.2 EXPANSIONS AND FACTORISATION

EXPANSIONS

(1) $a(b+c) = ab + ac$ or $(b+c)a = ab + ac$

(2) $a(b-c) = ab-ac$ or $(b-c)a = ab-ac$

(3) $(a+b)(c+d) = a(c+d) + b(c+d) = ac + ad + bc + bd$

Example

Expand the following.

(a) $2(4x+3)$ (b) $u(3x-1)$ (c) $(24a-1)a$ (d) $(y-2)(y+3)$

Solution

(a) $2(4x+3) = 2(4x) + 2(3) = 8x + 6$

(b) $u(3x-1) = u(3x)-u(1) = 3ux-u$

(c) $(24a-1)a = 24a^2-a$

(d) $(y-2)(y+3) = y^2 + 3y - 2y - 6 = y^2 + y - 6$

The Square of a Binomial

(i) $(a+b)^2 = a^2 + 2ab + b^2$ (ii) $(a-b)^2 = a^2 - 2ab + b^2$

We call these identities the **perfect square** identities. They are very useful in evaluating squares of numbers.

Example

Evaluate (a) 99^2 (b) 503^2

Solution

(a) $99^2 = (100-1)(100-1) = 10000-200+1 = 9801$

(b) $503^2 = (500+3)(500+3) = 250000 + 3000 + 9 = 253009$

FACTORISATION

Factorisation is the reverse process of expansion. Thus, $ab + ac = a(b+c)$.

Example

Factorise the following:

(a) $2x + 2y$ (b) $4xy + 12xz$ (c) $px - py + qx - qy$

(d) $x - y + xy - 1$ (e) $6x - 6y + 3ay - 3ax$

Solution

(a) $2x + 2y = 2(x + y)$ (b) $4xy + 12xz = 4x(y + 3z)$

(c) $px - py + qx - qy = p(x - y) + q(x - y) = (x - y)(p + q)$

(d) $x - y + xy - 1 = (xy - y) + (x - 1)$ (Rearrange and group)

$= y(x - 1) + 1(x - 1) = (x - 1)(y + 1)$

(e) $6x - 6y + 3ay - 3ax = 6(x - y) + 3a(y - x)$

$= 6(x - y) - 3a(x - y)$

$= (6 - 3a)(x - y)$

Note that $(x - y) = -(y - x)$!

The Difference of Two Squares

$$(a+b)(a-b) = a^2 - b^2$$

Example

Evaluate the following.

(a) $4^2 - 3^2$ (b) $25^2 - 16^2$

Solution

(a) $4^2 - 3^2 = (4 - 3)(4 + 3) = 1(7) = 7$

(b) $25^2 - 16^2 = (25 - 16)(25 + 16) = 9(41) = 369$

Example

Factorise the following:

(a) $25a^2 - 9b^2$ (b) $a^2 - b^2c^2$

Solution

(a) $25a^2 - 9b^2 = (5a)^2 - (3b)^2 = (5a - 3b)(5a + 3b)$

(b) $a^2 - b^2c^2 = a^2 - (bc)^2 = (a - bc)(a + bc)$

Factorising Quadratic Expressions

The following is the procedure for factorizing the expression $ax^2 + bx + c$.

(i) Multiply a by c to have ac.

(ii) Find the pair of integral factors p and q of ac whose sum or difference is b.

(iii) Substitute the middle term bx with the sum or difference of px and qx in $ax^2 + bx + c$.

(iv) Factorise the expression by grouping.

(v) When the coefficient of x^2 is 1, we can write the factors directly without taking pains to go through (iii) and (iv).

Example

Factorise:

(a) $2x^2 - x - 1$ (b) $x^2 - 4x - 21$

(c) $3 - 8y + 4y^2$ (d) $3x^2 + xy - 10y^2$

Solutions

(a) $2x^2 - x - 1 = 2x^2 - 2x + x - 1 = 2x(x-1) + 1(x-1)$
$= (x-1)(2x+1)$ (Factorising by grouping)

(b) $x^2 - 4x - 21 = (x-7)(x+3)$

(c) $3 - 8y + 4y^2 = 3 - 6y - 2y + 4y^2 = 3(1-2y) - 2y(1-2y)$
$= (1-2y)(3-2y)$

(d) $3x^2 + xy - 10y^2 = 3x^2 - 5xy + 6xy - 10y^2$
$= x(3x-5y) + 2y(3x-5y) = (x+2y)(3x-5y)$

Identities

An **identity is a statement**, which is true for all values of the variable. The right hand side substitutes the left hand side and vice versa.
Examples of identities are $(1+x)^2 = 1 + 2x + x^2$ and $4x^2-1 = (2x+1)(2x-1)$.

MULTIPLE CHOICE EXERCISE 2:2

1. The expression, which is a perfect square, is:
   [A] $(x+1)(x-1)$ [B] $x^2 + 2x + 1$ [C] $x^2 - y^2$ [D] $x^2 - 2x - 1$
2. The expression, which is not a difference of two squares, is:
   [A] $(x+1)(x-1)$ [B] $x^2 - 1$ [C] $x^2 - y^2$ [D] $(x-y)^2 = 6$
3. An identity among the following is:
   [A] $(x+1)(2x+3) = 2x^2 + 5x + 3$ [B] $(3p-1)(2p+1) = 3p^2 - 5p - 1$
   [C] $2y + 7 = 3y - 5$ [D] $x^2 + 2x + 1 = 6$
4. The statement, which is not an identity, is:
   [A] $(x+2)(x-1) = x^2 + x - 2$ [B] $(5x-1)(x+1) = 5x^2 + 4x - 1$
   [C] $(3x+1)(2x-1) = 6x^2 - x - 1$ [D] $(3x-1)(2x+1) = 3x^2 - 5x - 1$
5. When simplified, $5yx - 7xy + 4yx$ equals:
   [A] $9yx$ [B] $-9xy$ [C] $8xy$ [D] $2xy$
6. The simplified form of $6p + 7q - 8q - 5p$ is:

[A] $p-q$ [B] $q-p$ [C] $p+q$ [D] $11p-5q$

7. We can simplify $-7x + 8y - 2 + 9x - 10y + 4$ to have:
   [A] $2(x-y+1)$ [B] $2(x-y-1)$ [C] $2(x+y-1)$ [D] $2(x+y+1)$
8. Simplifying $15x - 12y + 8z - 14x + 12y - 8z$ leads to:
   [A] x [B] y [C] z [D] $x-y+z$
9. By simplification $x + (-x) + y$ is exactly:
   [A] $-y$ [B] y [C] $2x-y$ [D] $2x+y$
10. $32e + 6f - 12e + 4f$ can also be:
    [A] $38e+6f$ [B] $30e+8f$ [C] $28e+9f$ [D] $20e+10f$
11. When expanded $-2a(3a^2b + 4b^2)$ gives:
    [A] $-6ab^2-8a^2b$ [B] $-6ab^2-4ab^2$ [C] $-6ab^2+8a^2b$ [D] $-6a^3b-8ab^2$
12. On Simplification $13x - (2x - 4x - 3x)$ becomes:
    [A] $8x$ [B] $18x$ [C] $-8x$ [D] $-18x$
13. $9x - (5x - 3y) - y$ is equal to:
    [A] $4x - 2y$ [B] $4x + 2y$ [C] $5x - 2y$ [D] $5x + 2y$
14. $-2a - 5b - (8b - 5a) =$
    [A] $-8a+13b$ [B] $3a+3b$ [C] $3a-13b$ [D] $7a-13b$
15. $(2x + y) + (x - 2y)$ is the same as:
    [A] $3x+y$ [B] $x-3y$ [C] $x+3y$ [D] $3x-y$
16. $(2x + y) - (x - 2y)$ is equal to:
    [A] $3x-y$ [B] $x+3y$ [C] $x-3y$ [D] $3x+y$
17. $(2x - 3) - (2 - 3x)$ is equal to:
    [A] $5x-5$ [B] $5x-1$ [C] $x-5$ [D] $x-1$
18. Adding $(2x + y)$ and $(x - 2y)$ gives:
    [A] $3x+y$ [B] $x-3y$ [C] $x+3y$ [D] $3x-y$
19. Given the statement

    $x - 13y + 5z - 4m = x - (\qquad)$

    The expression required in the bracket is:

    [A] $-13y+5z-4m$ [B] $-13y+5z$

    [C] $-13y+5z-4x$ [D] $13y-5z+4m$
20. On expansion $(2x-5)(x-3)$ gives:
    [A] $x^2-11x-15$ [B] $2x^2-11x+15$ [C] $2x^2-5x-8$ [D] $x^2-5x+15$
21. Given that $p = 3 - 2y$ and $q = 4 + 3y$. The value of pq is:
    [A] $-6y^2-y-12$ [B] $6y^2-y-12$ [C] $-12+y+6y^2$ [D] $12+y-6y^2$
22. $(2x + y)(x - 2y)$ is equal to:

[A] $2x^2 - 2y^2$ [B] $2x^2 + 3xy + 2y^2$ [C] $2x^2 - 3xy - 2y^2$ [D] $2x^2 + 3xy - 2y^2$

23. $(2x - 1)(x + 2)$ is equal to:
[A] $2x^2 - 2$ [B] $2x^2 + x - 2$ [C] $2x^2 - x - 2$ [D] $2x^2 - 3x - 2$

24. On expansion $(4x - y)(x - 3y)$ becomes:
[A] $4x^2+13xy-3y^2$ [B] $6x^2-13xy+3y^2$ [C] $4x^2 - 13xy +3y^2$ [D] $6x^2+13xy-3y^2$

25. The product of $x - 1$ and $x + 1$ is:
[A] 2 [B] $2x$ [C] $x^2 + 2x - 1$ [D] $x^2 - 1$

26. If $(a + b)^2 = a^2 + 2ab + b^2$ the value of $(2a + 1)^2$ is:
[A] $4a^2 + 4a - 1$ [B] $4a^2 + 4a + 1$ [C] $4a^2 - 4a - 1$ [D] $4a^2 - 4a + 1$

27. The square of $x - 8$ is equal to:
[A] $x^2 - 16x - 64$ [B] $x^2 +16x - 64$ [C] $x^2 - 16x + 64$ [D] $x^2 - 32x + 64$

28. The coefficient of x in the expansion of $(x + 9)(x + 3)$ is:
[A] -12 [B] 12 [C] 3 [D] -3

29. The coefficients of x and x^2 in the expansion of $(x-3)^2$ are respectively:
[A] $-6, 1$ [B] 6, 1 [C] $-1, 6$ [D] $1, -6$

30. The coefficient of xy in the expansion of $(3x + 2y)(4x - 2y)$ is:
[A] -2 [B] -14 [C] 2 [D] 14

31. When factorised $3x(4 - y) - m(y - 4)$ becomes:
[A] $(3x + m)(4 - y)$ [B] $(3x - m)(4 - y)$ [C] $(3x +m)(y - 4)$ [D] $(3x - m)(y + 4)$

32. We can factorise the expression $x(a - c) + y(c - a)$ to obtain:
[A] $(a - c)(y + x)$ [B] $(a - c)(x - y)$ [C] $(a + c)(x - y)$ [D] $(a - c)(y - x)$

33. By factorising $m(2a - b) - 2n(b - 2a)$, the result is:
[A] $(2a - b)(2n - m)$ [B] $(2a - b)(m - 2n)$
[C] $(2a - b)(m + 2n)$ [D] $(2a - b)(m - 2n)$

34. The difference between the squares of the numbers 21 and 11 is:
[A] 20 [B] 100 [C] 220 [D] 320

35. The value of $13^2 - 12^2$ is:
[A] 25 [B] 5 [C] 1^2 [D] 125

36. $32x^3 - 8xy^2$ when factorised gives:
[A] $4(4x + y)(2x - y)$ [B] $(16x - y)(2x + y)$
[C] $8x(2x - y)$ [D] $8x(2x + y)(2x - y)$

37. By factorising $27p^2x^2 - 48y^2$ the result is:
[A] $3(3px - 4y)(3px + 4y)$ [B] $9(3px - 4y)^2$
[C] $9(px - 4y)(3px + 4y)$ [D] $3(3px - 4y)^2$

38. $(x - 2)(x + 3)$ are the factors of:
[A] $x^2 - 9$ [B] $x^2 - 6$ [C] $x^2 - x - 6$ [D] $x^2 + x - 6$

39. The result of factorising $x^2 + 4x - 192$ is:
[A] $(x - 4)(x + 48)$ [B] $(x + 48)(x + 4)$

[C] $(x-12)(x+16)$ [D] $(x-12)(x-16)$

40. When factorised, the expression $2x^2+x-15$ equals:
[A] $(2x+5)(x-3)$ [B] $(2x-5)(x+3)$
[C] $(2x-5)(x-3)$ [D] $(2x-3)(x+5)$

41. The quadratic $2e^2-3e+1$ when factorised, becomes:
[A] $(2e-1)(e-1)$ [B] $(e^2-3)(2e-1)$
[C] $(2e+3)(e-2)$ [D] $(2e-3)(e-1)$

42. Factorising $3a^2-11a+6$ leads to:
[A] $(3a-2)(a-3)$ [B] $(2a-2)(a-3)$
[C] $(3a-2)(a+3)$ [D] $(3a+2)(a-3)$

43. $2x^2-9x-45$, written as the product of two factors is:
[A] $(2x-9)(x-5)$ [B] $(2x-15)(x+3)$
[C] $(2x+15)(x-3)$ [D] $(2x-15)(x-3)$

44. On factorisation $6x^2+7x-20$ becomes:
[A] $(6x-5)(x+4)$ [B] $2(3x-5)(x+2)$
[C] $(3x+4)(2x-5)$ [D] $(3x-4)(2x+5)$

45. We can factorise the quadratic $3p^2+2p-1$ to have:
[A] $(3p+2)(p-1)$ [B] $(3p-1)(p+1)$
[C] $(3p+1)(p-1$ [D] $(3p-2)(p+1)$

46. Factorising the expression $2y^2+xy-3x^2$ gives rise to:
[A] $(x-y)(2y+3x)$ [B] $(2y-x)(2y+x)$
[C] $(3x-2y)(x-y)$ [D] $(2y+3x)(y-x)$

47. $6x^2+7xy-5y^2$ when factorised leads to:
[A] $(6x+5y)(x-y)$ [B] $(2x+5y)(3x-y)$
[C] $(3x+5y)(2x-y)$ [D] $(2x+y)(3x-5y)$

48. The result of factorising $mn-xy-nx+my$ is:
[A] $(n+y)(m-x)$ [B] $(n-y)(m+x)$
[C] $(x-m)(n-y)$ [D] $(n-y)(m-x)$

2.3 SIMPLE LINEAR EQUATIONS

Difference between Equations and Expressions

An expression is a meaningful combination of constants, operators, and variables representing numbers or quantities. Examples of expressions are $2x+3y-4$, $3x+2$ etc.

An **Equation** is a statement which states that two things are equal. Examples of equations are $3x+2=5$ and $2x+3y=4$.

A **simple linear equation** is an equation with one unknown whose power is one.

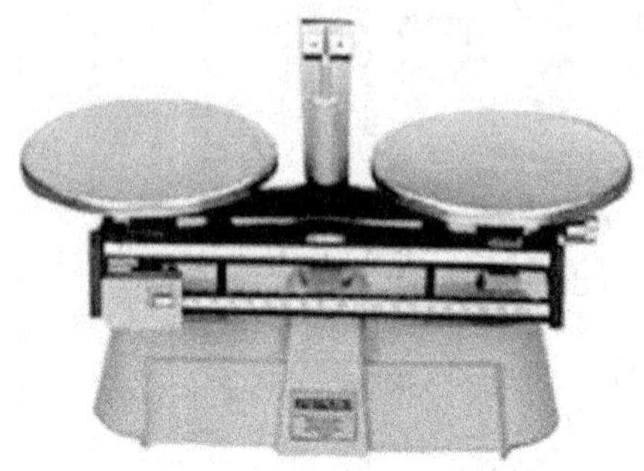

Equations are Balanced Systems.

Simple linear equations are classified as one step, two steps or multi-step depending on the number of steps required to solve them. However, solving simple linear equations requires the application of at least one of the principles in 1, 2 and 3 below.

1. Adding the same quantity to both sides does not destroy the balance.
2. Subtracting the same quantity from both sides does not destroy the balance.
3. Multiplying or dividing both sides by the same non-zero number does not destroy the balance.

Example

Solve the following equations.

(a) $1 - 2x = -7$ (b) $9 = 4m - 7$ (c) $7p = 8 + 3p$

(d) $\frac{v}{4} - 6 = 2$ (e) $5x - 2 = 3x + 4$ (f) $-7 - 4a = 48 + 7a$

Solutions

(a) $1 - 2x = -7$ (Subtract 1 from both sides)

$-2x = -8$ (Divide both sides by -2)

$x = 4$

(b) $9 = 4m - 7$ (Add 7 to both sides)

$16 = 4m$ (Divide both sides by 4)

$4 = m$ or $m = 4$

(c) $7p = 8 + 3p$

$4p = 8 \Rightarrow p = 2$

(d) $\frac{v}{4} - 6 = 2$

$\frac{v}{4} = 8 \quad \Rightarrow v = 32$

(e) $5x - 2 = 3x + 4$

$5x = 3x + 6$

$2x = 6 \Rightarrow x = 3$

(f) $-7 - 4a = 48 + 7a$

$-4a = 55 + 7a$

$-11a = 55 \Rightarrow a = -5$

Simple Linear Equations Involving Fractions and Decimals

When an equation involves fractions, multiply both sides by the LCM of the denominators to get rid of the fractions before solving. If it involves decimals, multiply both sides by a power of ten, to eliminate the decimals.

Examples

Solve the equation $0.17x - 10.966 = 1\frac{7}{20}x + 36.234$.

Solution

$$0.17x - 10.966 = 1\frac{7}{20}x + 36.234$$

$$0.17x - 10.966 = \frac{27}{20}x + 36.234$$

Multiply both sides by 1,000.

$$170x - 10966 = 1350x + 36234$$

$$-1180 = 47200$$

$$x = -40$$

Word Problems on Simple Linear Equations

Example

A girl has three times an amount of money as her friend. If their total sum is 700 FRS, find how much each of them has.

Solution

Let the girl's friend have *x* FRS, then the girl has 3*x* FRS.

$$\Rightarrow 3x + x = 700 \Rightarrow 4x = 700 \Rightarrow x = 175$$

Therefore, the girl has 3(175) = 525 FRS and her friend has 175 FRS.

MULTIPLE CHOICE EXERCISE 2.3

1. The letter x is an unknown in:
   [A] $3x + 5 = 0$ [B] $x^2 + x + 5$ [C] $y = x^2 + x + 5$ [D] $y = 3x + 5$
2. The letter x is a variable in:
   [A] $3x + 5 = 0$ [B] $2x^2 + x + 5$ [C] $(2x + 1)(x + 1)$ [D] $y = 3x + 5$
3. The additive inverse of $\frac{7}{2}$ is:
   [A] $\frac{7}{2}$ [B] $\frac{2}{7}$ [C] $-\frac{7}{2}$ [D] $-\frac{2}{7}$
4. The additive inverse of –14 is:
   [A] -14 [B] 14 [C] $\frac{1}{14}$ [D] $-\frac{1}{14}$
5. The multiplicative inverse of 12 is:
   [A] -12 [B] 12 [C] $-\frac{1}{12}$ [D] $\frac{1}{12}$
6. The multiplicative inverse of $\frac{2}{15}$ is:

[A] $-\frac{2}{15}$ [B] $\frac{15}{2}$ [C] $\frac{2}{15}$ [D] $-\frac{15}{2}$

7. Given that $8(x+8)=40$. The statement, which best interprets this equation is:
   [A] Eight times the sum of a number and 8 is 40
   [B] Eight times a number less than eight is 40
   [C] Eight less than a number is 40
   [D] The product of eight and a number is 40

8. The root of the equation $3x + 4x = 42$ is:
   [A] $x=4$ [B] $x=6$ [C] $x=8$ [D] $x=7$

9. The root of the equation $3n + 14 = 47$ is:
   [A] $n=8$ [B] $n=9$ [C] $n=10$ [D] $n=11$

10. The root of the equation $6y - 48 = 2y$ is:
   [A] $y=8$ [B] $n=9$ [C] $n=10$ [D] $n=12$

11. The value of x for which $2x + 8 = 0$ is:
   [A] -4 [B] 4 [C] -6 [D] 6

12. Given that $33 = 6y + 3$. The value of y must be:
   [A] 3 [B] 5 [C] 6 [D] 8

13. If $3x - 7 = 10$, then the value of x is:
   [A] $\frac{3}{17}$ [B] $-\frac{17}{3}$ [C] $\frac{17}{3}$ [D] $-\frac{3}{17}$

14. If $6x + 4 = -20$, then the value of x is:
   [A] $-\frac{3}{8}$ [B] $\frac{8}{3}$ [C] -4 [D] 4

15. The value of x in the equation $5x+1=31$ is:
   [A] 1 [B] 5 [C] 25 [D] 6

16. The value of y, which satisfies the equation, $4(y-4)=20$ is:
   [A] 1 [B] 24 [C] 6 [D] 9

17. The solution of $5(x-4)-4(x+1)=0$ is:
   [A] 16 [B] -24 [C] 24 [D] -16

18. The root of the equation
   $6(x - 4) + 3(x + 7) = 3$ is:
   [A] $\frac{3}{2}$ [B] $\frac{1}{3}$ [C] $\frac{1}{2}$ [D] $\frac{2}{3}$

19. $\frac{x-2}{3} = 8$, only if:
   [A] 26 [B] 22 [C] 24 [D] 19

20. When $\frac{x}{4} = \frac{5}{2}$, the value of x is:
   [A] 2 [B] 4 [C] 5 [D] 10

21. Given that $\frac{1}{3}(x+1)=6$ the value of x is:
   [A] 19 [B] 17 [C] 5 [D] 3

22. Given that $\frac{2}{x}=\frac{3}{6}$, the value of x is:

[A] 6 [B] 4 [C] 3 [D] 2

23. The only condition for which $\frac{5}{x+1}$ is equal to 4 is that:

[A] $x=4$ [B] $x=8$ [C] $x=\frac{1}{8}$ [D] $x=\frac{1}{4}$

24. Given that, $\frac{3-2y}{4}-\frac{2y}{6}$ the value of y is:

[A] $\frac{10}{9}$ [B] $\frac{9}{10}$ [C] 3 [D] −3

25. The root of the equation $\frac{2x+7}{6}+\frac{x-5}{3}=0$ is:

[A] $x=-\frac{3}{5}$ [B] $x=\frac{3}{4}$ [C] $x=\frac{1}{4}$ [D] $x=\frac{2}{5}$

26. Given that $\frac{3x-2}{6}-\frac{2x+7}{9}=2$. x is equal to:

[A] $x=-\frac{36}{5}$ [B] $x=\frac{36}{5}$ [C] $x=-\frac{16}{5}$ [D] $x=\frac{16}{5}$

27. The value of x, which satisfies the expression, $\frac{1}{x}+\frac{4}{3x}-\frac{5}{6x}+1=0$ is:

[A] $\frac{1}{6}$ [B] $\frac{1}{4}$ [C] $-\frac{3}{2}$ [D] $-\frac{7}{8}$

28. Using the relation $C=\frac{5}{9}(F-32)$, the value of F when $C=40$ is:

[A] 67 [B] 77 [C] 81 [D] 104

29. The value of t, which satisfies
$\frac{3t}{4}+\frac{1}{3}(21-t)=11$ is:

[A] $9\frac{3}{5}$ [B] $3\frac{9}{13}$ [C] 5 [D] $\frac{9}{13}$

30. If $8x-4=6x-10$, the value of $5x$ is:

[A] 7 [B] −15 [C] −3 [D] 3

31. The value of x, which satisfies the equation, $5(x-7)=7x-5$ is:

[A] $x=6$ [B] $x=-30$ [C] $x=-15$ [D] $x=-6$

32. If $2(x+1)=4x+3$, x equals:

[A] 2 [B] −2 [C] $\frac{1}{2}$ [D] $-\frac{1}{2}$

33. The value of v, which satisfies the equation, $\frac{12}{v}=\frac{15}{v+4}$ is:

[A] 10 [B] 12 [C] 14 [D] 16

34. The value of $3(p+7)$ for which $6p+5=4p+11$ is:
    [A] 15 [B] 20 [C] 25 [D] 30
35. The equation $\frac{2}{3}(x+5)=\frac{1}{4}(5x-3)$ has root:
    [A] $1\frac{1}{7}$ [B] 7 [C] 3 [D] $4\frac{3}{7}$
36. Given that $\frac{m}{3}+\frac{1}{2}=\frac{3}{4}+\frac{m}{4}$, the value of m is:
    [A] -3 [B] -2 [C] 2 [D] 3
37. The only condition for $0.6x-0.4$ to be equal to $1.2x+0.8$ is that:
    [A] $x=-2$ [B] $x=2$ [C] $x=-0.5$ [D] $x=-0.5$
38. Given that $0.9n-0.7=0.3n-0.1$, then n must be:
    [A] 4 [B] 3 [C] 2 [D] 1
39. A man is 23 years older than his son is this year. Given that his son will be 12 years old in ten years time, the man will be:
    [A] 25 years [B] 35 years [C] 33 years [D] 37 years
40. Abe bought two packets of sugar at x FRS each and four tins of milk at 400 FRS each. If the total cost is 3400 FRS, the price of a packet of sugar is:
    [A] 400 FRS [B] 1800 FRS [C] 900 FRS [D] 1600 FRS
41. 8 more than thrice a number is 35. The number is:
    [A] 27 [B] 43 [C] 9 [D] 14.3
42. A trader bought 400 liters of palm oil and sold $8x$ liters. If 160 liters are left, the number of liters he sold is:
    [A] 30 [B] 240 [C] 5 [D] 20
43. Given that $x=3$. The number, which we can add to $12x$ to make 57, is:
    [A] 21 [B] $\frac{19}{4}$ [C] 45 [D] 33
44. The sum of three consecutive numbers is 42. The largest of the numbers is:
    [A] 13 [B] 14 [C] 15 [D] 16
45. Three quarters of a certain number is 15. The number is:
    [A] 20 [B] 16 [C] 15 [D] 12
46. The ratio of $2x$ to $x+1$ is 5 : 3 only if x is:
    [A] $x=5$ [B] $x=4$ [C] $x=3$ [D] $x=2$
47. The ages of two people are in the ratio 5: 9. If the elder is 8 years older, the age of the younger is:
    [A] 45 years [B] 10 years [C] 8 years [D] 2 years

2.4 SIMULTANEOUS LINEAR EQUATIONS

1. If the simultaneous linear equations involve fractions or decimals, first multiply by the LCM or by a power of ten to eliminate the fractions or decimals respectively.
2. Algebraically we can solve any simultaneous linear equations in two unknowns using the substitution or the elimination method; however simultaneous linear equations with uniform coefficients are easier to solve using the method of elimination.

Example
Solve the following simultaneous linear equations using the substitution or elimination method. $\frac{x}{3} + 0.25y = \frac{1}{12}$ and $1.5x - \frac{y}{3} = -4$.

Solutions

$\frac{x}{3} + 0.25y = \frac{1}{12}$ ……………………………. ①

$1.5x - \frac{y}{3} = -4$ ……………………………. ②

Multiply ① by 12 and ② by 6 to eliminate the fractions and decimals.

$4x + 3y = 1$ ……………………………… ③

$9x - 2y = -24$ ……………………………… ④

We now solve the simultaneous equations using any of the methods we desire.

By the Method of Elimination
Multiply equation ③ by 2 and equation ④ by 3.

$8x + 6y = 2$ ……………………………… ⑤

$27x - 6y = -72$ ……………………………… ⑥

⑤+⑥: $35x = -70 \Rightarrow x = -2$

Substitute in ③: $4(-2) + 3y = 1 \Rightarrow y = 3$

By the Method of Substitution

From ③, $y = \frac{1-4x}{3}$ ……………………………… ⑦

Substitute in ④: $9x - 2\left(\frac{1-4x}{3}\right) = -24$

Multiply both sides by 3

$27x - 2(1 - 4x) = -72 \Rightarrow 35x - 2 = -72 \Rightarrow x = -2$.

Substitute in equation ⑦

$$y = \frac{1 - 4(-2)}{3} \Rightarrow y = 3$$

Simultaneous Linear Equations in Real Life

Example

A credit union gave Mr. Ngong 16000 francs consisting of 500 francs coins and 100 francs coins. The number of 100 francs coins is three times the number of 500 francs coins. How many of each type of coin did Mr. Ngong receive?

Solution

Let h = number of 100 francs coins and f = number of 500 francs coins

Then $100h + 500f = 16000 \Rightarrow h + 5f = 160$ ……………①

Also $h = 3f \Rightarrow h - 3f = 0$ ……………②

①-②: $8f = 160 \Rightarrow f = 20$

Substitute in ③: $h - 3(20) = 0 \Rightarrow h = 60$

∴ Number of 100 francs coins = 60 and number of 500 francs coins = 20.

MULTIPLE CHOICE EXERCISE 2.4

1. The pair of values of x and y which satisfy the simultaneous equations $x+y = 3$ and $3x - y = 1$ are:
   [A] (1,– 2) [B] (2,–1) [C] (1,2) [D] (–2,1)
2. Given that $x = 2y - 1$ and $2x = 2y - 1$. The value of x is:
   [A] – 9 [B] – 4 [C] 9 [D] 0
3. If $x = 2y - 1$ and $2x = 3y + 2$, then y equals:
   [A] 4 [B] 9 [C] –4 [D] –9
4. The roots of the simultaneous equations $x + y = 4$ and $2y - x = 5$ are:
   [A] (– 1,3) [B] (– 1,– 3) [C] (1,– 3) [D] (1,3)
5. The values of x and y that satisfy the simultaneous equations $2x + y = 7$ and $3x - 2y = 7$ are:
   [A] (– 1,3) [B] (3,1) [C] (–1,– 3) [D] (3,–1)
6. Given that $2p - m = 6$ and $2p + m = 1$. The value of $4p + 3m$ is:
   [A] 1 [B] 3 [C] 5 [D] 7
7. If $x + y = \frac{3}{2}$ and $x - y = \frac{5}{2}$, then $2y + x$ equals:
   [A] –2 [B] 1 [C] $\frac{1}{2}$ [D] –1
8. If $x + 2y = 1$ and $x - y = 2$, the value of $x + y$ is:
   [A] $1\frac{1}{3}$ [B] 1 [C] –1 [D] $-1\frac{1}{3}$

9. Given that $x + y = 7$ and $3x - y = 5$. When evaluated $\frac{y}{2} - 3$ gives:
[A] 3 [B] 1 [C] −1 [D] 4

10. If $2x + y = 7$ and $3x - y = 3$, then $7x$ is greater than 10 by:
[A] 1 [B] 7 [C] 3 [D] 10

11. Given the equations $4y - 5x = 14$ and $y = 3x$. The values of x and y are respectively:
[A] (2, 6) [B] (−2, −6) [C] (2, −6) [D] (−2, 6)

12. The values of x and y which satisfy the simultaneous equations $4x - y = 11$ and $5x + 2y = 4$ are:
[A] $x = 2, y = 3$ [B] $x = -2, y = -3$ [C] $x = -2, y = 3$ [D] $x = 2, y = -3$

13. If $3p - q = 6$ and $2p + q = 4$, then q is equal to:
[A] 0 [B] $\frac{1}{2}$ [C] $\frac{2}{3}$ [D] 1

14. The values of $x - y$, which satisfy the simultaneous equations $4x - 3y = 7$ and $3x - 2y = 5$ are:
[A] −3 [B] 3 [C] 2 [D] −2

15. $\frac{1}{3}x + y = 3$ and $x + \frac{1}{2}y = 4$ provided:
[A] $x = 3, y = 3$ [B] $x = 3, y = -3$ [C] $x = -3, y = 3$ [D] $x = -3, y = -3$

16. If the solutions of the pair of equations $2x + 3y = p$ and $3x - y = q$ are $x = -1$ and $y = 2$, the values of p and q are respectively:
[A] −4,5 [B] −5,4 [C] −4, −5 [D] 4, −5

17. An exercise book and a pencil cost 180 F. If the exercise book costs 140 F more than the pencil, then the pencil costs:
[A] 30 F [B] 25 F [C] 20 F [D] 15 F

18. Ambe is four times as old as Ndeh. If the sum of their ages is 20 years, the difference in their ages in years is:
[A] 16 [B] 12 [C] 8 [D] 4

19. 2 nuts and 3 bolts have a mass 28 g. 3 nuts and a bolt have a mass of 21 g. The mass of a bolt is:
[A] 4 g [B] 5 g [C] 6 g [D] 7 g

2.5 QUADRATIC EQUATIONS

The standard form of a quadratic equation is $ax^2 + bx + c = 0$, where a, b and c are constants and $a \neq 0$.

Solving Quadratic Equations

(a) *Factorisation method*

The fundamental theorem for solving the equation $(px+q)(rx+s)=0$ is that either $(px+q)=0$ or $(rx+s)=0 \Rightarrow x=-\frac{q}{p}$ or $x=-\frac{s}{r}$.

If $c=0$, the quadratic equation is $ax^2+bx=0$ which we can factorise and solve as follows.

$ax^2+bx=0 \Rightarrow x(ax+b)=0 \Rightarrow x=0 \text{ or } x=-\frac{b}{a}$.

If $b=0$ and $\frac{c}{a}\geq 0$, the quadratic equation is $ax^2+c=0$ which we can solve as follows.

$ax^2+c=0 \Leftrightarrow x^2=-\frac{c}{a} \Rightarrow x=\pm\sqrt{-\frac{c}{a}}$.

Example

Solve the following quadratic equations using the factorisation method.

(1) $x^2+9x+18=0$ (2) $6x^2+7x-3=0$

(3) $3x^2+4x=0$ (4) $4x^2-1=0$

Solution

(1)
$$x^2+9x+18=0$$
$$(x+3)(x+6)=0$$
$$\therefore (x+3)=0 \text{ or } (x+6)=0$$
$$x=-6 \text{ or } x=-3$$

(2)
$$6x^2+7x-3=0$$
$$(3x-1)(2x+3)=0$$
$$\therefore (3x-1)=0 \text{ or } (2x+3)=0$$
$$x=--\frac{3}{2} \text{ or } x=\frac{1}{3}$$

(3) $3x^2+4x=0 \Rightarrow x(3x+4)=0$
$$\Leftrightarrow x=0 \text{ or } x=-\frac{4}{3}$$

(4) $4x^2-1=0 \Leftrightarrow x^2=\frac{1}{4}$
$$\Rightarrow x=\pm\frac{1}{2}$$

The Quadratic Formula

If $ax^2+bx+c=0$, where a, b and c are constants and $a\neq 0$, then $x=\frac{-b\pm\sqrt{b^2-4ac}}{2a}$.

Example

Use the quadratic formula to solve the equation $2x^2 - 5x - 5 = 0$.

Solution

$x = \frac{-b \pm \sqrt{b^2 - 4ac}}{2a}$, $a = 2, b = -5$ and $c = -5$.

$\Rightarrow x = \frac{-(-5) \pm \sqrt{(-5)^2 - 4(2)(-5)}}{2(2)} = \frac{5 \pm \sqrt{65}}{4} \Rightarrow x = \frac{5 - \sqrt{65}}{4}$ or $= \frac{5 + \sqrt{65}}{4}$

Nature of roots of a Quadratic Equation

The **discriminant** of the expression $ax^2 + bx + c$ is $\Delta = b^2 - 4ac$.

1. If $\Delta > 0$, the equation has two real and distinct roots.
2. If $\Delta = 0$, the equation has repeated or equal real roots.
3. If $\Delta < 0$, the equation has no real roots i.e. the roots are complex or imaginary.

Worded Problems that lead to Quadratic Equations

Example

A man wants to buy a piece of land. The owner tells him that the area of the piece of land is 80 square metres and that the length is 2 metres longer than the width. The man hires you to work out the length and width for him. Go ahead and do your job.

Solution

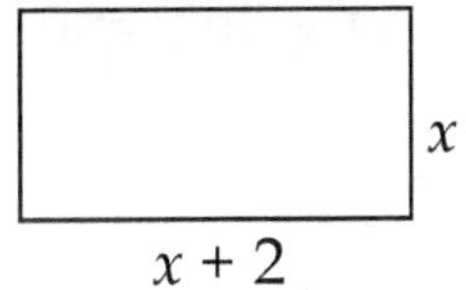

Let the width be x m, then the length will be $(x + 2)$ m.

$x(x + 2) = 80 \Rightarrow x^2 + 2x - 80 = 0$

$\Rightarrow x = 8$ or $x = -10$

Since $x > 0$, we discard $x = -10$. Clearly if the width is 8 m, the length will be 10 m, since $8 \times 10 = 80$.

MULTIPLE CHOICE EXERCISE 2:5

1. The expression, which is not a quadratic in x, is:
   [A] $2x^2 - 5x$ [B] $x(x - 5)$ [C] $x^2 - 5$ [D] $5(x - 1)$
2. The quadratic equation whose roots are $x = -2$ and $x = 7$ is:
   [A] $x^2 + 2x - 7 = 0$ [B] $x^2 - 2x + 7 = 0$
   [C] $x^2 - 5x - 14 = 0$ [D] $x^2 + 5x - 14 = 0$
3. A quadratic equation has roots $-\frac{2}{3}$ and $-\frac{1}{4}$. The required equation is:

[A] $x^2 - \frac{11}{12}x + 2 = 0$ [B] $12x^2 - 3x + 2 = 0$
[C] $12x^2 - 11x + 2 = 0$ [D] $12x^2 - 11x - 2 = 0$

4. The only quadratic equation with roots $-\frac{1}{2}$ and 2 is:
[A] $3x^2 - 3x + 2 = 0$ [B] $3x^2 + 3x + 2 = 0$
[C] $2x^2 - 3x + 2 = 0$ [D] $2x^2 - 3x + 2 = 0$

5. The quadratic equation whose roots are 3 and $\frac{2}{3}$ is:
[A] $3x^2 - 11x + 6 = 0$ [B] $x^2 - 11x + 6 = 0$
[C] $3x^2 - 11x + 2 = 0$ [D] $x^2 - 11x - 2 = 0$

6. Given that the roots of a quadratic equation are $\frac{1}{4}$ and 3 then the quadratic equation is:
[A] $4x^2 - 13x + 3 = 0$ [B] $4x^2 - 13x - 3 = 0$
[C] $4x^2 + 13x - 3 = 0$ [D] $4x^2 + 13x + 3 = 0$

7. The equation whose roots are 4 and –5 is:
[A] $x^2 - x - 20 = 0$ [B] $x^2 + x + 20 = 0$
[C] $x^2 - x + 20 = 0$ [D] $x^2 + x - 20 = 0$

8. The equation whose roots are $\frac{2}{3}$ and $-\frac{1}{4}$ is:
[A] $12x^2 + 11x + 2 = 0$ [B] $12x^2 - 11x + 2 = 0$
[C] $12x^2 - 5x - 2 = 0$ [D] $12x^2 - 11x - 2 = 0$

9. The roots of a quadratic equation in x are $-m$ and $2n$. The equation is:
[A] $x^2 + x(m - 2n) - 2mn = 0$ [B] $x^2 - x(m - 2n) - 2mn = 0$
[C] $x^2 - x(m - 2n) + 2mn = 0$ [D] $x^2 + x(m - 2n) + 2mn = 0$

10. The equation whose roots are – 8 and 5 is:
[A] $x^2 + 3x + 40 = 0$ [B] $x^2 - 3x - 40 = 0$
[C] $x^2 + 3x - 40 = 0$ [D] $x^2 - 3x + 40 = 0$

11. If $x^2 + 15x + 50 = ax^2 + bx + c = 0$, it is not true to say that:
[A] $x = 5$ [B] $x = 10$ [C] $x + 10 = 0$ [D] $bc = 750$

12. If $2x^2 + kx - 14 = (x+2)(2x-7)$ then, the value of k is:
[A] −3 [B] 5 [C] 9 [D] 11

13. Given that $5x^2 + 4x + 3 = a + b(x+1) + cx(x+1)$. The value of the constants a, b, and c are respectively:
[A] 7, 4, 5 [B] 4, −1, 5 [C] −1, 5, 4 [D] 7, 5, 4

14. The equation $(x+2)(x-7) = 0$ has roots:
[A] −2 and 7 [B] 2 and – 7 [C] −2 and – 7 [D] 2 and 7

15. If the roots of the equation
$(3x - 1)(x + 2) = 0$ are p and q, then, the value of $p + q$ is:
[A] $2\frac{1}{2}$ [B] $1\frac{2}{3}$ [C] $-1\frac{2}{3}$ [D] $-2\frac{1}{2}$

16. The values of x, which satisfy, $x^2 + 2x + 1 = 25$ are:
[A] −6, −4 [B] 6, −4 [C] 6, 4 [D] −6, 4

17. The values of x, which satisfy the equation, $x^2 - 2x - 3 = 0$ are:
[A] (−3, 1) [B] (−1, 3) [C] (−3, 1) [D] (−1, −3)

18. The smaller value of x for which $x^2 - 3x + 2 = 0$ is:
[A] 1 [B] 2 [C] − 1 [D] − 2

19. $6x^2 - 7x - 5 = 0$, only if:
[A] $x = \frac{1}{2}$ or $-2\frac{1}{2}$ [B] $x = \frac{1}{3}$ or $2\frac{1}{2}$ [C] $x = 1\frac{2}{3}$ or $-\frac{1}{2}$ [D] $x = -1\frac{2}{3}$ or $\frac{1}{2}$

20. $2x^2 - 3x - 2 = 0$ is true if and only if:
[A] $x = -2$ or $\frac{1}{2}$ [B] $x = 1$ or 8 [C] $x = -\frac{1}{2}$ or 2 [D] $x = -1$ or 2

21. $2a^2 - 3a - 27 = 0 \Leftrightarrow a =$:
[A] $-3, -\frac{9}{2}$ [B] $-\frac{2}{3}, 9$ [C] $3, \frac{9}{2}$ [D] $-3, \frac{9}{2}$

22. The equation $3x^2 + 25x - 18 = 0$ has roots:
[A] −3, 2 [B] $-9, \frac{2}{3}$ [C] $-\frac{3}{2}, 9$ [D] −2, 3

23. The sum of the roots of the equation $2x^2 + 3x - 9 = 0$ is:
[A] −18 [B] −6 [C] $\frac{9}{2}$ [D] $-\frac{3}{2}$

24. The two values of x that satisfy the equation $5x^2 - 4x - 1 = 0$ are:
[A] $1, -\frac{1}{5}$ [B] $-1, -\frac{1}{5}$ [C] $-1, \frac{1}{5}$ [D] $1, \frac{1}{5}$

25. The solutions of the equation $3 + 5x - 2x^2 = 0$ are:
[A] $-\frac{1}{2}, -3$ [B] 2,3 [C] −2,3 [D] $-\frac{1}{2}, 3$

26. Given that $10 - 3x - x^2 = 0$. The values of x are:
[A] $x = 2$ or $x = -5$ [B] $x = -2$ or $x = 5$
[C] $x = -1$ or $x = 10$ [D] $x = 2$ or $x = 5$

27. The equation $3a + 10 = a^2$ gives rise to the roots:
[A] $a = 5$ or $a = 2$ [B] $a = -5$ or $a = 2$
[C] $a = -5$ or $a = -2$ [D] $a = 5$ or $a = -2$

28. One of the roots of the equation $6x^2 = 5 - 7x$ is:
[A] $-\frac{1}{2}$ [B] $-\frac{1}{3}$ [C] $\frac{1}{2}$ [D] $2\frac{1}{2}$

29. The values of y, which satisfy the equation $3y^2 = 3y$ are:
[A] $y = -3$ or $y = 9$ [B] $y = 0$ or $y = 9$
[C] $y = -3$ or $y = 3$ [D] $y = 3$ or $y = 9$

30. The equation $7y^2 = 27y$ has roots:
[A] $y = 3$ or $y = 7$ [B] $y = 0$ or $y = 7$
[C] $y = 0$ or $y = \frac{3}{7}$ [D] $y = 0$ or $y = 9$

31. A root of the equation $x^2 + 6x = 0$ is:
[A] 0 [B] 6 [C] 2 [D] 3
32. If $x^2 - k^2 = 0$ where k is an integer, the truth about the roots of the equation is:
[A] The two roots are equal.
[B] The sum of the two roots is zero.
[C] The difference of the two roots is $2x$.
[D] The sum of the two roots is zero and the difference of the two roots is $2x$.
33. The value of k, which makes the expression $m^2 - 8m + k$ a perfect square, is:
[A] 2 [B] 4 [C] 8 [D] 16
34. For $x^2 - 6x$ to be a perfect square:
[A] add 9 [B] add 36 [C] 1 [D] add 3
35. The number to the expression $x^2 - 8x$ to make a perfect square is:
[A] 36 [B] 9 [C] 25 [D] 16
36. On subtracting five times a certain integer from twice the square of the integer, the result is 63. The integer is:
[A] 21 [B] 7 [C] 9 [D] 4
37. The area of a rectangle is the product of its length and breadth. The length and breadth of a rectangle are $(x–3)$ cm and $(x–5)$ cm. The area of the rectangle is 24 cm^2 only if:
[A] $x = -9$ or -1 [B] $x = 9$ or 1 [C] $x = 9$ or -1 [D] $x = -9$ or 1
38. The nature of the roots of the quadratic equation $x^2 - 3x + 2 = 0$ is that the roots are:
[A] Real and equal [B] Real and distinct
[C] Imaginary [D] Imaginary and equal
39. The sum of the squares of two natural numbers is 29. One of the numbers is three more than the other is. The larger of the numbers is:
[A] –5 [B] –2 [C] 2 [D] 5
40. 120 soldiers are standing in rows. There are 2 more soldiers in each row than there are rows. The number of rows is:
[A] 8 [B] 10 [C] 12 [D] 55
41. The difference between two numbers is 6 and the difference between their squares is 132. The numbers are:
[A] 8,14 [B] –8,–14 [C] –8,14 [D] 8,–14
42. A man uses 50 m of fencing to fence a garden at his back yard with his house as one side of the fence. If the area of the garden is 300 square metres, the length of the garden is:
[A] 6 [B] 10 [C] 15 [D] 25

2.4 ALGEBRAIC FRACTIONS

Simplifying Algebraic Fractions

The same rules used in manipulating numerical fractions are used in manipulating algebraic fractions.

1. To simplify fractions first factorise the numerators and denominators completely, and then cancel out any common factors.
2. To divide by a fraction multiply by the reciprocal of the fraction.
3. To add and subtract fractions, find the LCM of the denominators and use it to evaluate the sums and differences, and then simplify the single fraction to its lowest terms.

Example

Evaluate and simplify the following.

1. $\frac{9p^2}{6p^2-3pq}$ 2. $\frac{5m+5n}{5n^2-5m^2}$ 3. $\frac{3-3a}{6ab-6b}$ 4. $\frac{x^2+2x-15}{2x^2-12x+18}$

5. $\frac{4y^2-1}{9y-3y^2} \div \frac{2y^2-7y-4}{y^2-7y+12}$. 6. $\frac{3}{3x+1} - \frac{2}{2x+3}$ 7. $\frac{x+1}{2} - \frac{x+2}{3} + \frac{x-4}{4}$

Solution

1. $$\frac{9p^2}{6p^2-3pq} = \frac{\cancel{9}^{3}\,\cancel{p^2}^{p}}{\cancel{3}_{1}\,\cancel{p}_{1}(2p-q)} = \frac{3p}{2p-q}$$

2. $$\frac{5m+5n}{5n^2-5m^2} = \frac{\cancel{5}^{1}\,\cancel{(m+n)}^{1}}{\cancel{5}_{1}\,\cancel{(n+m)}_{1}(n-m)} = \frac{1}{n-m}$$

3. $$\frac{3-3a}{6ab-6b} = \frac{\cancel{3}^{1}\,\cancel{(1-a)}^{-1}}{\cancel{6}_{2}\,b\,\cancel{(a-1)}_{1}} = -\frac{1}{2b}$$ (Note that $a-1=-(a-1)$).

4. $$\frac{x^2+2x-15}{2x^2-12x+18} = \frac{(x+5)\,\cancel{(x-3)}^{1}}{2\,\cancel{(x-3)}_{1}(x-3)} = \frac{x+5}{2(x-3)}$$

5. $\frac{4y^2-1}{9y-3y^2} \div \frac{2y^2-7y-4}{y^2-7y+12} = \frac{(2y-1)(2y+1)}{3y(3-y)} \cdot \frac{(y-3)(y-4)}{(2y+1)(y-4)} = \frac{1-2y}{3y}$

6. $\frac{3}{3x+1} - \frac{2}{2x+3} = \frac{3(2x+3)-2(3x+1)}{(3x+1)(2x+3)} = \frac{6x+9-6x-2}{(3x+1)(2x+3)} = \frac{7}{(3x+1)(2x+3)}$

7. $\frac{x+1}{2} - \frac{x+2}{3} + \frac{x-4}{4} = \frac{6(x+1)-4(x+2)+3(x-4)}{12} = \frac{6x+6-4x-8+3x-12}{12} = \frac{5x-14}{12}$

UNDEFINED EXPRESSIONS

To find the values of x for which an algebraic fraction is undefined, factorise and cancel any common factors, then equate the denominator to zero and solve this equation.

Example

Find the value(s) of x for which each of the expression $\frac{2x^2-3x+1}{6x^2+5x+1}$ is undefined.

Solutions

$$\frac{2x^2-3x+1}{6x^2+5x+1} = \frac{(2x-1)(x-1)}{(2x+1)(3x+1)}$$

For the expression to be undefined, $(2x+1)(3x+1) = 0$

$\Rightarrow x = -\frac{1}{2}$ or $x = -\frac{1}{3}$.

MULTIPLE CHOICE EXERCISE 2:6

1. As a single fraction, $\frac{x-3}{6} - \frac{x+3}{5}$ equals:

[A] $\frac{-x+33}{30}$ [B] $-\frac{x+33}{30}$ [C] $\frac{x-33}{30}$ [D] $\frac{x+33}{30}$

2. When $\frac{t+2}{2} + \frac{t+3}{3}$ is simplified, the result is:

[A] $\frac{5t+5}{6}$ [B] $\frac{5t+12}{6}$ [C] $\frac{2t+5}{6}$ [D] $\frac{t^2+6}{6}$

3. $\frac{x-2}{6}-\frac{x-7}{4}$ is equal to:

[A] $\frac{8-5x}{6}$ [B] $\frac{17-5x}{6}$ [C] $\frac{x-17}{12}$ [D] $\frac{17-x}{12}$

4. When simplified, $\frac{1}{u}+\frac{1}{v}=\frac{1}{f}$ equals:

[A] $\frac{vf^2+uf^2-uv}{fuv}$ [B] $\frac{vf+uf^2-uv}{fuv}$ [C] $\frac{vf+uf-uv}{fuv}$ [D] $\frac{vf+uf-uv}{u}$

5. The result of simplifying $\frac{x^2-y^2}{x+y}$ is:

[A] $x+y$ [B] x^2+y^2 [C] $x-y$ [D] $(x-y)^2$

6. When simplified $\frac{2-18m^2}{1+3m}$ becomes:

[A] $(1+3m)$ [B] $2(1+3m^2)$ [C] $2(1+3m)$ [D] $2(1-3m^2)$

7. In its simplest form $\frac{x^2-3x+2}{x^2-4}$ is:

[A] $\frac{x+1}{x+2}$ [B] $\frac{x+2}{x-2}$ [C] $\frac{x+1}{x-2}$ [D] $\frac{x-1}{x+2}$

8. As a single fraction $\frac{5}{1-x}+\frac{2}{1+x}$ is:

[A] $\frac{7+3x}{1-x^2}$ [B] $\frac{7+3x}{(1-x)^2}$ [C] $\frac{7+3x}{(1+x)^2}$ [D] $\frac{7-3x}{1-x^2}$

9. $\frac{3}{6r}+\frac{3}{4r}$ as a single fraction is:

[A] $\frac{1}{12r}$ [B] $\frac{12}{r}$ [C] $\frac{1}{6r}$ [D] $\frac{5}{4r}$

10. The simplified form of $\frac{5}{x-y}-\frac{4}{y-x}$ is:

[A] $\frac{9}{x-y}$ [B] $\frac{9}{y-x}$ [C] $\frac{1}{x-y}$ [D] $\frac{9}{y-x}$

11. After expressing as a single fraction, $\frac{x}{x-2}-\frac{x+2}{x+3}$ becomes:

[A] $\frac{3x-4}{(x-2)(x+3)}$ [B] $\frac{2x^2+3x-4}{(x-2)(x+3)}$ [C] $\frac{2}{(x-2)(x+3)}$ [D] $\frac{3x+4}{(x-2)(x+3)}$

12. Simplifying $\frac{4}{x+1}-\frac{3}{x-1}$ leads to:

[A] $\frac{x+7}{x^2-1}$ [B] $\frac{x-1}{x^2+1}$ [C] $\frac{x-7}{x^2-1}$ [D] $\frac{x-11}{x^2-1}$

13. $\frac{1}{x-3}-\frac{3(x-1)}{x^2-9}$ as a single fraction is:

[A] $\frac{x-1}{x-3}$ [B] $-\frac{2}{x+3}$ [C] $\frac{x-1}{x+3}$ [D] $\frac{4x}{x^2-9}$

14. In its simplest form, $\frac{2x-1}{3}-\frac{x+3}{2}$ is:

[A] $\frac{x+7}{6}$ [B] $\frac{x+8}{6}$ [C] $\frac{x-11}{6}$ [D] $\frac{x-4}{6}$

15. By simplifying $\frac{1}{1-x}+\frac{2}{1+x}$, the result is:

[A] $\frac{x-3}{1-x^2}$ [B] $\frac{x-3}{1+x^2}$ [C] $\frac{3-x}{1-x^2}$ [D] $\frac{3+x}{1+x}$

16. Simplifying $\frac{1}{x-3}-\frac{2}{x+4}$, gives rise to:

[A] $\frac{1-x}{(x-3)(x+4)}$ [B] $\frac{7-x}{(x-3)(x+4)}$ [C] $\frac{10-x}{(x-3)(x+4)}$ [D] $\frac{-(x+2)}{(x-3)(x+4)}$

17. The result of simplifying $1-\frac{12-3x^2}{2x^2-8}$ is:

[A] $\frac{2-x}{x-2}$ [B] $\frac{3x}{4}$ [C] $\frac{3}{2}$ [D] $\frac{5}{2}$

18. $4-\frac{y-x}{x}$, expressed as a single fraction is:

[A] $\frac{5x-y}{x}$ [B] $\frac{3x-y}{x}$ [C] $\frac{4-y+x}{x}$ [D] $\frac{4-y-x}{x}$

19. When simplified $\frac{3x-2}{x-5}-\frac{2x+3}{2(x-5)}$ equals:

[A] $\frac{1}{x-5}$ [B] $\frac{1}{2(x-5)}$ [C] $\frac{4x-7}{2(x-5)}$ [D] $\frac{4x-7}{x-5}$

20. $\frac{2}{a+b}-\frac{b}{a^2-b^2}$ can be simplified to obtain:

[A] $\frac{3}{a+b}$ [B] $\frac{a-3b}{a^2-b^2}$ [C] $\frac{3a+b}{a^2-b^2}$ [D] $\frac{2a-3b}{a^2-b^2}$

21. The fraction that is equal to $\frac{4}{a-3}-\frac{1}{a+2}$ is:

[A] $\frac{3a+11}{(a+3)(a-2)}$ [B] $\frac{3a+11}{(a-3)(a+2)}$ [C] $\frac{3a-11}{(a-3)(a+2)}$ [D] $\frac{3a+11}{(a-3)(a-2)}$

22. The expression $\frac{a-b}{x-y}$ is undefined when:

[A] $a=b$ [B] $x=y$ [C] $a-b=0$ [D] $\frac{a-b}{x-y}=0$

23. The quotient $\frac{2x+1}{x^2+6x+5}$ is impossible when:

[A] $x = 1$ [B] $x = 5$ [C] $x = 1$ or $x = 5$ [D] $x = -1$ or $x = -5$

24. The restricted domain of $\frac{2x^2+3x+1}{6x^2-x-1}$ is:

[A] $x = -\frac{1}{3}$ or $\frac{1}{2}$ [B] $x = \frac{1}{3}$ or $-\frac{1}{2}$ [C] $x = -\frac{1}{3}$ or $-\frac{1}{2}$ [D] $x = \frac{1}{3}$ or $\frac{1}{2}$

25. A value of y, which makes the expression $\frac{y+2}{y^2-3y-10}$ undefined, is:

[A] $y = 10$ [B] $y = 2$ [C] $y = 3$ [D] $y = 5$

26. The values of x, which renders the expression $\frac{x-5}{x^2-2x-3}$ undefined, is:

[A] $3, -2$ [B] $-1, -3$ [C] $-1, 3$ [D] $1, -3$

27. The value of x for which the expression $\frac{x^2+15x+50}{x-5}$ is undefined is:

[A] -10 [B] -5 [C] 0 [D] 5

28. The values of x, which make the expression $\frac{6x-1}{x^2+4x-5}$ undefined are:

[A] $+4$ and $+1$ [B] -5 and $+1$ [C] $+4$ or -1 [D] $+5$ or -1

29. The pair of values of x for which $\frac{1}{2x^2-13x+15}$ is undefined are:

[A] 5 or $\frac{3}{2}$ [B] 1 or $\frac{15}{13}$ [C] 2 or 15 [D] 13 or 15

30. The pair of values of x for which the expression $\frac{3x-2}{4x^2+9x-9}$ is undefined is:

[A] $-\frac{3}{4}$ or 3 [B] $-\frac{2}{3}$ or -3 [C] $\frac{2}{3}$ or 3 [D] $\frac{3}{4}$ or -3

31. The restricted domain of $\frac{2y-16}{y^2-11y+24}$ is:

[A] 8 [B] -3 [C] 3 [D] 3 or 8

32. $\frac{2x^2-9x-5}{6x^2+7x+2}$ is impossible when:

[A] $\frac{2}{3}$ [B] $-\frac{2}{3}$ [C] 5 [D] -5

2.5 TRANSPOSITION OF FORMULAE

1. To make a subject of a formula, solve for the subject as with equations.
2. If there is a square root sign, first square both sides to eliminate it.
3. If the formula contains a quadratic, factorise or use the quadratic formula to find the two expressions of the subject.

Example

Make x the subject of each of the following formulae

(a) $r - \frac{m}{x} = p^2$ (b) $\sqrt{\left(\frac{a}{x} - b\right)} = c$

Solutions

(a)

$$r - \frac{m}{x} = p^2$$

$$rx - m = p^2 x$$

$$rx - p^2 x - m = 0$$

$$rx - p^2 x = m$$

$$x(r - p^2) = m \Leftrightarrow x = \frac{m}{r - p^2}$$

(b)

$$\sqrt{\left(\frac{a}{x} - b\right)} = c$$

Squaring both sides

$$\frac{a}{x} - b = c^2$$

$$a - bx = c^2 x$$

$$a = (b + c^2)x$$

$$x = \frac{a}{b+c^2}$$

Example

Make x the subject of the formula $2x^2 + 7xyz - 15y^2z^2 = 0$. Find the values of x for which $y = 2$ and $z = -7$.

Solution

$$2x^2 + 7xyz - 15y^2z^2 = 0$$

$$2x^2 + 10xyz - 3xyz - 15y^2z^2 = 0$$

$$2x(x + 5yz) - 3yz(x + 5yz) = 0$$

$$(2x - 3yz)(x + 5yz) = 0$$

$$x = \frac{3}{2}yz \text{ or } x = -5yz$$

When $y = 2$ and $z = -7$

$$x = \frac{3}{2}(2)(-7) \text{ or } x = -5(2)(-7)$$

$$\Leftrightarrow x = -21 \text{ or } x = 70$$

Alternatively, the quadratic formula $x = \frac{-b \pm \sqrt{b^2 - 4ac}}{2a}$ may be used.

Thus, $a = 2, b = 7yz, c = -15y^2z^2$

$$x = \frac{-7yz \pm \sqrt{(7yz)^2 - 4(2)(-15y^2z^2)}}{2(2)} \quad x = \frac{-7yz \pm \sqrt{169y^2z^2}}{2(2)}$$

$$x = \frac{3yz}{2} \text{ or } x = -5yz$$

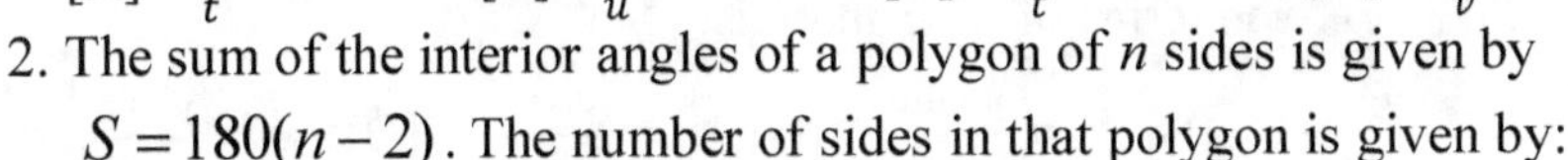

1. Given that $v = u + at$. In terms of *u*, *v* and *t*, *a* will be:

   [A] $\frac{u+v}{t}$ [B] $\frac{v+t}{u}$ [C] $\frac{v-u}{t}$ [D] $\frac{u+t}{v}$

2. The sum of the interior angles of a polygon of *n* sides is given by $S = 180(n-2)$. The number of sides in that polygon is given by:

   [A] $n = \frac{s}{180} - 2$ [B] $n = \frac{s}{180}$ [C] $n = \frac{2s}{180}$ [D] $n = \frac{s}{180} + 2$

3. The volume *V* of a cylinder is given by $V = \pi r^2 h$ where *r* and *h* are the radius and height of the cylinder respectively. The height *h* is given by:

   [A] $h = \pi r^2 V$ [B] $h = \frac{V}{\pi r^2}$ [C] $h = \frac{\pi r^2}{V}$ [D] $h = \frac{\pi}{r^2 V}$

4. As a subject of the formula $A = 2\pi rh$, *r* is:

   [A] $\frac{A}{2\pi h}$ [B] $\frac{\pi}{2Ah}$ [C] $\frac{\pi h}{2A}$ [D] $2\pi Ah$

5. The volume of a cone is given by the formula $V = \frac{1}{3}\pi r^2 h$. In terms of *V*, *r* and π, the height *h* is:

   [A] $\frac{\pi r^2}{3V}$ [B] $\frac{3Vr^2}{\pi}$ [C] $\frac{3V}{\pi r^2}$ [D] $\frac{\pi r^2 V}{3}$

6. In the temperature conversion formula $F = \frac{9}{5}C + 32$. *C* is given by:

   [A] $\frac{5F-160}{9}$ [B] $\frac{5F+160}{9}$ [C] $\frac{9F+160}{5}$ [D] $\frac{9F-160}{5}$

7. If $I = \frac{PRT}{100}$, the value of *T* when $P = 450$, $R = 12$ and $I = 90$ is:

   [A] $\frac{3}{5}$ [B] $\frac{5}{6}$ [C] $\frac{5}{3}$ [D] 15

8. If $y = \frac{a+p}{a-p}$, then :

   [A] $\frac{2a-y}{a+y} = p$ [B] $\frac{ay-1}{y+1} = p$ [C] $\frac{a(y-1)}{y+1} = p$ [D] $\frac{2y-1}{y-1} = p$

9. If $h(m+n) = m(h+r)$, *h* in terms of *m*, *n* and *r* is:

   [A] $h = \frac{mr}{2m+n}$ [B] $h = \frac{mr}{2m-n}$ [C] $h = \frac{m+r}{n}$ [D] $h = \frac{mr}{n_1}$

10. If *q* is made the subject of the relation $t = \sqrt{\frac{pq}{r} - r^3 q}$, the relation will now be:

    [A] $q = \frac{t^2}{p-r^4}$ [B] $q = \frac{rt^2}{p-r^4}$ [C] $q = \frac{rt}{p-r^4}$ [D] $q = \frac{p-r^4}{rt^2}$

11. Making S the subject of the formula $V = \dfrac{K}{\sqrt{T-S}}$ gives:

[A] $S = T - \frac{K^2}{V^2}$ [B] $S = \frac{K^2}{V^2} - T$ [C] $S = T - \frac{V^2}{K^2}$ [D] $S = T\left(\frac{V^2 - K^2}{V^2}\right)$

12. If $y = \sqrt{ax - b}$, x in terms of y, a and b is:

[A] $x = x = \frac{y^2 - b}{a}$ [B] $x = \frac{y+b}{a}$ [C] $x = \frac{y-b}{a}$ [D] $x = \frac{y^2+b}{a}$

13. $k = m\sqrt{\frac{t-p}{r}}$ as subject of the formula, t is equal to:

[A] $\frac{rk^2+p}{m^2}$ [B] $\frac{rk^2-pm^2}{m^2}$ [C] $\frac{rk^2-p}{m^2}$ [D] $\frac{rk^2-p^2}{m^2}$

14. Given that $U = \frac{T}{5} - \frac{2R}{Q}$, then T expressed in terms of U, R and Q is:

[A] $T = 5U + \frac{2R}{Q}$ [B] $T = 5U + \frac{10R}{Q}$ [C] $T = 5\left(U - \frac{2R}{Q}\right)$ [D] $T = 5U - \frac{2R}{Q}$

2.6 THE REMAINDER AND FACTOR THEOREMS

Remainder theorem

The **remainder theorem** states that, when a polynomial $f(x)$ is divided by $ax + b$, the remainder is $f\left(-\frac{b}{a}\right)$

Factor theorem

The **factor theorem** states that, if the remainder when the polynomial $f(x)$ is divided by $ax + b$ is zero, then $ax + b$ is a factor of $f(x)$. Conversely, if $f\left(-\frac{b}{a}\right) = 0$ then $ax + b$ is a factor of $f(x)$.

Example

Find the remainder when $f(x) = 6x^3 - 5x^2 - 17x + 7$ is divided by $2x + 3$.

Solution

$$f\left(-\frac{3}{2}\right) = 6\left(-\frac{3}{2}\right)^3 - 5\left(-\frac{3}{2}\right)^2 - 17\left(-\frac{3}{2}\right) = -\frac{81}{4} + \frac{45}{4} + \frac{51}{2} + 7 = 1$$

Therefore, the remainder is 1.

Example

Determine the factors of $f(x) = 2x^3 + x^2 - 7x - 6$.

Solution

By trial and error method, $f(-1) = 2(-1)^3 + (-1)^2 - 7(-1) = 0$.
Therefore, $(x + 1)$ is a factor of $f(x)$.
$f(2) = 2(2)^3 + (2)^2 - 7(2) = 0$. Therefore, $(x - 2)$ is a factor of $f(x)$.
$\Rightarrow f(x) = (x + 1)(x - 2)(ax + k)$
Since the coefficient of x^3 and the independent terms are 2 and −6 respectively, then $(ax + k) = (2x + 3)$.
$\Rightarrow f(x) = (x + 1)(x - 2)(2x + 3)$

MULTIPLE CHOICE EXERCISE 2:8

1. The factor theorem states that if the remainder when $f(x)$ is divided by $ax + b$ is 0 then:
   [A] $f\left(\frac{a}{b}\right)$ [B] $f\left(-\frac{a}{b}\right)$ [C] $f\left(-\frac{b}{a}\right)$ [D] $f\left(\frac{b}{a}\right)$
2. The remainder theorem states that when a polynomial $f(x)$ is divided by $ax + b$ the remainder is:
   [A] $f\left(\frac{a}{b}\right)$ [B] $f\left(-\frac{a}{b}\right)$ [C] $f\left(\frac{b}{a}\right)$ [D] $f\left(-\frac{b}{a}\right)$
3. If $f(x)$ is divided by $ax - 1$ the remainder is:
   [A] $f(a)$ [B] $f(-a)$ [C] $f\left(-\frac{1}{a}\right)$ [D] $f\left(\frac{1}{a}\right)$
4. The remainder when $x^3 + 2x^2 + 1$ is divided by x is:
   [A] x [B] 1 [C] 2 [D] −1
5. $f(x) = -x^2 + kx - 6$ has a factor $(x + 2)$. The value of k is:
   [A] −3 [B] −7 [C] −5 [D] 7
6. Given that $f(x) = 2x^2 + 3x - 2$. It is true to say that:
   [A] $(x - 2)$ is a factor of $f(x)$.
   [B] $(x + 2)$ is a factor of $f(x)$.
   [C] $f(x)$ leaves a remainder −2 when divided by $(x + 2)$.
   [D] $(2x + 1)$ is a factor of $f(x)$.
7. The remainder when $x^3 - 2x^2 - x - 2$ is divided by $x + 1$ is:
   [A] −2 [B] 4 [C] −4 [D] 2
8. If $f(x) = x^3 - 3x - 4$. The remainder when $f(x)$ is divided by $x - 4$ is:
   [A] −56 [B] 48 [C] −80 [D] 72
9. The remainder when $x^3 + 3x - 4$ is divided by $x + 1$ is:
   [A] −2 [B] −8 [C] 0 [D] −7
10. Given that $f(x) = 2x^3 - 3x^2 - 3x + 11$. The remainder when $f(x)$ is divided by $2x + 1$ is:
    [A] 1.5 [B] 2.5 [C] −1.5 [D] −2.5
11. The remainder when $3x^3 - x^2 - 2x + 13$ is divided by $x + 2$ is:
    [A] 4 [B] −3 [C] −4 [D] 3

12. Given that when divided by $x+1$ and $x+2$ the expression ax^2+bx+3 leaves remainders 6 and 9 respectively. The values of a and b are:
    [A] $a=-2, b=1$ [B] $a=-3, b=0$ [C] $a=0, b=-3$ [D] $a=2, b=-1$
13. The remainder when x^3+3x^2-5x-6 is divided by $x+2$ is:
    [A] -12 [B] 8 [C] 4 [D] 12
14. $x-1$ and $x-2$ are both factors of x^3+ax^2+bx-6 when:
    [A] $a=-6, b=11$ [B] $a=6, b=11$
    [C] $a=6, b=-11$ [D] $a=-6, b=-11$
15. If $(x+2)$ is a factor of x^3+kx^2-2x+4. The value of k is:
    [A] 2 [B] -2 [C] 1 [D] 0
16. x^3-3x^2+6x-2 has remainder 2 when divided by:
    [A] $x-1$ [B] $x+1$ [C] $x+2$ [D] $2x-1$
17. x^3-3x^2+2x-6 has a factor:
    [A] $x-4$ [B] $x-2$ [C] $x-3$ [D] $x+3$
18. Given that $f(x)=3x^3+4x^2-3x-4$. One of the factors of $f(x)$ is:
    [A] $3x-4$ [B] $4x+7$ [C] $x-1$ [D] $x-4$
19. A factor of x^3+2x^2-5x-6 is:
    [A] $x+2$ [B] $x-1$ [C] $x+1$ [D] $x-2$
20. A factor of $x^3+3x^2-4x-12$ is:
    [A] $x-4$ [B] $x+4$ [C] $x-3$ [D] $x+3$
21. $x-2$ is a factor of:
    [A] $x^3-3x^2-4x+12$ [B] $x^3+3x^2+4x+12$
    [C] $x^3-3x^2+4x+12$ [D] $x^3+3x^2-4x+12$

2.7 LINEAR AND QUADRATIC INEQUALITIES

A statement that two real quantities or expressions are not equal is called an **inequality**.
The following symbols are used for inequalities.

Symbol	Meaning
$<$	is less than
$>$	is greater than
$\leq$	is less than or equal to
$\geq$	is greater than or equal to

Ordering

$\forall a, b, c \in \mathbb{R}$,

1. Either $a < b$ or $a > b$ or $a = b$. 2. $a < b \Leftrightarrow b > a$.
3. If $a < b$ and $b < c$ then $a < c$. This is known as the **transitive property**.
4. If $a < b$ and $c > 0$ then $ac < bc$ and if $a < b$ and $c < 0$ then $ac > bc$.

Representation of inequalities

On a real number line, if $a < b$, the point a is to the left of the point b.
Also, on a real number line, an open circle o is used for < and > while a fill-in circle • is used for $\leq$ and $\geq$.

Open Intervals

An open interval is denoted as $a < x < b$ or $]a, b[$ or (a, b) and is represented as below. The boundary points a and b are not included.

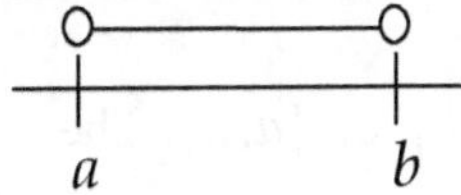

or

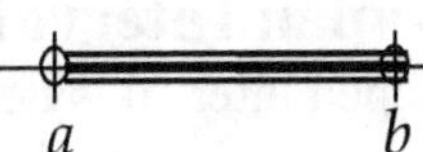

Example
The following represent $5 < x < 12$, $]5, 12[$ or $(5, 12)$.

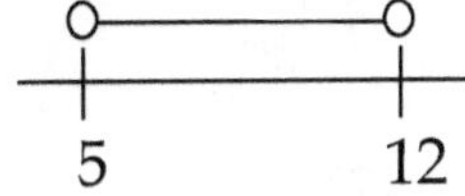

or

Closed Intervals

A closed interval is denoted as $a \leq x \leq b$ or $[a, b]$. The boundary points a and b are included.

or

Example
The following represent $5 \leq x \leq 12$ or $[5, 12]$.

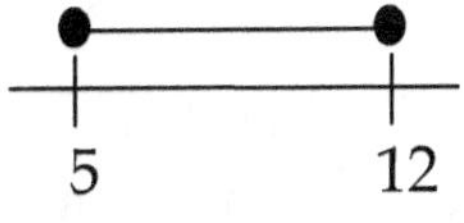

or

Half Open or Half Closed Intervals

Open-Closed Interval

An open-closed interval is denoted as $a < x \leq b,\ (a, b]$ or $]a, b]$. The boundary point a is not included but b is included.

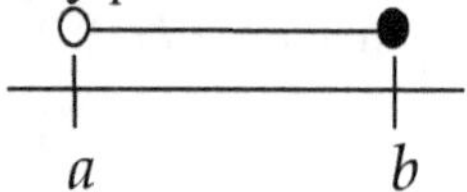

or

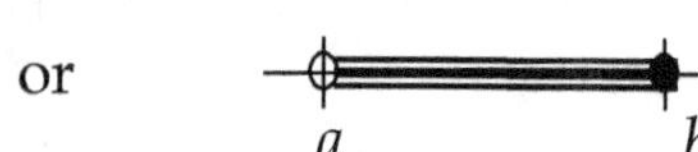

Example

The following represent $5 < x \leq 12,\ (5, 12]$ or $]5, 12]$.

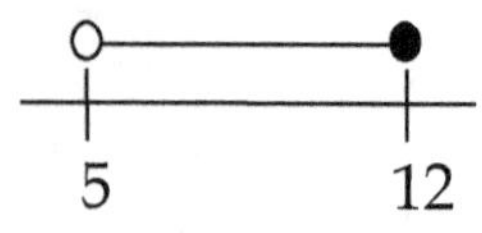

Closed-open Interval

A closed-open interval is denoted as $a \leq x < b$ or $[a, b[$ or $[a, b)$. The boundary point a is included but b is not included.

or

Example

The following represent $5 \leq x < 12$ or $[5, 12[$ or $[5, 12)$.

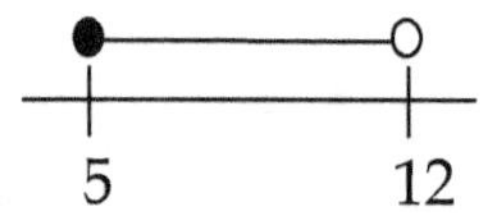

or

Conditional Inequalities or Inequations

Conditional inequalities (or inequations) are inequalities which contain at least one unknown and whose truth or falsity depends on the range of values of the unknowns. For instance, the inequality $x + 2 < 5$ is an inequation but the inequality $-5 < 8$ is not an inequation.

Solving Inequations

1. All the rules for solving equations apply when solving inequalities.
2. The only difference is that if both sides of an inequality are multiplied or divided by any negative real quantity, the inequality changes sense from $<$ to $>$ or $\leq$ to $\geq$ and vice versa.

Linear Inequalities

Example

Solve the following inequalities and represent the solution on a number line.

1. $5 - 2x \leq 13$ 2. $\frac{2}{3}n + 7 > 11$

Solution

1. $5 - 2x \leq 13$

$-2x \leq 8$

$x \geq -4$

−4

2. $\frac{2}{3}n + 7 < 11$

$\frac{2}{3}n < 4$

$2n < 12$

$n < 6$

6

Compound Inequalities

Example

Solve the following inequalities and represent your result on a number line.

(a) $3 \leq 6 + 3x \leq 9$ (b) $3x - 4 < x \leq 5x + 12$

Solution

(a) $3 \leq 6 + 3x \leq 9$

Subtract 6 from each expression.

$-3 \leq 3x \leq 3$

Divide through by 3.

$-1 \leq x \leq 1$

−1 1

(b) $3x - 4 < x \leq 5x + 12$.

$\Rightarrow 3x - 4 < x$ and $x \leq 5x + 12$

$2x - 4 < 0$ $-4x \leq 12$

$2x < 4$ $x \geq -3$

$x < 2$

We can then combined the results to have $-3 \leq x < 2$.

−3 2

Absolute Value Inequalities

If $|n| \leq k$ then, $-k \leq n \leq k$.

Example

Solve the inequality $|3x + 2| \leq 5$ and represent your solution on a real number line.

Solution

$|3x + 2| \leq 5$

$-5 \leq 3x + 2 \leq 5$

$-7 \leq 3x \leq 3$

$-\frac{7}{3} \le x \le 1$

$-\frac{7}{3}$ 1

Quadratic Inequations

Example

Solve the following inequations and represent your solution on a number line.

(a) $x^2 - 5x + 6 < 0$ (b) $x^2 + 2x - 3 \ge 0$

Solution

(a) $x^2 - 5x + 6 < 0$

$(x-2)(x-3) < 0$

Critical values are 2 and 3

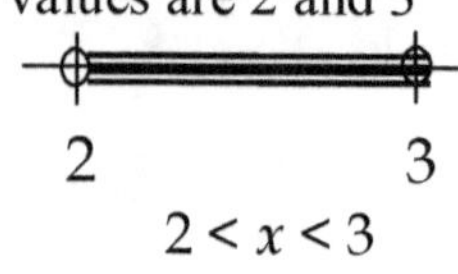

$2 < x < 3$

(b) $x^2 + 2x - 3 \ge 0$

$(x+3)(x-1) \ge 0$

Critical values are –3 and 1.

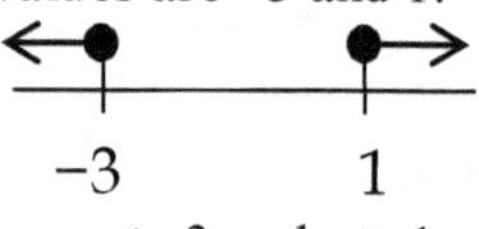

$x \le -3$ and $x \ge 1$

Solution set = $\{x: x \le -3 \text{ or } x \ge 1, x \in \mathbb{R}\}$

Absolute Inequalities

An **absolute inequality** is an inequality, which is always true for all real values of the variables involved. Examples are $x^2 \ge 0, x \in \mathbb{R}$; $(x \pm 1)^2 \ge 0, x \in \mathbb{R}$; $(x \pm y)^2 \ge 0, x, y \in \mathbb{R}$.

MULTIPLE CHOICE EXERCISE 2:9

1. The solution of the inequality $3m + 3 > 9$ is:
   [A] $m > 2$ [B] $m > 3$ [C] $m > 4$ [D] $m > 6$
2. If x is positive, the range of values of x for which $4 + 3x < 10$ is:
   [A] $0 < x < 2$ [B] $x < 2$ [C] $1 < x < 2$ [D] $0 > x > 2$
3. The inequality $\frac{1}{3}(2x-1) < 5$ is true only if:
   [A] $x < -5$ [B] $x > -5$ [C] $x < 7$ [D] $x > 8$
4. p and q are two positive real numbers such that $p > 2q$. The inequality which is not true is:

[A] $-p<-2q$ [B] $-p>2q$ [C] $-p<2q$ [D] $-p<\frac{1}{2}q$

5. The solution of the inequality $(y-3)<\frac{y}{3}$ is:

   [A] $y>-\frac{9}{2}$ [B] $y<3$ [C] $y>4$ [D] $y<\frac{9}{2}$

6. The solution of the inequality $3x-8\geq 5x$ is:

   [A] $x\geq 4$ [B] $x\geq 1$ [C] $x\geq -4$ [D] $x\geq -1$

7. The range of values of x which satisfy the inequality $2x+3<5x$ are:

   [A] $x>1$ [B] $x<\frac{3}{7}$ [C] $x<\frac{3}{7}$ [D] $x>-1$

8. The smallest whole number which satisfies the inequality $9-2x<5x-12$ is:

   [A] 1 [B] 2 [C] 3 [D] 4

9. A distance d metres which is more than 18 m, but not more than 23 m can be represented by:

   [A] $18\leq d\leq 23$ [B] $18<d\leq 23$ [C] $18\leq d<23$ [D] $d<18$ or $d>23$

10. Nfor had x oranges. He ate 2 and shared the remainder equally with Ngala. In terms of x, the inequality which represents the information that Ngala's share is at least 5 oranges is:

    [A] $\frac{x}{2}-2\leq 5$ [B] $\frac{x}{2}-2\geq 5$ [C] $\frac{(x-2)}{2}\geq 5$ [D] $\frac{(x-2)}{2}\leq 5$

11. The range of values of x for which $\frac{1}{2}(4x+2)-(x-5)\leq\frac{1}{4}(3x-1)$ is:

    [A] $x\geq 25$ [B] $x\leq 25$ [C] $x\geq -25$ [D] $x\leq -25$

12. The number line below which represents the solution to the inequality $\frac{x}{3}-\frac{(x-3)}{2}<1$ is:

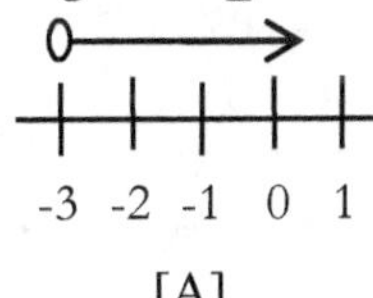

-1 0 1 2 3

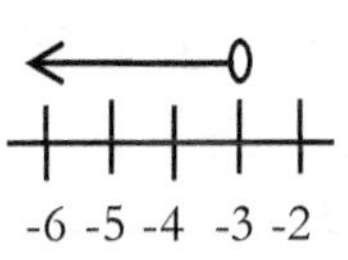

1 2 3 4 5

[A] [B] [C] [D]

13. If x varies over the set of real numbers, the inequality illustrated in the number line below is:

    [A] $-3<x\leq 2$ [B] $-3\leq x<2$ [C] $-3<x<2$ [D] $-3\leq x\leq 2$

−4 −3 −2 −1 0 1 2 3 4 5

14. If x is a real number the inequality represented in Figure 16:3 is:

    [A] $\{x: -5<x\leq 3\}$ [B] $\{x: -5\leq x\leq 3\}$

[C] $\{x: -5 \leq x < 3\}$ [D] $\{x: -5 < x < 3\}$

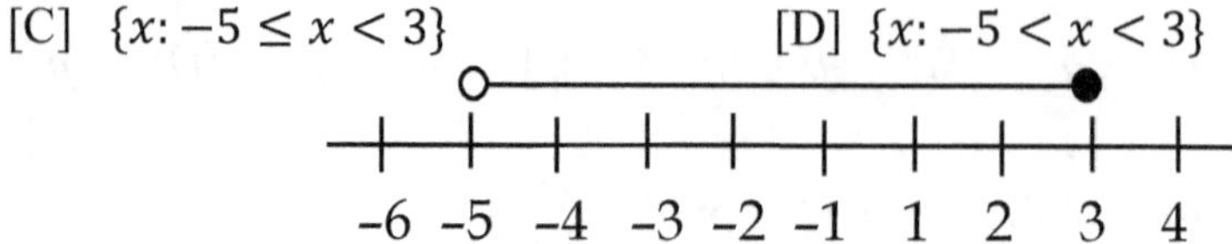

15. If $x \in \mathbb{R}$, the inequality more illustrated in the number line below is:
[A] $x < 4$ [B] $x > -2$ [C] $-2 < x \leq 4$ [D] $-2 \leq x < 4$

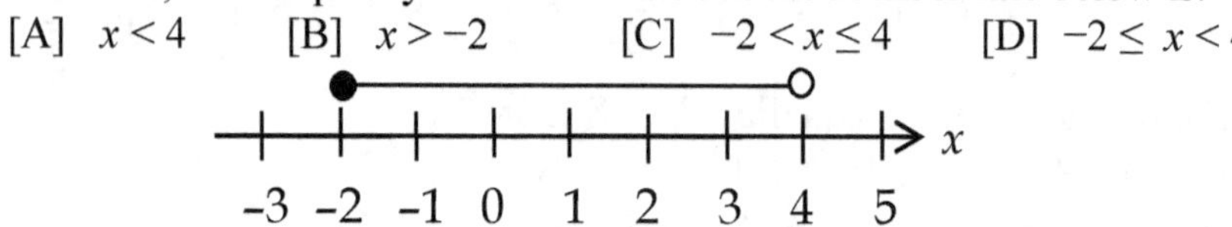

16. If x is real, the inequality more illustrated in the number line below is:
[A] $x < 4$ [B] $x > -2$ [C] $-2 < x \leq 4$ [D] $-2 \leq x < 4$

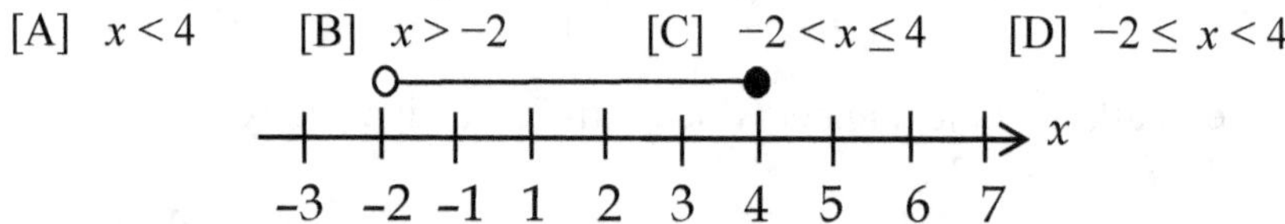

17. If x varies over the set of real numbers, the inequality illustrated below is:
[A] $-2 \leq x < 3$ [B] $-2 < x \leq 3$ [C] $-2 < x < 3$ [D] $-2 \leq x \leq 3$

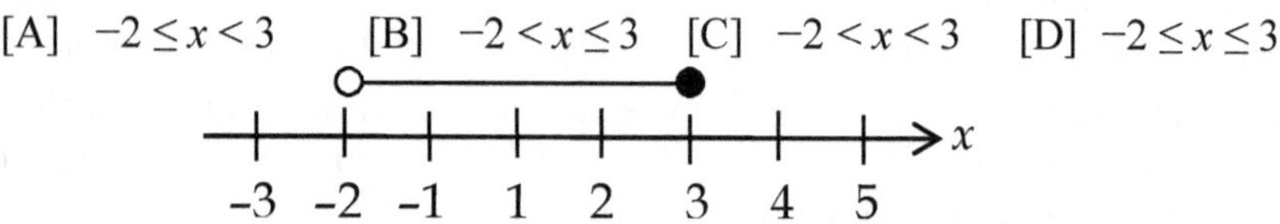

18. The pairs of inequalities represented on the number line below are:
[A] $x < -2$ and $x \geq 1$ [B] $x \leq -2$ and $x > 1$
[C] $x \leq -2$ and $x < 1$ [D] $x < -2$ and $x > 1$

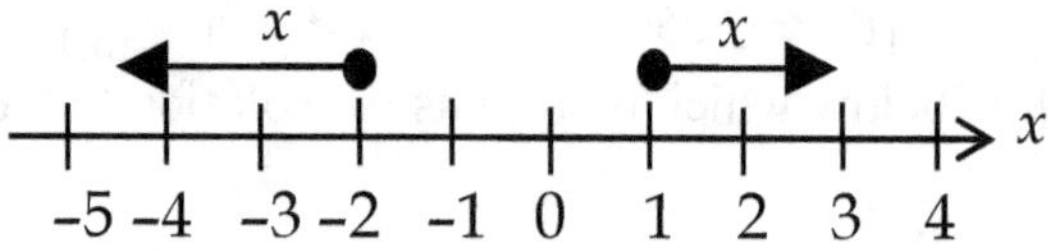

19. The number line below which represents the inequality $2 \leq x < 9$ is:

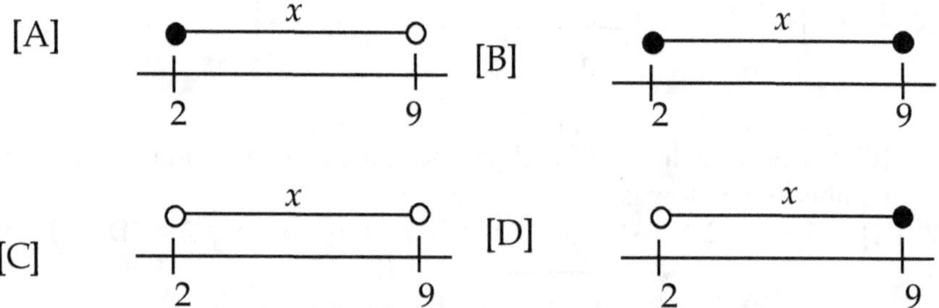

20. The number line below represents:
[A] $0 < x < -7$ [B] $-7 < x < 1$ [C] $-7 < x \leq -1$ [D] $-7 \leq x < -1$

21. Given that, $x \in \mathbb{R}$. An absolute inequality among the following is:
[A] $|3x + 2| \leq 5$ [B] $(x-1)^2 \geq (x+1)^2$
[C] $(x-1)^2 \geq 0$ [D] $-2 < x < 0$
22. The inequalities represented on the number line below are:
[A] $-2 < x \leq 0$ or $x > 3$ [B] $-2 \leq x < 0$ or $x \geq 3$
[C] $-2 \leq x < 0$ or $x > 3$ [D] $-2 \leq x \leq 0$ or $x > 3$

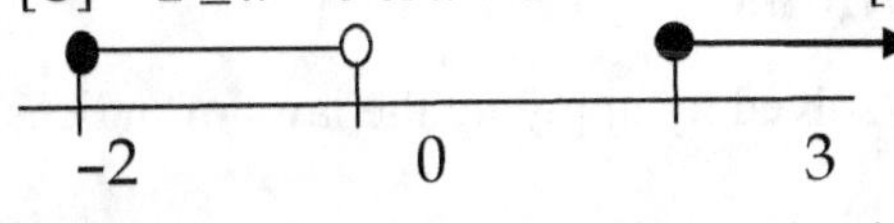

23. Given that, $x \in \mathbb{R}$. An absolute value inequality among the following is:
[A] $|7x-16| \geq 0$ [B] $(x+3)^2 \geq (x-3)^2$ [C] $(x-3)^2 \geq 0$ [D] $-4 < x+2 < 10$
24. A conditional inequality is an inequality which is:
[A] always true [B] sometimes true [C] always false [D] sometimes false

2.8 INDICES, LOGARITHMS AND SURDS

THEORY OF INDICES

b^p ← power, index, exponent, or logarithm
b ← base

Laws of Indices

(1) $b^m \times b^n = b^{m+n}$ (2) $b^m \div b^n = b^{m-n}$ (3) $b^0 = 1$

(4) $b^1 = b$ (5) $b^{-n} = \dfrac{1}{b^n}$ (6) $(b^m)^n = (b^n)^m = b^{mn}$

(7) $b^{\frac{1}{n}} = \sqrt[n]{b}$ (8) $b^{\frac{m}{n}} = \sqrt[n]{b^m} = \left(\sqrt[n]{b}\right)^m$

Example

Simplify (a) $\dfrac{(2p)^3(3p)^2}{36p^4}$ (b) $\left(\dfrac{625}{144}\right)^{-\frac{3}{2}}$

Solution

(a) $\frac{(2p)^3(3p)^2}{36p^4} = \frac{2^3 \times 3^2 \times p^3 \times p^2}{2^2 \times 3^2 \times p^4}$

$$= 2^{3-2} \times 3^{2-2} \times p^{3+2-4} = 2^1 \times 3^0 \times p^1 = 2 \times 1 \times p = 2p$$

(b) $\left(\frac{625}{144}\right)^{-\frac{3}{2}} = \left(\frac{144}{625}\right)^{\frac{3}{2}} = \left(\frac{12}{25}\right)^3 = \frac{1728}{15625}$

Exponential or Index Equations

Exponential or index equations are solved by applying the laws of indices.

Example

Find the value of x for which $64^x = 16^{2x+1}$.

Solution

$$64^x = 16^{2x+1} \Leftrightarrow 2^{6x} = 2^{4(2x+1)}$$

Equating exponents,

$6x = 4(2x + 1) \Longrightarrow 6x = 8x + 4 \Longrightarrow -2x = 4$ and $x = -2$

LOGARITHMS

Definition of Logarithms

The logarithm of a positive number n to a base b is the power p to which b must be raised to give the number n.

Symbolically, $\log_b n = p \Leftrightarrow n = b^n, n \in \mathbb{R}$

N.B: The logarithm of negative numbers and zero do not exist.

Laws of Logarithms

(1) $\log_b xy = \log_b x + \log_b y$ (2) $\log_b (\frac{x}{y}) = \log_b x - \log_b y$

(3) $\log_b x^n = n \log_b x$ (4) $\log_b b = 1$ (5) $\log_b 1 = 0$

(6) $b^{\log_b a} = a$ (7) $\log_{a^n} b = \frac{1}{n} \log_a b$

(8) **Change of base formula:** $\log_b x = \frac{\log_a x}{\log_a b}$

Example

Without using tables or calculators, simplify the following.

1. $\log_{10}\left(\frac{15}{8}\right)-2\log_{10}\left(\frac{5}{9}\right)+\log_{10}\left(\frac{400}{243}\right)$

2. $\frac{\log_{10}6-\log_{10}3}{\log_{10}8-\log_{10}4}\div\frac{\log_{10}5}{\log_{10}0.2}$

3. $\log_{10}\sqrt{35}+\log_{10}\sqrt{2}-\log_{10}\sqrt{7}$

Solution

1. $\log_{10}\left(\frac{15}{8}\right)-2\log_{10}\left(\frac{5}{9}\right)+\log_{10}\left(\frac{400}{243}\right)=\log_{10}\left(\frac{15}{8}\right)\left(\frac{400}{243}\right)-\log_{10}\left(\frac{5}{9}\right)^2$

$$=\log_{10}\left(\frac{15}{8}\right)\left(\frac{400}{243}\right)\left(\frac{81}{25}\right)=\log_{10}10=1$$

2. $\frac{\log_{10}6-\log_{10}3}{\log_{10}8-\log_{10}4}\div\frac{\log_{10}5}{\log_{10}0.2}=\frac{\log_{10}\left(\frac{6}{3}\right)}{\log_{10}\left(\frac{8}{4}\right)}\times\frac{\log_{10}0.2}{\log_{10}5}$

$$=\frac{\log_{10}2}{\log_{10}2}\times\frac{\log_{10}\left(\frac{1}{5}\right)}{\log_{10}5}=\frac{\log_{10}5^{-1}}{\log_{10}5}=\frac{-1\log_{10}5}{\log_{10}5}=-1$$

3. $\log_{10}\sqrt{35}+\log_{10}\sqrt{2}-\log_{10}\sqrt{7}=\log_{10}\left(\frac{(35)(2)}{(7)}\right)^{\frac{1}{2}}=\frac{1}{2}\log_{10}10=\frac{1}{2}$

Example

Given that $\log_{10}3=0.4771$ and $\log_{10}4=0.6021$. Find the value of $\log_4 3$ to three significant figures.

Solution

$$\log_4 3=\frac{\log_{10}3}{\log_{10}4}=\frac{0.4771}{0.6021}=0.792$$

SURDS

Laws of Surds

1. $\sqrt[n]{ab}=\sqrt[n]{a}\times\sqrt[n]{b}$

2. $\sqrt[n]{\frac{a}{b}}=\frac{\sqrt[n]{a}}{\sqrt[n]{b}}$

3. $\sqrt[n]{a^m}=\left(\sqrt[n]{a}\right)^m$

4. $\sqrt[n]{a^n} = a$ 5. $\sqrt[mn]{a} = \sqrt[n]{\sqrt[m]{a}}$ 6. $\sqrt{a^0} = 1$

Note that $\sqrt[n]{a} + \sqrt[n]{b} \neq \sqrt[n]{a+b}$ and $\sqrt[n]{a} - \sqrt[n]{b} \neq \sqrt[n]{a-b}$

The Conjugate of a Surd

The **conjugate** of $a + \sqrt{b}$ is $a - \sqrt{b}$.

$$(a + \sqrt{b})(a - \sqrt{b}) = a^2 - (\sqrt{b})^2 = a^2 - b$$

Rationalizing the Denominator

Rationalizing the denominator is the process of removing the radical sign from the denominator of a surd expression by multiplying both numerator and denominator by the conjugate.

Example

Simplify (a) $\frac{2}{\sqrt{3}}$ (b) $\frac{1}{\sqrt{5}}$ (c) $\frac{\sqrt{7}}{2\sqrt{7}-5}$

Solution

(a) $\frac{2}{\sqrt{3}} = \frac{2}{\sqrt{3}} \times \frac{-\sqrt{3}}{-\sqrt{3}} = \frac{2\sqrt{3}}{3}$ (b) $\frac{1}{\sqrt{5}} = \frac{1}{\sqrt{5}} \times \frac{-\sqrt{5}}{-\sqrt{5}} = \frac{\sqrt{5}}{5}$

(c) $\frac{\sqrt{7}}{2\sqrt{7}-5} = \left(\frac{\sqrt{7}}{2\sqrt{7}-5}\right)\left(\frac{2\sqrt{7}+5}{2\sqrt{7}+5}\right) = \frac{14+5\sqrt{7}}{28-25} = \frac{14+5\sqrt{7}}{3}$

Surd Equations

Example

Solve the following equations

(a) $\sqrt{2x-1} = 3$ (b) $\sqrt{4x+1} = x-5$

Solution

(a) $\sqrt{2x-1} = 3$

Squaring both sides

$2x - 1 = 9$

$2x = 10$

$x = 5$

(b) $\sqrt{4x+1} = x - 5$

Squaring both sides

$4x + 1 = x^2 - 10x + 25$

$x^2 - 14x + 24 = 0$

$(x - 12)(x - 2) = 0$

$x = 12$ or $x = 2$

MULTIPLE CHOICE EXERCISE 2:10

1. The value $(2^3)^2$ is:
   [A] 16 [B] 32 [C] 36 [D] 64
2. The value $2^2 + 3^3$ is:
   [A] 13 [B] 25 [C] 31 [D] 36
3. The value $64^{\frac{1}{3}}$ is:
   [A] 16 [B] 8 [C] 4 [D] 2
4. 5^4 has the value of is:
   [A] 9 [B] 20 [C] 125 [D] 625
5. The value $2^0 - 2^{-2}$ is:
   [A] 1 [B] $-\frac{1}{4}$ [C] $\frac{3}{4}$ [D] $\frac{1}{4}$
6. $\frac{2^3 \times 2^4}{2^2}$ is equal to:
   [A] 2^4 [B] 2^3 [C] 2^2 [D] 2^5
7. The value $\left(\frac{196}{225}\right)^{-\frac{1}{2}}$ is:
   [A] $\frac{17}{14}$ [B] $\frac{15}{14}$ [C] $\frac{14}{15}$ [D] $\frac{14}{17}$
8. On simplification $16^{\frac{1}{2}}\left(4^{-1} + 5^0\right)$ gives:
   [A] 5 [B] $5\frac{1}{2}$ [C] $4\frac{1}{2}$ [D] $4\frac{1}{4}$
9. $(0.7)^3$ equals:
   [A] 2.1 [B] 0.49 [C] 3.43 [D] 0.343
10. When simplified $\left(27^2\right)^{\frac{1}{3}}$ equals:
    [A] 81 [B] 6 [C] 9 [D] 8
11. After evaluating $36^{\frac{1}{2}} \times 64^{-\frac{1}{2}} \times 5^0$ equals:
    [A] $\frac{3}{4}$ [B] $\frac{1}{24}$ [C] $\frac{2}{3}$ [D] $1\frac{1}{2}$
12. When $\frac{9^{-\frac{1}{2}}}{27^{\frac{2}{3}}}$ is simplified the result is:
    [A] $\frac{1}{2}$ [B] $\frac{1}{9}$ [C] $\frac{1}{18}$ [D] $\frac{1}{27}$

13. Simplifying $125^{-\frac{1}{3}} \times 49^{-\frac{1}{2}} \times 10^{0}$ gives:

[A] 350 [B] 35 [C] $\frac{1}{35}$ [D] $\frac{1}{350}$

14. $\left(\frac{1}{4}\right)^{-1\frac{1}{2}}$ when simplified is:

[A] $\frac{1}{8}$ [B] $\frac{1}{4}$ [C] 8 [D] 4

15. On evaluation $\left(\frac{16}{81}\right)^{\frac{1}{4}}$ becomes:

[A] $\frac{8}{27}$ [B] $\frac{1}{3}$ [C] $\frac{4}{9}$ [D] $\frac{2}{3}$

16. The result of simplifying $\frac{4^{-\frac{1}{2}} \times 16^{\frac{3}{4}}}{4^{\frac{1}{2}}}$ is:

[A] $\frac{1}{4}$ [B] 0 [C] 1 [D] 2

17. On evaluation, the result of $\frac{27^{\frac{1}{3}}}{16^{-\frac{1}{4}}}$ is:

[A] 6 [B] 2 [C] 4 [D] 3

18. $16^{\frac{5}{4}} \times 2^{-3} \times 3^{0}$ is equal to:

[A] 20 [B] 2 [C] 4 [D] 10

19. The result of evaluating $0.027^{-\frac{1}{2}}$ is:

[A] $3\frac{1}{3}$ [B] 3 [C] $\frac{3}{10}$ [D] $\frac{1}{3}$

20. $\frac{8^{\frac{2}{3}} \times 27^{-\frac{1}{3}}}{64^{\frac{1}{3}}}$ simplifies to:

[A] $\frac{1}{3}$ [B] $\frac{1}{9}$ [C] $\frac{16}{3}$ [D] $\frac{27}{8}$

21. After evaluating $5\frac{2}{3} \times \left(\frac{2}{3}\right)^{2} \div \left(1\frac{1}{2}\right)^{-1}$, the result is:

[A] $\frac{12}{5}$ [B] $\frac{8}{5}$ [C] $3\frac{3}{5}$ [D] $4\frac{1}{8}$

22. The result of simplifying $\frac{2^{\frac{1}{2}} \times 8^{\frac{1}{2}}}{4}$:

[A] 1 [B] 2 [C] 4 [D] 16

23. $\left(\frac{1}{4}\right)^{-1\frac{1}{2}}$ is equal to:

[A] 8 [B] 4 [C] $\frac{1}{4}$ [D] $\frac{1}{16}$

24. The result of evaluating $\left(\frac{16}{81}\right)^{-\frac{3}{4}} \times \left(\frac{100}{81}\right)$ is:

[A] $\frac{80}{243}$ [B] $\frac{20}{27}$ [C] $\frac{25}{6}$ [D] $\frac{15}{4}$

25. $\left(3a^2\right)^3$ is equal to:

[A] $3a^6$ [B] $9a^6$ [C] $27a^2$ [D] $27a^6$

26. $a^2 \times b \times a^4 \times b^2$ simplifies to:

[A] a^6b^2 [B] a^8b^2 [C] a^3b^3 [D] a^6b^3

27. When simplified $\frac{x^3y^4z^7}{x^2y^6z^7}$ to:

[A] $\frac{x}{y^2}$ [B] $\frac{y^2}{x}$ [C] $\frac{x^2}{y}$ [D] $\frac{1}{y}$

28. $\left(m^2n^5\right) \div \left(m^3n^4\right)$ equals:

[A] mn^{-1} [B] $m^{-1}n$ [C] m^5n^9 [D] mn

29. When $10a^6$ is divided by $5a^3$ the result is:

[A] $3a^3$ [B] $2a$ [C] a^3 [D] $2a^3$

30. If $2^x \times 3^2 = 144$, the value of x is:

[A] 7 [B] 5 [C] 4 [D] 8

31. If $x^2 \times 3^2 \times 1^2 = 144$, the value of x is:

[A] – 4 [B] 2 [C] –2 [D] 16

32. When $3^x = 81$, the value of x is:

[A] 2 [B] 3 [C] 4 [D] 27

33. The value of x for which $3^x = 243$ is:

[A] 6 [B] 5 [C] 4 [D] 3

34. If $3^x + 6 = 87$, the value of x is:

[A] 1 [B] 2 [C] 3 [D] 4

35. If $3^{2x} = 27$ the value of x is:
[A] 1 [B] 1.5 [C] 4.5 [D] 18

36. The value of t for which $\frac{64}{27} = \left(\frac{3}{4}\right)^{t-1}$ is:
[A] −4 [B] 2 [C] 4 [D] −2

37. Given that $27^{(1+x)} = 9$. The value of x is:
[A] −3 [B] $-\frac{1}{3}$ [C] $\frac{1}{3}$ [D] 2

38. If $16(4)^{2x} = \left(\frac{1}{2}\right)^{x}$, the value of x is:
[A] −3 [B] $-\frac{4}{5}$ [C] $-\frac{4}{3}$ [D] $\frac{4}{3}$

39. $\left(\frac{1}{4}\right)^{2-y} = 1$, the value of y is:
[A] −2 [B] $-\frac{1}{2}$ [C] $\frac{1}{2}$ [D] 2

40. The solution of the equation $2\sqrt{x} = 4$ is:
[A] −2 [B] 2 [C] 4 [D] 6

41. The value of x for which $2^{-6x} = 8^{(1-x)}$ is true is:
[A] $-\frac{7}{3}$ [B] $\frac{1}{3}$ [C] −1 [D] $\frac{7}{9}$

42. The value of n which satisfies the equation $\frac{3^{(1-n)}}{9^{-2n}} = \frac{1}{9}$ is:
[A] $-\frac{3}{2}$ [B] $\frac{1}{3}$ [C] −3 [D] 1

43. When simplified $\frac{1}{4}\left(2^{n} - 2^{n+2}\right)$ becomes:
[A] $2^{n-1} - 2^{n}$ [B] $2^{n-2}(1 - 2^{n})$ [C] $2^{n+2} - 2$ [D] 2^{n}

44. When $56x^{-4} \div 14x^{-8}$ is simplified the result is:
[A] $2x^{-12}$ [B] $4x^{-4}$ [C] $4x^{+4}$ [D] $4x^{-3}$

45. Given that $81 \times 2^{2n-2} = k$, the value of $\sqrt{k}$ which satisfies this equation is:
[A] 4.5×2^{n} [B] 4.5×2^{2n} [C] $9 \times 2^{n-1}$ [D] 9×2^{n}

46. The word that is not another name for logarithm is:
[A] Power [B] exponent [C] Base [D] index

47. If $\log_{10} 5.444 = 0.7359$, then $\log_{10} 54440$ is:
[A] 3.7359 [B] 4.7359 [C] 5.7359 [D] 6.7359
48. If $\log_{10} x = 2.8765$, the value of x is:
[A] Greater than 100 [B] between 1 and 10
[C] less than 10 [D] between 10 and 100
49. The value of $\log_{10} 6 + \log_{10} 45 - \log_{10} 27$ is:
[A] 0 [B] 1 [C] 1.1738 [D] 10
50. On simplification $\frac{\log\sqrt{8}}{\log 8}$ is equal to:
[A] $\frac{1}{3}$ [B] $\frac{1}{3}\log\sqrt{8}$ [C] $\frac{1}{3}\log\sqrt{2}$ [D] $\frac{1}{2}$

Simplifying $\frac{\log 27^{\frac{1}{3}}}{\log 81}$ gives:
[A] 14 [B] $\frac{3}{8}$ [C] $\frac{1}{2}$ [D] $\frac{3}{4}$
51. The value of $\log_{10} 25 + \log_{10} 32 - \log_{10} 8$ is:
[A] 0.2 [B] 2 [C] 100 [D] 409
52. The value of $2\log_3 6 + \log_3 16$ is:
[A] $4 - \log_3 2$ [B] $3 + \log_3 2$ [C] $2 + 6\log_3 2$ [D] $3 - \log_3 2$
53. On evaluation, $\log_{10} 4 + \log_{10} 25$ becomes:
[A] 1 [B] 2 [C] 3 [D] 4
54. $\log_3 9 + \log_3 15 - \log_3 5$ is equal to:
[A] $\log_3 19$ [B] log3 [C] 3 [D] 1
55. Simplifying $\log_7 8 - \log_7 2 + \log_7 4$, gives:
[A] 0 [B] 2 [C] $2\log_2 7$ [D] $4\log_2 7$
56. The value of $\log_{10} 5 + \log_{10} 20$ is:
[A] 2 [B] 3 [C] 4 [D] 5
57. If $\log_{10} 2 = 0.3010$ and $\log_{10} 2^y = 1.8062$. The value of y to the nearest whole number is:
[A] 4 [B] 6 [C] 5 [D] 2
58. If $3\log_{10} a = \log_{10} 64$, the value of a is:
[A] 4 [B] 6 [C] 8 [D] 16
59. Given that $\log p = 2\log x + 3\log q$. The correct expression of p in terms of x and q is:
[A] $p = 6xq$ [B] $p = x^2q^3$ [C] $p = x^2 + q^3$ [D] $p = 2x + 3q$
60. The solution of the equation $\log_8 x - 4\log_8 x = 2$ is:

[A] $\frac{1}{4}$ [B] $\frac{1}{2}$ [C] 4 [D] 2

61. $7^{x-1} = \log_5 5$, then x is equal to:
[A] 1 [B] 7 [C] -1 [D] -7

62. Given that $\frac{1}{3}\log_{10} p = 1$. The value of p is:
[A] 3 [B] 10 [C] 100 [D] 1000

63. $\log_2 a = \log_8 4$ only if a is equal to:
[A] $2^{\frac{1}{2}}$ [B] $2^{\frac{2}{3}}$ [C] $4^{\frac{2}{3}}$ [D] $4^{\frac{1}{3}}$

64. If $\log_a x = p$, then in terms of a and p, x is equal to:
[A] a^p [B] $\frac{a}{p}$ [C] p^a [D] ap

65. The value of p for which $\frac{1}{2}\log_{10} p = 1$ is true is:
[A] 10^{-1} [B] 10^3 [C] 10^2 [D] 10^1

66. Given that $\log_4 x = -3$. The value of x is:
[A] $\frac{1}{81}$ [B] $\frac{1}{64}$ [C] 64 [D] 81

67. On simplification $\frac{4\sqrt{18}}{\sqrt{8}}$ becomes:
[A] 2 [B] 3 [C] 6 [D] 12

68. The value of $\sqrt{96} + \sqrt{54} - \sqrt{24}$ is:
[A] $\sqrt{6}$ [B] $2\sqrt{6}$ [C] $3\sqrt{6}$ [D] $5\sqrt{6}$

69. $\sqrt{32} - \sqrt{98} + 5\sqrt{2}$ is equal to
[A] $\frac{1}{2}\sqrt{2}$ [B] $2\sqrt{2}$ [C] $3\sqrt{2}$ [D] $4\sqrt{2}$

70. The value of $3\sqrt{12} + 10\sqrt{3} - \frac{6}{\sqrt{3}}$ is:
[A] $7\sqrt{3}$ [B] $10\sqrt{3}$ [C] $14\sqrt{3}$ [D] $18\sqrt{3}$

71. Given that $\frac{1}{\sqrt{2}} = 0.7071$. Then, $\frac{3\sqrt{2}}{2}$ is greater than $\frac{1}{\sqrt{2}}$ by:
[A] -3 [B] -1.4142 [C] 1.4142 [D] 3

72. Evaluating $\sqrt{20} \times \left(\sqrt{5}\right)^3$ gives:
[A] 10 [B] 20 [C] 25 [D] 50

73. If $K\sqrt{28} + \sqrt{63} - \sqrt{7} = 0$ the value of K is:
[A] -2 [B] -1 [C] 1 [D] 2

74. $\sqrt{128} + \sqrt{18} - \sqrt{k} = 7\sqrt{2}$, then k must be:
[A] 8 [B] 16 [C] 32 [D] 48

75. When the denominator is rationalized, $\frac{10}{\sqrt{32}}$ becomes:

[A] $\frac{5}{4}\sqrt{2}$ [B] $\frac{4}{5}\sqrt{2}$ [C] $\frac{5}{16}\sqrt{2}$ [D] $\frac{16}{5}\sqrt{2}$

76. The number $\frac{6}{\sqrt{2}}$ is equal to:

[A] $4\sqrt{2}$ [B] $3\sqrt{2}$ [C] $2\sqrt{2}$ [D] 2

2.9 ELEMENTARY SEQUENCES

The nth Term of a Sequence

Example

Write down the 6th term U_6 of a sequence whose n^{th} term is given by $U_n = n(n+1)$, where $n \in \mathbb{N}$.

Solution

$$U_n = n(n+1)$$
$$\Rightarrow U_6 = 6(6+1) = 42$$

Example

Find the nth term of the sequences 1, 4,7,10, 13, 16 …

.

Solution

(a) 1, 4,7,10, 13, 16 …

$1 = 3(1) - 2$
$4 = 3(2) - 2$
$7 = 3(3) - 2$
$10 = 3(4) - 2$
$13 = 3(5) - 2$
$16 = 3(6) - 2$
$\therefore Un = 3n - 2$

Sum of the First n Terms of a Sequence

Example

The sum S_n of the first n terms, of a sequence is given by $S_n = 3n^2 - 2n$, where $n \in \mathbb{N}$. Find (a) S_1, S_2 and S_{10}. (b) U_1, U_2 and U_{10}.

Solution

$S_n = 3n^2 - 2n$

(a) $S_1 = 3(1)^2 - 2(1) = 1$
$S_2 = 3(2)^2 - 2(2) = 8$
$S_{10} = 3(10)^2 - 2(10) = 280$

(b) $U_1 = S_1 = 1$
$U_2 = S_2 - S_1 = 8 - 1 = 7$
$U_{10} = S_{10} - S_9$
$S_9 = 3(9)^2 - 2(9) = 225$
$U_{10} = 280 - 225 = 55$

ARITHMETIC PROGRESSION (A.P.)

The nth term (General Term) of an A.P.

The general term of an A.P. is given by $U_n = a + (n-1)d$.

Example
Find the 200th term of the A.P. 4, 2, 0, –2, –4…

Solution
$U_n = a + (n-1)d,\ a = 4, d = -2$ and $n = 200$
$\Rightarrow U_{200} = 4 + (200-1)(-2) = -394$

Sum of the first n terms of an A.P.

The sum of the first *n* terms of an A.P. is given by

$$S_n = \frac{n}{2}\{2a + (n-1)d\} \text{ or } S_n = \frac{n}{2}(a + l)$$

Example
Find the sum of the first 20 terms of the A.P. 3, 7, 11, 15.

Solution
$a = 3, d = 4, n = 20, S_n = \frac{n}{2}\{2a + (n-1)d\}$
$\Rightarrow S_{20} = \frac{20}{2}\{2(3) + (20-1)4\} = 10\{6 + 19(4)\} = 820$

The Arithmetic Mean

The arithmetic mean of two numbers x_1 and x_2 is given by $\bar{x} = \frac{x_1 + x_2}{2}$.

The arithmetic mean of a set of *n* numbers $x_1, x_2, x_3, \ldots, x_n$ is given by $\bar{x} = \frac{x_1 + x_2 + x_3 + \cdots + x_n}{n}$.

Example
Find the arithmetic mean of 13 and 17.

Solution

$$\bar{x} = \frac{a+b}{2} = \frac{13+17}{2} = 15$$

GEOMETRIC PROGRESSION (G.P.)

The nth term of a G.P.

The nth term of a G.P. is given by $U_n = ar^{n-1}$

Example

The nth term of the G.P. 2, 4, 8, 16… is 2048. Find the value of *n*.

Solution

$U_n = ar^{n-1}, a = 2, r = 2$ and $U_n = 2048 \Rightarrow 2048 = 2(2)^{n-1} \Rightarrow n = 11.$

The Sum of n terms of a G.P.

The sum of the first *n* terms of a G.P. is given by

$$S_n = \frac{a(r^n - 1)}{(r-1)} \text{ or } S_n = \frac{a(1-r^n)}{(1-r)}$$

Example

Find the sum of the first 20 terms of the G.P. 3, 6, 12, 24, …

Solution

$a = 3, r = 2 \text{ and } n = 20 \Rightarrow S_n = \frac{a(r^n-1)}{r-1} = \frac{3(2^{20}-1)}{2-1} = 3145725.$

Sum to infinity, S_∞

The **sum to infinity** of the G.P. is denoted by S_∞ and given by

$$S_\infty = \frac{a}{(1-r)}, \quad |r| < 1$$

Example

Find the sum to infinity of the G.P. whose first term is 81 and whose common ratio is $\frac{1}{3}$.

Solution

$a = 81, r = \frac{1}{3} \Rightarrow S_\infty = \frac{a}{1-r} = \frac{81}{1-\frac{1}{3}} = 81\left(\frac{3}{2}\right) = \frac{243}{2}$

Geometric Mean

The **geometric mean** of two numbers a and c, denoted by GM is given by

$$GM = \pm\sqrt{ac}$$

The geometric mean of a set of n numbers $x_1, x_2, x_3,..,x_n$ is given by

$$GM = \pm\sqrt[n]{x_1 x_2 \ldots x_n}$$

Example

Find the geometric mean of 6 and $13\frac{1}{2}$.

Solution

$$GM = \pm\sqrt{ac} = \pm\sqrt{6\left(\frac{27}{2}\right)} \pm\sqrt{81} = \pm 9$$

Example

Insert 2 numbers between –6 and $20\frac{1}{4}$ so that the sequence will be a G.P.

Solution

Consecutive terms of the G.P. are $-6, -6r, -6r^2, 20\frac{1}{4}$.

$$\Rightarrow -\frac{6r}{6} = \frac{20\frac{1}{4}}{-6r^2} \Rightarrow r^3 = -\frac{81}{24} \text{ and } r = -\frac{3}{2}$$

Therefore, the 2 numbers are $-6\left(-\frac{3}{2}\right)$ and $-6\left(-\frac{3}{2}\right)^2$ i.e. 9 and $-\frac{27}{2}$ in that order.

MULTIPLE CHOICE EXERCISE 2:11

1. The next 2 terms in the sequence 1, 2, 4, 7, 11, 16… are:
   [A] 17, 29 [B] 29, 24 [C] 22, 29 [D] 29, 40
2. The next term in the sequence 1, 4, 9, 16,…is:
   [A] 20 [B] 25 [C] 23 [D] 27
3. The next term in the sequence 2, 5, 11, 23, 47…is:
   [A] 95 [B] 93 [C] 71 [D] 27
4. In the sequence 1,3,7,15,31 the number that must be added to 31 to give the next term is:
   [A] 4 [B] 8 [C] 16 [D] 32
5. The number represented by * in the arithmetic progression 14, – 3, *, – 37 is:

[A] 11 [B] –14 [C] 17 [D] – 20

6. The 4th term of an *A.P.* whose first term is 2 and whose common difference is 0.5 is:
   [A] 3 [B] $\frac{7}{2}$ [C] $\frac{11}{2}$ [D] 5
7. The first term of an *A.P.* is equal to twice the common difference *d*. In terms of *d* the 5th term of the *A.P.* is:
   [A] 4*d* [B] 5*d* [C] 6*d* [D] $a + 5d$
8. The 9th term of the Arithmetic progression 18, 12, 6, 0,–6,... is:
   [A] –54 [B] –30 [C] 30 [D] 42
9. The eleventh term of an *A.P.* is 25 and its first term is 3. Its common difference is:
   [A] $1\frac{9}{10}$ [B] $2\frac{1}{5}$ [C] $2\frac{4}{5}$ [D] $2\frac{1}{2}$
10. If the first term of an *A.P.* is 4 and the 5th term is 12, then the mean of the first five terms is:
    [A] 4 [B] 6 [C] 8 [D] 10
11. The nth term U_n of the *A.P.* 11, 4, –3… is:
    [A] $U_n = 19 + 7n$ [B] $U_n = 19 - 7n$ [C] $U_n = 18 - 7n$ [D] $U_n = 18 + 7n$
12. The nth term *Un* of the sequence 4, 10, 16,… is:
    [A] $2(3n - 1)$ [B] $2(2 + 3^{n-1})$ [C] $2^2 + 2$ [D] $2(3n + 1)$
13. In an *A.P.*, the first term is 2, and the sum of the first and the 6th terms is $16\frac{1}{2}$. The 4th term is:
    [A] $5\frac{1}{2}$ [B] $9\frac{1}{2}$ [C] 8 [D] 7
14. The sum of the 1st and 2nd terms of an *A.P.* is 4 and the 10th term is 19. The sum of the 5th and 6th terms is:
    [A] 11 [B] 22 [C] 21 [D] 20
15. It is observed that, if $1 + 3 + 5 + 7 + 9 + 11 + 13 + 15 = p^2$, then the value of *p* is:
    [A] 6 [B] 7 [C] 8 [D] 9

$$1 + 3 = 2^2$$
$$1 + 3 + 5 = 3^2$$
$$1 + 3 + 5 + 7 = 4^2$$

16. The common ratio of a *G.P.* is 2. If the 5th term is greater than the first by 45, the 5th term will be:
    [A] 3 [B] 6 [C] 45 [D] 48
17. The fifth and seventh terms of the geometric progression $-2, -3, -\frac{9}{2}, -\frac{27}{4}\cdots$ are respectively:
    [A] $-\frac{81}{16}, -\frac{729}{32}$ [B] $-\frac{8}{81}, \frac{72}{18}$ [C] $-\frac{21}{8}, \frac{32}{618}$ [D] $-\frac{27}{16}, -\frac{79}{18}$

18. The 6th term of a geometric progression is $-\frac{2}{27}$ and the first term is 18. The common ratio is:
    [A] $-\frac{1}{2}$ [B] $-\frac{1}{3}$ [C] $\frac{1}{4}$ [D] 3
19. The common ratio of the $G.P$ log 3, log 9, log 81,... is:
    [A] 1 [B] 2 [C] 3 [D] 6
20. The 16th term of the $G.P.$ 2,6,18,... is given by:
    [A] 2×3^{12} [B] 2×3^{13} [C] 2×3^{15} [D] 2×3^{16}
21. The sum of the first 5 terms of the $G.P.$ 2,6,18,...is:
    [A] 121 [B] 243 [C] 242 [D] 130
22. If the second and fourth terms of a $G.P$ are 8 and 32 respectively the sum of the first four terms will be:
    [A] 28 [B] 40 [C] 48 [D] 60
23. If the second and 5th terms of a $G.P.$ are -6 and 48 respectively, the sum of the first four terms is:
    [A] -45 [B] -15 [C] 15 [D] 45
24. The nth term of a sequence is represented by $3\times2^{(2-n)}$. The first three terms of the sequence are respectively:
    [A] $\frac{3}{2}, 3, 6$ [B] $6, 3, \frac{3}{2}$ [C] $\frac{3}{2}, 3, \frac{1}{3}$ [D] $\frac{2}{3}, 3, \frac{8}{3}$
25. The nth term of a sequence is $2^{(2n-1)}$. If $U_n = 2^9$, n is:
    [A] 3 [B] 4 [C] 5 [D] 6
26. The next two terms of the sequence 1,5,14,30,55,... are respectively:
    [A] 61,110 [B] 67,116 [C] 81,140 [D] 91,140
27. The nth term of a sequence is given by $(-1)^{(n-2)}$. The sum of the second and third terms is:
    [A] 0 [B] 1 [C] 2 [D] 6
28. The sum of the first n terms of a sequence is given by $S_n = 17n - 3n^2$. The fourth term of the sequence is:
    [A] 20 [B] -10 [C] -4 [D] 10
29. The nth term of a sequence is given as $\frac{n^2+n}{2}$. The 7th term of the sequence is:
    [A] 36 [B] 28 [C] 21 [D] 14
30. The common difference in the sequence 10, 2, -6, -14, . . .is:
    [A] 10 [B] 8 [C] -16 [D] -8
31. The next three terms and the rule that describe the sequence 10, 20, 40, 80 . . . . are:
    [A] 82, 84, 86; start with 10 and add 2 repeatedly.
    [B] 90, 100, 110; start with 10 and add 10 repeatedly.
    [C] 320, 1280, 5120; start with 10 and multiply by 4 repeatedly.
    [D] 160, 320, 640; start with 10 and multiply by 2 repeatedly.

32. The next three terms in the sequence 3,12,21,30,... are:
    [A] 40, 50, 60 [B] 38, 46, 54 [C] 39, 48, 57 [D] 36, 32, 39
33. The next four terms in the sequence −4, −1, 2, 5... are:
    [A] 5, 8, 11, 14 [B] 8, 11, 14, 17 [C] 3, 6, 9, 12 [D] 0, 8, 11, 14
34. If n points are marked on a circle, where n is a whole number greater than 1, $\frac{1}{2}n^2 - \frac{1}{2}n$ segments can be drawn to connect these points. The number of segments that can be drawn if 8 points are marked on the circle is:
    [A] 28 [B] 32 [C] 60 [D] 56
35. If the pattern continues, the number of squares in the 8th diagram is:
    [A] 16 [B] 18 [C] 14 [D] 15

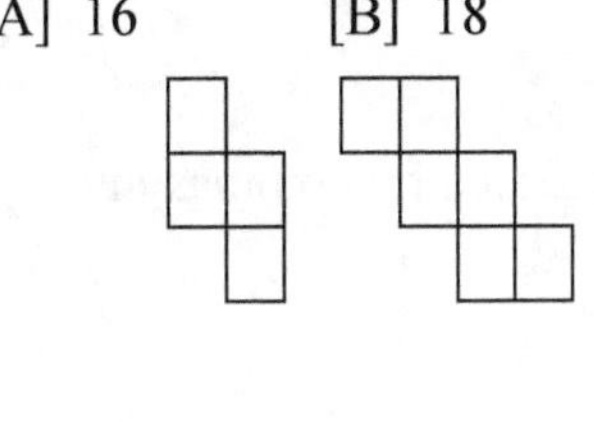

36. The number of dots in the ninth figure is:
    [A] 25 dots [B] 27 dots [C] 19 dots [D] 26 dots

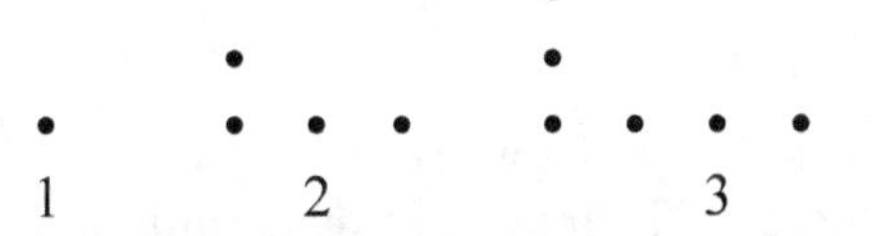

37. The diagram below represents a five by five square of black and white floor tiles. The numbers of black tiles which can be added to the existing pattern to make it six by six are:
    [A] 8 [B] 6 [C] 5 [D] 4

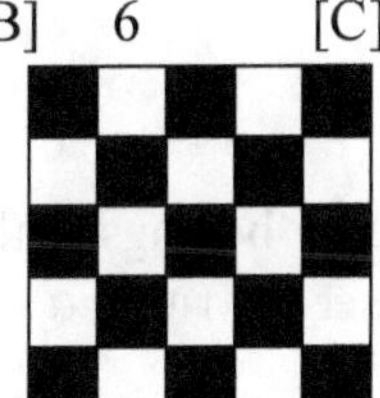

CHAPTER 3

SETS, BINARY OPERATIONS AND LOGIC

3.1 SET THEORY AND LANGUAGE

A set is a well defined collection of objects. Sets are denoted by capital letters$A, B, C \cdots$.

Definition of Sets

1. *The Roster Method*

This is done by listing the elements of a set, separated by commas and enclosed in braces. For instance, $\{a, e, i, o, u\}$.

2. *The Rule Definition Method*

In this case a rule, which qualifies any element as a member of the set, is stated or defined. If C is the set {North, West, South, East} then we can write C = cardinal points.

3. *Set Builder Notation Method*

$\{x{:}x$ is a day of the week$\}$ or $\{x/x$ is a day of the week$\}$ is the set {Sunday, Monday, Tuesday, Wednesday, Thursday, Friday, Saturday}. $P = \{(x, y){:}\ y = 2x + 5\}$ is the set of all points (x, y) which satisfy the line $y = 2x + 5$. $\{x{:}\ 0 \leq x \leq 3\}$ is the set of all points which lie between the points 0 and 3 inclusively.

Membership Notation

Let $V = \{x / x$ is an English vowel$\}$, then a belongs to the set V, or a is a member of V or a is an element of the set V written $a \in V$. b is not an element of V is written $b \notin V$.

The Cardinality of a Set

The cardinality, cardinal number or the order of a set is the number of elements the set. The cardinality of $V = \{a, e, i, o, u\}$ is $n\ (V) = 5$.

TYPES OF SETS

A **singleton or unit set** is a set whose cardinality is 1.

A **doubleton or Pair set** is a set whose cardinality is 2.
A **trebleton** is a set whose cardinality is 3.
An **empty** or **null set** denoted by Ø or { } is a set with no elements.
A **finite set** is a set which has an exact countable number of elements.
An **infinite set** is a set whose elements are inexhaustible.

The Universal Set

The universal set denoted by $\mathscr{E}$ or U is a set which contains all the elements under consideration in a particular situation.

Subsets

If all the elements of a set A are found in another set B, then A is said to be a subset of B, written $A \subset B$ or $B \supset A$. If A is not a subset of B this is written $A \not\subset B$. For instance if $X = \{1,2,3,4,5,6,7,8\}$ and $Y = \{2,5,7\}$, then $X \subset Y$ or $Y \supset X$.

Equality of Sets

Two sets X and Y are equal if they have the same elements. i.e. $X \subset Y$ and $Y \subset X$. By implication $n(X) = n(Y)$.
Note that *repeating elements of a set does not change the set.*
Thus, $\{2, 4, 6, 8\} = \{8, 2, 6, 4, 6, 8\}$

Equivalent Sets

If two sets X and Y are such that $n(X) = n(Y)$ but $X \not\subset Y$ then X and Y are said to be equivalent denoted by $X \sim Y$.

Proper and Improper Subsets

If two non equivalent non empty sets X and Y are such that all the elements of X are found in Y, then X is a **proper subset** of Y and Y is a **superset** of X. On the other hand if two sets X and Y are such that $X = \emptyset$ or $X = Y$ then X is an improper subset of Y.

Intersection of Sets

The intersection of the sets $A, B, C, \cdots$ denoted by $A \cap B \cap C \cdots$ is the set, which consist of the elements common to all the sets $A, B, C, \cdots$. Using set builder notation $A \cap B \cap C \cdots = \{x : x \in A, x \in B, x \in C \cdots\}$. $A \cap B$ is read as '*A* intersection *B*' or 'the intersection of *A* and *B*' or '*A* cap *B*'.

Union of Sets

The union of the sets $A, B, C, \cdots$ denoted by $A \cup B \cup C \cdots$ is the set, which consist of all the elements of the sets $A, B, C, \cdots$ put together. Using set builder notation $A \cup B \cup C \cdots = \{x : x \in A \text{ or } x \in B \text{ or } x \in C \cdots\}$. $A \cup B$ is read 'A union B' or the union of 'A and B' or 'A cup B'.

Disjoint Sets

If two or more sets have no elements in common, they are said to be **disjoint**. $A, B, C, \cdots$ are disjoint $\Leftrightarrow A \cap B \cap C \cdots = \emptyset$ or $n(A \cap B \cap C \cdots) = 0$.

The Complement of a Set $C_{\mathscr{E}}^{A}$ or A'

The complement of a set A denoted by A' or $C_{\mathscr{E}}^{A}$ is the set containing all the elements of the universal set $\mathscr{E}$ that do not belong to the set A.

Example

$\mathscr{E} = \{0,1,2,3,4,5,6,7,8,9,10,11,12\}$, $A = \{2, 4, 6, 8, 10, 12\}$,
$B = \{1, 2, 3, 6, 9\}$ and $C = \{4, 5, 7, 8\}$, find (i) $A \cap B$ (ii) $A \cup B$ (iii) A'
(iv) State in symbolic language the relationship between B and C.

Solution

(i) $A \cap B = \{2,6\}$ (ii) $A \cup B = \{1, 2, 3, 4, 6, 8, 9, 10, 12\}$
(iii) $A' = \{0, 1, 3, 5, 7, 9, 11\}$ (iv) $B \cap C = \emptyset$ or $n(B \cap C) = 0$.

Representation of Sets-Venn Diagrams

We use Venn diagrams to represent relationships between sets as illustrate below.

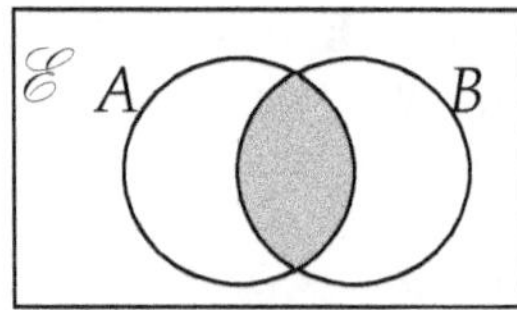

$A \cap B$ is shaded

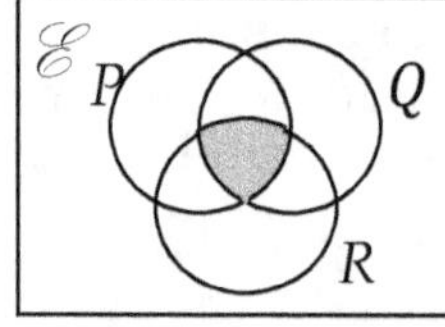

$P \cap Q \cap R$ is shaded

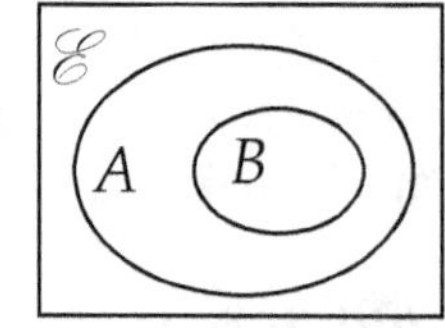

$A \subset B$

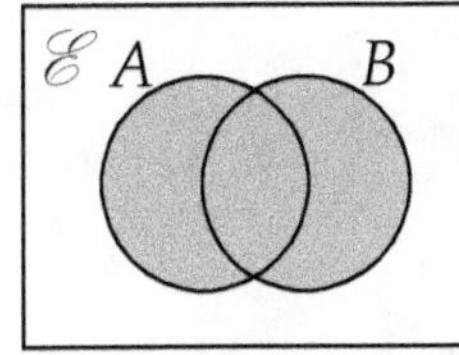

$A \cup B$ is shaded

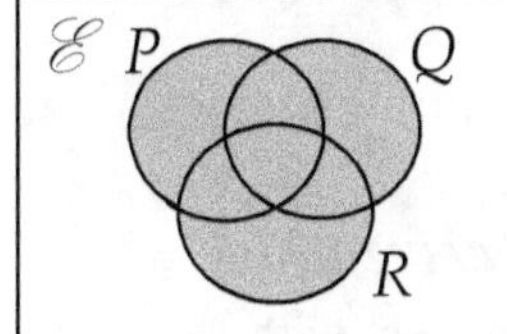

$P \cup Q \cup R$ is shaded

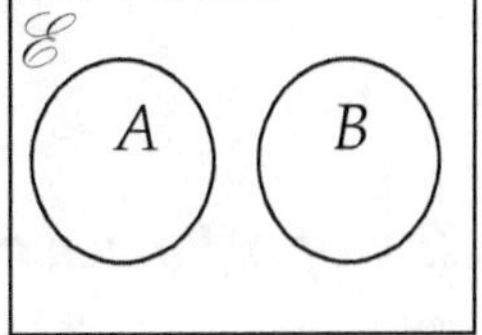

$A \cap B = \varnothing$

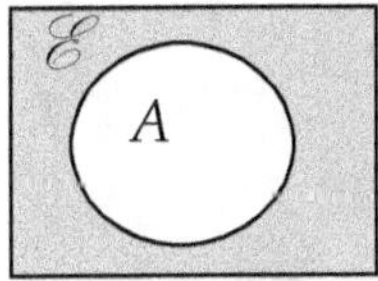

A' is shaded

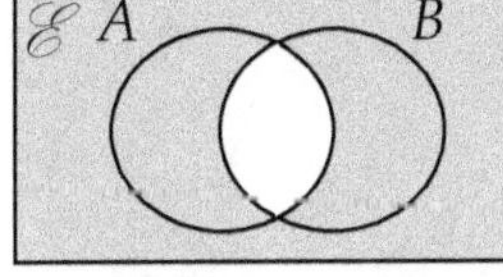

$(A \cap B)'$ is shaded

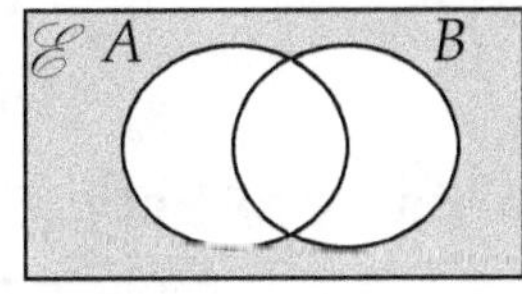

$(A \cup B)'$ is shaded

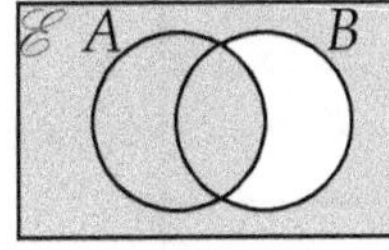

$A \cup B'$ is shaded

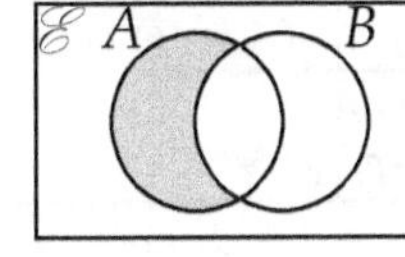

$A \cap B'$ is shaded

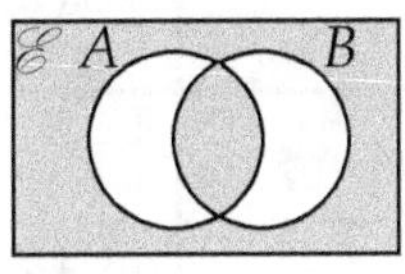

$(A \cap B) \cup (A \cup B)'$ is shaded

Example

$\mathscr{E} = \{1,2,3,4,5,6,7,8,9,10\}$, $A = \{1, 3, 4, 5, 7\}$, $B = \{3, 4, 8, 9\}$

Draw a Venn diagram to illustrate the relationship between these sets. Using this Venn diagram, list the elements of

(i) $A \cap B$ (ii) $A \cup B$ (iii) B'.

Solution

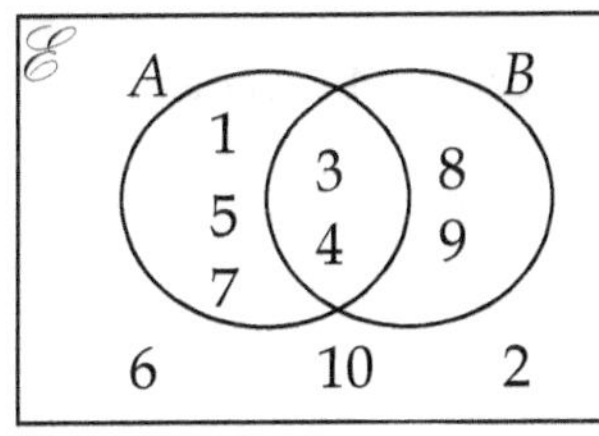

(i) $A \cap B = \{3, 4\}$

(ii) $A \cup B = \{1, 3, 4, 5, 7, 8, 9\}$

(iii) $B' = \{1, 2, 5, 6, 7, 10\}$

The Power or Derived Set of a Set

The power or derived set $P(A)$ of a set A is a set, which consists of all the subsets of A.

Example

If $A = \{a, b, c\}$, find the power set of A.

Solution

$P(A) = \{\emptyset, \{a\}, \{b\}, \{c\}, \{a, b\}, \{a, c\}, \{b, c\}, A\}$.

Cardinality of a Power or Derived Set

The cardinality of the power set of a set with n elements is 2^n.

Excluding the improper subsets, the cardinality of the power set of a set with n elements is $2^n - 2$.

Algebraic Laws of Sets

1	Idempotent Laws	$A \cup A = A \cap A = A$
2	Associative Laws	$A \cup (B \cup C) = (A \cup B) \cup C$ $A \cap (B \cap C) = (A \cap B) \cap C$
3	Commutative Laws	$A \cup B = B \cup A$ $A \cap B = B \cap A$
4	Distributive Laws	$A \cup (B \cap C) = (A \cup B) \cap (A \cup C)$ $A \cap (B \cup C) = (A \cap B) \cup (A \cap C)$
5	De-Morgan's Laws	$(A \cup B)' = A' \cap B'$ $(A \cap B)' = A' \cup B'$
6	Complement Laws	$A \cup A' = \mathscr{E}, A \cap A' = \emptyset$ $\mathscr{E}' = \emptyset, \emptyset' = \mathscr{E}, (A')' = A$
7	Identity Laws	$A \cup \emptyset = A, A \cap \emptyset = \emptyset$ $A \cup \mathscr{E} = \mathscr{E}, A \cap \mathscr{E} = A$

Cardinality Logic

Example

A newsagent sells three papers; the Post, the Messenger and the Standard. Customers buy one of the three papers. It is found that:

80 customers buy the Post,
70 customers buy the Messenger,
60 customers buy the Standard,
21 customers buy the Post and the Messenger,
14 customers buy the Messenger and the Standard,
16 customers buy the Standard and the Post,
6 customers buy all three newspapers.

Draw a Venn diagram to illustrate this information. Hence find the number of customers who buy only (a) The Post. (b) The Messenger. (c) The standard. (d) Find also the total number of customers for the three papers.

Solution

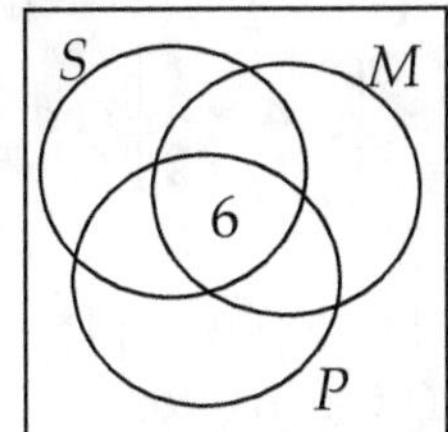

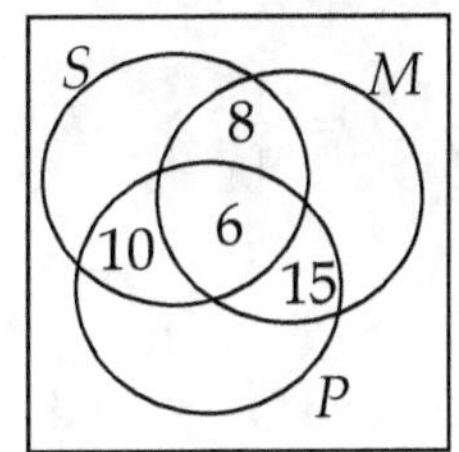

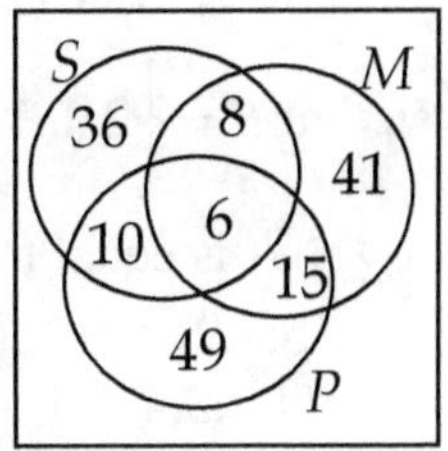

(a) Post customers only = $n(M \cup S)' = 49$.
(b) Messenger customers only = $n(P \cup S)' = 41$.
(c) Standard customers only = $n(M \cup P)' = 36$.
(d) Total number of customers = $n(M \cup P \cup S)$
$= 36 + 41 + 49 + 10 + 8 + 15 + 6 = 165$

MULTIPLE CHOICE EXERCISE 3:1

1. Let $P = \{2, 4, 6, 8, 10\}$, $Q = \{1, 3, 5, 7\}$, $R = \{1,2,3,4\}$ and $S = \{10, 6, 2, 8, 4\}$. It is true to say that:
   [A] P and Q are equivalent sets. [B] Q and R are equivalent sets
   [C] P and S are equivalent sets [D] Q and R are equal sets
2. Let $P = \{2, 4, 6, 8, 10\}$, $Q = \{1, 3, 5, 7\}$, $R = \{1,2,3,4\}$ and $S = \{10, 6, 2, 8, 4\}$. It is not true to say that:
   [A] P and Q are equivalent sets [B] Q and R are equivalent sets
   [C] P and S are equal sets [D] P and $P \cap S$ are equal sets.
3. Given that $H = \emptyset$. It is true to say that:
   [A] $0 \in H$ [B] $n(H) = \emptyset$ [C] $H = \{0\}$ [D] $n(H) = 0$
4. Given the universal set $\mathscr{E} = \{x: 0 < x < 10, x \in Z\}$. The complement of the set $P = \{x: x \in \mathscr{E}$, x is not divisible by 4$\}$ is:
   [A] $\{4\}$ [B] $\{4,8\}$ [C] $\{1,2,3\}$ [D] $\{1,2,3,5,6,7,9\}$
5. If $A = \{a, b, c\}$, $B = \{a, b, c, d, e\}$ and $C = \{a, b, c, d, e, f\}$, $(A \cup B) \cap (A \cup C)$ is equal to:
   [A] $\{a, b, c, d\}$ [B] $\{a, b, c, d, e\}$ [C] $\{a, b, c, d, e, f\}$ [D] $\{a, b, c\}$
6. Let J be the set of positive integers. If $H = \{x: x^2 < 3, x \neq 0\}$, then
   [A] $H = \{1\}$ [B] H is an infinite set [C] $H = \{0,1\}$ [D] $H = \{\ \}$
7. Given that $P = \left\{2, 1, 3, 9, \frac{1}{2}\right\}$, $Q = \left\{1, 2\frac{1}{2}, 3, 7\right\}$, $R = \left\{5, 4, 2\frac{1}{2}\right\}$ and $P \cup Q \cup R$ is equal to:

[A] $\left\{5,4,2\frac{1}{2}\right\}$ [B] $\{1,2,3,4,5,6,7\}$ [C] $\{1,9\}$ [D] $\left\{\frac{1}{2},1,2,2\frac{1}{2},3,4,5,7,9\right\}$

8. Given that $P=\left\{2,1,3,9,\frac{1}{2}\right\}, Q=\left\{1,2\frac{1}{2},3,7\right\}$, $R=\left\{5,4,2\frac{1}{2}\right\}$ and $P\cap Q\cap R$ is equal to:

   [A] $\{5,7,9\}$ [B] $\emptyset$ [C] $\{1,3,7\}$ [D] $\{4\}$

9. Let $\mathscr{E}=\{1,2,3,4\}$, $P=\{2,3\}$ and $Q=\{2,3,4\}$. $P\cap Q$ is equal to:

   [A] $\{1,2,3\}$ [B] $\{1,3,4\}$ [C] $\{2,3\}$ [D] $\{1,3\}$

10. $S=\{1,2,3,4,5,6\}$, $T=\{2,4,5,7\}$ and $R=\{1,4,5\}$. Then $(S\cap T)\cup R$ is:

    [A] $\{1,4,5\}$ [B] $\{2,4,5\}$ [C] $\{2,3,4,5\}$ [D] $\{1,2,4,5\}$

11. If $R=\{2,4,6,7\}$ and $S=\{1,2,4,8\}$ then $R\cup S$ equals:

    [A] $\{1,2,4,6,7,8\}$ [B] $\{1,2,4,7,8\}$ [C] $\{1,4,7,8\}$ [D] $\{2,6,7\}$

12. If $P=\{3,5,6\}$ and $Q=\{4,5,6\}$ then $P\cap Q$ equals:

    [A] $\{3,6\}$ [B] $\{4,5\}$ [C] $\{4,6\}$ [D] $\{5,6\}$

13. If $S\subset R$, then:

    [A] $S\cap R=R$ [B] $S\cap R=S$ [C] $S\cup R=S'$ [D] $S\cup R=S$

14. If $P=\{3,7,11,13\}$, $Q=\{2,4,8,16\}$. It follows that:

    [A] $(P\cap Q)'=\{2,3,4,13\}$ [B] $n(P\cup Q)=4$ [C] $P\cup Q=\emptyset$ [D] $P\cap Q=\emptyset$

15. Given that $P=\{b, d, e, f\}$ and $Q=\{a, c, f, g\}$ are subsets of the universal set $U=\{a, b, c, d, e, f, g\}$. $P'\cap Q$ is equal to:

    [A] $\{a, c\}$ [B] $\{a, c, d, g\}$ [C] $\{c, d, g\}$ [D] $\{a, c, g\}$

16.

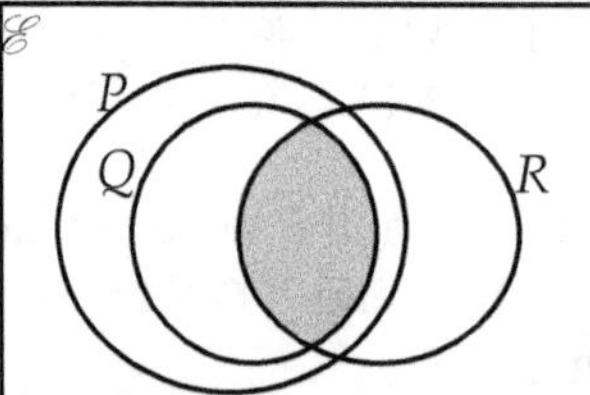

In the Venn diagram above, the shaded region is described by

[A] $Q\cap R$ [B] $Q\subset R$ [C] $P\cap Q\cup R$ [D] $P\cup Q\cap R$

17. Given the universal set $\mathscr{E}=\{1,2,3,4,5\}$, $P=\{1,2\}$, $Q=\{2,3,4\}$, then $(P\cap Q)'$ is:

    [A] $\{2\}$ [B] $\{5\}$ [C] $\{1,3,4,5\}$ [D] $\{1,2,3,5\}$

18.

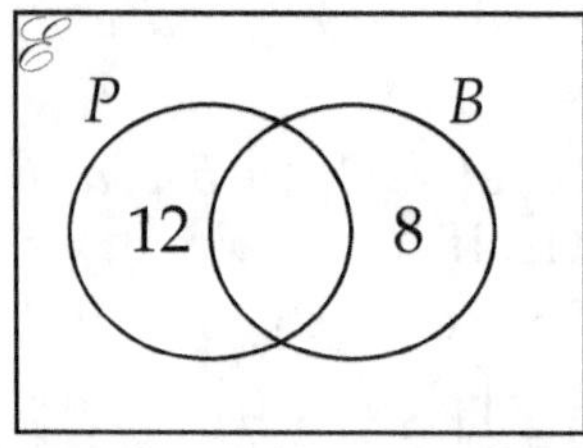

In the above Venn diagram Figure 20:14, given that $\frac{4}{9}$ of the total number of students offering Physics (P) or Biology (B) offer Physics only. The number of students offering both subjects is:

[A] 27 [B] 20 [C] 7 [D] 18

19. In the following Venn diagram, $n(P \cap Q)$ is:

[A] 1 [B] 2 [C] 4 [D] 6

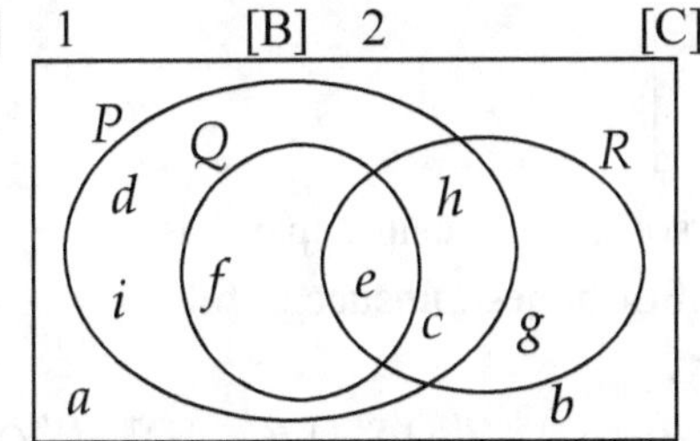

20. In the following Venn diagram, $Q' \cap R$ is equal to:

[A] $\{e\}$ [B] $\{c, h\}$ [C] $\{c, g, h\}$ [D] $\{c, e, g, h\}$

21. In the following Venn diagram, the shaded portion is:

[A] $P' \cap Q$ [B] $(R \cap Q) \cup (P' \cap R)$ [C] $P' \cap Q \cap R$ [D] $(P \cup Q)' \cap R$

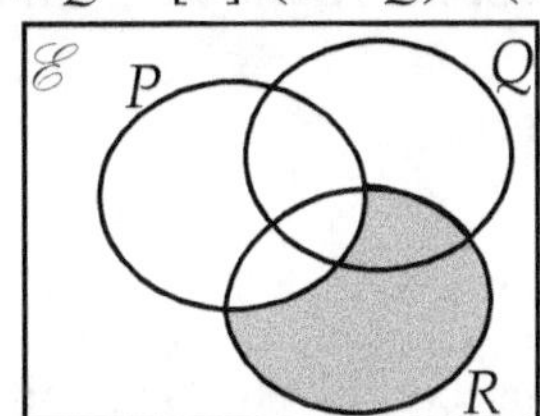

22. In a class of 80 students, every student study Economics or Geography. If 65 students study Economics and 50 study Geography, the number of students who study both subjects is:

[A] 15 [B] 30 [C] 35 [D] 45

23. A and B are two sets. The number of elements in $A \cup B$ is 49. The number in A is 22 and the number in B is 34. The number of elements in $A \cap B$ is:

[A] 7 [B] 27 [C] 15 [D] 12

24. If $n(P) = 21$, $n(R) = 33$ and $n(P \cup R) = 46$. $n(P \cap R)$ is equal to:

[A] 8 [B] 34 [C] 58 [D] 100

25. The following Venn diagram shows the number of students who studied physics P, chemistry C, and mathematics M in a certain school. How many students studied at least two of the three subjects?

[A] 165 [B] 160 [C] 155 [D] 135

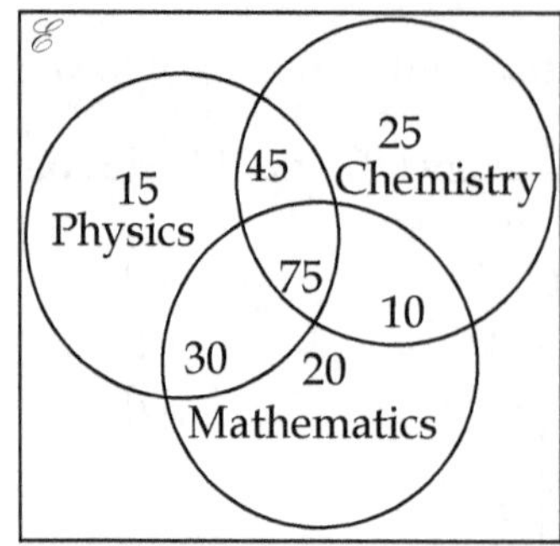

26. Let $\mathscr{E}$ = All men, H = honest persons, B = Businesspersons, S = Successful persons. The set of dishonest, unsuccessful businesspersons is represented by:
    [A] $H \cap S' \cap B$ [B] $H \cap S \cap B$ [C] $H' \cup S' \cup B$ [D] $H' \cap S' \cap B$
27. Let M = mammals, V = vertebrates, I = insects, H = horned animals. The statement, 'All mammals are vertebrates' is denoted by:
    [A] $M \cap V \neq \emptyset$ [B] $M \subset V$ [C] $M \cup V = \emptyset$ [D] $M \cup H \neq \emptyset$
28. Let M = mammals, V = vertebrates, I = insects, H = horned animals. $V \cap I = \emptyset$ in ordinary English means:
    [A] Vertebrates and insects are empty sets.
    [B] Some vertebrates are insects
    [C] Vertebrates cannot join with insects.
    [D] No insect is a vertebrate.
29. Let M = mammals, V = vertebrates, I = insects, H = horned animals. Using set notation the statement 'Some mammals are horned animals', can be written as:
    [A] $H \cap M \neq \emptyset$ [B] $M \cap H = \emptyset$ [C] $M \cup H \neq \emptyset$ [D] $M \cup H = \emptyset$
30. The statement $A \subset B$ is the same as:
    [A] $A \cap B = B$ [B] $A \cap B = A$ [C] $A \cup B = B$ [D] $A \cup B = A$

3.2 BINARY OPERATIONS

Properties of Operations

In an operation $*$ defined on a set S

1. S is **closed** under $*$ iff $\forall a, b \in S \Rightarrow a * b \in S$.
2. The operation has an **identity or neutral element** e iff $\forall a \in S, a * e = e * a = a$.
3. The **inverse** a^{-1} of a is that element such that, $a * a^{-1} = a^{-1} * a = e$.
4. The operation$*$ is **commutative**, iff $\forall a, b \in S, a * b = b * a$.
5. The operation$*$ is **associative**, iff $\forall a, b, c \in S, (a * b) * c = a * (b * c$.

Addition and multiplication are commutative and associative on the set $\mathbb{R}$ but Subtraction and division are not.

Operation Tables

On operation tables,

1. Symmetry of elements about the leading diagonal indicates that the operation is commutative.
2. The identity element is the element on the intersection of the row and column whose elements are in the same order as the header row and header column.
3. Any two elements which head the row and column which intersect on the identity element are inverses of each other.

$\circ$	a	b	c	d
a	b	d	a	c
b	d	c	b	a
c	a	b	c	d
d	c	a	d	b

(i)

$\circ$	a	b	c	d
a	b	d	a	c
b	d	c	b	a
c	a	b	c	d
d	c	a	d	b

(ii)

The tables above show that:

(a) The operation is commutative.

(b) The identity element is c.

(c) $a^{-1} = d$, $d^{-1} = a$, b and c are **self inverse** $\Rightarrow b^{-1} = b$ and $c^{-1} = c$.

Residue Classes

In arithmetic modulo m,

1. Any integer $n = qm + r \Rightarrow n = r \bmod m$ where the remainder (residue) $r \in \{0, 1, 2 \cdots (m-1)\}$.
2. **Equivalent numbers** are integers which have the same residue and are said to be **congruent modulo m**.

Example

The operation $*$ is defined over the set $S = \{0, 1, 2, 3, 4\}$ as multiplication modulo 5. Find $2 * 4$.

Solution

$2 * 4 = \frac{2(4)}{5} = 1R3 = 3$

ELEMENTARY GROUP THEORY

A non empty set, S, together with a binary operation $*$ forms a group $(S,*)$ if the set is **closed**, has an **identity element**, every element has an **inverse** and the operation is **associative** over S.
The mnemonic *CL-ID-IN-AS* may help as a mental aid.

An Abelian or commutative group is a group, for which the **commutative law** also holds.
The mnemonic *CL-ID-IN-AS-CO* may be useful as a mental aid.

MULTIPLE CHOICE EXERCISE 3:2

1. An operation * is defined on $\mathbb{Z}$ the set of integers, by $a*b = a^2b - b^2a$. 3*5 is equal to:
   [A] –45 [B] –30 [C] 45 [D] 30
2. An operation * is defined on $\mathbb{Z}$ the set of integers, by $a*b = a^2b - b^2a$. The operation * is:
   [A] Commutative because $a * b \neq b * a, \forall a, b \in \mathbb{Z}$.
   [B] Not commutative because $a * b \neq b * a, \forall a, b \in \mathbb{Z}$.
   [C] Commutative because $a * b = b * a, \forall a, b \in \mathbb{Z}$.
   [D] Not commutative because $a * b = b * a, \forall a, b \in \mathbb{Z}$.
3 An operation * is defined on the set $\mathbb{R}$, of real numbers, as $x * y = xy - (x + y)$. The value of 2*3 is:
   [A] 6 [B] –5 [C] 1 [D] 5
4. An operation * is defined on the set $\mathbb{R}$, of real numbers, as $x * y = xy - (x + y)$. Given that $x*2 = 8$, the value of x is:
   [A] 6 [B] –6 [C] –10 [D] 10
5. An operation $*$ is defined on $\mathbb{Z}$ as $\forall a, b \in \mathbb{Z}, a * b = a + b - 4$. The identity element for the operation is:
   [A] 1 [B] –1 [C] –4 [D] 4
6. An operation $*$ is defined on $\mathbb{Z}$ as as $\forall a, b \in \mathbb{Z}, a * b = a + b - 4$. The inverse of 5 is:
   [A] $\frac{1}{5}$ [B] –5 [C] 3 [D] –3

7. The operation $*$ is defined on $\mathbb{Q}$, the set of rational numbers, by

$a * b = \dfrac{a+3b}{2}$. Given that $5*u = 20$, the value of u is:

[A] $u = \dfrac{35}{3}$ [B] $u = \dfrac{8}{3}$ [C] $u = \dfrac{5}{3}$ [D] $u = \dfrac{1}{3}$

8. If the operation $a \sim b$ gives the numerical difference between a and b, e.g. 8~5 =3 and 7~11=4. The operation table which best represents the operation ~ for the set $S = \{0, 4, 8, 12\}$ is:

[A]

~	0	4	8	12
0	0	8	4	12
4	4	0	4	8
8	8	4	0	4
12	12	4	8	0

[B]

~	0	4	8	12
0	0	4	8	12
4	4	0	4	8
8	8	4	0	4
12	12	8	4	0

[C]

~	0	4	8	12
0	12	8	4	0
4	8	4	0	4
8	4	0	12	8
12	0	8	4	12

[D]

~	0	4	8	12
0	12	8	4	0
4	8	4	0	4
8	4	0	12	8
12	0	4	8	12

9. If the operation $a \sim b$ gives the numerical difference between a and b. For instance, 8~5 =3 and 7~11=4. The identity element is:

[A] 12 [B] 8 [C] 4 [D] 0

10. The operation * is defined on $\mathbb{R}$, the set of real numbers, by $a * b = a + b + 2ab$. Given that $x * x = 12$, the possible value(s) of x are:

[A] –3 and 2 [B] –3 and –2 [C] 3 and –2 [D] 3 and 2

11. The operation * is defined on the set $\mathbb{N}$ of natural numbers by $x * y$ is the absolute value of x–y. e.g. 2*2=0 , 5*8=3 etc. The identity element is:

[A] 0 [B] 1 [C] –1 [D] there is no identity element

12. The operation * is defined on the set $\mathbb{N}$ of natural numbers by $x * y$ is the absolute value of $x - y$. e.g. $2 * 2 = 0$, $5 * 8 = 3$ etc. The inverse of each element is:

[A] Its reciprocal. [B] The element itself.

[C] The negation of the element. [D] The square of the element.

13. The operation $*$ is:

[A] Not associative because $a * b \neq b * a, \forall a, b \in \mathbb{N}$.

[B] Associative because $(a * b) * c = a * (b * c), \forall a, b, c \in \mathbb{N}$.

[C] Not associative because $(a * b) * c \neq a * (b * c), \forall a, b, c \in \mathbb{N}$.

[D] Not associative because $(a * b) * c = a * (b * c), \forall a, b, c \in \mathbb{N}$.

14. The table below is a combination table for the set $S = \{a, b, c, d\}$ under a given operation$*$. The identity element is:

[A] a [B] b [C] c [D] d

$*$	a	b	c	d
a	b	c	a	d
b	c	d	b	a
c	a	b	c	d
d	d	a	d	c

15. The table above is a combination table for the set $S = \{a, b, c, d\}$ under a given operation$*$. For this operation, it is true to say that:
[A] $a^{-1} = b,\ b^{-1} = c,\ c^{-1} = d,\ d^{-1} = c$ [B] $a^{-1} = d,\ b^{-1} = c,\ c^{-1} = a,\ d^{-1} = b$
[C] $a^{-1} = c,\ b^{-1} = d,\ c^{-1} = a,\ d^{-1} = b$ [D] $a^{-1} = b,\ b^{-1} = a,\ c^{-1} = c,\ d^{-1} = d$

16. Table 20:7 below is a combination table for the set $S = \{a, b, c, d\}$ under a given operation$*$. The operation $*$ is:
[A] Not associative because $a * b \neq b * a, \forall a, b \in S$.
[B] Not associative because $(a * b) * c = a * (b * c), \forall a, b, c \in S$.
[C] Not associative because, $(a * b) * c \neq a * (b * c), \forall a, b, c \in S$.
[D] Associative because, $(a * b) * c = a * (b * c), \forall a, b, c \in S$.

17. A binary operation $*$ is defined on $\mathbb{R}$ by $x*y = xy + 2x$. $-3*4$ is equal to:
[A] -4 [B] -18 [C] -12 [D] 8

18. A binary operation $*$ is defined on $\mathbb{R}$ by $x*y = xy + 2x$. The operation $*$ is:
[A] Not commutative because $a * b \neq b * a, \forall a, b \in \mathbb{R}$.
[B] Commutative because, $a * b \neq b * a, \forall a, b \in \mathbb{R}$.
[C] Not commutative because, $a * b = b * a, \forall a, b \in \mathbb{R}$.
[D] Commutative because, $a * b = b * a, \forall a, b \in \mathbb{R}$.

19. The operation $*$ on the set of real numbers is defined by $a * b = \dfrac{8a + 5b}{3}$.
The value of b for which $2 * b = -6$ is:
[A] $\frac{34}{5}$ [B] $-\frac{34}{5}$ [C] $-\frac{2}{5}$ [D] $\frac{2}{5}$

20. The operation ∇ is defined on the set $\{0, 2, 4, 6, 8\}$ by $x \nabla y = (x - y)^2$. The property or properties valid for the operation is/are:
[A] closure [B] commutativity
[C] associativity [D] closure, commutativity and associativity.

21. A binary operation is defined on $\mathbb{R}$ the set of real numbers as follows $a*b = a^2 + ab + b^2$. The value of $-3*2$ is:
[A] -6 [B] -12 [C] 4 [D] 7

22. The operation $*$ on the set $\mathbb{R}$ of real numbers is defined by
$a * b = \dfrac{a^2 + b^2}{2ab}$. The value of $3*-4$ is:
[A] -25 [B] $-\frac{24}{25}$ [C] -24 [D] $-\frac{25}{24}$

23. The operation * on the set $\mathbb{R}$ of real numbers is defined by

$a * b = \frac{a^2 + b^2}{2ab}$. Given that $2*x = 1$, the value of x is:

[A] 4 [B] -4 [C] -2 [D] 2

3.3 LOGIC

Statements or Propositions

A **statement** or **proposition** is a sentence that is either true or false but not both. The truthfulness or falsity of a statement is called its **truth value**. On truth tables, the truth value of a true statement is denoted by T or 1, while that of a false statement is denoted by F or 0.

A **closed statement** is a statement concerning a definite object.
An **open statement** is a statement that does not concern a definite object.

The open part in an open statement is called the **variable.**
The set of all the possible values which the variable can take is called the **domain** or the **replacement set** of the variable.

Examples

State the domain and the truth set for each of the following statements.
(a) $5x + 2 = 13$ (b) $7y < 42$ (c) $(x + 1)(x - 2) = 0$

Solution

(a) The domain is $\mathbb{Q}$ and the truth set is $\left\{x: x = \frac{11}{5}, x \in \mathbb{Q}\right\}$.
(b) The domain is $\mathbb{R}$ and the truth set is $\{y : y < 6, y \in \mathbb{R}\}$.
(c) The domain is $\mathbb{Z}$ and the truth set is $\{-1, 2\}$.

Negation, $\sim p$

The negation of a statement p denoted by $\sim p$ or $\neg p$ or p' and read 'not p' is the statement formed from p by inserting the word "not" into p or placing one of the phrases "it is false that" or "it is not true that" before the statement p.

If p is true $\sim p$ is false; if p is false $\sim p$ is true.

Example

State the negation of the following statements.
q: It is raining.
r: All Cameroonians speak both English and French.
s: 5 is a multiple of 3.
t: $2x + 1 = 0$, for all values of x.

Solution

$\sim q$: It is not raining.

$\sim r$: It is not true that all Cameroonians speak both English and French.

$\sim s$: 5 is not a multiple of 3.

$\sim t$: It is not true that $2x + 1 = 0$, for all values of x.

Composite Statement

A **conjunction** $p \wedge q$ is a composite statement made by combining two statements p and q with the use of the preposition "and".

$p \wedge q$ is true iff p is true and q is true. $p \wedge q$ is false iff p or q is false.

p	q	$p \wedge q$
T	T	T
T	F	F
F	T	F
F	F	F

or

p	q	$p \wedge q$
1	1	1
1	0	0
0	1	0
0	0	0

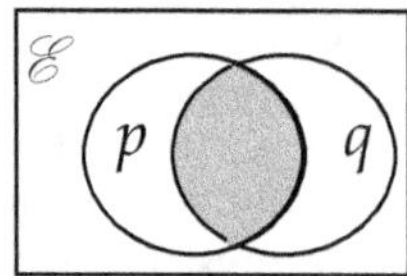

$p \wedge q$ is shaded

A **disjunction** is a composite statement $p \vee q$ made by combining two statements p and q with the use of the "or" or 'either…or'.

In logic "or" is used in the sense of "and/or".

If $p \vee q$ is true then p is true or q is true or both p and q are true.

If $p \vee q$ is false then both p and q are false.

p	q	$p \vee q$
T	T	T
T	F	T
F	T	T
F	F	F

or

p	q	$p \vee q$
1	1	1
1	0	1
0	1	1
0	0	0

Example

Given the statements x: Ngwa ate rice and y: Ngwa ate beans.

Make (a) a conjunction involving x and y.

(b) a disjunction involving x and y.

Solution

(a) $x \wedge y$: Ngwa ate rice and beans.

(b) $x \vee y$: Ngwa ate rice or beans.

Conditional Statement, $p \Rightarrow q$ or $p \rightarrow q$

A **Conditional Statement** $p \rightarrow q$ is a composite statement obtained by joining two statements p and q in such a way that if the first statement p is true, the second statement q must be true. $p \rightarrow q$ or $p \Rightarrow q$ is read as "*p* implies *q*", "*p* is sufficient for *q*", "*p* only if *q*" or "*q* is necessary for *p* ". $p \Rightarrow q$ called an **implication** and is used in theorems instead of $p \rightarrow q$.

$p \rightarrow q$ is always true except p is true and q is false.

p	q	$p \rightarrow q$
T	T	T
T	F	F
F	T	T
F	F	T

or

p	q	$p \rightarrow q$
1	1	1
1	0	0
0	1	1
0	0	1

Example

Which of the following statements is false?

p: If Cameroon is in Africa, then $3 \times 2 = 7$.

q: If Cameroon is in Europe, then $3 \times 2 = 6$.

r: If Cameroon is in Africa, then $3 \times 2 = 6$.

s: If Cameroon is in Europe, then $3 \times 2 = 7$.

Solution

Only p is false and q, r and s are all true.

Biconditional Statement, $p \Leftrightarrow q$ or $p \leftrightarrow q$

A **Biconditional Statement** is a proposition involving two statements, one of which is true if, and only if, the other is true.

The biconditional statement $p \Leftrightarrow q$ or $p \leftrightarrow q$ can be read in the following ways.

(i) p is a necessary and sufficient condition for q

(ii) p implies and is implied by q

(iii) p if and only if q sometimes written as p iff q.

For a true biconditional statement, either both statements are true or both are false, otherwise the statement is false.

We can summarize this fundamental property on a truth table as follows;

p	q	$p \leftrightarrow q$
T	T	T
T	F	F
F	T	F
F	F	T

or

p	q	$p \leftrightarrow q$
1	1	1
1	0	0
0	1	0
0	0	1

Example

Which of the following statements is true?

p: Cameroon is in Africa $\leftrightarrow$ $3 \times 2 = 7$.

q: Cameroon is in Europe $\leftrightarrow 3 \times 2 = 6$.

r: Cameroon is in Africa $\leftrightarrow 3 \times 2 = 6$.

s: Cameroon is in Europe $\leftrightarrow 3 \times 2 = 7$.

Solution

The biconditional statements r and s are true and p and q are false.

Connectors

The logical symbols ~, ∧, ∨, → and ↔ are called connectors. Their respective set algebra equivalents are ', ∩, ∪, ⊂ and =. ~ is called a unitary connector because it affects only one statement while ∧, ∨, → and ↔ are called binary connectors because they combine two statements.

Logical Equivalence

Two statements p and q are **logically equivalent** if and only if they have the same truth tables. The statement 'p and q are logically equivalent' is written symbolically as $p \equiv q$.

Example

Draw truth tables showing the following:

(a) $p \wedge q$ (b) $p \vee q$ (c) $q \wedge p$ (d) $q \vee p$.

Hence deduce that:

(i) $p \wedge q \equiv q \wedge p$ (ii) $p \vee q \equiv q \vee p$

Solution

(a)

p	q	$p \wedge q$
T	T	T
T	F	F
F	T	F
F	F	F

(b)

p	q	$p \vee q$
T	T	T
T	F	T
F	T	T
F	F	F

(c)

p	q	$q \wedge p$
T	T	T
T	F	F
F	T	F
F	F	F

(d)

p	q	$q \vee p$
T	T	T
T	F	T
F	T	T
F	F	F

We can see that the truth tables for $p \wedge q$ and $q \wedge p$ are the same and the truth tables for $p \vee q$ and $q \vee p$ are the same.
Therefore, (i) $p \wedge q \equiv q \wedge p$ (ii) $p \vee q \equiv q \vee p$

De Morgan's Laws of Logic

The De Morgan's laws of logic state that:
(i) $\sim p \vee \sim q \equiv \sim (p \wedge q)$ (ii) $\sim p \wedge \sim q \equiv \sim (p \vee q)$
Compare these with the De Morgan laws of sets.

Example
Draw the truth tables for:
(a) $\sim p \vee \sim q$ (b) $\sim (p \wedge q)$ (c) $\sim p \wedge \sim q$ (d) $\sim (p \vee q)$.
Hence, deduce that the De Morgan laws of logic are true.

Solution

(a)

p	q	$\sim p$	$\sim q$	$\sim p \vee \sim q$
T	T	F	F	F
T	F	F	T	T
F	T	T	F	T
F	F	T	T	T

(b)

p	q	$p \wedge q$	$\sim (p \wedge q)$
T	T	T	F
T	F	F	T
F	T	F	T
F	F	F	T

(c)

p	q	$\sim p$	$\sim q$	$\sim p \wedge \sim q$
T	T	F	F	F
T	F	F	T	F
F	T	T	F	F
F	F	T	T	T

(d)

p	q	$p \vee q$	$\sim (p \vee q)$
T	T	T	F
T	F	F	F
F	T	F	F
F	F	F	T

The truth tables for $\sim p \wedge \sim q$ and $\sim (p \vee q)$ are the same and the truth tables for $\sim p \vee \sim q$ and $\sim (p \wedge q)$ are the same Hence the De Morgan's laws are true.

Tautologies

A tautology is a proposition that is always true. For instance, the proposition "p or not p" is a tautology.

p	$\sim p$	$p \vee \sim p$
T	F	T
F	T	T

Contradictions

A contradiction is a proposition that is always false. For instance, the proposition "p and not p" is a contradiction.

p	$\sim p$	$p \wedge \sim p$
T	F	F
F	T	F

Quantifiers

Quantifiers are words such as; for all, for every, for some, for each, for any, there exist, for at least etc which suggest the idea of quantity.

Universal Quantifier, ∀

Example: $\forall x \in \mathbb{R}, x^2 \geq 0$. This is read "for all real values of x, x^2 is greater than or equal to zero.

Existential Quantifier

Example: Let $A = \{1,2,3,4,5\}$, then we can write $\exists x \in A$: x is a multiple of 2 read "there exists, at least one element of A which is a multiple of 2".

Unitary Existential Quantifier

Example: Let $A = \{1,2,3,4,5\}$, then we can write $\exists! \, x \in A$: x is a multiple of 3 read "there exists one and only one" element of A which is a multiple of 3".

Example

In each of the following cases write statements using a quantifier and explain each in English.

(1) Let $A = \{1, 2, 3, 4, 5\}$ and $p(x)$: $0 \leq x - 1 < 5$.

(2) $\mathbb{R}$ is the set of real numbers and $p(a)$: $a \times 1 = 1 \times a = 0$.

(3) $\mathbb{R}$ is the set of real numbers and $p(a)$: $a + 0 = 0 + a = a$.

(4) Let $A = \{1, 2, 3, 4, 5\}$ and $p(x)$: $x - 1$ is odd.

Solution

(1) $\forall x \in A, p(x)$ or $(\forall x \in A)\, p(x)$ means "for all values of x belonging to the set A, $0 \leq x - 1 < 5$.

(2) $\exists!\, a \in \mathbb{R}, p(a)$ or $(\exists!\, a \in \mathbb{R}), p(a)$ means "there exist one and only one element a belonging to $\mathbb{R}$, such that $a \times 1 = 1 \times a = 0$".

(3) $\forall a \in \mathbb{R}, p(a)$ or $(\forall a \in \mathbb{R}), p(a)$ means "for all real values of a, $a + 0 = 0 + a = a$.

(4) $\exists x \in A, p(x)\ or\ (\exists x \in A), p(x)$ means "there exist at least one element x belonging to the set A, such that $x - 1$ is odd".

Syllogisms

A **syllogism** is an argument made up of statements in one of the following four forms:

(a) The **Universal affirmative** is of the form: All A's are B's.
Example p: all cows eat grass.

(b) The **Universal negative** is of the form: No A's are B's.
Example q: No historians are chemists.

(c) A **particular affirmative** is of the form: Some A's are B's.
Example r: Some polygons are rectangles.

(d) A **particular negative** is of the form: Some A's are not B's.
Example s: Some students are not studious.

The common nouns, such as cows, grass, historians, chemists etc are the terms of the syllogism.

Hypotheses and Conclusion

Consider the following statements.

p: All even numbers are divisible by 2

q: 14 is an even number

Therefore,

r: 14 is divisible by 2.

Statements such as p and q are called the **premises** or **hypotheses** while r is called the **conclusion**.

It follows that if the premises or hypotheses are true then the conclusion is bound to be true.

MULTIPLE CHOICE EXERCISE 3:3

1. Given the following truth table the correct relationship between the statements p and q is:

p	q
T	F
F	T

[A] $p \Rightarrow q$ [B] $p \Leftrightarrow q$
[C] $\sim p = q$ [D] $p = q$

2. The closed statement among the following is:
[A] $4 + 4 = 8$ [B] He ate rice [C] $2x - 1 \leq 0$ [D] $2x - 1 = 0$
3. The open statement among the following is:
[A] $7 + 7 = 14$ [B] Her makeup is good
[C] $3 + 4 \leq 0$ [D] Ngwa is a footballer
4. Which of the following is not a statement?
[A] $3 + 4 \leq 0$ [B] Roses are red
[C] Tse is sick [D] Read your Bible everyday
5. A statement among the following is:
[A] $5 + 8$ [B] Hit the iron while hot
[C] $ax^2 + bx + c$ [D] $x = -b \pm \frac{\sqrt{b^2 - 4ac}}{2}$
6. The truth set for the statement $X = \{x: 2x-1 \leq 7, x \in \mathbb{N}\}$ is:
[A] $X = \{0, 1, 2, 3, 4\}$ [B] $X = \{x: x < 4, x \in \mathbb{N}\}$
[C] $X = \{x: 0 \leq x < 4, x \in \mathbb{N}\}$ [D] $X = \{x: 0 \leq x \leq 4, x \in \mathbb{N}\}$
7. A statement among the following is:
[A] To God be the glory. [B] 10-3
[C] Stand up when the visitor arrives [D] $5x > -20$
8. If p: Bih is in Bamenda and q: Bamenda is in Cameroon. Then:
[A] $q \Rightarrow p$ [B] $\sim p \Rightarrow q$ [C] $p \Rightarrow q$ [D] $\sim q \Rightarrow p$
9. Given that, p: ΔABC is equilateral and q: ΔABC is equiangular. Then:
[A] $\sim p \Rightarrow q$ [B] $\sim q \Leftrightarrow p$ [C] $p \Leftrightarrow q$ [D] $\sim p \Rightarrow q$
10. The truth table for $p \wedge q$ is:

[A]

p	q	$p \wedge q$
T	T	T
T	F	F
F	T	F
F	F	F

[B]

p	q	$p \wedge q$
T	T	T
T	F	T
F	T	T
F	F	F

[C]

p	q	$p \wedge q$
T	T	T
T	F	F
F	T	T
F	F	T

[D]

p	q	$p \wedge q$
T	T	T
T	F	T
F	T	F
F	F	F

11. The truth table for $p \vee q$ is:

[A]

p	q	$p \vee q$
T	T	T
T	F	F
F	T	F
F	F	F

[B]

p	q	$p \vee q$
T	T	T
T	F	T
F	T	T
F	F	F

[C]

p	q	$p \vee q$
T	T	T
T	F	F
F	T	T
F	F	T

[D]

p	q	$p \vee q$
T	T	T
T	F	T
F	T	F
F	F	F

12. Given the following statements and the triangle below.
$p: a + b = \theta,\ q: c + \theta = 180°$ and $r: a + b + c = 180°$
The arguments which proves that the exterior angle of a triangle is equal to the sum of the two interior opposite angles is:

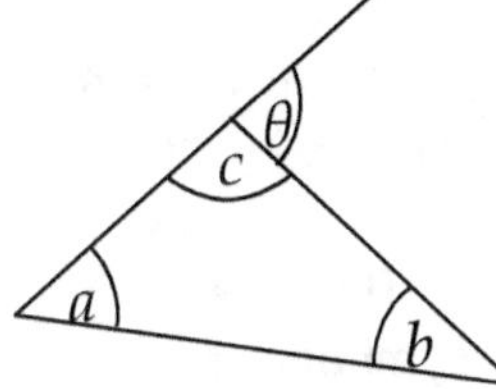

[A] $q \Rightarrow r \Rightarrow p$
[B] $r \Rightarrow p \Rightarrow q$
[C] $r \Rightarrow q \Rightarrow p$
[D] $q \Rightarrow p \Rightarrow r$

13. Which of the following is a composite statement?
[A] He studied Mathematics at E.N.S. [B] The kola nut has been eaten.
[C] He can drive. [D] There is no chalk in the school.

14. Which of the following statements is not a composite statement?
[A] Bamenda is in the NW region of Cameroon.
[B] Ngala was dead and buried.
[C] Fonjong eats coco yams.
[D] The sun is shining when it is raining.

15. The truth table that represents a tautology is:

p	q	$p \wedge q$
T	T	T
T	F	T
F	T	T
F	F	T

[A]

p	q	$p \wedge q$
T	T	T
T	F	F
F	T	F
F	F	F

[B]

p	q	$p \wedge q$
T	T	T
T	F	F
F	T	F
F	F	T

[C]

p	q	$p \wedge q$
T	T	F
T	F	F
F	T	F
F	F	F

[D]

16. The truth table that represents a contradiction is:

p	q	$p \wedge q$
T	T	T
T	F	T
F	T	T
F	F	T

[A]

p	q	$p \wedge q$
T	T	T
T	F	F
F	T	F
F	F	F

[B]

p	q	$p \wedge q$
T	T	T
T	F	F
F	T	F
F	F	T

[C]

p	q	$p \wedge q$
T	T	F
T	F	F
F	T	F
F	F	F

[D]

17. An inclusive disjunction among the following is:
    [A] Nursing requires a mastery of physics or chemistry.
    [B] At 6 a.m. I shall go to Bamenda or Douala.
    [C] I shall be sleeping or eating.
    [D] In high school, I shall offer LS1 or LA1.
18. An exclusive disjunction among the following is:
    [A] $-5 < 0$ or $5 > 2$. [B] $4 > 1$ or $4 < 6$.
    [C] ΔABC is equilateral or a right-angled triangle. [D] $0 < x \leq 20$ or $x > 3$.
19. The symbol which is not a logical connector is:
    [A] $\wedge$ [B] $>$ [C] $\vee$ [D] $\longrightarrow$
20. The unitary connector among the following logical symbols is:
    [A] $\sim$ [B] $>$ [C] $\vee$ [D] $\longrightarrow$
21. The statement which is not a syllogism is:
    [A] All cows eat grass. [B] No historians are chemists.
    [C] I love yams. [D] Some polygons are rectangles.
22. The negation of the statement "All Anglophones speak English" is:
    [A] Some Anglophones speak English.
    [B] Some Anglophones do not speak English.
    [C] No Anglophone speaks English.
    [D] All Anglophones do not speak English.
23. The negation of the statement "All x are not y" is:
    [A] Some x are y. [B] Some x are not y.
    [C] No x are y. [D] All x are not y.
24. The negation of the statement "Not all that glitters is gold" is:
    [A] All that glitters is not gold. [B] Some of what glitters is gold.
    [C] Nothing that glitters is gold. [D] All that glitters is gold

CHAPTER 4

RELATIONS AND FUNCTIONS

4.1 BINARY RELATIONS

A **relation** $\Re$ is a set of ordered pairs or a rule that assigns an element $x \in A$ to another element $y \in A$ (relations in a set) or an element $x \in A$ to another element $y \in B$ (relations from one set to another).
A **mathematical relation** or **number relation** is relation between sets of numbers. The statement 'a relates b' is denoted by $a\Re b$ or(a, b).

Ways of Defining Relations

We can define relations using a rule, a formula, ordered pairs, set builder notation, a table of values, a graph or an arrow (pappy) diagram.

The Cartesian product of Two Sets

The Cartesian product $A \times B$, read as "A cross B" is defined as

$$A \times B = \{(a, b): a \in A \text{ and } b \in B\}$$

Note!

1. A relation is a subset of a Cartesian product.
2. $A \times B \neq B \times A$ and $(a, b) \neq (b, a)$
3. We should not confuse (a, b) with the same notation used to represent the open interval $a < x < b$.

Example

Given that $A = \{1,2,3\}$ and $B = \{x, y\}$. Find (a) $A \times B$ (b) $B \times A$.

Solution

(a) $A \times B = \{(1, x), (1, y), (2, x), (2, y), (3, x), (3, y)\}$
(b) $B \times A = \{(x, 1), (x, 2), (x, 3), (y, 1), (y, 2), (y, 3)\}$

Example

A relation "is a factor of" is defined from the set $X = \{2, 3, 4\}$ to another set $Y = \{1, 2, 3, 4, 5, 6, 7, 8\}$. Illustrate this relation,
(a) As a set of ordered pairs. (b) Diagrammatically.

Solution

(a) $\Re = \{(2,2), (2,4), (2,6), (2,8), (3,3), (3,6), (4,4), (4,8)\}$

(b)

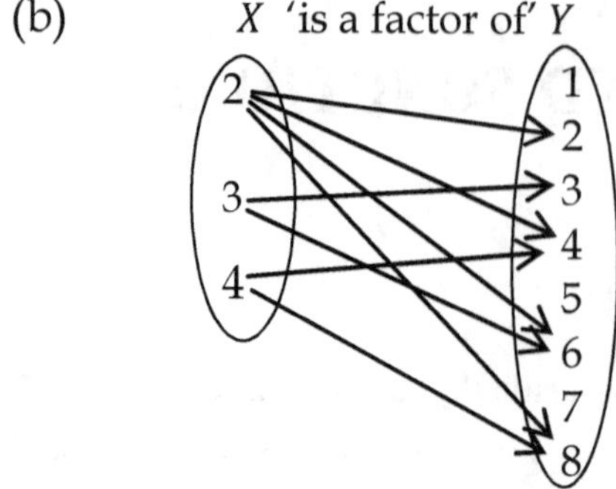

The set of elements of the codomain which are images of the domain is called the **range**.

Linear Relation

A linear relation is a relation which represents a straight line equation $y = mx + c$, where m and c are constants.

Example

Let $\mathfrak{R}$ be the relation "is half of" defined from the set $A = \{1,2,3,4\}$ to $B = \{2,4,6,8\}$. Draw a graph of this relation.

Solution

The graph (i) is made of discrete points. If the relation were defined on $\mathbb{R}$ the graph will be continuous as in (ii).

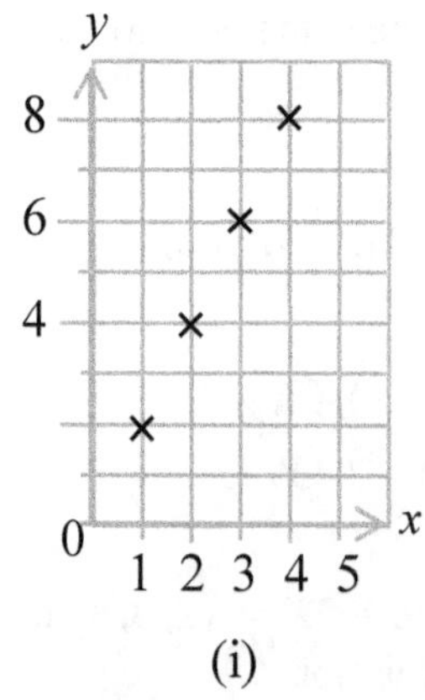

(i)

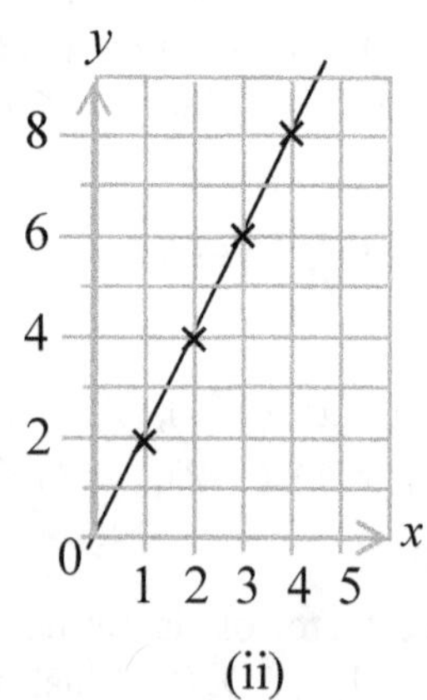

(ii)

Inverse Relation

The inverse $\mathfrak{R}^{-1}$ of a relation $\mathfrak{R}$ from a set A to B is the opposite relation from B to A. A relation is **self inverse** if the inverse relation is the relation itself. For instance the relation 'is a cousin of' is self inverse.

Example

Find the inverse of each of the following relations

(a) $\Re = \{(-4,4),(-2,2),(0,0),(2,-2),(4,-4)\}$ (b)

x	0	1	2	3	4
y	0	2	4	6	8

Solution

(a) $\Re = \{(4,-4),(2,-2),(0,0),(-2,2),(-4,4)\}$ (b)

x	0	2	4	6	8
y	0	1	2	3	4

Relations in a Set

Example

A relation is defined on that set $X = \{1,2,3,4,5,6\}$ as 'is a factor of'. Represent this relation on an arrow diagram.

Solution

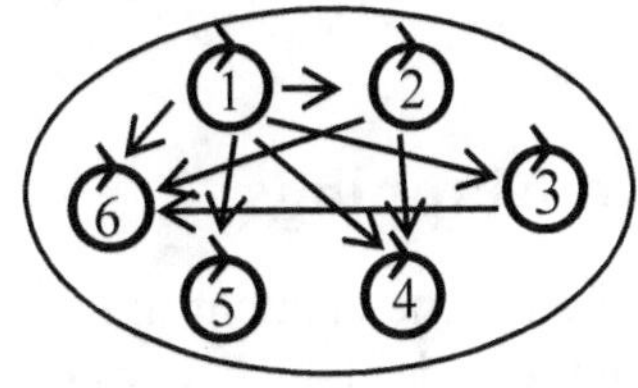

The loops ⥀ round each of the numbers indicates that each number is a factor of itself.

Properties of Relations in a Set

A relation $\Re$ in a set A is said to be:

(a) **Reflexive** if every element in A is related to itself. i.e. $\forall x \in A, x\Re x$.
(b) **Symmetric** if $\forall x, y \in A, x\Re y \Rightarrow y\Re x$.
(c) **Transitive** if $\forall x, y, z \in A, x\Re y$ and $y\Re z \Rightarrow x\Re z$.
(d) **Anti-symmetric** if $\forall x, y \in A, x\Re y$ and $y\Re x \Rightarrow x = y$.

Equivalence relations

An **equivalence relation (RST relation)** in a set A is a relation $\Re$, which is reflexive, symmetric and transitive.

Order Relations

An **order relation (RAT relation)** in a set A is a relation $\Re$, which is reflexive, anti-symmetric and transitive.

4.2 MAPPINGS AND FUNCTIONS

The Concept of a Mapping and a Function

A **mapping** is a relation in which every element, x of the domain A is related to one and only one element y of the codomain B.

By implication, on arrow diagrams one and only one arrow leaves each element in the domain or equivalently, on a graph a vertical line cannot be drawn to cut more than one point.

A **function** is a mapping which involves the set of numbers.

Function Notation

The two common notations used for functions are:

(a) $f(x) = 5x + 4$ or $y = 5x + 4$. (b) $f: x \mapsto 5x + 4$.

$\mapsto$ is used to map elements of two sets as in (b) and $\rightarrow$ is used to map one set to another. For example $f: A \rightarrow B$.

Classification of Functions and Mappings

A **one-one mapping** is a mapping in which one and only one element in the domain A is mapped to one and only one element in the codomain B. A **one-one function** is called a bijection or bijective function. For instance;

Let $A = \{1,2,3,4\}$ and $B = \{2,3,4,5\}$. The function $f: A \rightarrow B, f: x \longmapsto x + 1$, shown below is injective and surjective.

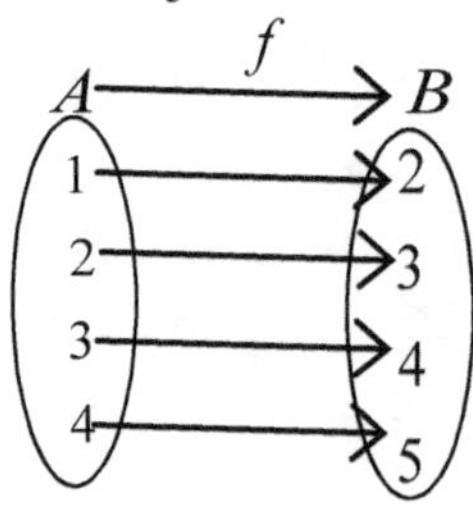

An "**onto**" mapping is a mapping in which all the elements of the codomain are images of elements of the domain.
An **on-to function** is called a surjection or surjective function.

For instance; let $A = \{-3, -2, -1, 0, 1, 2, 3\}$ and $B = \{0, 1, 4, 9\}$. The relation $f\!:\!A \longrightarrow B$, defined by $f: x \longmapsto x$ shown below is surjective.

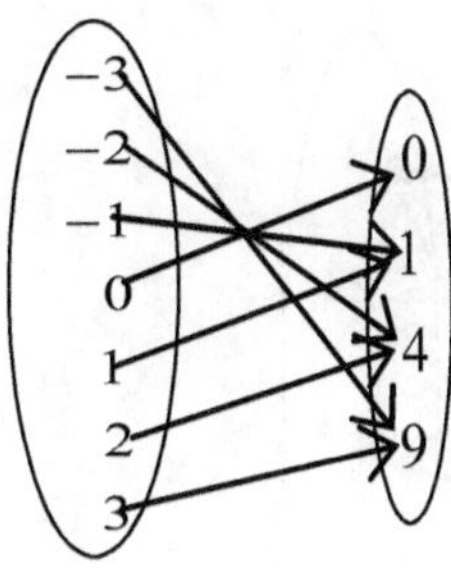

For a surjection, the cardinality of the codomain B is always less than or equal to that of the domain A.

$$n(B) \le n(A)$$

An "**into**" mapping is a mapping in which not all the elements of the codomain are images of elements of the domain.
An **in-to function** is called an injection or injective function.

Let $A = \{1,2,3,4\}$ and $B = \{1,2,3,4,5,6\}$. The function $f : A \to B$, defined by $f : x \mapsto x+1$, shown below is injective.

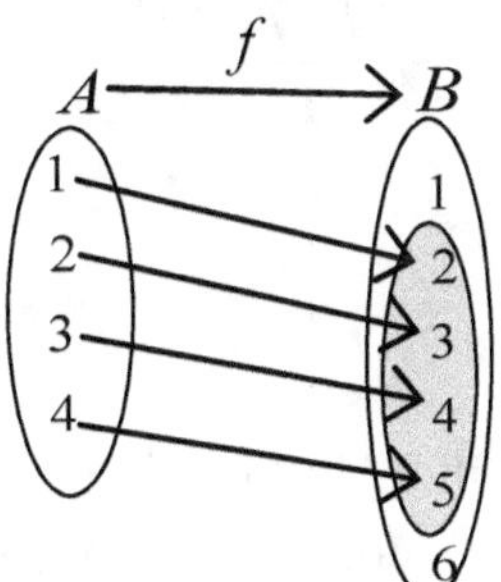

For an injection, the cardinality of the domain A is always less than or equal to that of the codomain B.

$$n(A) \le n(B)$$

A **many-one mapping** is a mapping in which two or more elements in the domain give rise to the same image in the codomain.

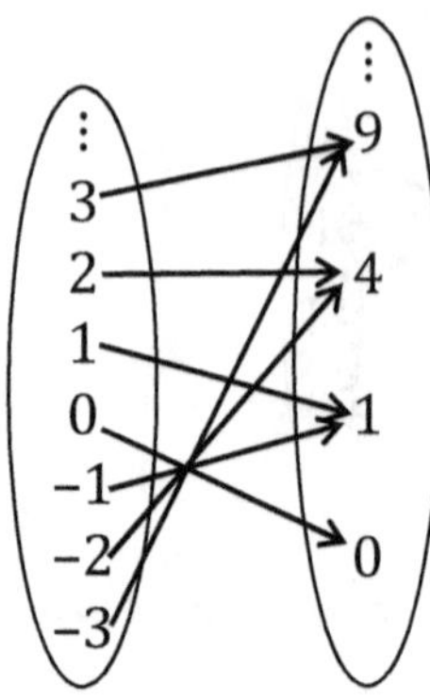

Many-one function

Flow Charts

Example

Draw a flow chart to represent the function $f: x \mapsto \frac{9-7x}{6}$.

Solution

$$x \rightarrow \boxed{\times(-7)} \xrightarrow{-7x} \boxed{+9} \xrightarrow{9-7x} \boxed{\div 6} \rightarrow \frac{9-7x}{6}$$

Inverse Function

The inverse f^{-1} of a bijective function f is a function that performs the reverse process of what the function f does. For instance, if $g: x \mapsto x-2$ then the inverse of g denoted by g^{-1} is $g^{-1}: x \mapsto x+2$.

Finding the Inverse of a function

To find the inverse f^{-1} of a function $f: x \mapsto f(x)$ write $y = f(x)$ and solve for x, then substitute x for y to have $f^{-1}(x)$.

Example

Given that $f: x \longmapsto \frac{2y-7}{5}$, find the inverse of f.

Solution

Let $\frac{2y-7}{5} = x \Rightarrow y = \frac{5x+7}{2} \Rightarrow f^{-1}: x \mapsto \frac{5x+7}{2}$.

Alternatively; draw a flow chart of the function and used it to draw the corresponding inverse flow chart as shown below.

$$x \rightarrow \boxed{\times 2} \xrightarrow{2x} \boxed{-7} \xrightarrow{2x-7} \boxed{\div 5} \rightarrow \frac{2x-7}{5}$$

$$\frac{5x+7}{2} \leftarrow \boxed{\div 2} \xleftarrow{5x+7} \boxed{+7} \xleftarrow{5x} \boxed{\times 5} \leftarrow x$$

$$\Rightarrow f^{-1}: x \mapsto \frac{5x+7}{2}$$

Composite Functions

A composite function such as $f \circ g \circ h$ by convention is operated from right to left. The following figure shows $h = f \circ g(x) = 2x + 5$, the result of composing set $A = \{1,2,3,4,5,6\}$ using the function $g: x \mapsto 2x$ followed by $f: x \mapsto x + 5$.

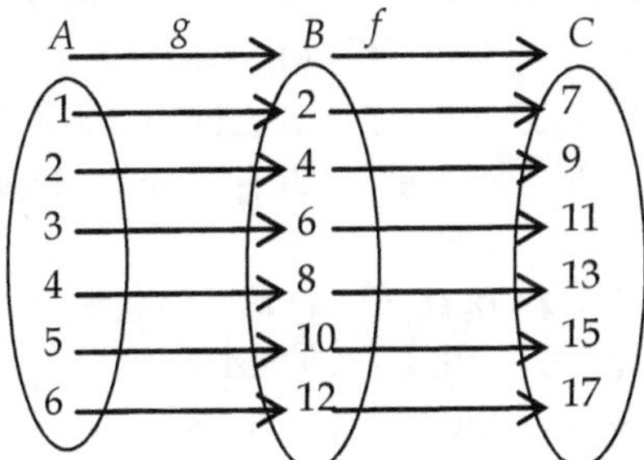

An **identity function** denoted by $I: x \mapsto x$ is a function which maps every element to itself.

Example

The functions f and g are defined on $\mathbb{R}$, the set of real numbers, by

$f: x \mapsto x+4, \quad g: x \mapsto \frac{1}{x+2}, x \neq -2$. Express $f \circ g$ in the form

$f \circ g: x \mapsto \cdots$ Hence evaluate $f \circ g(2)$.

Solution

$$f \circ g(x) = f\left(\frac{1}{x+2}\right) = \frac{1}{x+2} + 4 = \frac{4x+9}{x+2} \Rightarrow f \circ g: x \mapsto \frac{4x+9}{x+2}$$

$$\Rightarrow f \circ g(2) = \frac{4(2)+9}{2+2} = \frac{17}{4}$$

Restricted Domain and Restricted Function

A **restricted function** is a function whose domain and/or codomain are reduced to one of its subset. For instance, we can limit the domain of definition of the function $f: \mathbb{R} \to \mathbb{R}$, $f: x \longmapsto x^2$ which has no inverse to $f: \mathbb{R}^+ \to \mathbb{R}^+$ so as to enable it to have an inverse. The set $\mathbb{R}^+$ is called a **restricted domain**.

MULTIPLE CHOICE EXERCISE 4:1

1. Given that $A = \{1,2,3,4,5\}$ and $B = \{-1,0,1,\ldots,12\}$. A relation $\Re$ is defined from A to B as $a \,\Re\, b$ means $b=3a-4$ e.g. $_1\,\Re\,_{-1}$ since, $-1=3(1)-4$. The set of ordered pairs (a, b) of the relation R is:
   [A] $\{(1,-1),(2,2),(3,5),(4,8),(5,11)\}$ [B] $\{(-1,1),(2,2),(5,3),(8,4),(11,5)\}$
   [C] $\{(2,2),(3,5)\}$ [D] $\{(2,2),(5,3)\}$
2. Given that $A = \{1,2,3,4,5\}$ and $B = \{-1,0,1,\ldots,12\}$. A relation $\Re$ is defined from A to B as $a \,\Re\, b$ means $b=3a-4$ e.g. $_1\,\Re\,_{-1}$ since, $-1=3(1)-4$. It is true that:
   [A] $\Re$ is an onto relation since, $\forall a\in A,\ b\in B,\ (a,b)\in A\times B$.
   [B] $\Re$ is an onto relation since, $\forall a\in A,\ b\in B,\ (a,b)\notin A\times B$.
   [C] $\Re$ is an onto relation since, $\forall a\in A,\ b\in B,\ (a,b)\supset A\times B$.
   [D] $\Re$ is an onto relation since, $\forall a\in A,\ b\in B,\ (a,b)\subset A\times B$.
3. The properties which both satisfy the relation 'is less than' are:
   [A] reflexive and transitive. [B] reflexive and symmetric.
   [C] anti-symmetric and transitive [D] symmetric and transitive.
4. The following figure, shows a mapping involving two sets A and B described by the relation:

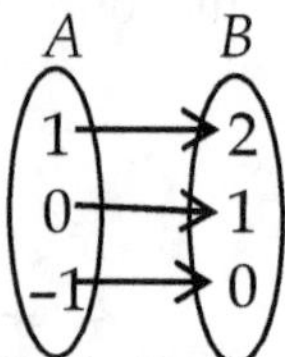

   [A] $x\mapsto x$
   [B] $x\mapsto x-1$
   [C] $x\mapsto x+$
   [D] $x\mapsto 2x+\mathbf{1}$
5. The ages of five children Cha, Abe, Eme, Bep and Dah, are such that Cha is older than Abe but younger than Eme. Eme is younger than Bep.

Dah is older than Bep. The correct arrangement of the Children in the ascending order of their ages is:

[A] Dah, Bep, Eme, Cha, Abe [B] Abe, Cha, Eme, Bep, Dah

[C] Eme, Cha, Abe, Dah, Bep [D] Eme, Bep, Dah, Abe, Cha

6. A relation is defined on the set $X = \{1, 2, 3, 4, 5, 6\}$ as 'is a factor of'. The diagrams among the following, which represents this relation is:

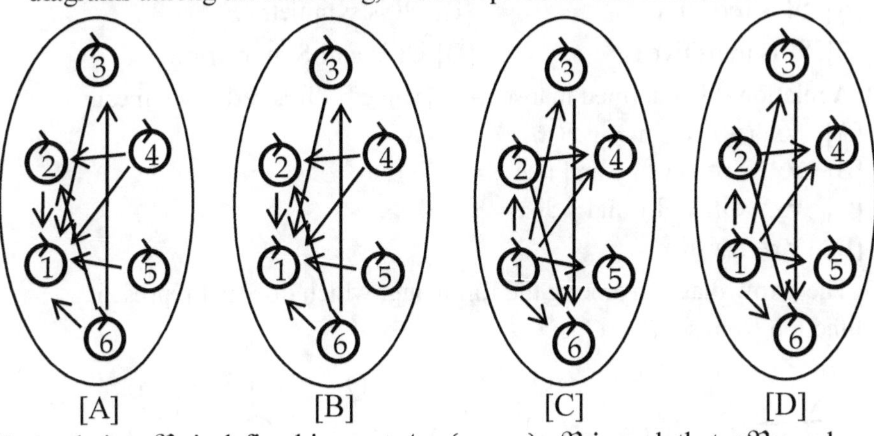

7. A relation $\Re$ is defined in a set $A = \{x, y, z\}$. $\Re$ is such that $x\Re y$ and $y\Re z \Rightarrow x\Re z$. Therefore:

[A] $\Re$ is reflexive. [B] $\Re$ is symmetric.

[C] $\Re$ is transitive. [D] $\Re$ is anti-Symmetric.

8. A relation $\Re$ is defined in a set $A = \{x, y, z\}$. If $\Re$ is said to be anti-symmetric. This means that:

[A] $\forall x, y \in A, x \Re y \Rightarrow y \Re x$. [B] $\forall x, y, z \in A, x \Re y$ and $y \Re z \Rightarrow x \Re z$.

[C] $\forall x, y \in A, x \Re y$ and $y \Re x \Rightarrow x = y$. [D] $\forall x \in A, x \Re x$.

9. A relation $\Re$ is defined in a set $A = \{x, y, z\}$. $\Re$ is such that, $x \Re x, \forall x \in A$. Therefore:

[A] $\Re$ is reflexive. [B] $\Re$ is symmetric.

[C] $\Re$ is transitive. [D] $\Re$ is anti-Symmetric.

10. A relation $\Re$ is defined in a set $A = \{x, y, z\}$. $\Re$ is symmetric means:

[A] $\forall x, y \in A, x \Re y \Rightarrow y \Re x$.

[B] $\forall x, y, z \in A, x \Re y$ and $y \Re z \Rightarrow x \Re z$.

[C] $\forall x, y \in A, x \Re y$ and $y \Re x \Rightarrow x = y$.

[D] $\forall x \in A, x \Re x$.

11. A relation $\Re$ is defined in a set $A = \{x, y, z\}$. $\Re$ is such that, $x \Re y \Rightarrow y \Re x, \forall x, y \in A$. Therefore, $\Re$ is:

[A] $\Re$ is reflexive. [B] $\Re$ is symmetric.

[C] $\Re$ is transitive. [D] $\Re$ is anti-Symmetric.

12. A relation $\Re$ is defined in a set $A = \{x, y, z\}$. $\Re$ is transitive means:

[A] $\forall x, y \in A, x \,\Re\, y \Rightarrow y \,\Re\, x$.
[B] $\forall x, y, z \in A, x \,\Re\, y$ and $y \,\Re\, z \Rightarrow x \,\Re\, z$.
[C] $\forall x, y \in A, x \,\Re\, y$ and $y \,\Re\, x \Rightarrow x = y$.
[D] $\forall x \in A, x \,\Re\, x$.

13. A relation $\Re$ is defined in a set $A = \{x, y, z\}$. $\Re$ is such that, $x \,\Re\, y$ and $y \,\Re\, x \Rightarrow x = y$, $\forall x, y \in A$. Therefore, $\Re$ is:
[A] $\Re$ is reflexive. [B] $\Re$ is symmetric.
[C] $\Re$ is transitive. [D] $\Re$ is anti-Symmetric.

14. A relation $\Re$ is defined in a set $A = \{x, y, z\}$. $\Re$ is reflexive means:
[A] $\forall x, y \in A, x \,\Re\, y \Rightarrow y \,\Re\, x$.
[B] $\forall x, y, z \in A, x \,\Re\, y$ and $y \,\Re\, z \Rightarrow x \,\Re\, z$.
[C] $\forall x, y \in A, x \,\Re\, y$ and $y \,\Re\, x \Rightarrow x = y$.
[D] $\forall x \in A, x \,\Re\, x$.

15. The arrow diagram among the following, which does not represent a function from set P to Q is:

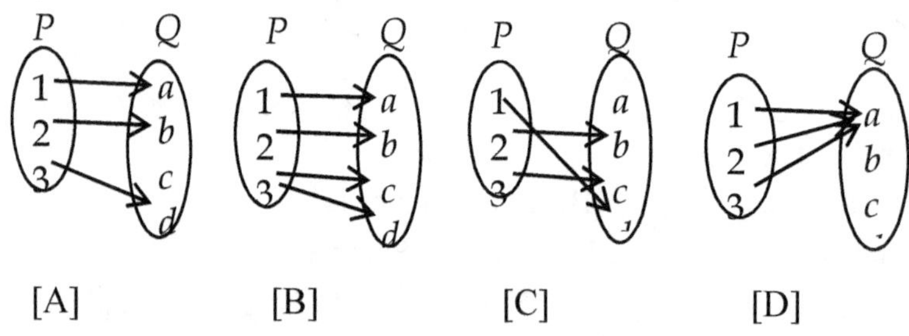

16. A function f: $\mathbb{R} \to \mathbb{R}$ is defined by $f : x \mapsto 2(x^2 + 1)$. The elements of the domain whose image is 10 are:
[A] -1 or 3 [B] 1 or 3 [C] -1 or -3 [D] -2 or 2

17. The function f: $\mathbb{R} \to \mathbb{R}$ is defined by $f : x \mapsto \begin{cases} -x \text{ when } x < -1 \\ 1 \text{ when } -1 \leq x < 1 \\ 2x - 1 \text{ when } x > 1 \end{cases}$

The graph of f among the following is:

[A]
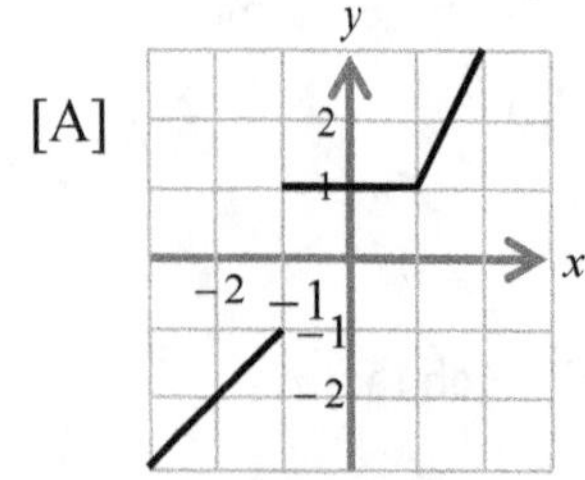

[B]
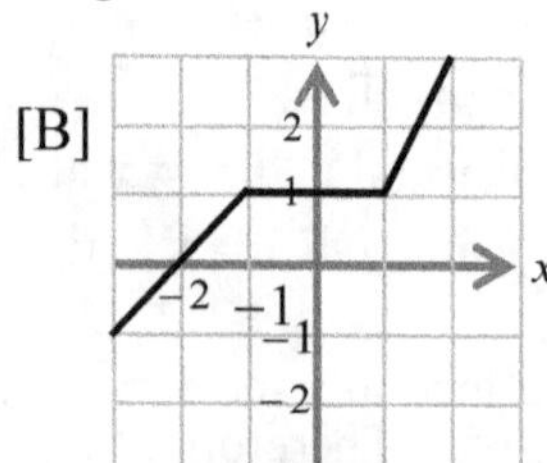

[C] 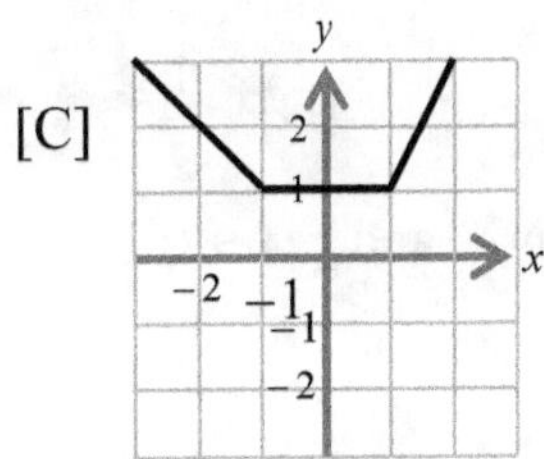

[D] 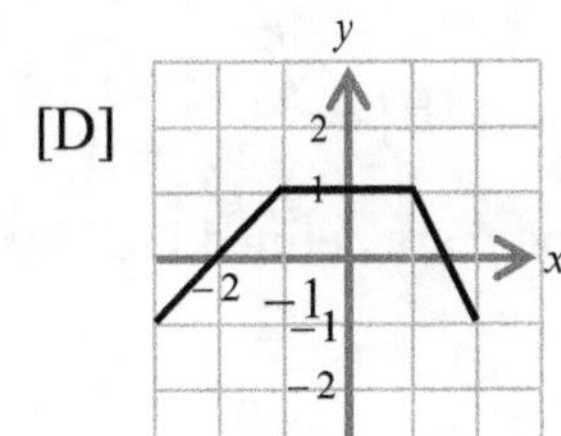

18. A function $f: \mathbb{R} \longrightarrow \mathbb{R}$ is defined by $f: x \mapsto x + 2$. $f\left(\frac{1}{2}\right)$ is equal to:

[A] $\frac{5}{4}$ [B] 1 [C] $\frac{3}{2}$ [D] $\frac{5}{2}$

19. A function $f: \mathbb{R} \longrightarrow \mathbb{R}$ is defined by $f: x \mapsto x + 2$. $f \circ f(x)$ is:

[A] 6 [B] 8 [C] 12 [D] 4

20. A function $f: \mathbb{R} \longrightarrow \mathbb{R}$ is defined by $f: x \mapsto x + 2$. If $f\left(\frac{3}{a}\right) = f \circ f(a)$, the value of a must be:

[A] $-\frac{3}{2}$ [B] $\frac{3}{2}$ [C] 1 [D] 3

21. The function g is defined in $\mathbb{R}$, the set of real numbers, by $g: x \mapsto x + 4$

[A] $g^{-1}: x \mapsto \frac{4}{x}$ [B] $g^{-1}: x \mapsto x - 4$ [C] $g^{-1}: x \mapsto \frac{x}{4}$ [D] $g^{-1}: x \mapsto \frac{1}{x-4}$

22. The function f is defined in $\mathbb{R}$, the set of real numbers, by $f: x \frac{1}{x+2}, x \neq -2$. $f(-4)$ is equal to:

[A] $-\frac{1}{2}$ [B] $\frac{1}{2}$ [C] $-\frac{1}{6}$ [D] $\frac{1}{6}$

23. The functions f and g are defined in $\mathbb{R}$, the set of real numbers, by $f: x \mapsto \frac{1}{x+2}, x \neq 2$ and $g: x \mapsto x + 4$. $g \circ f(2)$ is equal to:

[A] $\frac{3}{2}$ [B] $\frac{13}{2}$ [C] $\frac{17}{4}$ [D] $\frac{5}{2}$

24. Given the function f defined in the set $\mathbb{R}$ of real numbers by $f: x \mapsto x^2 - 3x + 2$, $f(-3)$ is equal to:

[A] 2 [B] 20 [C] 7 [D] 5

25. Given the function f defined in the set $\mathbb{R}$ of real numbers by $f: x \mapsto x^2 - 3x + 2$. The values of x for which $f(x) = 0$ are:

[A] 2 and 1 [B] −2 and −1 [C] 2 and −1 [D] −2 and 1

26. The functions f and g are defined in $\mathbb{R}$ by $f: x \mapsto \sin x^\circ$ and $g: x \mapsto 2x$. $f \circ g(45)$ is equal to:

[A] 0 [B] $\frac{\sqrt{2}}{2}$ [C] 1 [D] $\frac{1}{2}$

27. The functions f and g are defined in $\mathbb{R}$ by $f: x \mapsto \sin x^{\circ}$ and $g: x \mapsto 2x$. $g \circ f(30)$ is equal to:

[A] 0 [B] $\frac{\sqrt{2}}{2}$ [C] 1 [D] $\frac{1}{2}$

28. The function f is defined on $\mathbb{Z}$ the set of integers, by $f: x \mapsto 1\text{-}2x$. $f(-4)$ is equal to:

[A] -7 [B] 7 [C] -9 [D] 9

29. The functions f and g are defined on $\mathbb{Z}$ the set of integers, by $f: x \mapsto 1-2x$ and $g: x \mapsto 5x - k$. Given that $f \circ g(x) = g \circ f(x)$, the value of k is:

[A] -4 [B] 4 [C] $-\frac{4}{3}$ [D] $\frac{4}{3}$

30. The number of simple functions that make up the composite function $f: x \mapsto (2x+5)^2$ is:

[A] 2 [B] 3 [C] 4 [D] 5

31. A function is defined on $\mathbb{Z}$ the set of integers as $f: x \mapsto 3 + x$. The image of -4 is:

[A] -1 [B] 1 [C] 4 [D] -4

32. The relation from set P to set Q in the arrow diagram below is:

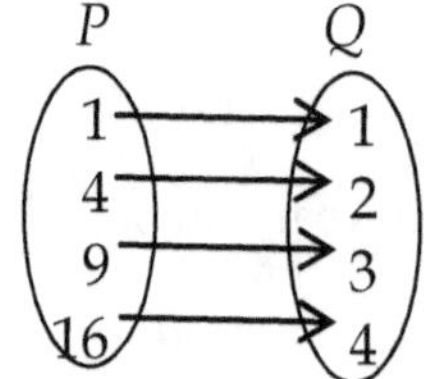

[A] 'is the square root of'
[B] 'is double'
[C] 'is the square of'
[D] 'is 4 times

33. Given that $f: x \mapsto \frac{1}{x-2}$, then $f^{-1}: x \mapsto$:

[A] $\frac{1}{x} - 2$ [B] $\frac{1}{x} + 2$ [C] $\frac{1}{x+2}$ [D] $\frac{1}{x-2}$

33. Given that $g(x) = x^2 - 6, x \in \mathbb{Z}$, then the value of m when $g(2m) = 10$ is:

[A] $\sqrt{2}$ [B] 4 [C] -2 or 2 [D] 0 or 4

34.

x	-2	-1	0	1
y	-1	2	1	0

The correct graph and assertion about the relation in the above table is:

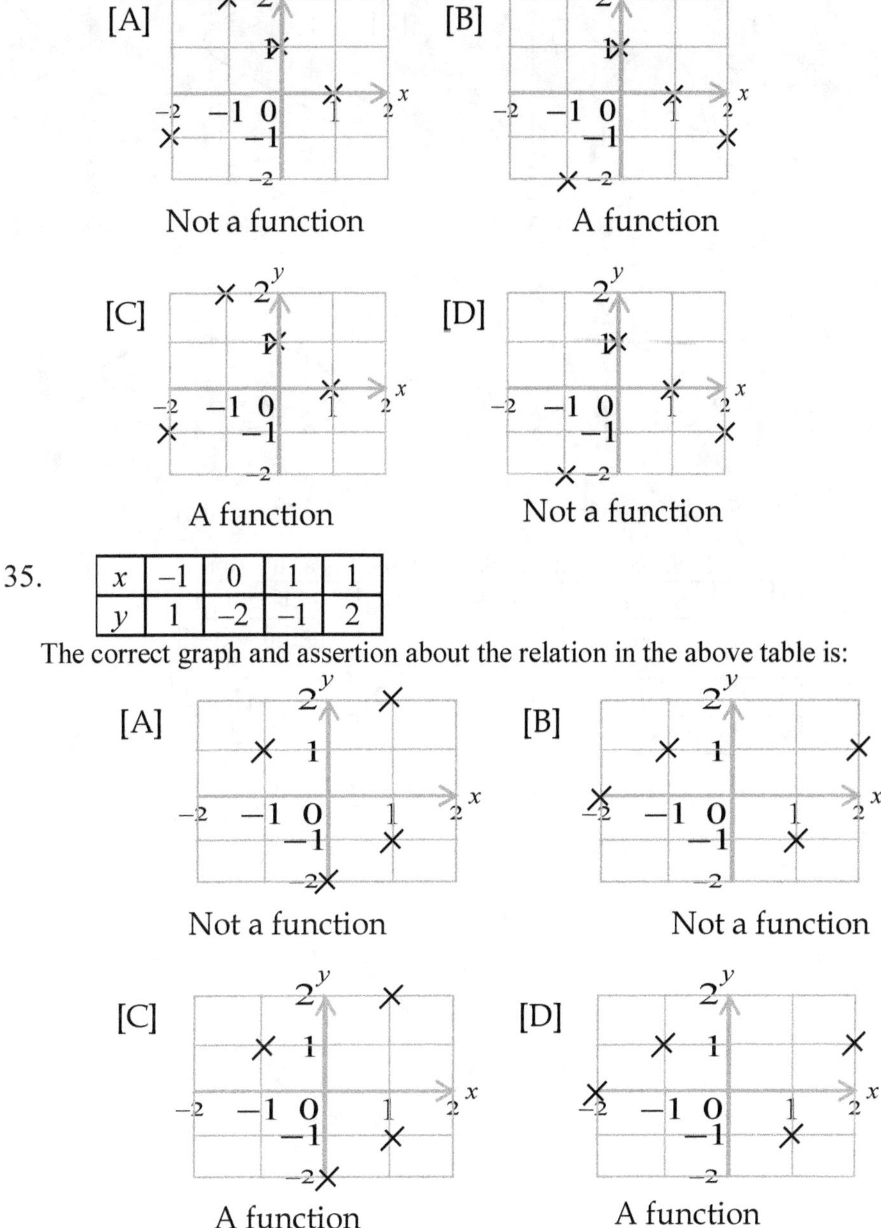

[A] Not a function

[B] A function

[C] A function

[D] Not a function

35.

x	-1	0	1	1
y	1	-2	-1	2

The correct graph and assertion about the relation in the above table is:

[A] Not a function

[B] Not a function

[C] A function

[D] A function

36. The graph which represents a function is:

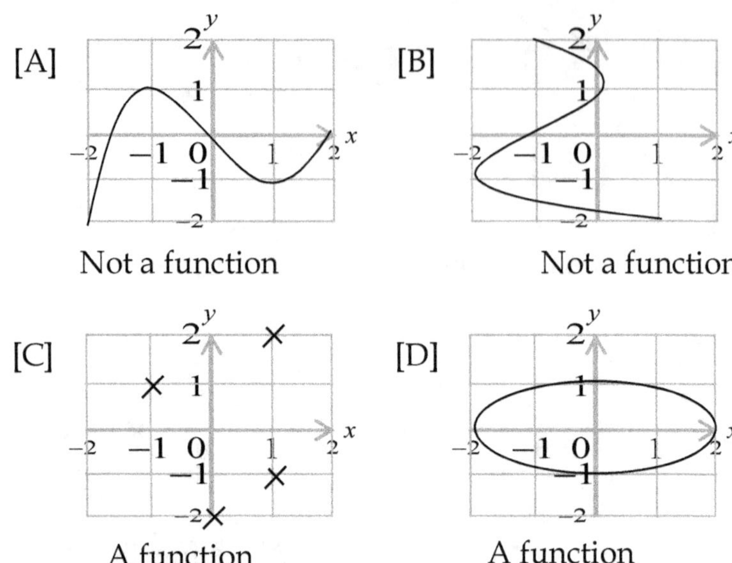

[A] Not a function   [B] Not a function

[C] A function   [D] A function

37. The table which represents a relation which is a function is:

[A]

x	–2	–1	0	1
y	–1	2	1	–1

[B]

x	–2	–1	0	1
y	–2	0	–1	–2

[C]

x	–2	–1	0	1
y	–2	–1	0	–1

[D]

x	–2	–1	0	–2
y	–1	2	1	–2

CHAPTER 5

EUCLIDEAN GEOMETRY

5.1 POINTS, LINES AND ANGLES

Definition of terms

A Point: A point is a sizeless representation of position.

• A Point

A Line: A line is a sizeless path of a series of points.

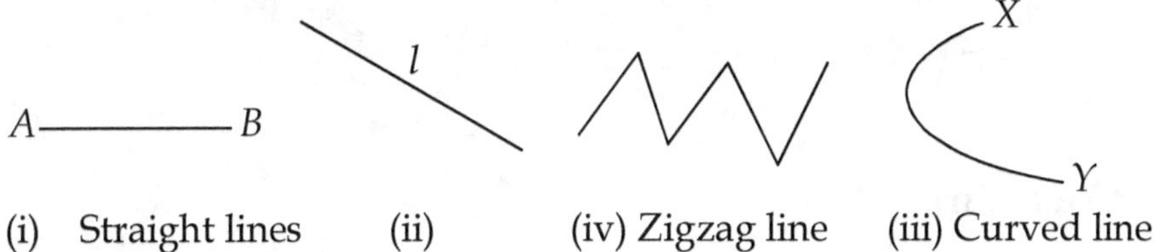

(i) Straight lines (ii) (iv) Zigzag line (iii) Curved line

A line containing the points A and B is denoted by (AB).
A **straight line** is the shortest distance line between two points.

Line Segments and Rays

A **line segment** is part of a line consisting of two points, called endpoints, and all the points that lie between these endpoints.
A **ray** is a line which has a definite beginning but is endless in one direction.

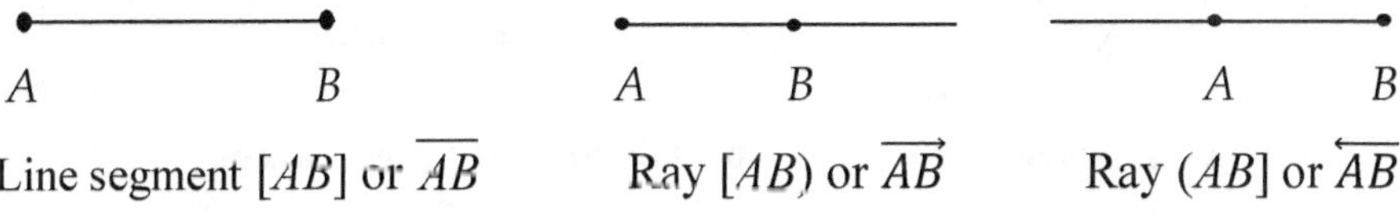

Line segment $[AB]$ or $\overline{AB}$ Ray $[AB)$ or $\overrightarrow{AB}$ Ray $(AB]$ or $\overleftarrow{AB}$

ANGLES

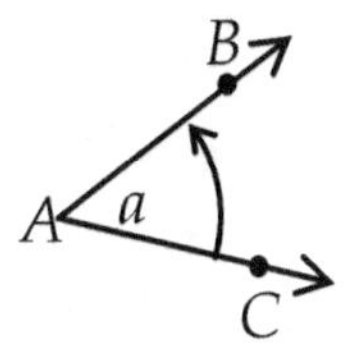

The following are all ways of denoting the angle on the left.
Angle A, angle a, $\hat{a}$, $\angle BAC$, $\angle CAB$, $B\hat{A}C$, $C\hat{A}B$, angle BAC, or angle CAB.

Types of Angles

1. A full turn or a revolution

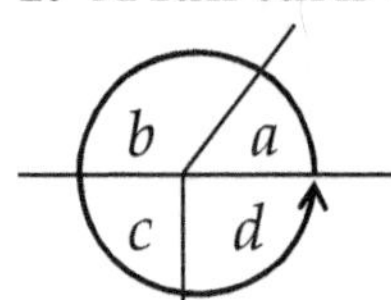

A **full turn** or a **revolution** (or **angles at a point)** is an angle which is equal to 360°.

Angles at a point sum-up to 360°.

$\Rightarrow a + b + c + d = 360°$

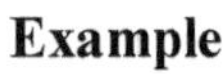
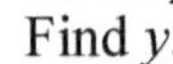

Example

Find y.

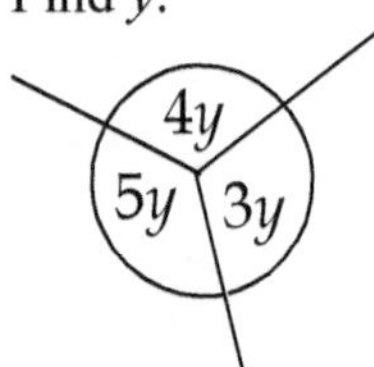

Solution

$$5y + 4y + 3y = 360°$$
$$\Rightarrow 12y = 360°$$
$$y = 30°$$

2. A Half Turn

A half turn is equal to 180° or adjacent angles on a straight line sum-up to 180°.

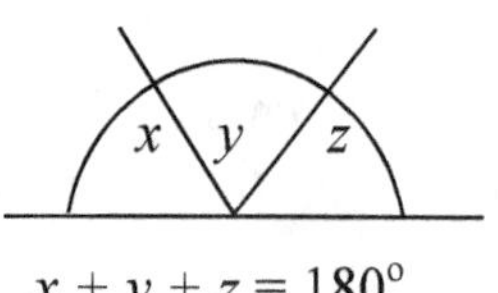

$x + y + z = 180°$

Example

Find the value of x.

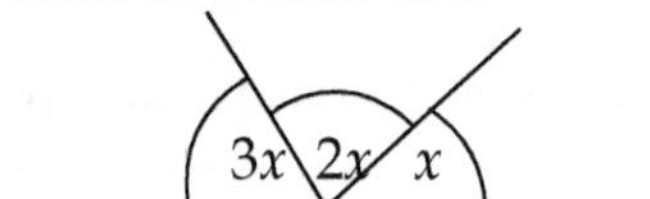

Solution

$$3x + 2x + x = 180$$
$$6x = 180°$$
$$\Rightarrow x = 30°$$

Two angles, whose sum is 180°, are said to be **supplementary**.

3. A Quarter Turn or a Right Angle

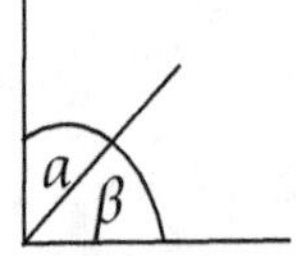

A **quarter turn** or **right-angle** is an angle which is equal to 90°.Two angles whose sum is 90° are said to be **complementary**.

$\Rightarrow a + \beta = 90°$ °

Example

Use the figure below to write down an equation that satisfies p and q. Given also that $p - q = 40°$, find the values of p and q.

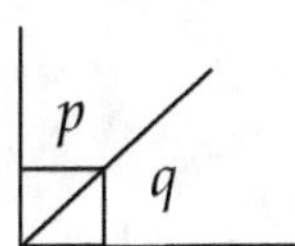

Solution

The required equation is $p + q = 90^{\circ}$.........①

Since $p - q = 40^{\circ}$.........②

① + ②: $2p = 130^{\circ} \Rightarrow p = 65^{\circ}$

Substitute in ①, $65 + q = 90^{\circ} \Rightarrow q = 25^{\circ}$

4. **An acute angle** is an angle, whose value is less than 90°.
5. **An obtuse angle** is an angle, whose value lies between 90° and 180°.
6. **A reflex angle** is an angle, which lies between 180° and 360°

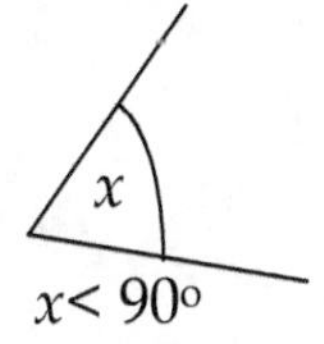

$x < 90^{\circ}$

Acute angle

$90^{\circ} < y < 180^{\circ}$

Obtuse angle

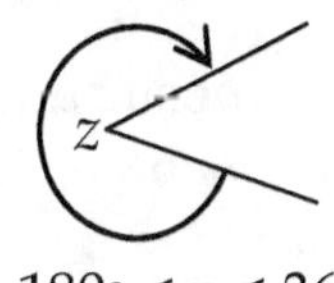

$180^{\circ} < z < 360^{\circ}$

Reflex angle

Adjacent and Non-Adjacent angles

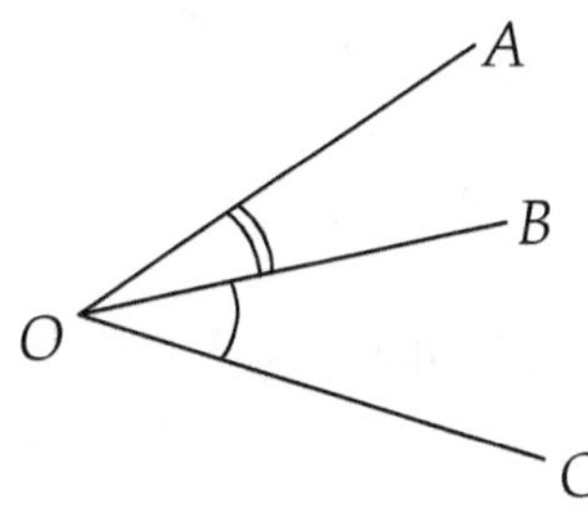

(a) *Adjacent angles*

$\angle AOB$ and $\angle BOC$ are **adjacent angles** because the lines forming them all intersect at exactly one point.
For $\angle AOB$ and $\angle BOC$ to be adjacent, $\angle AOB + \angle BOC = \angle AOC$.

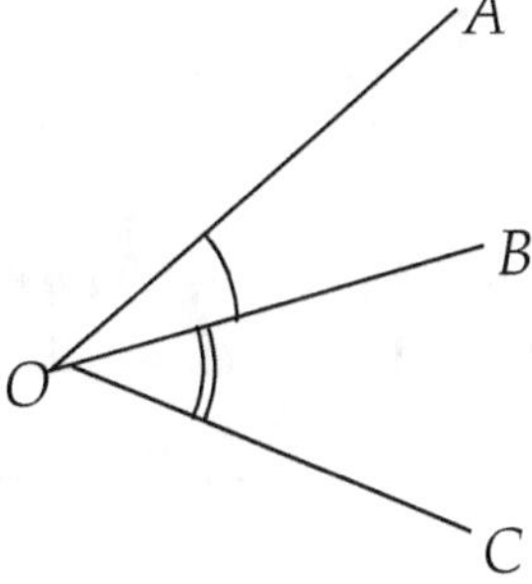

(b) *Non-Adjacent Angles*

$\angle AOB$ and $\angle BOC$ are not **adjacent angles** because the lines forming them do not all intersect at exactly one point.
$\angle AOB + \angle BOC \neq \angle AOC$

Example

Given that in the following figure, *POQ* and *QOR* are adjacent angles and that $\angle POR$ is equal to 59°, find the size of x.

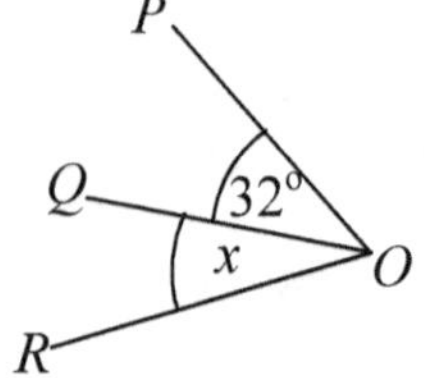

Solution
$\angle x = \angle POR - \angle POQ = 59° - 32 = 27°$

Angles between Intersecting Lines

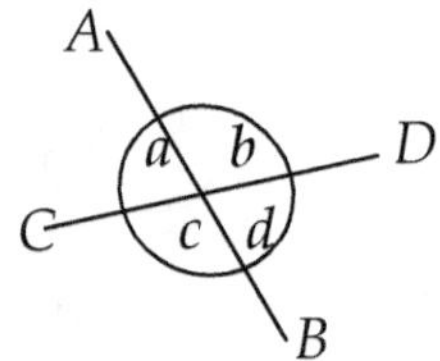

Vertically opposite angles are equal.
i.e. $a = d$ and $b = c$.

Perpendicular Lines

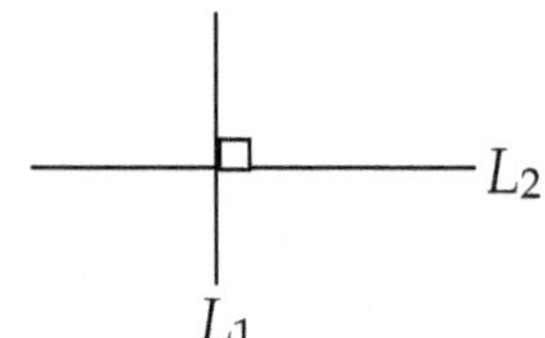

If the angle between two intersecting lines is 90°, the lines are said to be perpendicular.

$$L_1 \perp L_2$$

Parallel Lines and Transversals

A **transversal** is a line, which intersects two or more parallel lines.

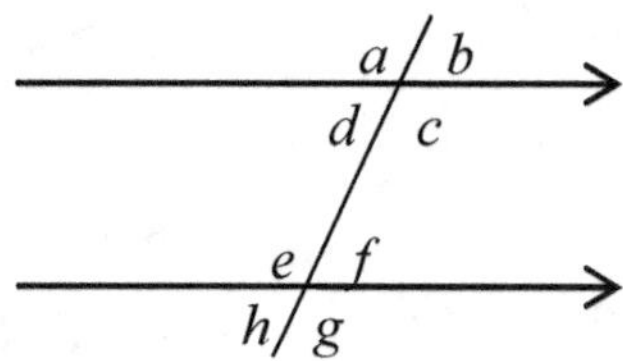

When a transversal intersects parallel lines,

Pairs of alternate interior (Z-) angles are equal.	$d = f, \quad c = e$
Pairs of alternate exterior angles are equal.	$a = g, \ b = h$
Pairs of corresponding angles are equal.	$a = e, \ b = f, \quad d = h, \ c = g$
Pairs of co-interior angles are supplementary.	$d + e = 180°, \ c + f = 180°$

Simply put, when a transversal intersects parallel lines all the acute angles, are equal and all the obtuse angles are equal.

Example

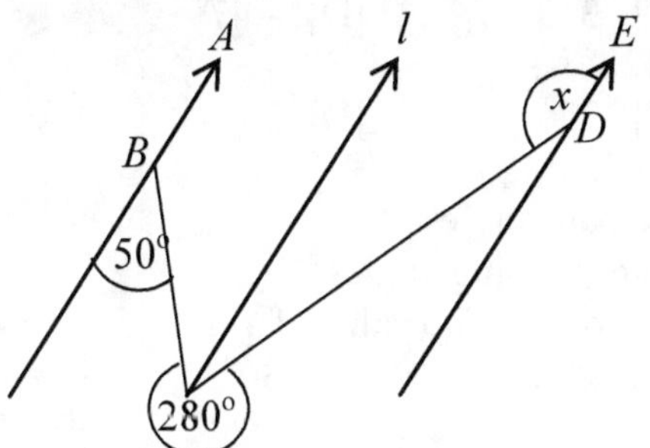

In the figure on the left, AB, l and DE are parallel. Find x.

Solution

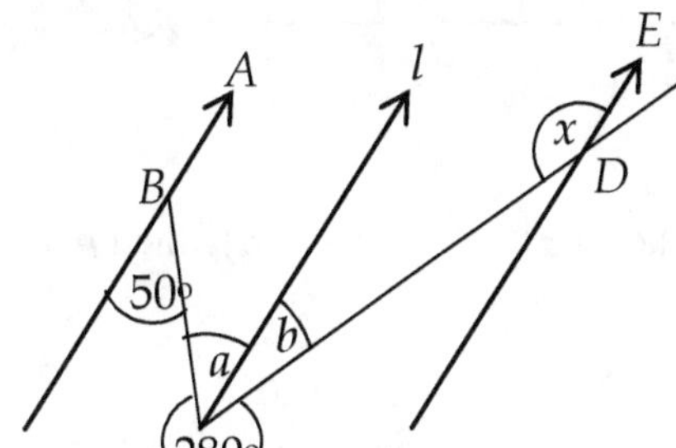

$a = 50°$ [alternate angles]
$a + b + 280° = 360°$ [angles at a point]
$\Rightarrow 50° + b + 280° = 360°$
$\Rightarrow b = 30°$
$b + x = 180°$ [co-interior angles]
$\Rightarrow 30° + x = 180°$
$\Rightarrow x = 150°$

MULTIPLE CHOICE EXERCISE 5:1

1. 141° is an example of:
   [A] an acute angle [B] an obtuse angle
   [C] a reflex angle [D] an alternate angle
2. An angle which is between 180° and 360° is called:
   [A] a complementary angle [B] an acute angle
   [C] an obtuse angle [D] a reflex angle
3. The angle which is not a reflex angle is:
   [A] 317° [B] 258° [C] 193° [D] 116°
4. The angle θ shown in the figure below is:
   [A] acute [B] right [C] obtuse D] reflex

   θ

5. A reflex angle is:

[A] $< 90^{\circ}$ [B] $> 90^{\circ}$ but $< 180^{\circ}$ [C] $> 180^{\circ}$ but $< 360^{\circ}$ [D] equal to 90°

6. The complement of 47° is
   [A] 133° [B] 43° [C] 313° [D] -47°
7. The angle which is supplementary to 128° is:
   [A] 52° [B] 42° [C] 32° [D] 22°
8. Two angles whose sum is 180° are called:
   [A] complementary angles [B] alternate angles
   [C] supplementary angles [D] corresponding angles
9. In figure (i) below, AO is perpendicular to OB. The value of x is:
   [A] 75° [B] 15° [C] 22.5° [D] 30°

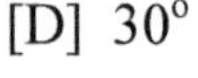

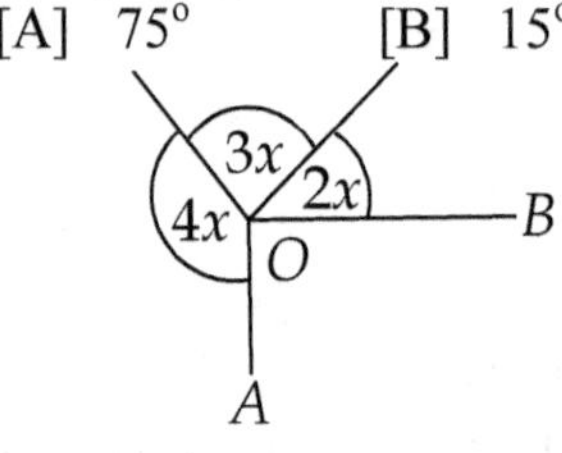

(i)

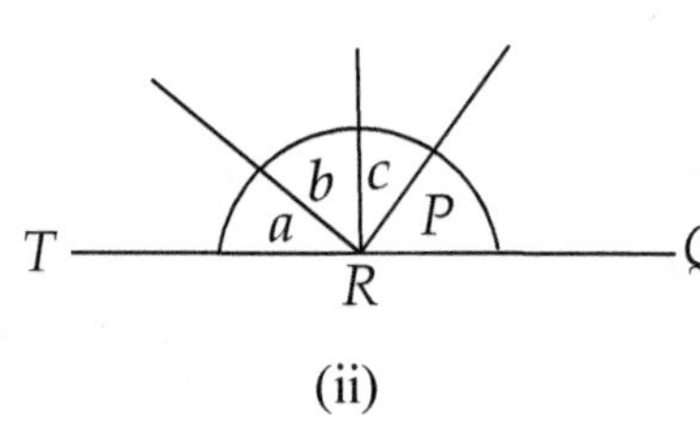

(ii)

10. In figure (ii) above, TRQ is a straight line. If $P = \frac{1}{3}(a+b+c)$, then p equals:
    [A] 45° [B] 60° [C] 90° [D] 120°
11. In figure (i) below, AB, CD and XY are straight lines intersecting at W. The value of $\angle CWX$ is:
    [A] 80° [B] 100° [C] 120° [D] 140°

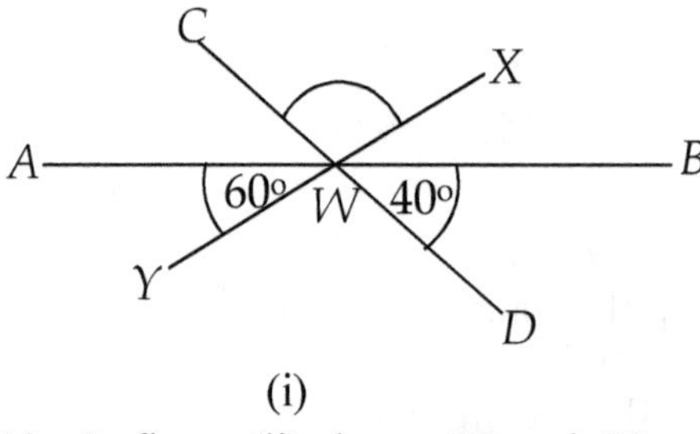

(i)

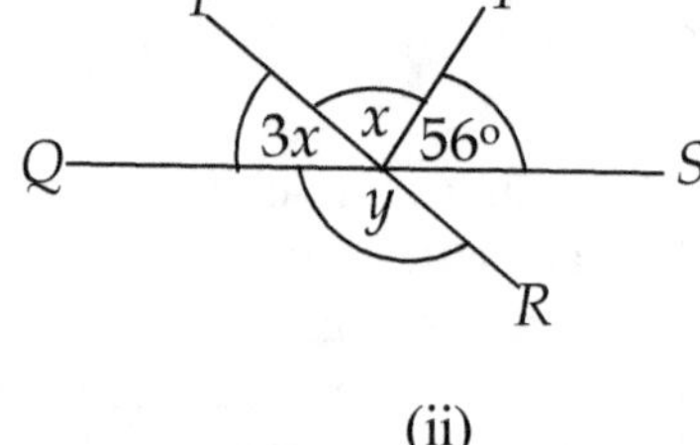

(ii)

12. In figure (ii) above, PR and QS are straight lines. The value of the angle y is:
    [A] 30° [B] 56° [C] 87° [D] 93°
13. In figure (i) below, XY and PQ intersect at O. Angle $TOP = 64^{\circ}$, implies that m is equal to:
    [A] 93° [B] 96° [C] 116° [D] 151°

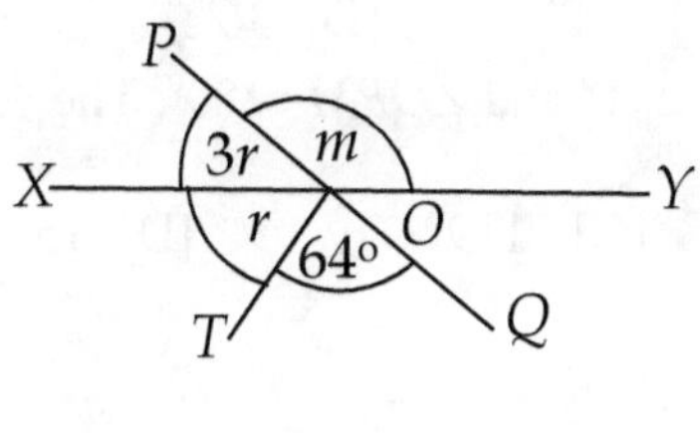

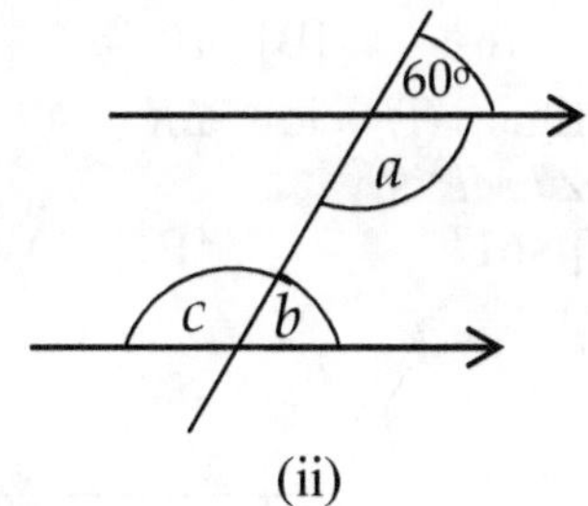

(i) (ii)

14. The angle a, shown figure (ii) above is equal to:
[A] 60° [B] 120° [C] 30° [D] 90°
15. The angle b shown figure (ii) above is equal to:
[A] 60° [B] 120° [C] 30° [D] 90°
16. The angle c shown figure (ii) above is equal to:
[A] 60° [B] 120° [C] 30° [D] 90°
17. In figure (i) below, it is true that:
[A] $a = d$ [B] $a = b$ [C] $e = b$ [D] $a = c$

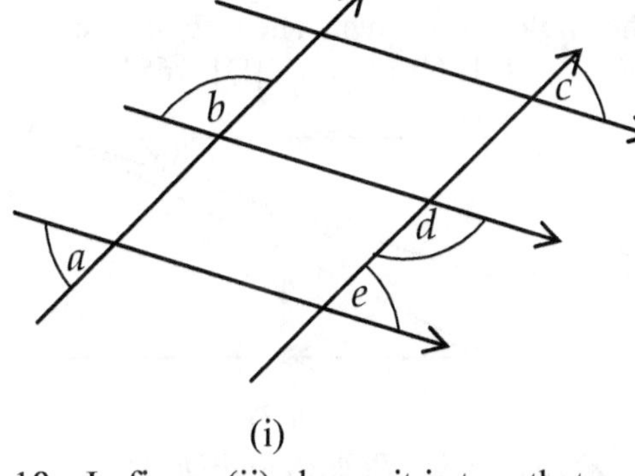

(i) (ii)

18. In figure (ii) above, it is true that:
[A] $q = p+r$ [B] $p + q + r = 180°$ [C] $q = r - p$ [D] $q = 360° - p - r$.
19. In figure (i) below, the relation between x and y is:
[A] $x = y$ [B] $x = y + 180°$ [C] $x = y - 180°$ [D] $x + y = 180°$.

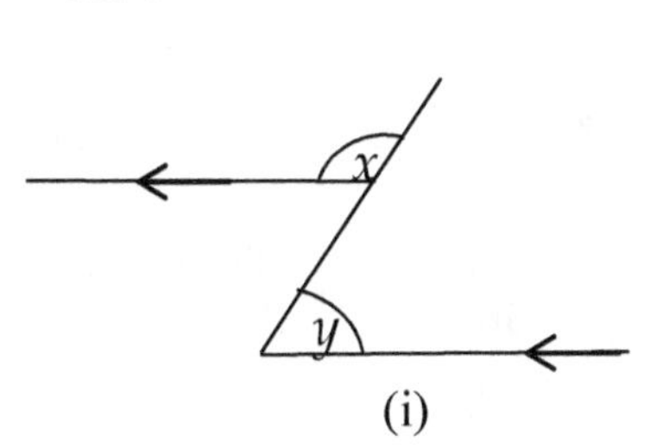

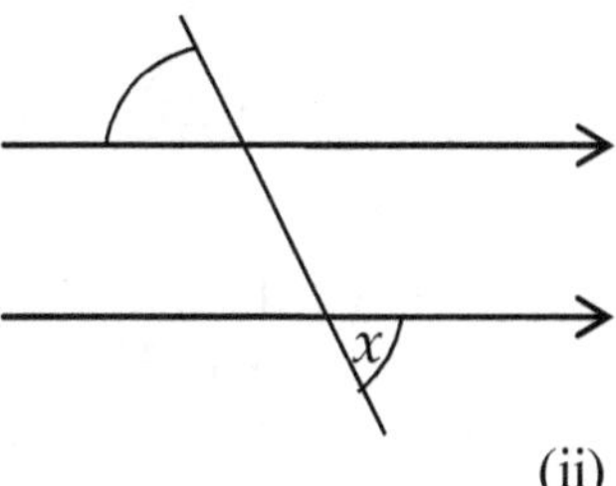

(i) (ii)

20. The value of x in figure (ii) above is:

[A] 76° [B] 104° [C] 14° [D] 36°

21. In figure (i) below, $LK \parallel PQ$. $\angle KLM = 241^\circ$ and $\angle QPM = 89^\circ$. The value of $\angle LMP$ is:
[A] 61° [B] 30° [C] 119° [D] 150°

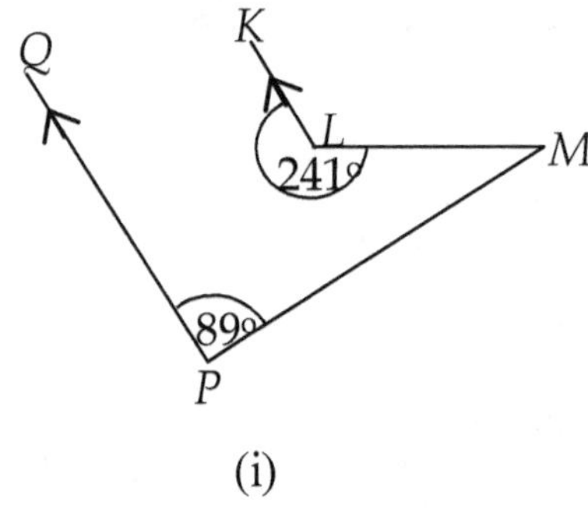

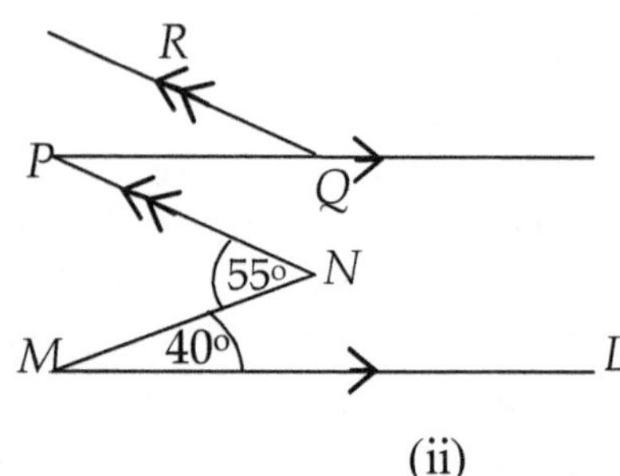

(i) (ii)

22. In figure (ii) above, $ML \parallel PQ$ and $NP \parallel QR$. Given that $\angle LMN = 40^\circ$ and $\angle MNP = 55^\circ$ then $\angle RQP$ equals:
[A] 15° [B] 25° [C] 35° [D] 40°

23. In figure (i) below, $PQ \parallel RS$ and the angles are shown. The size of x is:
[A] 145° [B] 150° [C] 155° [D] 165°

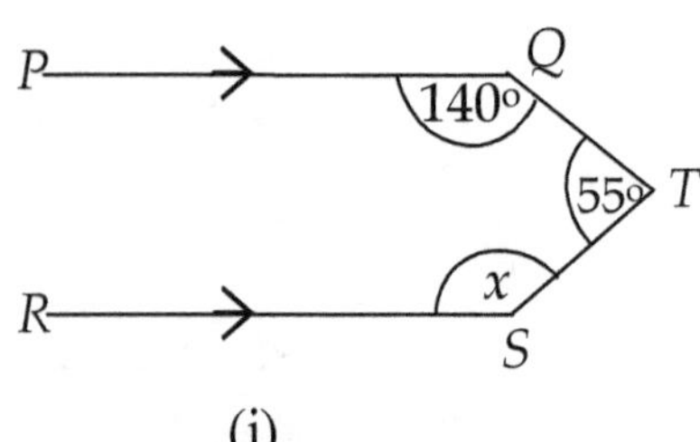

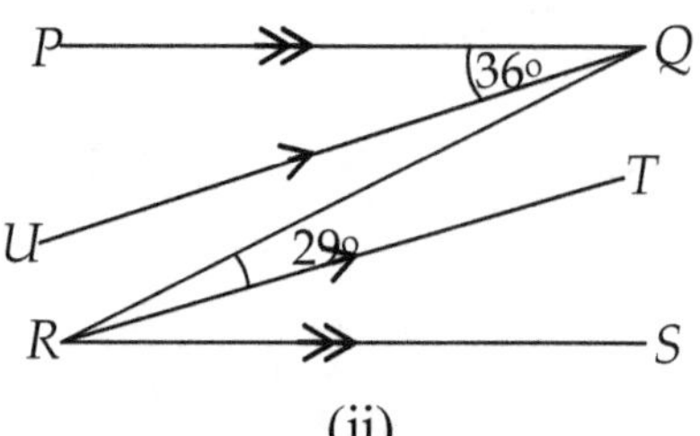

(i) (ii)

24. In figure (ii) above, $\angle PQU = 36^\circ$, $\angle QRT = 29^\circ$, $PQ \parallel RS$ and $UQ \parallel RT$. $\angle PQR$ should be:
[A] 94° [B] 65° [C] 61° [D] 54°

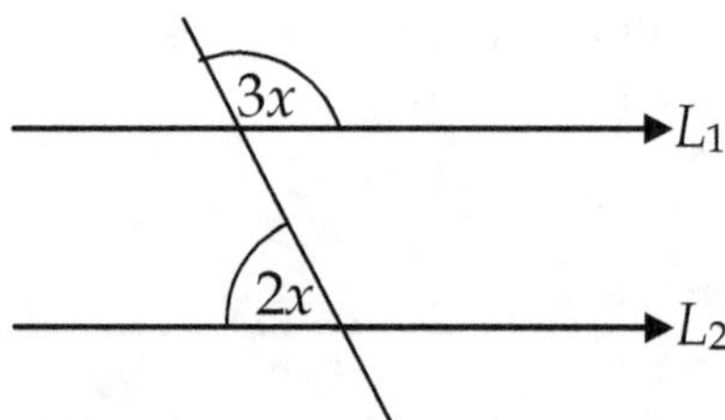

25. In the figure above, the lines L_L and L_2 are parallel. The value of x is:
[A] 45° [B] 30° [C] 36° [D] 60°

26. The diagram which shows an impossible situation is:

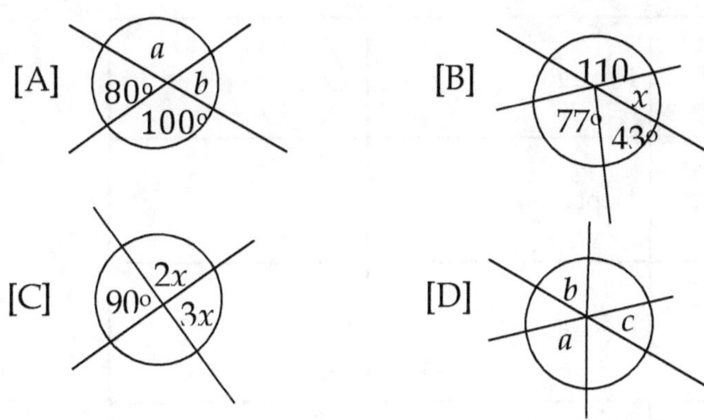

27. The fraction of the circle which has been shaded in figure (i) below is:
[A] $\frac{7}{9}$ [B] $\frac{2}{9}$ [C] $\frac{3}{8}$ [D] $\frac{5}{8}$

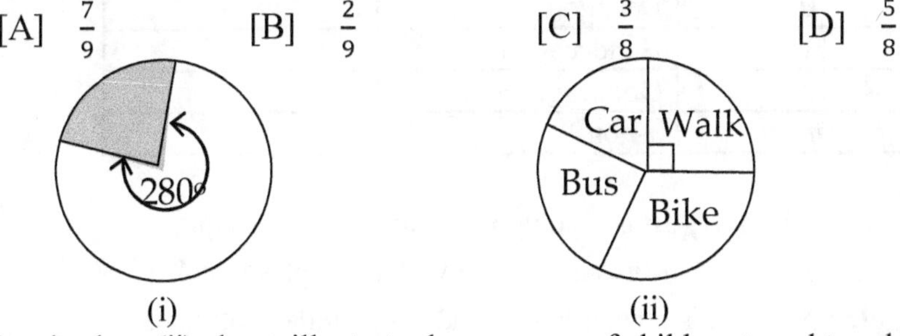

(i) (ii)

28. The pie chart (ii) above illustrates how groups of children travel to school. The percentage of children who walk to school is:
[A] 90% [B] 20% [C] 25% [D] 10%

5.2 SIMPLE PLANE FIGURES

What is a Polygon?

A **polygon** is a plane figure with three or more sides.

Polygons are named according to the number of sides they contain.

Number of sides	Name of polygon	Diagram
3	Triangle	
4	Quadrilateral	
5	Pentagon	

6	Hexagon	
7	Heptagon	
8	Octagon	
9	Nonagon	
10	Decagon	
11	Hendecagon	
12	Dodecagon	
n	n-gon	

An **equilateral polygon** is a polygon with all the sides equal.

An **equiangular polygon** is a polygon with all angles equal.

A **regular polygon** is an equilateral and equiangular polygon.

An **irregular polygon** is one with at least one of its sides or angles different.

A **convex polygon** is a one with none of its interior angles greater than 180°.

A **concave** or **re-entrant polygon** is a polygon with at least one of its interior angles greater than 180°.

Quadrilaterals and Their Properties

A quadrilateral is a four sided plane figure.

Name and Shape	Properties
Trapezium	Two parallel sides.
Kite	Two pairs of adjacent sides are equal. Diagonals intersect each other at right angles.

Name and Shape	Properties
Parallelogram	Opposite sides are equal and parallel. Opposite angles are equal. Adjacent angles are supplementary (sum up to 180°) Diagonals bisect each other. Diagonals bisect the parallelogram.
Rhombus	All the four sides are equal. Opposite sides are parallel. Opposite angles are equal. Diagonals bisect each other at right angles. Diagonals bisect the rhombus.
Rectangle	Opposite sides are equal and parallel. Each angle is equal to 90°. Diagonals bisect each other. Diagonals are equal in length. Diagonals bisect the rectangle.
Square	All sides are equal. Opposite sides are parallel. All angles are equal, each equal to 90°. Diagonals bisect each other at right angles. Diagonals bisect the square. Diagonals are equal in length.

Relationship between Quadrilaterals

The relationship between quadrilaterals can be illustrated using the quadrilateral family tree shown on the right or a Venn diagram shown below.

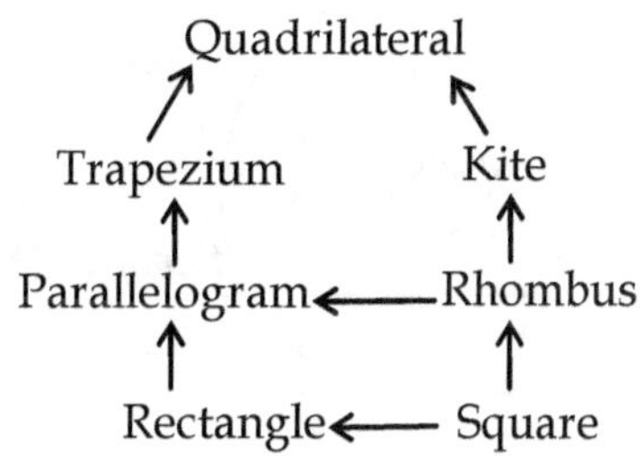

Let $\mathscr{E}$ = All quadrilaterals
P = All parallelograms
K = All kites
R_h = All rhombuses
R_e = All rectangles
S = All squares
T = All trapeziums

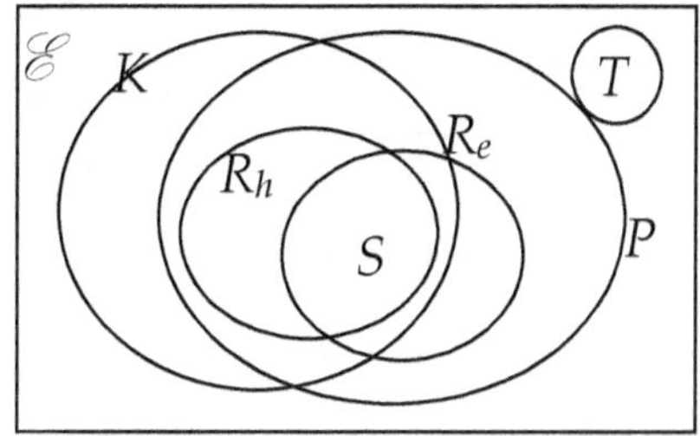

Types of Triangles

Triangles are classified in two ways;

(a) ***By the Measures of their Angles***

(i) 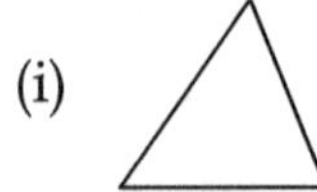 (ii) 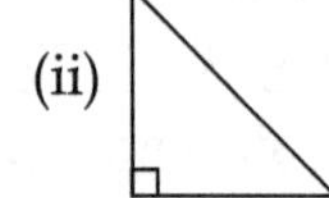 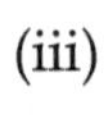 (iii) 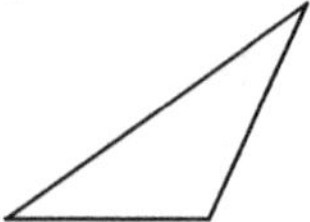

(1) **An Acute angle triangle** (i) has each angle less than 90°.
(2) A **right-angled triangle** (ii) has one of its angles equal to 90°.
(3) **Obtuse angle triangle** (iii) has one of its angles between 90° and 180°.

(b) ***By the Measures of their Sides***

(iv) 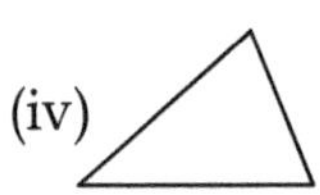 (v) 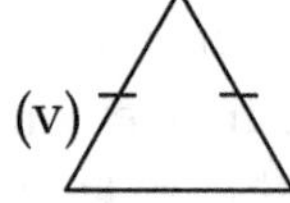 (vi) 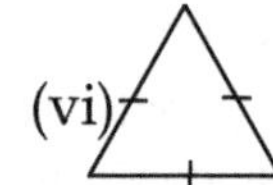

(1) **A Scalene triangle** (iv) has no sides equal.
(2) **An Isosceles triangle** (v) has two sides equal.
(3) **An Equilateral triangle** (vi) has all the sides equal.

Vocabulary Associated with Circles

A circle is a plane figure bounded by points, which are equidistant from a fixed point called the **centre** of the circle.

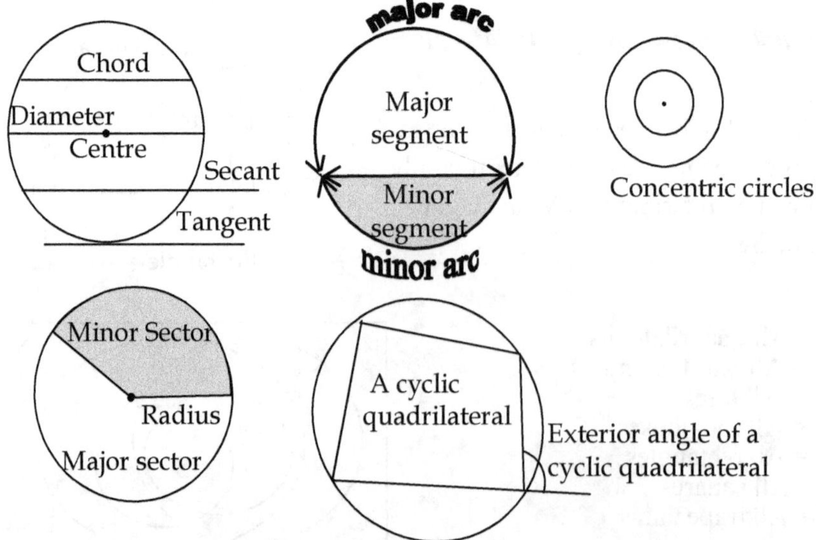

The **circumference** of a circle denoted by C is the distance round the circle.
The **radius** denoted by r is the distance from the centre of a circle to a point on the circle.
A chord is a line segment joining any two points on the circle.
The **diameter** the longest chord of a circle always passes through the centre and its length is twice that of the radius. It is usually denoted by d.
A **tangent** is a straight line that intersects or touches the circle at exactly one point and is always perpendicular to the radius at the point of contact.
A **secant** is a straight line that intersecting a circle at two points.
An **arc** is part of the circumference. The longer of the arcs is called the **major arc** while the shorter is called the **minor arc**.
A **sector** is a portion of a circle bounded by two radii and an arc of the circle. The larger of the sectors is called the **major sector** while the smaller is called the **minor sector**.
A **segment** is a portion of a circle bounded by a chord of the circle and an arc of the circle. The smaller segment is called the **minor segment** while the larger segment is called the **major segment**.
Concentric circles are circles with the same centre.
A **cyclic quadrilateral** is a quadrilateral which is inscribed in a circle.

MULTIPLE CHOICE EXERCISE 5:2

1. The statement which is always true of a rhombus is:
   [A] All the angles are complementary
   [B] All the sides are equal
   [C] The adjacent angles are equal
   [D] All the angles are equal
2. The plane shape which is not a quadrilateral is:
   [A] A kite [B] a rhombus [C] A pentagon [D] a parallelogram
3. The plane shape which is not an example of a quadrilateral is:
   [A] Square [B] Trapezium [C] Rhombus [D] Triangle
4. The statement which is not true of a parallelogram is:
   [A] It has more than 4 sides [B] Opposite sides are parallel.
   [C] It has exactly 4 sides. [D] The sum of its angles is 360°.
5. A seven sided plane figure is called:
   [A] An octagon [B] a pentagon [C] A hexagon [D] a heptagon
6. A polygon with all its interior angles less than 180° is definitely:
   [A] a convex polygon [B] a regular polygon
   [C] a re-entrant polygon [D] a quadrilateral
7. A triangle with vertices $(-4, 4)$, $(4,4)$, $(0,-1)$ is:

[A] Right-angled [B] equilateral [C] Isosceles [D] scalene

8. The largest angle of any triangle:
   [A] Must always be an acute angle.
   [B] Can sometimes be an acute angle.
   [C] Can never be a right-angle.
   [D] Must always be an obtuse angle.
9. A quadrilateral with one pair of sides equal is:
   [A] A rhombus [B] a parallelogram [C] A rectangle [D] a trapezium
10. The assertion about a rhombus which may not be true is:
    [A] The opposite angles are equal.
    [B] The diagonals bisect the angles through which they pass.
    [C] The diagonals are equal.
    [D] Opposite sides are equal.
11. A quadrilateral whose diagonals bisect at right angles is:
    [A] A rectangle [B] a parallelogram [C] A trapezium [D] a rhombus
12. The property/properties which do not characterise a rectangle is/are:
    [A] The diagonals bisect at right angles
    [B] Opposite sides are equal and parallel
    [C] Each of its angles is a right angle
    [D] Opposite sides are equal and parallel and each angles is 90°.
13. In the following figure, the value of the angle marked y is:
    [A] 28° [B] 62° [C] 118° [D] 152°

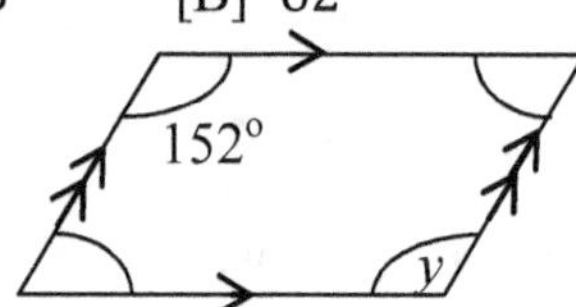

14. The value of the angle marked x in figure (i) below is:
    [A] 140° [B] 130° [C] 110° [D] 70°

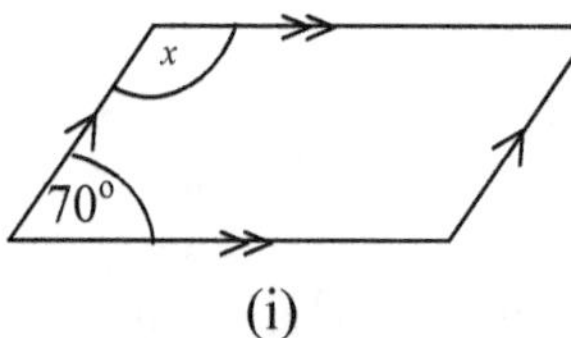

(i)

x
85°
y

(ii)

15. The value of the angle marked y in figure (ii) above is:
    [A] 270° [B] 210° [C] 190° [D] 95°
16. In figure (ii) above, the sum of x and y is:
    [A] 270° [B] 210° [C] 190° [D] 95°
17. The values of x, y, and z in figure (i) below are respectively:
    [A] 130°, 50°, 130° [B] 140°, 40°, 140°
    [C] 150°, 30°, 150° [D] 120°, 60°, 120°

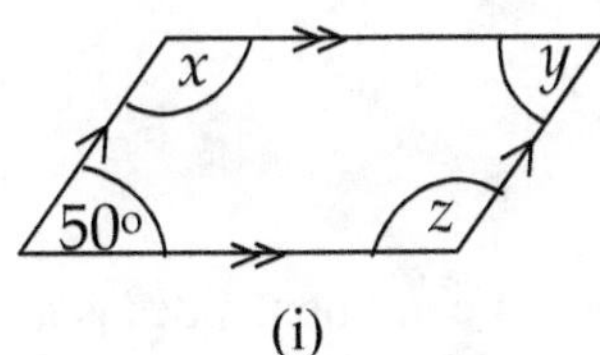

(i)

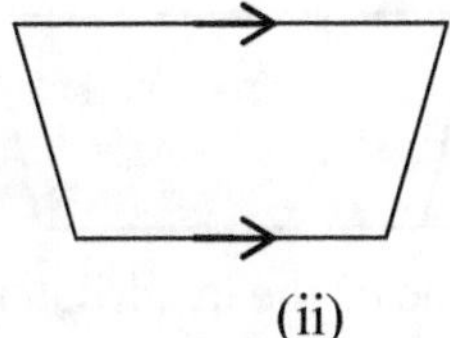

(ii)

18. The name given to the quadrilateral in figure (ii) above is:
    [A] A parallelogram [B] a trapezium [C] A rhombus [D] a rectangle
19. Let $\mathscr{E}$ = {x:x is a polygon}
    A = {x:x is a regular polygon}
    B = {x:x is a quadrilateral}
    Then an element of $A \cap B$ is a:
    [A] square [B] rhombus [C] trapezium [D] rectangle
20. Let $\mathscr{E}$ = {x:x is a polygon}
    A = {x:x is a regular polygon}
    B = {x:x is a quadrilateral}
    Then an element of $A' \cap B$ can never be a:
    [A] square [B] trapezium [C] rhombus [D] rectangle
21. The sum of the angles of a square is:
    [A] 90° [B] 120° [C] 180° [D] 360°
22. The shaded portion in figure (i) belowis called a:
    [A] minor segment [B] major segment
    [C] minor sector [D] major sector

(i)

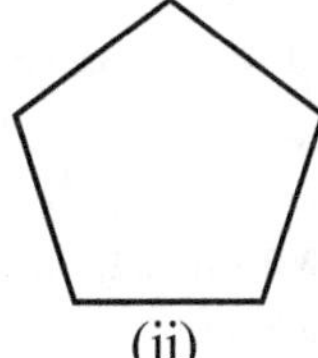

(ii)

23. By the number of sides the polygon in figure (ii) above is the:
    [A] octagon [B] pentagon [C] quadrilateral [D] hexagon
24. The polygon which has the shape of the Cameroon flag is the:
    [A] octagon [B] pentagon [C] hexagon [D] quadrilateral
25. A shape that has two more sides than a football field is the:
    [A] pentagonal [B] octagonal [C] hexagonal [D] decagonal
26. A polygon has 4 more sides than a rectangle. The polygon is:
    [A] pentagon [B] decagon [C] hexagon [D] octagon
27. The real name of the plane figure below and some of its possible names are:
    [A] parallelogram; quadrilateral, rhombus.
    [B] Rhombus; quadrilateral, trapezium.
    [C] Trapezium; quadrilateral, polygon.
    [D] Quadrilateral, parallelogram, polygon.

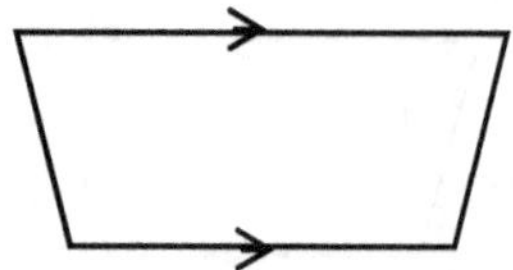

28. The best name of the quadrilateral which has 4 congruent sides is a:
[A] parallelogram [B] rhombus [C] trapezoid [D] rectangle
29. A triangle has sides 3, 5, 8 and angles 25°, 85°, 70°. By the measure of its angles and its sides the triangle is:
[A] Isosceles, obtuse [C] isosceles, acute
[B] Scalene, obtuse [D] scalene, acute
30. The diagram among the following which is a polygon is:

[A]

[B]

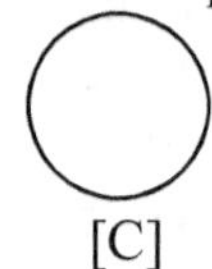
[C]

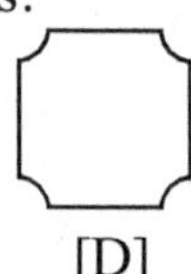
[D]

31. The property which makes a rhombus different from every other parallelogram is:
[A] All sides are equal.
[B] Opposite sides are parallel.
[C] Opposite angles are equal.
[D] Diagonals bisect each other at right angles.
32. The property which makes a square a unique rectangle is:
[A] All sides are equal.
[B] Opposite sides are parallel.
[C] Opposite angles are equal.
[D] Diagonals bisect each other at right angles.
33. The property which makes a square a unique rhombus is:
[A] All sides are equal.
[B] Opposite sides are parallel.
[C] Opposite angles are equal.
[D] Diagonals bisect each other at right angles.
34. Among the properties, the property which makes a rectangle a special parallelogram is:
[A] Opposite sides are equal. [B] Opposite sides are parallel.
[C] Diagonals bisect each other. [D] Diagonals are equal in length.

5.3 SIMPLE SOLID FIGURES

Prisms

A prism is a solid with uniform polygonal cross-sectional area.
Prisms are named from the nature of their cross-sections.

Nature of cross-section	Name of Prism	Diagram of Prism
Triangle	Triangular Prism	
Square	Square Prism or cube	
Rectangle	Rectangular Prism or cuboid	
Pentagon	Pentagonal Prism	
Hexagon	Hexagonal Prism	

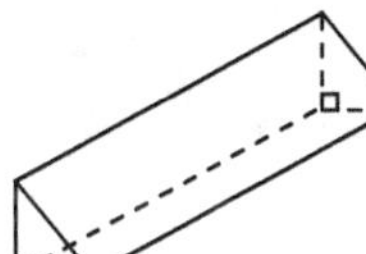

A **right-angled triangular prism** is a prism whose cross-section is a right-angled triangle.

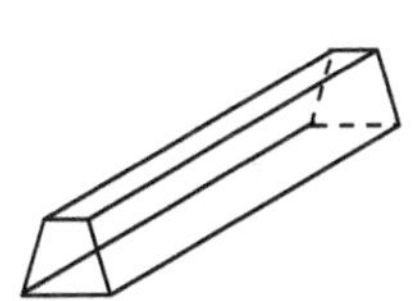

A **trapezoidal prism** is a prism whose cross-section is a trapezium.

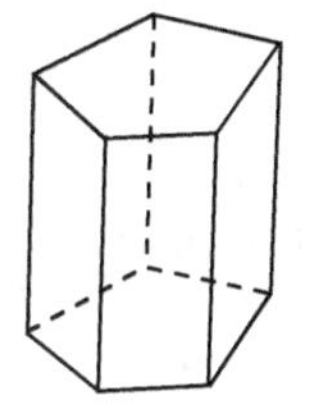

Right prism

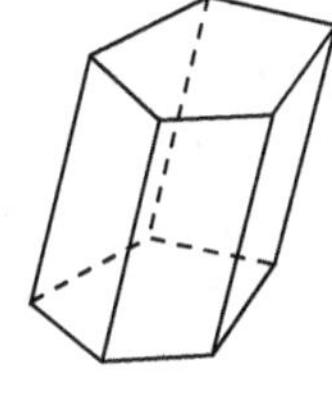

Oblique prism

A prism is called a **right prism** if the lateral surfaces are made up of rectangles; otherwise, it is called an oblique prism.

Cylinders

A cylinder is a solid with a uniform circular cross-section. When it is closed at both ends, it is described as a **solid** cylinder. The curved surface of a cylinder is called the **lateral surface** of the cylinder.

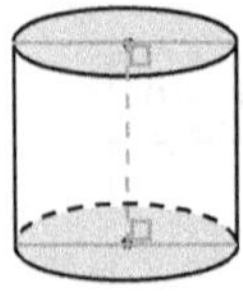

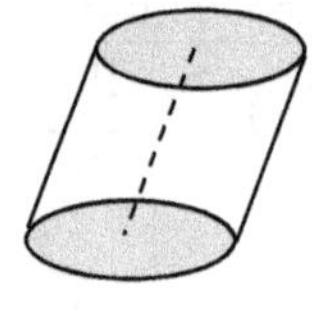

When the two circular faces of a cylinder are parallel and any line joining corresponding points are perpendicular, it is called as a **right circular** cylinder otherwise; it is called an oblique cylinder.

Right circular **Oblique cylinder**

Pyramids

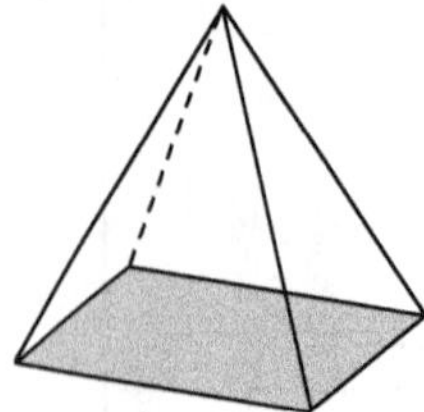

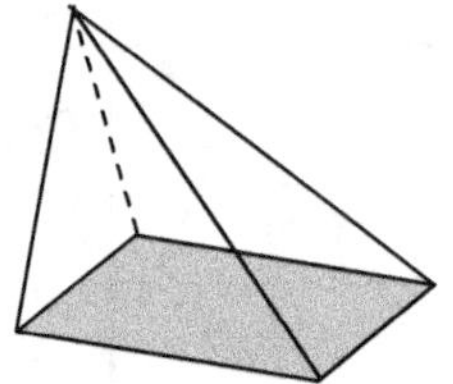

Right pyramid **Oblique pyramid**

A **pyramid** is a solid with triangular faces and a polygonal base. A pyramid is called a **right pyramid** if the line connecting the apex and the center of the base is perpendicular to the base; otherwise, it is called an oblique **pyramid**. Pyramids derive their names from the nature of their bases.

Nature of Base	Name of Pyramid
Triangle	Triangular Pyramid
Square	Square pyramid
Rectangle	Rectangular pyramid
Pentagon	Pentagonal pyramid
Hexagon	Hexagonal pyramid

Cones

A **cone** is a solid whose base is circular and whose lateral surface reduces gradually to the vertex.

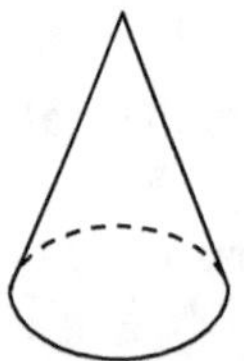

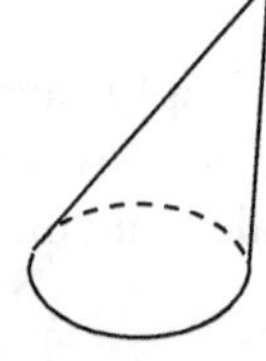

A cone is called a **right cone** if the line connecting the apex and the center of the base is perpendicular to the base; otherwise, it is called an oblique cone.

Right cone **Oblique cone**

Frustums

When the smaller end of a cone or pyramid is cut off through its cross-section by a plane parallel to the base the portion left is called a **frustum**.

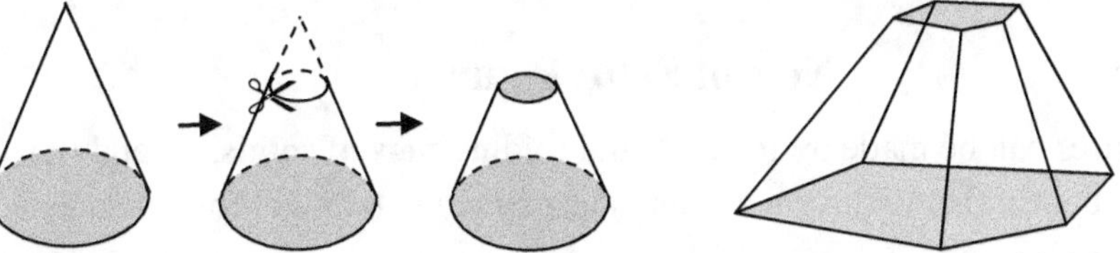

Conical frustum   Pentagonal pyramidal frustum

Frustums are named from the nature of their bases.

Polyhedrons

Polyhedrons are solid figures whose faces are made up of polygons. The faces of a regular polyhedron are equal in size and shape. There are only five regular polyhedrons.

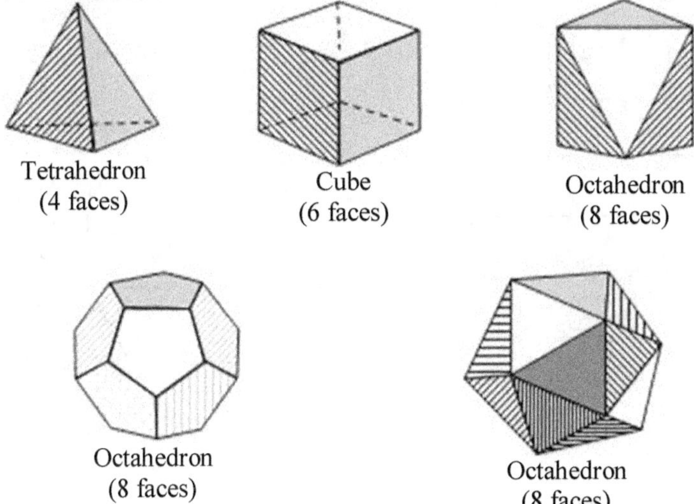

The Sphere and the Hemisphere

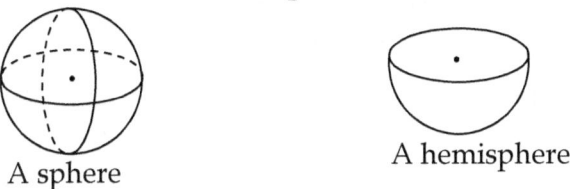

A sphere is a solid in which all the points on the surface are a fixed distance from a particular fixed point called the centre. A hemisphere is half of a sphere.

Nets of Solid Figures

Solid figures can be made by forming and folding **nets of solids.** (i) and (ii) will form cubes, (iii) will form a triangular prism and (iv) will form a square pyramid.

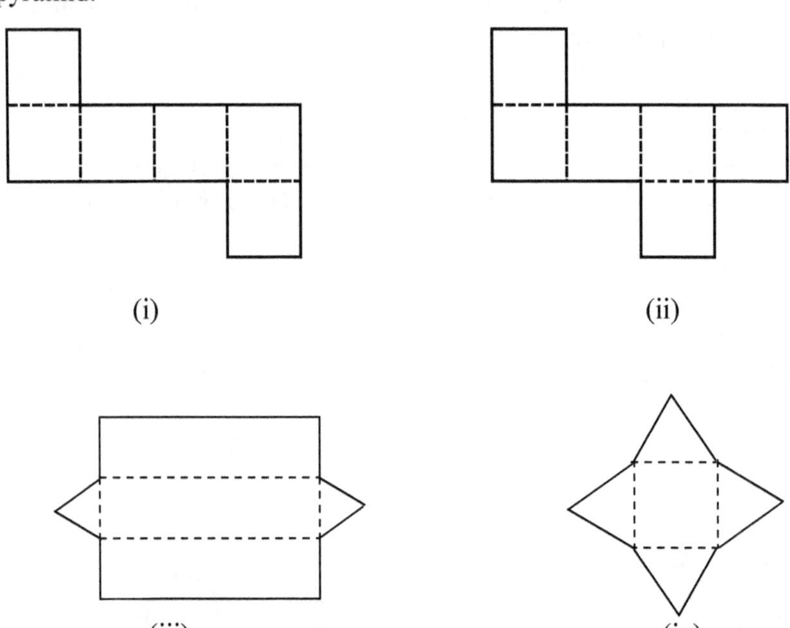

(i) (ii)

(iii) (iv)

MULTIPLE CHOICE EXERCISE 5:3

1. The following net is a net of:

[A] A tetrahedron [B] a pyramid
[C] A cone [D] a triangular prism

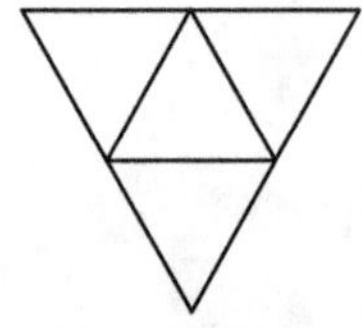

2. A prism is a solid figure with:

[A] Regular faces [B] Uniform cross-sectional area.
[C] Triangular faces [D] A square base and regular triangular faces.

3. Among the following, the figure that is certainly not a prism is:

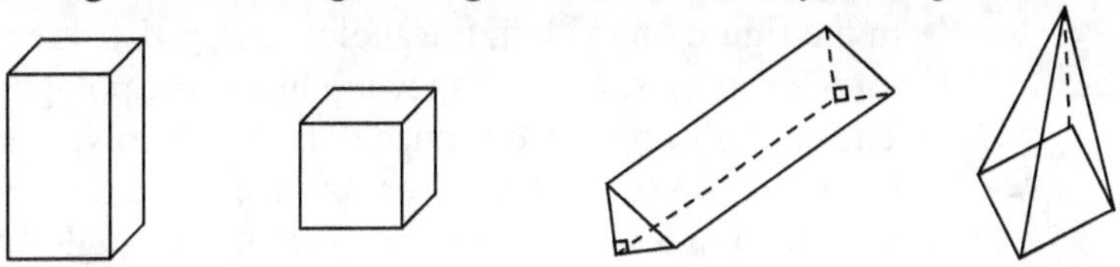

[A] [B] [C] [D]

4. The following Figure is called:

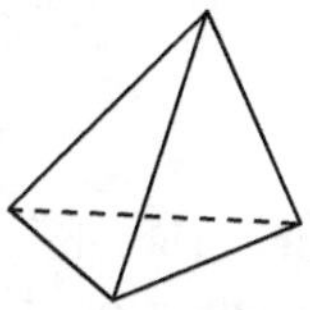

[A] A rhombus [B] a triangular prism
[C] A triangular pyramid [D] a cone

5. The solid formed by the following net is:

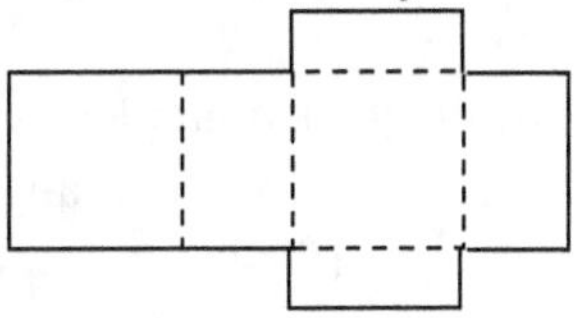

[A] Hexagonal prism
[B] rectangular pyramid
[C] Hexagonal pyramid
[D] rectangular prism

6. The figure which has one rectangular base and four lateral triangular surfaces is:

[A] Square pyramid [B] rectangular pyramid
[C] Cone [D] rectangular prism

7. A solid with two parallel and congruent bases *cannot* be:

[A] Cone [B] prism [C] cylinder [D] cube

8. None of the diagrams in the figure below are drawn to scale.

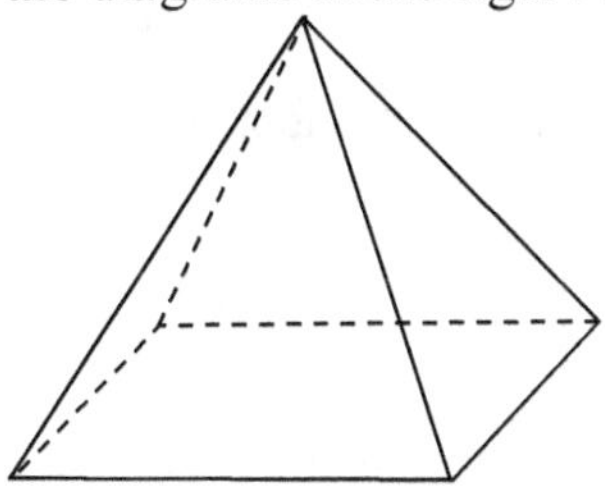

The net in the following which corresponds to the figure above is:

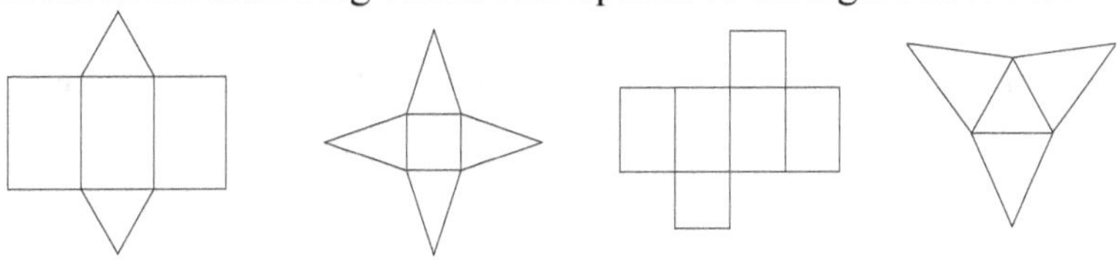

[A] [B] [C] [D]

5.4 POLYGON THEOREMS

Interior and exterior angles of polygons

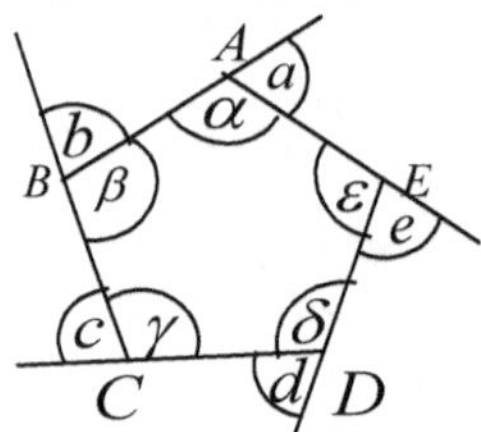

In the figure on the left, the angles α, β, γ, δ, ε, are **interior angles**, because they are inside the polygon. The angles a, b, c, d, e formed outside the polygons when AE, BA, CB, DC, ED are produced are the **exterior angles** of the polygon. An interior angle and the corresponding exterior angle are supplementary.

Chasles' Theorem

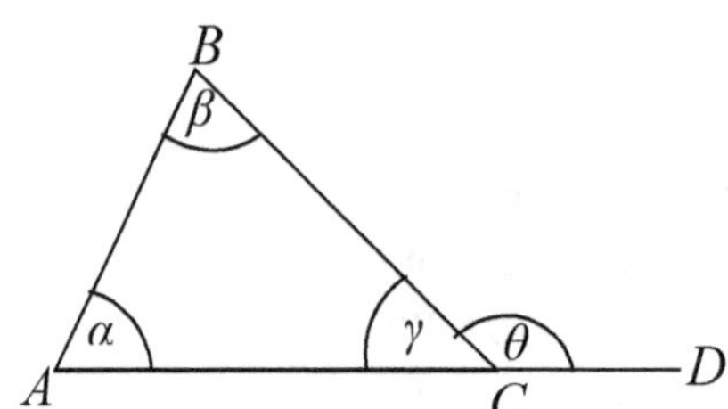

1. The exterior angle of a triangle is equal to the sum of the two interior opposite angles. i.e $\theta = \alpha + \beta$

2. The sum of the interior angles of a triangle is 180° (or two right angles). i.e. $\alpha + \beta + \gamma = 180°$

Polygon Theorems

In a convex polygon with n sides,

(1) The sum of the interior angles is $(2n - 4)$ right angles or $(n - 2)180°$.

(2) The sum of the exterior angles is 360° no matter the value of n.

Example

Each angle of a regular polygon is 170°. Find the number of sides of the polygon.

Solution

$170n = (n - 2)180 \Rightarrow n = 36.$

Or $(180 - 170)n = 360 \Rightarrow n = 36$. Therefore, the polygon has 36 sides.

Example

One angle of a hexagon is 140° and 5 angles are equal. Find the value of each of the 5 angles.

Solution

Let the value of each of the 5 angles be p.

$$\Rightarrow 5p + 140 = (6 - 2)180 \Rightarrow p = 116°$$

Note!!
It is easier to use the theorem on sum of exterior angles of a polygon than that on the sum of interior angles, though both lead to the same answer.

MULTIPLE CHOICE EXERCISE 5:4

1. In the following figure, $AB \parallel CD$. The size of the angle marked x is:
   [A] 103° [B] 93° [C] 62° [D] 52°

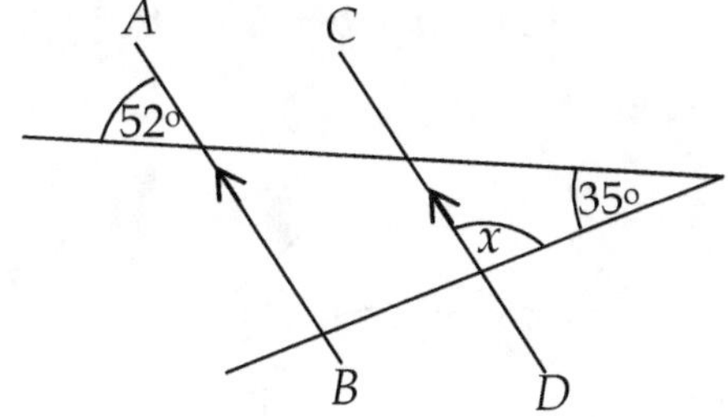

2. In the following figure, the value of x is:

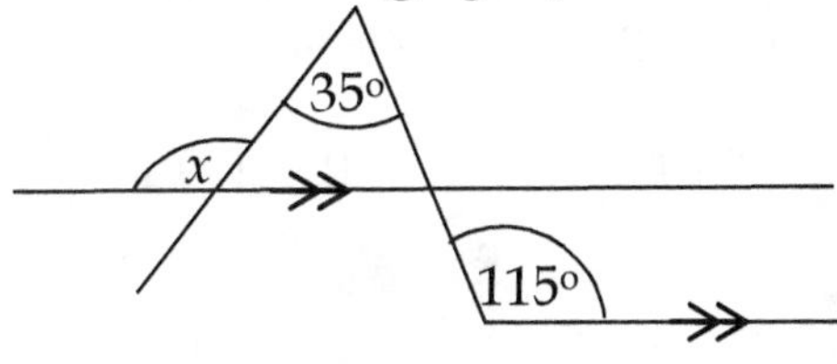

   [A] 35° [B] 80° [C] 100° [D] 115°

3. In the following figure, $PQ \parallel ST$. The value of x is:
   [A] 82° [B] 108° [C] 124° [D] 164°

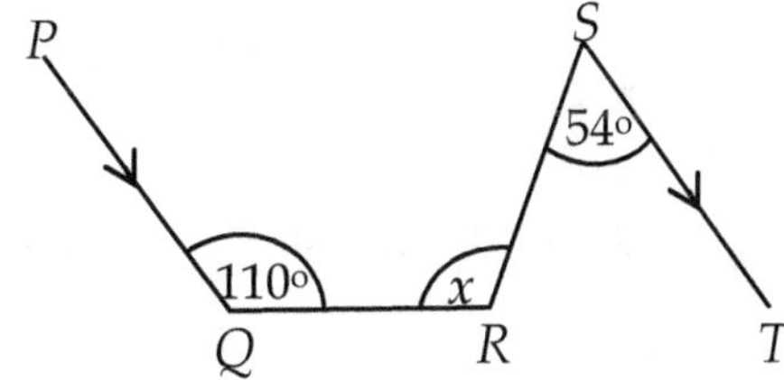

4. The number of sides in a regular polygon whose interior angle is 135° is:
   [A] 7 [B] 8 [C] 10 [D] 12
5. The angles of a pentagon are $x°, 2x°, (x + 60)°, (x + 10)°, x°, (x - 10)°$.
   The value of x is:
   [A] 80° [B] 75° [C] 60° [D] 40°

6. The number of sides in a regular polygon with each of its interior angles equal to 108° is:
   [A] 4 [B] 5 [C] 6 [D] 7
7. The number of sides in a regular polygon with each of its interior angles equal to 140° is:
   [A] 7 [B] 8 [C] 9 [D] 10
8. The number of sides in a regular polygon with each of its interior angles 120° is:
   [A] 4 [B] 6 [C] 7 [D] 8
9. In figure (i) below, the true relation is:
   [A] $a + b + x = 180°$ [B] $a = b + x$
   [C] $a - b = 180° - x$ [D] $a + b = x + 180°$

(i)

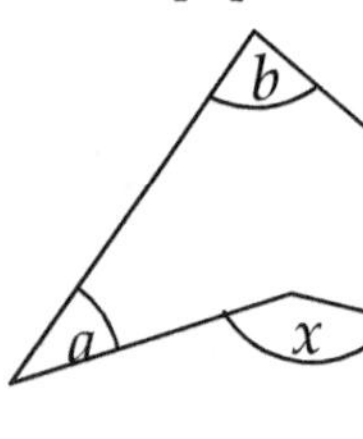

(ii)

10. In figure (ii) above, x is equal to:
   [A] $a + b + c$ [B] $360° - (a + b + c)$
   [C] $a + b + c + 180°$ [D] $360° - a + b + c$
11. In figure (i) below, y is equal to:
   [A] 80° [B] 70° [C] 40° [D] 100°

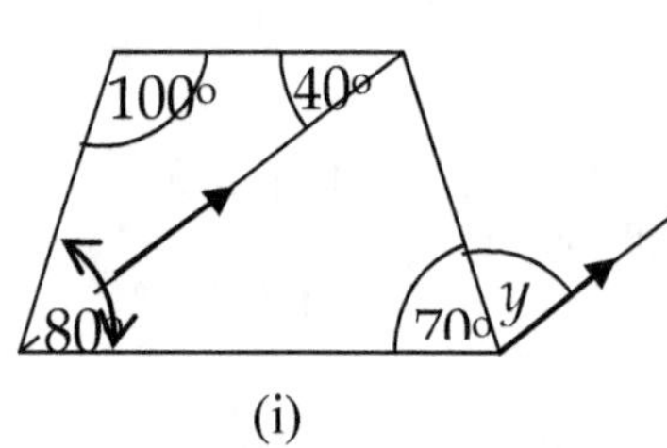

(i)

(ii)

12. In figure (ii) above, the size of angle ACB is:
   [A] 40° [B] 50° [C] 60° [D] 80°
13. In figure (i) below, $|PQ| = |PR| = RS$ and $\angle RPS = 32°$. The value of $\angle QPR$ is:

(i)

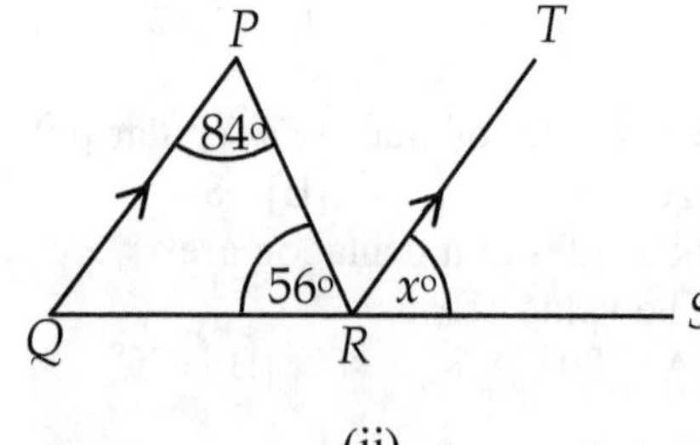

(ii)

[A] 64° [B] 52° [C] 32° [D] 26°

14. In figure (ii) above, QRS is a straight line, $QP \| RT$, $\angle PQR = 56°$, $\angle QPR = 84°$, $\angle TRS = x°$. The value of x is:
    [A] 28° [B] 40° [C] 44° [D] 84°

15. In figure (i) below, ABC is a triangle, BC is produced to D, $|AB| = |AC|$, $\angle BAC = 50°$. The value of $\angle ACD$ is:
    [A] 115° [B] 65° [C] 60° [D] 50°

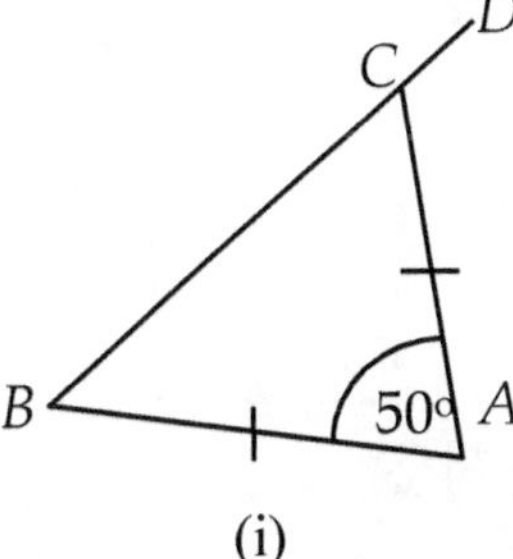

(i)

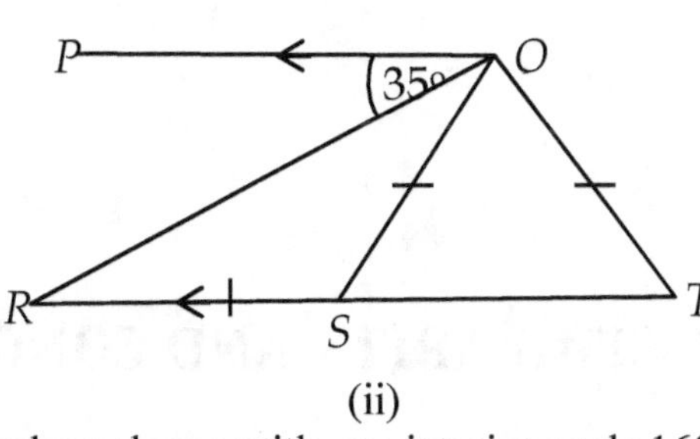

(ii)

16. The number of sides in a regular polygon with one interior angle 160° is:
    [A] 10 [B] 36 [C] 18 [D] 20

17. In figure (ii) above, PQ is parallel to RST. $\angle PQR = 35°$ and $|RS| = |SQ| = |TQ|$. The size of $\angle STQ$ is:
    [A] 35° [B] 40° [C] 70° [D] 110°

18. In figure (i) below, $WXYZ$ is a rhombus and $\angle WYZ = 20°$
    The value of angle XZY is:
    [A] 20° [B] 30° [C] 60° [D] 70°

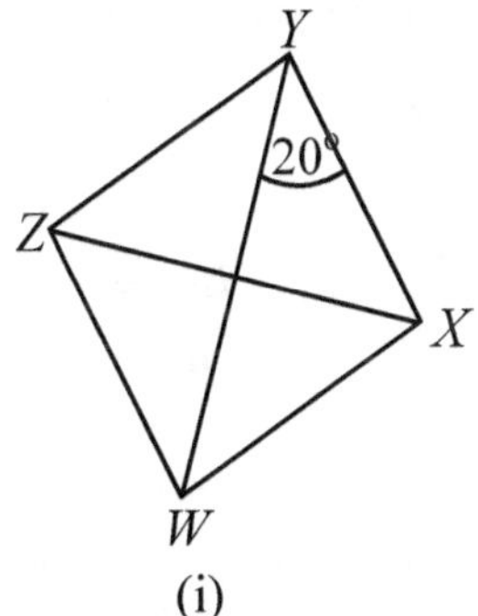

(i)

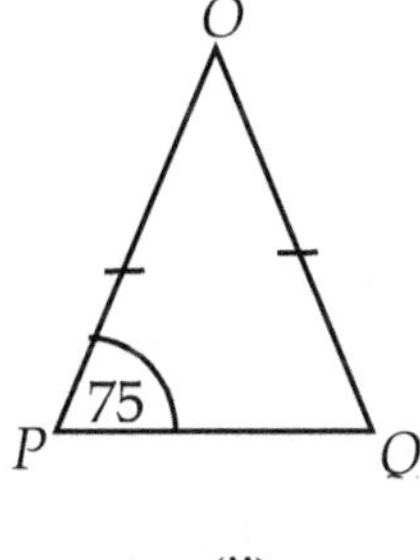

(ii)

19. A regular polygon centre O can be sub-divided into isosceles triangles identical to triangle POQ in figure (ii) above. The number of such triangles in the polygon is:
    [A] 12 [B] 10 [C] 9 [D] 8

20. The value of angle t in the following figure is:
    [A] 115° [B] 120° [C] 125° [D] 145°

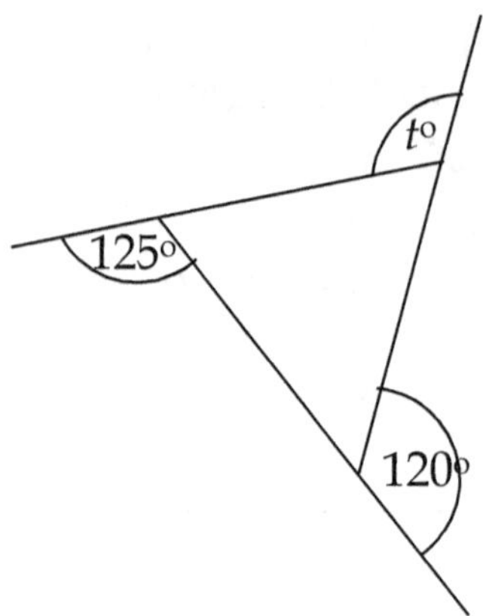

5.5 SIMILARITY AND CONGRUENCY

Congruent Figures

Congruent figures are figures that have the same shape and the same size. The statement 'A is congruent to B' is written $A \equiv B$.

Congruent Triangles

Conditions for Triangles to be Congruent

1. If the three sides of one triangle are equal to the three sides of the other then they are congruent by **side-side-side** abbreviated **SSS**.

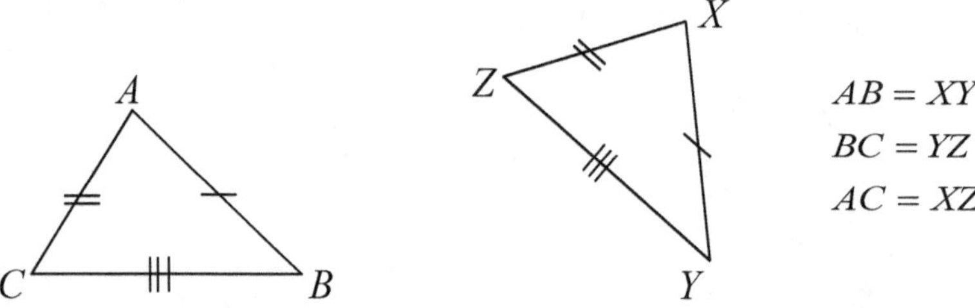

2. If two sides of one triangle are equal to two sides of the other and the included angles are equal then they are congruent by **side -included angle-side** abbreviated **SAS**.

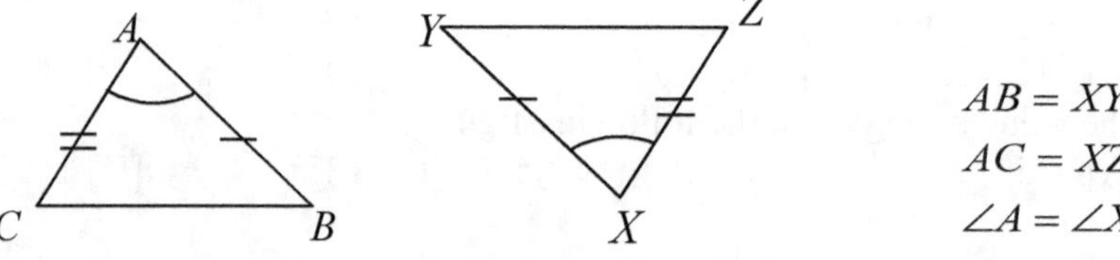

3. If two triangles are such that two angles of one are equal to two angles of the other and the included sides are equal then they are congruent by **angle-side-angle** abbreviated **ASA.**

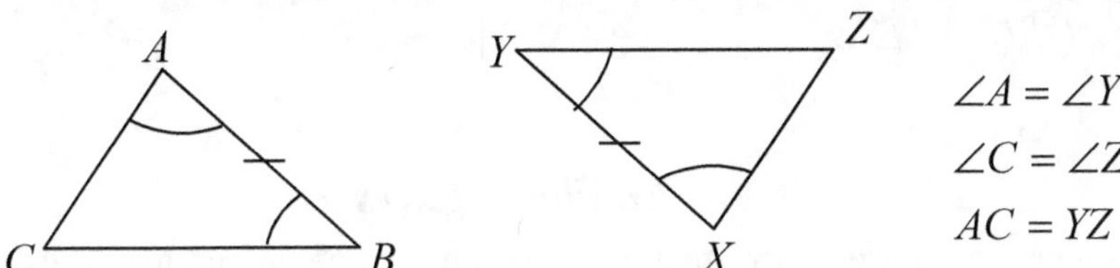

4. If two right-angled triangles have equal hypotenuse and one arm of another is equal to one arm of the other, then they are congruent by **right angle-hypotenuse- side** abbreviated **RHS**

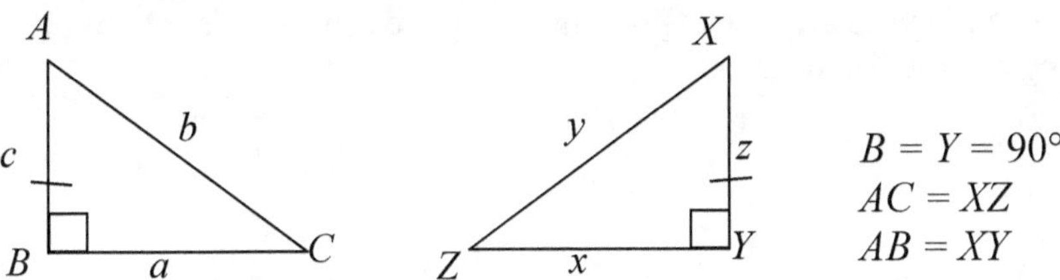

Example

In the following figure, $PR = PS$ and Q and T are the mid-points of PR and PS respectively. Determine the pairs of triangles, which are congruent giving arguments and reasons leading to your answer.

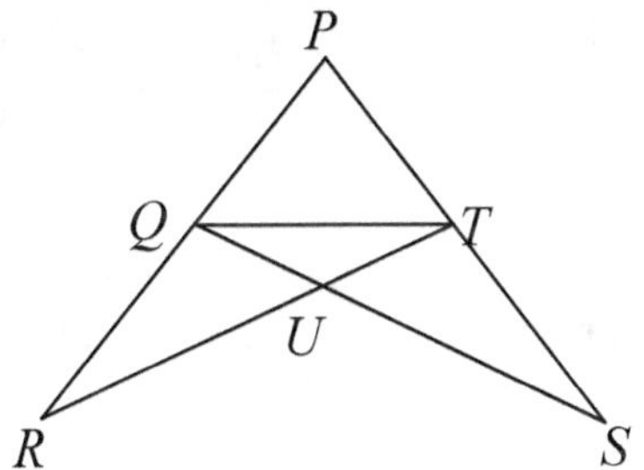

Solution

$PS = PR$ [Given]

$PQ = PT$ [P and Q are mid-points of PR and PS]

$\angle RPS$ is common to ΔPRT and ΔPSQ [Shown on diagram]

$\therefore \Delta PRT \equiv \Delta PSQ$ [SAS]

$RT = SQ$ [$\Delta PRT \equiv \Delta PSQ$ as proven above]

$RQ = ST$ [P and Q are mid-points of PR and PS]

QT is common to ΔQRT and ΔTSQ [Shown on diagram]

$\therefore \Delta QRT \equiv \Delta TSQ$ [SSS]

$\angle QUR = \angle TUS$	[Vertically opposite angles]
$\angle QRU = \angle TSU$	[$\Delta QRT \equiv \Delta TSQ$]
$\angle RQU = \angle RTS$	[Sum of $\angle$s of Δ]
$\therefore \Delta QRU \equiv \Delta TSU$	[ASA]

Similar Figures

Similar figures are figures that have the same shape but not necessarily the same size. The statement '*A* is similar to *B*' is written *A///B*.

Conditions for Triangles to be Similar

1. If two triangles are equiangular, then they are similar by **angle-angle-angle** abbreviated **AAA**.

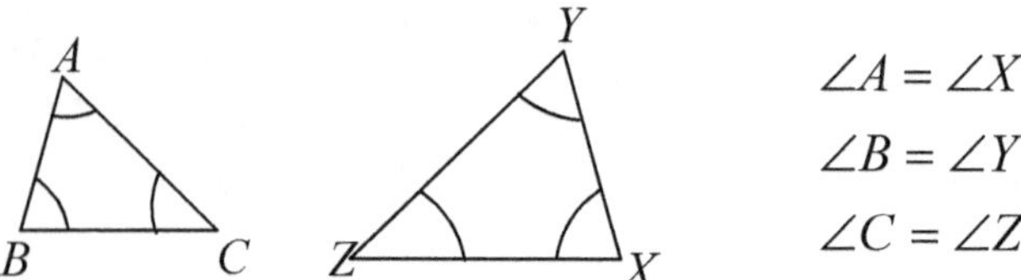

2. If the corresponding sides of two triangles are in a common ratio then they are similar by **side- side-side** abbreviated **SSS**.

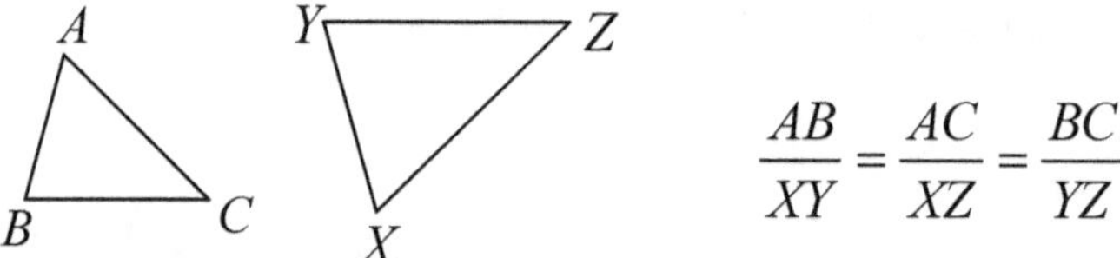

3. If an angle of one triangle is equal to an angle of another triangle and the sides containing this angle are in a common ratio then they are similar by **side- included angle-side** abbreviated **SAS**.

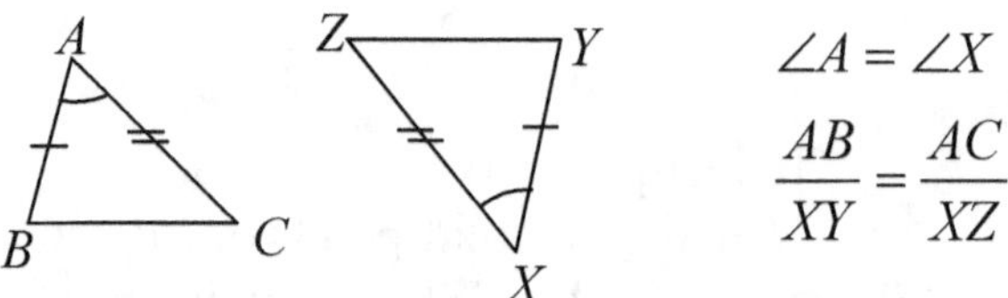

Example

In the figure below, $PQ \parallel ST$ calculate the value of y.

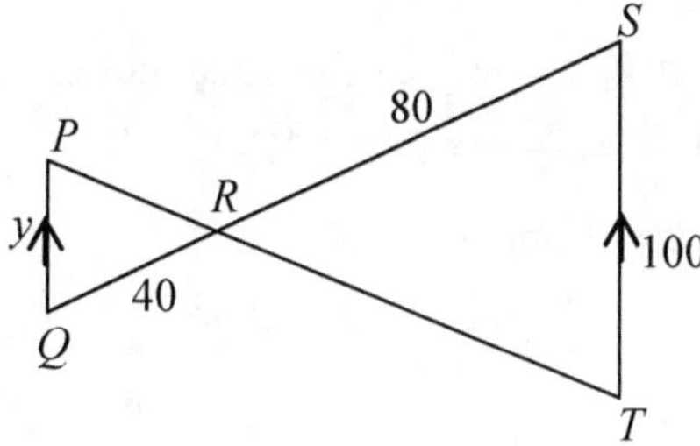

Solution

Since $PQ \parallel ST$, $\Delta RST \sim \Delta RQP \Rightarrow \frac{y}{100} = \frac{40}{80} \Rightarrow y = 50$

Ratio of Areas of Similar Figures

If the ratio of corresponding sides of two similar plane figures is $m\!:\!n$, then the ratio of their areas is $m^3 : n^3$.

Example

The ratio of the corresponding sides of two similar triangles is 1:3. Calculate the area of the larger triangle given that the area of the smaller triangle is 8 cm^2.

Solution

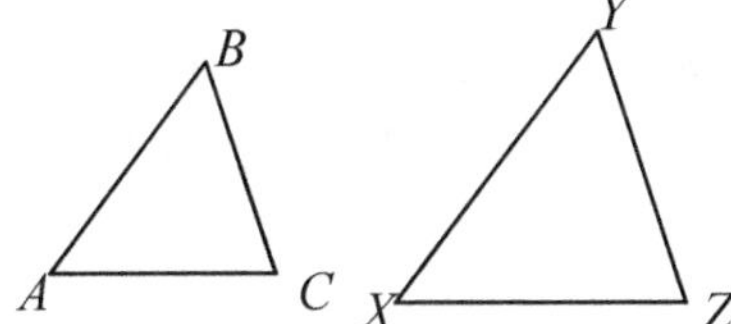

$$\frac{\text{Area of } \Delta XYZ}{\text{Area of } \Delta ABC} = \frac{3^2}{1^2}$$

$$\begin{aligned} \Rightarrow \text{Area of } \Delta XYZ &= 9 \times \text{Area of } \Delta ABC \\ &= 9 \times 8 \\ &= 72 \text{ cm}^2 \end{aligned}$$

Ratio of Volumes of Similar Figures

If the ratio of corresponding sides of two similar solid figures is $m\!:\!n$, then the ratio of their volumes is $m^3 : n^3$.

Example

A cube, whose volume is 20 cm^3, is enlarged such that its volume now is V cm^3. Given that the scale factor of the enlargement is 4. Find the value of V.

Solution

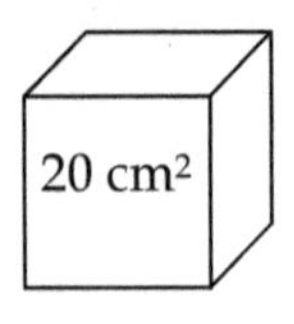

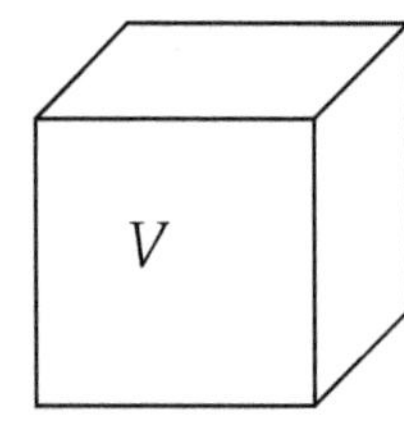

Let V_0 = volume of original cube.

Then, $\frac{V}{V_0} = \frac{k^3}{1} \Rightarrow V = k^3 V_0$

$k = 4$ and $V_0 = 20$ cm

$\Rightarrow V = 4^3 \times 20 = 1280 \text{ cm}^3$

MULTIPLE CHOICE EXERCISE 5:5

1. The pair of triangles among which is definitely congruent is:

[A] 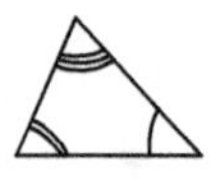 

[B] 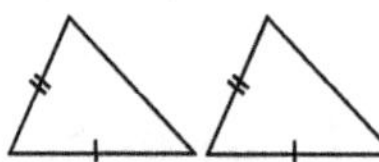

[C] 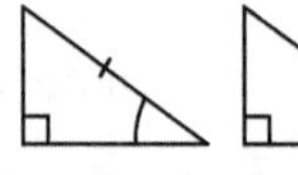

[D] 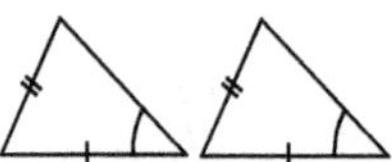

2. The pair of triangles in the figure below which is definitely congruent is:

[A] 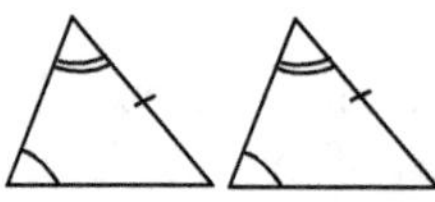

[B] 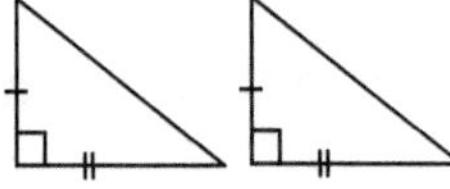

[C] 

[D] 

3. In three triangles *PQR*, *DEF* and *XYZ*: triangles *PQR* and *DEF* are equiangular but not congruent; triangles *DEF* and *XYZ* are congruent. It follows that triangles:
   [A] *PQR* and *DEF* are equal in area
   [B] *PQR* and *XYZ* are congruent
   [C] *PQR* and *XYZ* are equal in area
   [D] *PQR* and *XYZ* are similar

4. The pair of triangles in the figure below (not drawn to scale) which is similar is:

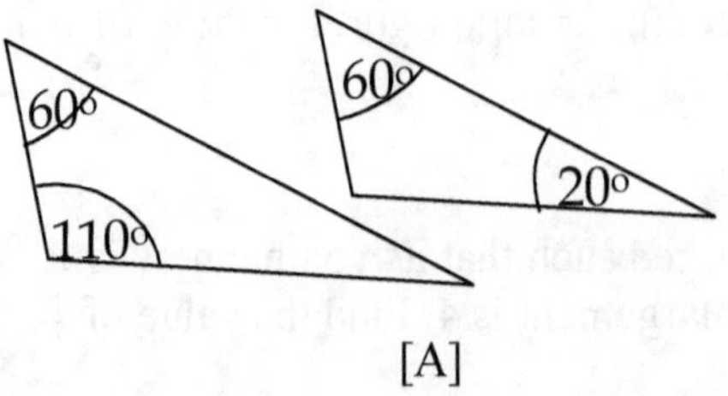

[A]

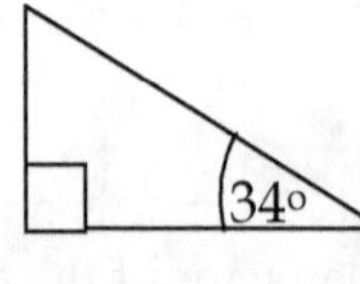

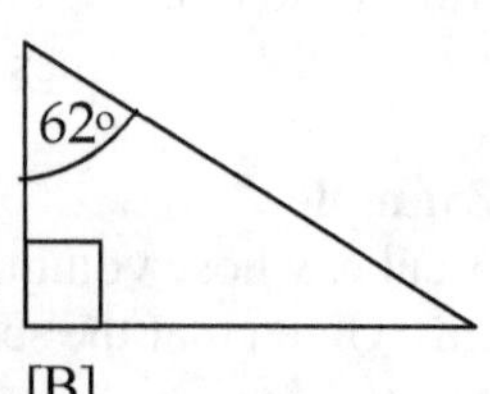

[B]

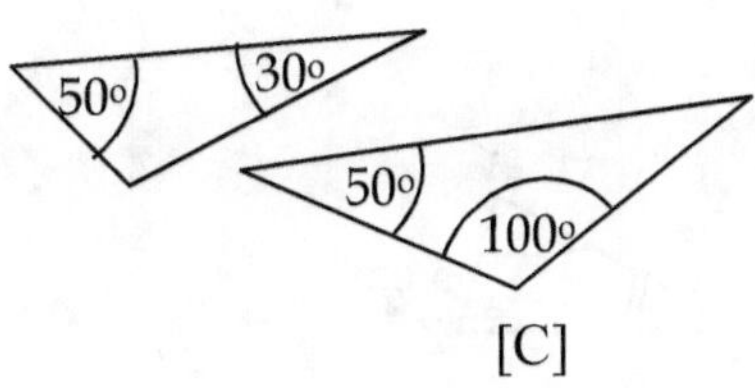

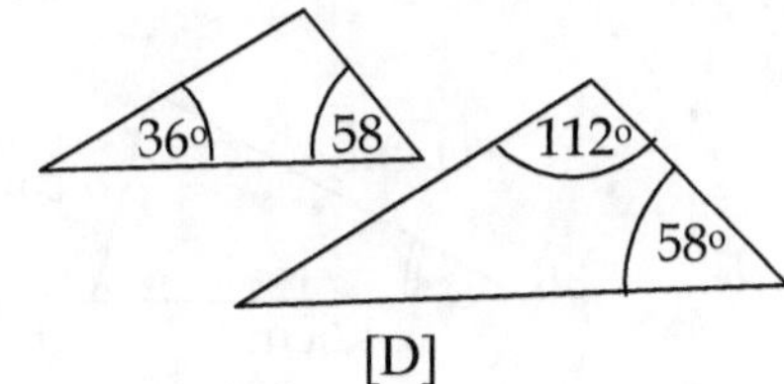

[C] [D]

5. The pair of triangles in the figure below which is similar is:

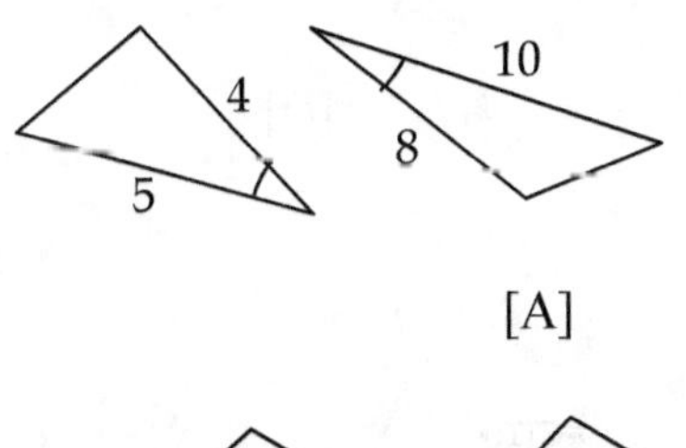

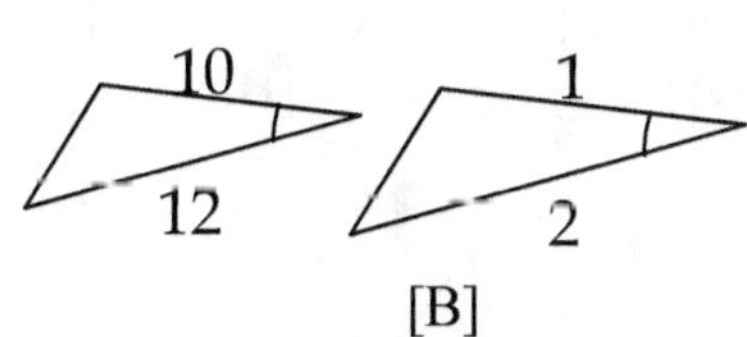

[A] [B]

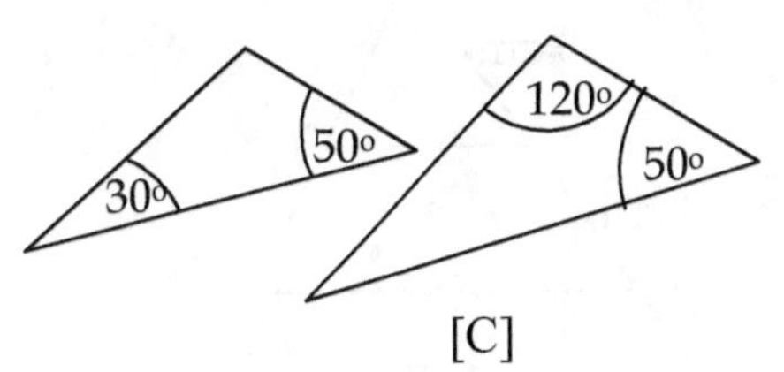

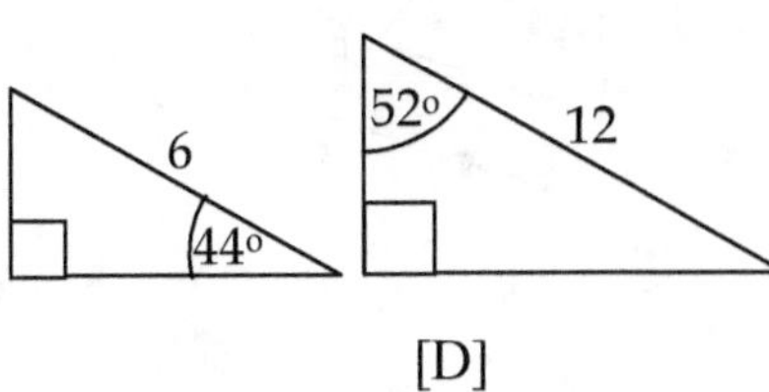

[C] [D]

6. Given that the triangles in the figure below are similar. It follows that:

[A] $\frac{AC}{XY}=\frac{YZ}{XZ}$ [B] $\frac{AC}{XY}=\frac{BC}{YZ}$ [C] $\frac{BC}{AB}=\frac{YZ}{XZ}$ [D] $\frac{BC}{AB}=\frac{XZ}{YZ}$

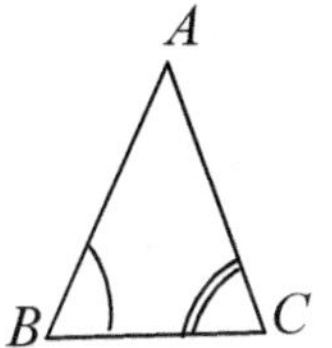

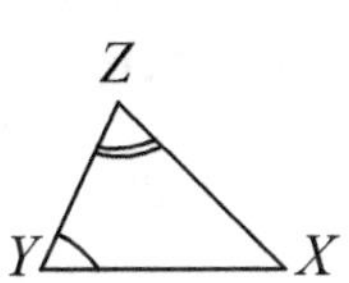

7. In the figure below, if $\frac{AB}{XY}=\frac{AC}{XZ}$ and $\angle B=\angle Y$. Then:

[A] $\frac{AB}{XY}=\frac{BC}{YZ}$ [B] $\Delta A=\Delta X$ [C] $\Delta C=\Delta Z$

[D] None of the above is necessarily true.

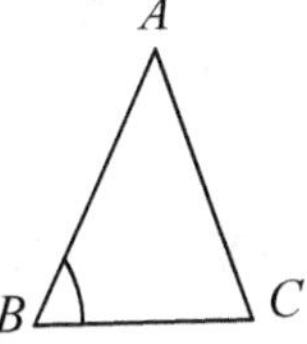

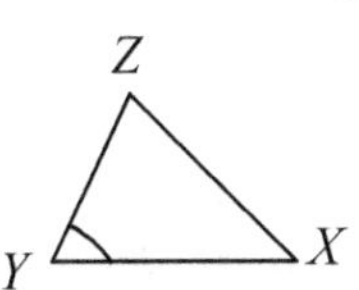

8. In the figure below, $\angle A=\angle X$ and $\angle B=\angle Y$. Hence XY is equal to:

[A] $6\frac{7}{8}$ cm [B] $17\frac{3}{5}$ cm [C] $19\frac{1}{5}$ cm [D] $8\frac{1}{2}$ cm

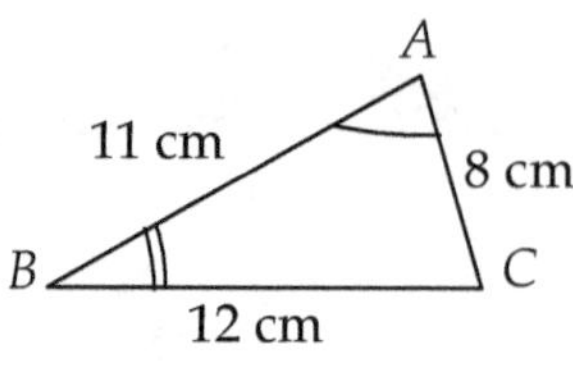

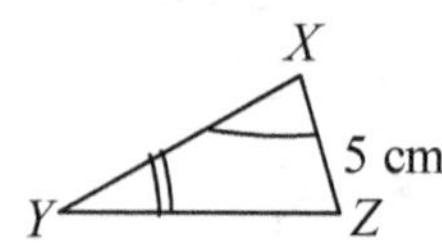

9. In figure (i) below, $PS = 8$ cm and $QS = 2$ cm. Hence $\frac{ST}{QR}$ is equal to:

[A] $\frac{1}{4}$ [B] $\frac{4}{1}$ [C] $\frac{4}{5}$ [D] $\frac{5}{4}$

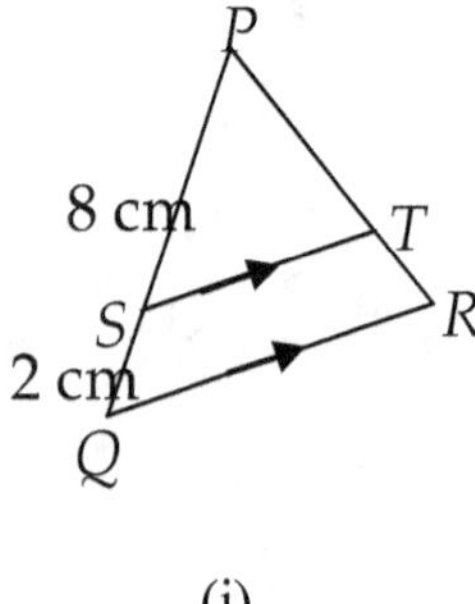

(i)

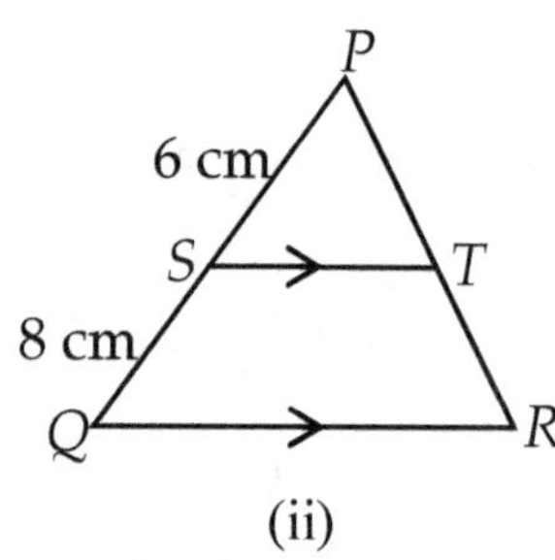

(ii)

10. In figure (ii) above, ST and QR are parallel. $|PS| = 6$ cm, $|SQ| = 8$ cm, $|PR| = 18\frac{2}{3}$ cm. $|PT|$ is equal to:

[A] 7 cm [B] 8 cm [C] $8\frac{2}{3}$ cm [D] 10 cm

11. In figure (i) below, $|AB|$=12 cm, $|AE|$=8 cm, $|DC|$=9 cm and $AB \parallel DC$. The length $|EC|$ is:
[A] 10 cm [B] 9 cm [C] 8 cm [D] 6 cm

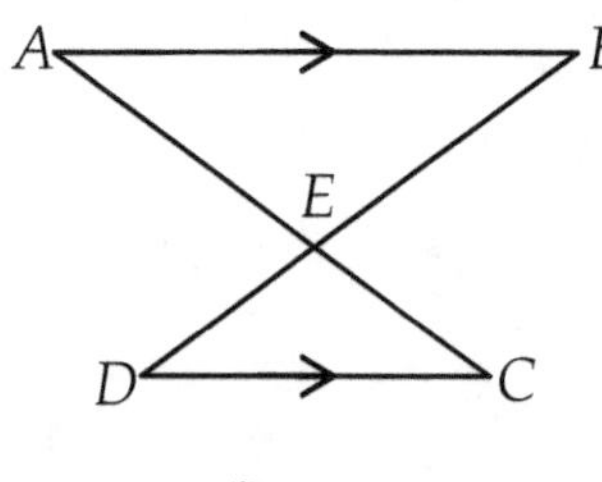

(i)

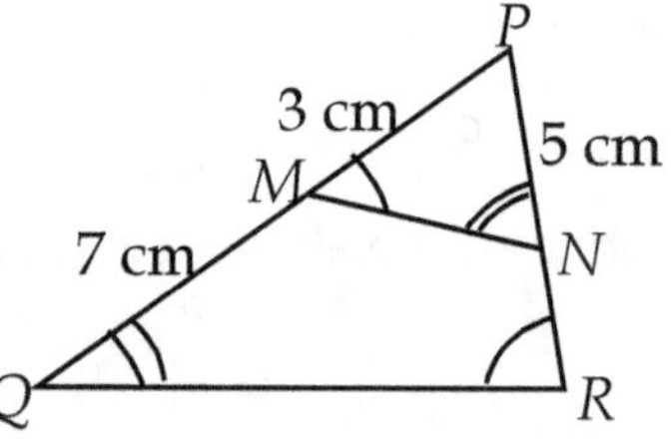

(ii)

12. In figure (ii) above, $\angle PMN = \angle PRQ$ and $\angle PNM = \angle PQR$. If $|PM|$=3 cm, $|MQ|$=7 cm and $|PN|$=5 cm, $|NR|$=

[A] 1 cm [B] 3 cm [C] $3\frac{1}{2}$ cm [D] 5 cm

13. In figure (i) below, $EF \| QR$, $PE = 2$ cm, $EQ = 4$ cm and $FR = 6$ cm. x should be:
[A] 2 cm [B] 3 cm [C] 4 cm [D] 6 cm

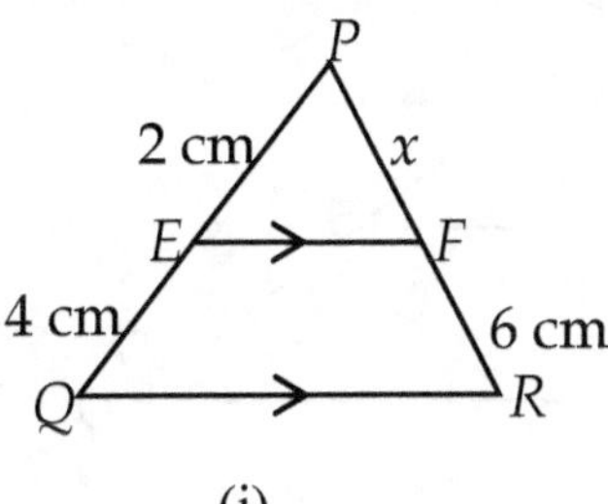

(i)

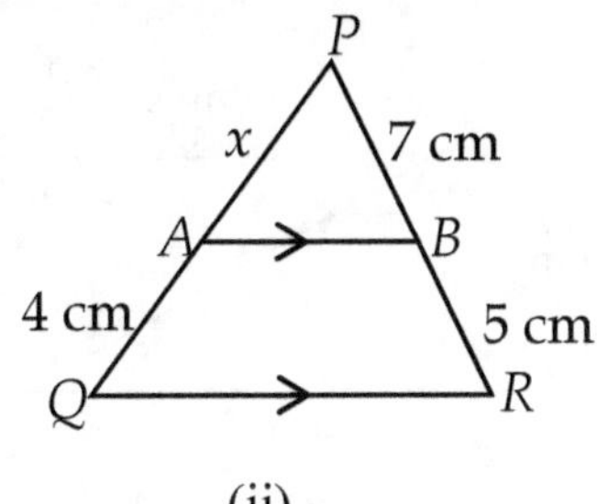

(ii)

14. The value of x in figure (ii) above is:
[A] 6.8 cm [B] 6.6 cm [C] 6.5 cm [D] 5.6 cm

15. In the figure below, XY is parallel to BC and AB is parallel to YZ. Hence:

[A] $\angle ABC = \angle ZYC$ [B] $\frac{YZ}{ZC} = \frac{AC}{BC}$

[C] ΔABC is similar to ΔZYC [D] $\frac{ZC}{AC} = \frac{YZ}{AB}$

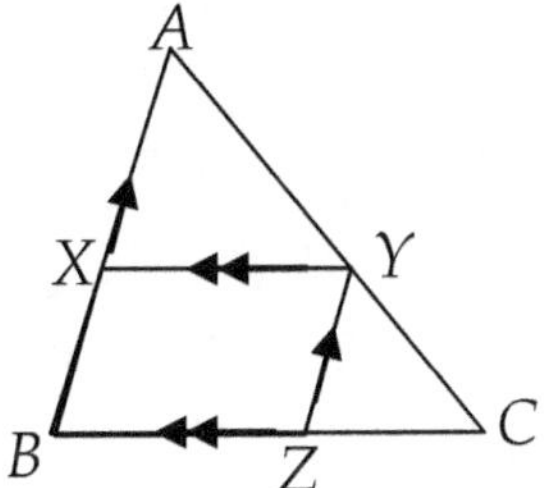

16. In the figure (i) below, AB is parallel to DC, $AB = 3$ cm and $DC = 5$ cm. Hence $\frac{XD}{XB}$ is equal to:

[A] $\frac{3}{5}$ [B] $\frac{5}{3}$ [C] $\frac{5}{8}$ [D] $\frac{8}{5}$

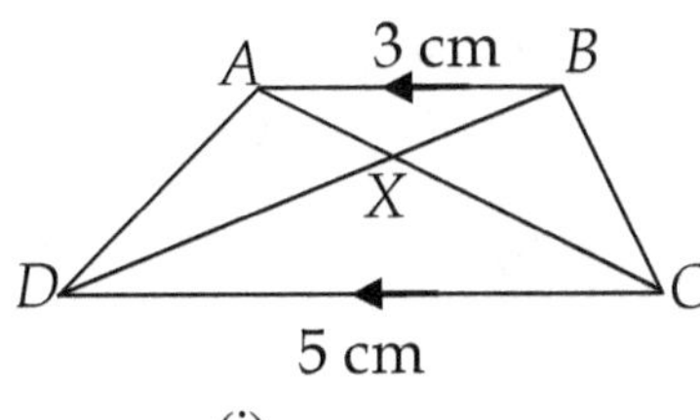

(i)

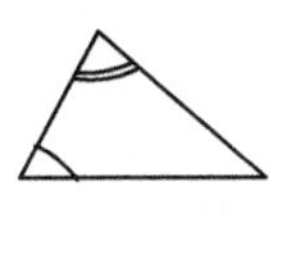
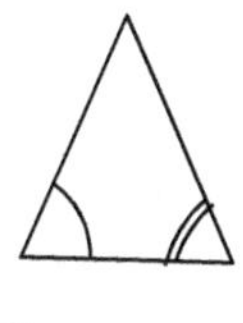
(ii)

17. The triangles in figure (ii) above are:

[A] congruent [B] similar [C] identical [D] none of the above

18. The ratio of the areas of the two triangles figure (i) below is:
[A] 1:6 [B] 1:2 [C] 1:4 [D] 1:9

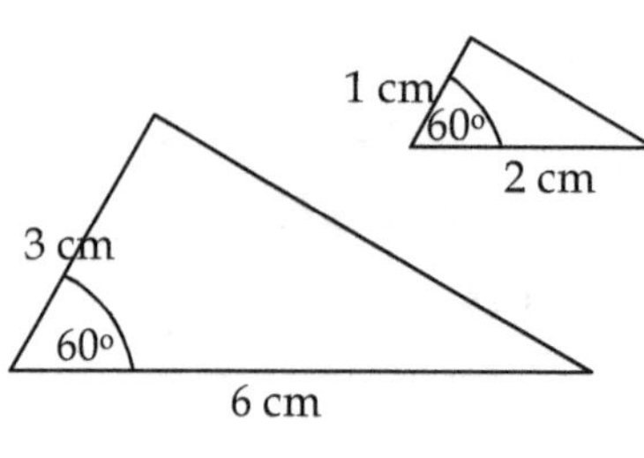

(i)

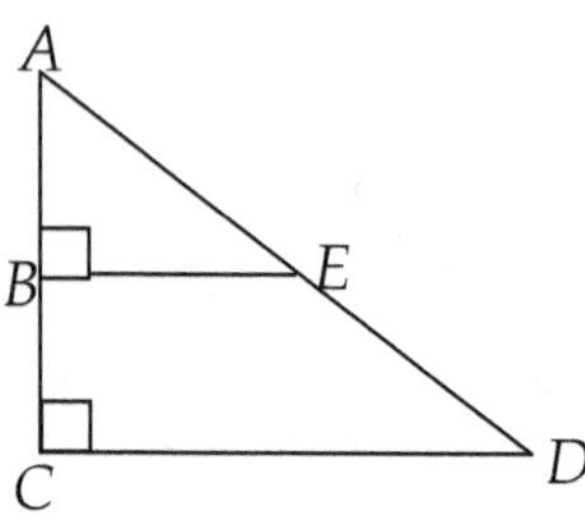

(ii)

19. Figure (ii) above shows a right-angled triangle ACD. $AB = 6$ cm, $AC = 8$ cm, $CD = 6$ cm and BE is parallel to CD. The value of the length BE is:
[A] 9 [B] 6 [C] 4.5 [D] 4

20. The figure below, shows two right-angled triangles OMN and OPQ. Given $\frac{OM}{OP} = \frac{1}{4}$, MN is parallel to PQ, $OM = 1$ cm, $ON = 2$ cm. The area of POQ in cm^2 is:
[A] 20 [B] 16 [C] 12 [D] 8

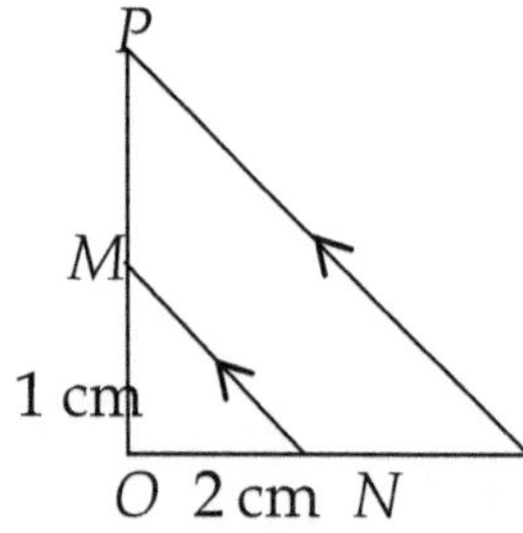

21. On a map drawn to a scale of 2 cm representing 1 km, the area represented by a square of side 4 cm is:
[A] 2 km^2 [B] 4 km [C] 4 km^2 [D] 1 km^2

22. Two similar cylinders have heights of 3 cm and 6 cm respectively. The ratio of their volumes is:
[A] 1:4 [B] 1:8 [C] 2:5 [D] 1:2

23. In figure (i) below, triangle ABC is similar to triangle AED and AB=16 cm, AE= 8 cm and AC=14 cm. The value of the length of the side marked x is:
[A] 7 cm [B] $\frac{80}{7}$ cm [C] $\frac{70}{8}$ cm [D] 6 cm

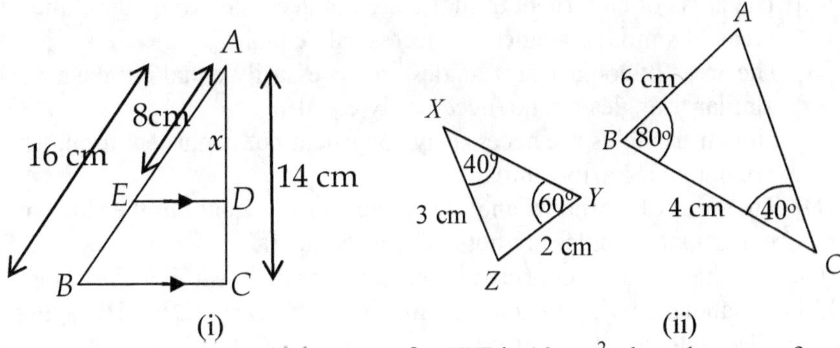

24. In Figure (ii) above, if the area of ΔXYZ is 10 cm^2, then the area of ΔABC is:
[A] 160 cm^2 [B] 40 cm^2 [C] 90 cm^2 [D] insufficient information.
25. In the figure below, ΔABC is similar to ΔDEF. Given that the area of ΔABC is 20 cm^2, then the area of ΔDEF is:
[A] 10 cm^2 [B] 5 cm^2 [C] 8 cm^2 [D] None of the above

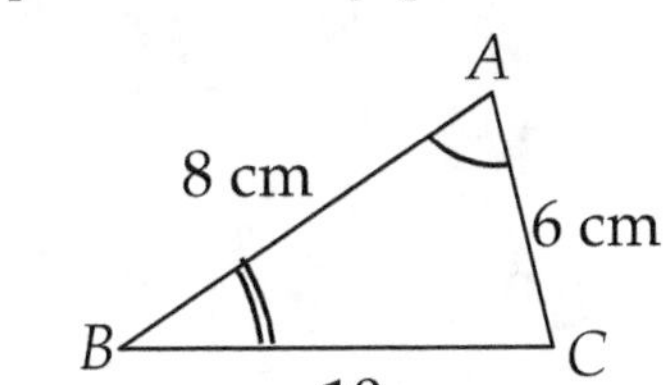

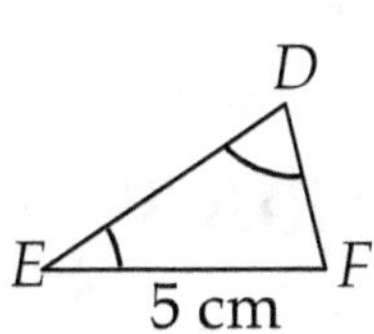

26. In the figure below, $\angle A = \angle X$ and $\angle B = \angle Y$. ΔABC has an area of 36 cm^2 and ΔXYZ has an area of 4 cm^2. If $AB = 4$ cm, then XY is equal to:
[A] $\frac{3}{4}$ [B] $\frac{4}{3}$ [C] $\frac{4}{4}$ [D] $\frac{9}{4}$

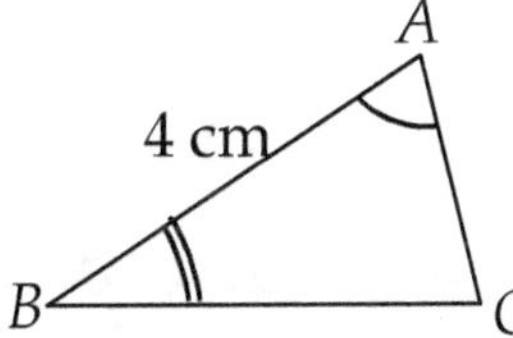

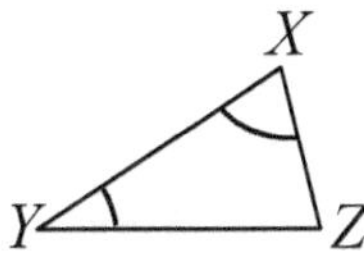

27. Two buckets A and B, identical in shape, are such that the dimensions of A are three times as large as the corresponding dimensions of B. The ratio of the volumes of A:B is:
[A] 1:9 [B] 27:1 [C] 9:1 [D] 1:27
28. The pairs of triangles PQR, XYZ are congruent is:
[A] $XY = PQ, XZ = QR, \angle X = \angle Q$ [B] $XY = QR, YZ = PR, \angle Y = \angle P$
[C] $\angle Y = \angle P, \angle Z = \angle Q, XZ = PQ$ [D] $\angle Z = \angle P, \angle Y = \angle Q, XY = PR$
29. Similar triangles differ from congruent triangles in that:

[A] The areas of congruent triangles are not necessarily equal but the areas of similar triangles are necessarily equal.
[B] The areas of congruent triangles are necessarily equal but the areas of similar triangles are not necessarily equal.
[C] Similar triangles are necessarily congruent but congruent triangles are not necessarily similar.
[D] The sides of similar triangles are necessarily equal but the sides of congruent triangles are not necessarily equal.

30. The set which consist entirely of elements which are similar figures is:
[A] Triangles [B] Quadrilaterals [C] Circles [D] Hexagons

31. The false statement (s) is/are:
[A] All Similar objects have the same shape but not necessarily the same size.
[B] All similar objects are congruent.
[C] All congruent objects are similar.
[D] All similar objects are congruent and have the same shape but not necessarily the same size.

5.6 CONSTRUCTIONS AND LOCI

CONSTRUCTIONS

We normally do constructions with pencil, ruler and a pair of compasses, and unless otherwise stated never use any other device. After a construction do not erase the construction lines.

Constructing a Triangle *ABC* with Sides of Given Length

To construct a triangle ABC with sides of length $AB = 9$ cm, $BC = 6$ cm and $AC = 4$ cm.

(i) Draw a line longer than 9 cm and mark the point A on it.
(ii) With open compass measure 9 cm from your ruler and with centre A draw an arc at B.
(iii) With compass, measure 6 cm and with centre B, draw an arc on one side of the line.
(iv) With compass, measure 4 cm and with centre A, draw another arc to cut the first. Mark their point of intersection C.
(v) Using ruler and pencil join AC and BC

Note!! Construction lines do not end at the vertices *A*, *B*, and *C*.

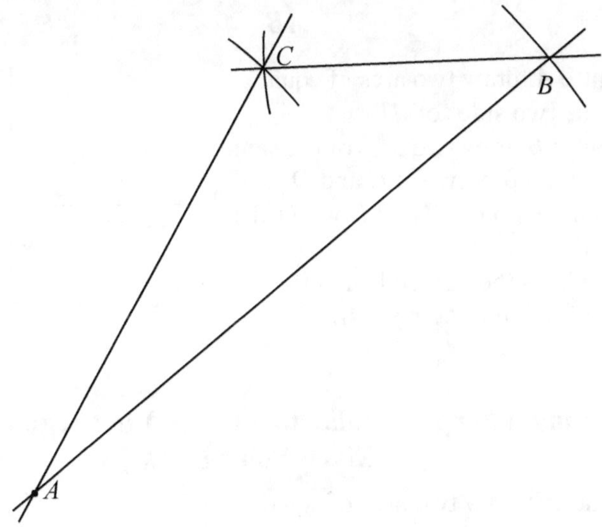

Height, Median and Perpendicular Bisector

The **altitude** (or **height**) of a triangle is the perpendicular distance from one side of the triangle to the opposite vertex.

The **perpendicular bisector** or **mediator** of a side of a triangle is a line that is perpendicular to the side and passes through its midpoint.

The **median** of a triangle is a line from the midpoint of a side of the triangle to the opposite vertex.

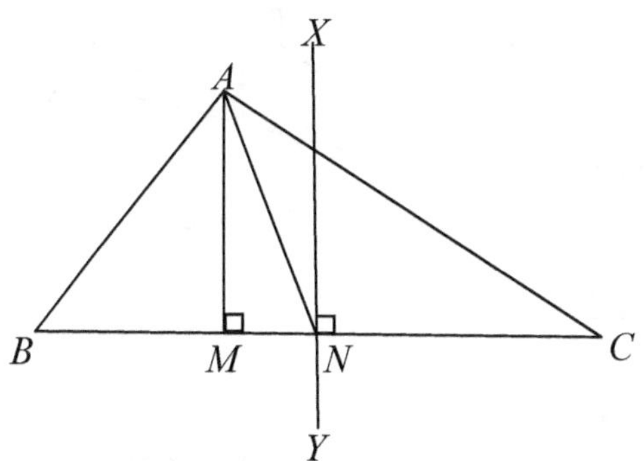

The figure on the left shows the altitude (or height) *AM*, the median *AN* and the perpendicular bisector (or mediator) *XY* of the triangle *ABC*, with respect to the side *BC*.

Constructing a Perpendicular Bisector of a Given Line Segment *AB*

(i) With centre *A*, draw two arcs of equal radii on the two sides of *AB*.
(ii) With centre *B* draw two arcs of the same radii to cut the first two at *C* and *D*.
(iii) Now join the points *C* and *D* with ruler and pencil.
The line *CD* is the perpendicular bisector of the line segment *AB*.

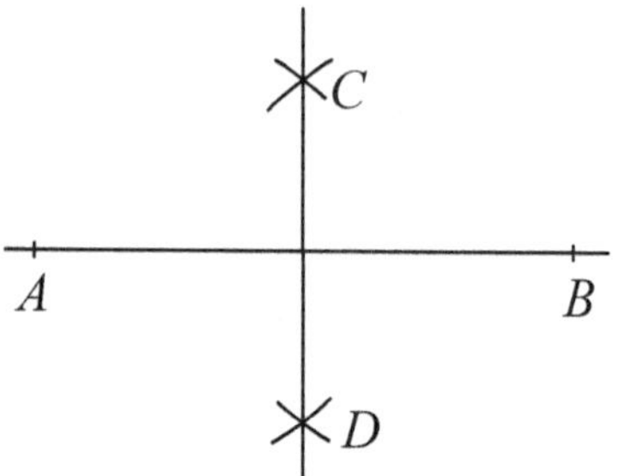

Constructing a Perpendicular to a Given Line Segment, from a Given Point *P*

(i) With centre *P* draw two arcs of equal radii to cut *AB* at two points *C* and *D*.
(ii) With centres *C* and *D* draw two arcs of equal radii to intersect at *Q* on the opposite side of *P*.
(iii) Now join *PQ*.
PQ is perpendicular to *AB*.

Note that the construction is the same event if *P* lies on *AB*.

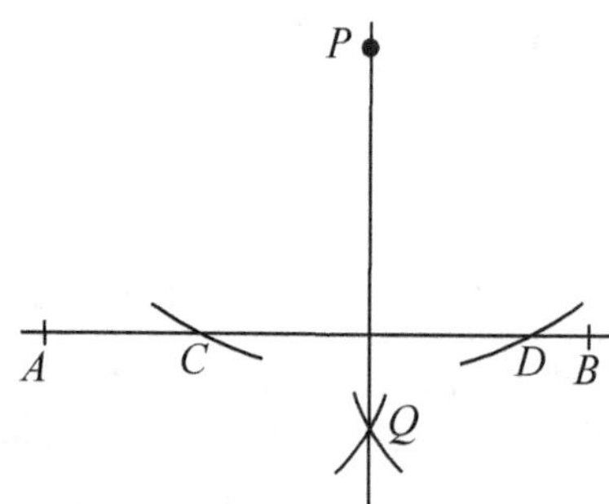

Bisecting a Given Angle *PAR*

(i) With centre *A* draw two arcs of equal radii to cut the adjacent sides to the given angle at two points *B* and *C*.
(ii) With centres *B* and *C* draw two arcs of equal radii to intersect at *D*.
(iii) Join *A* and *D* with ruler and pencil.
AD is the bisector of the angle *PAR*.

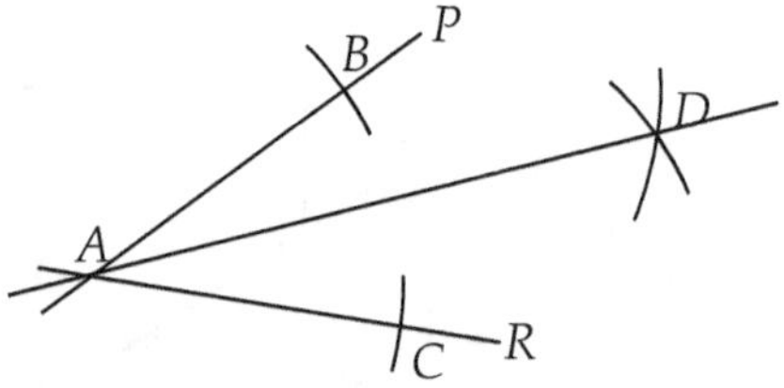

Constructing an Angle of 60°

To construct an angle of 60°, use the properties of an equilateral triangle as follows.

(i) With centre *A* draw two arcs of equal radii one to cut the line segment *AB* at *B* and the other on one side of *AB*.

(ii) With centre B and the same radius, draw another arc to intersect the first at C.

(iii) Join AC. The angle BAC is 60°.

After constructing an angle of 60°, a bonus is obtained by the 120° angle constructed.

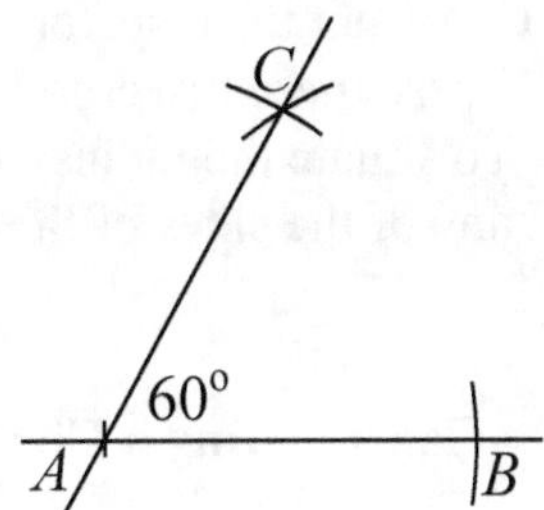

To construct an angle of 45° simply construct an angle of 90° and bisect it.

To construct an angle of 30° simply construct an angle of 60° and bisect it.

Trisecting a Right Angle *ABC*

(i) With centre B draw a large arc to cut AB at A and BC at C.

(ii) With centre C draw an arc of the same radius to cut the large arc at D.

(iii) With centre A draw an arc of the same radius to cut the large arc at E.

(iv) Join BD and BE.

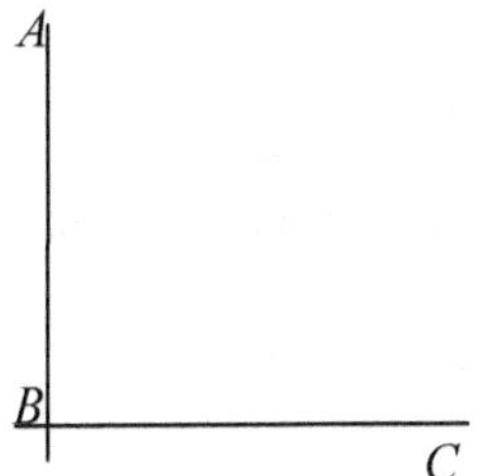

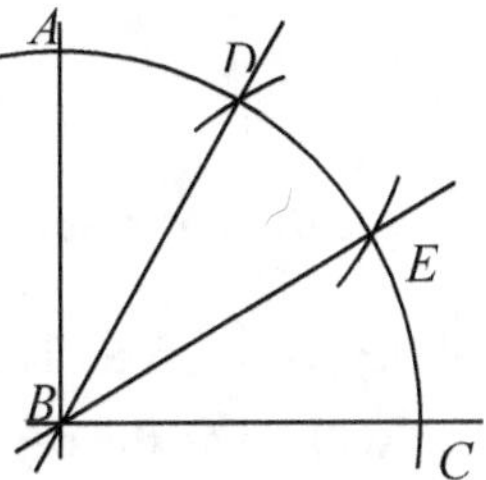

Constructing an Inscribed Circle of a Given Triangle *ABC*

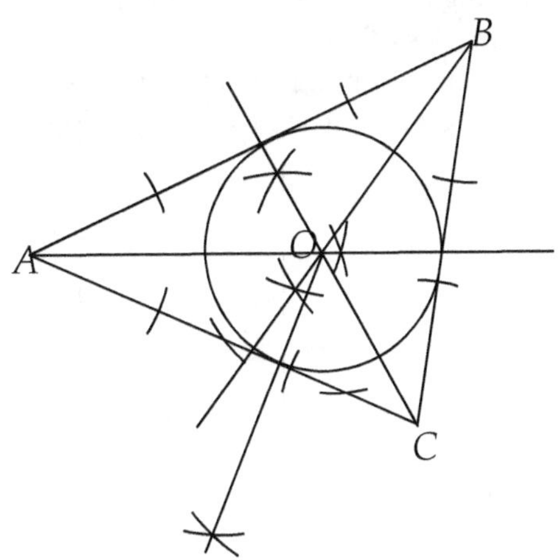

(i) Construct the bisectors of each of the angles of the triangle *ABC*.
(ii) Where these bisectors intersect is the centre *O* of the circle.
(iii) To situate the radius of the circle, construct a perpendicular from *O* to any of the sides of the triangle and draw the circle.

Constructing a Circumscribed Circle of a Given Triangle *ABC*

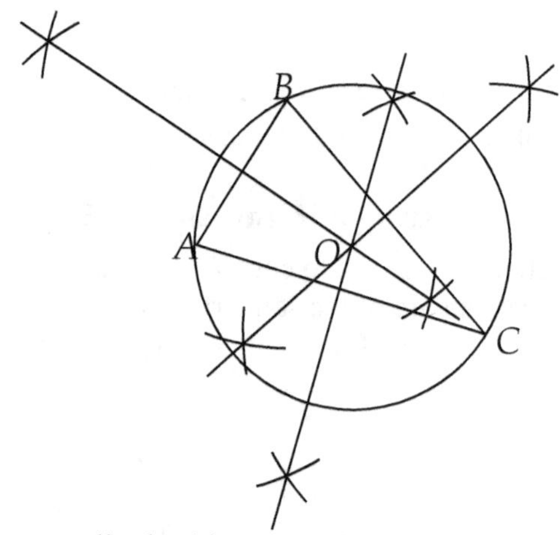

(i) Construct the perpendicular bisectors of each of the sides of the triangle *ABC*.
(ii) Where these bisectors intersect is the centre *O* of the circle.
(iii) Draw the circle with *O* as the centre.

LOCI

A locus is the path of a moving point subject to some restrictions.
The following are some examples of loci.

(1) The locus of a point which is moving such that its distance from a fixed point is always constant is a circle. (See figure (i) below).

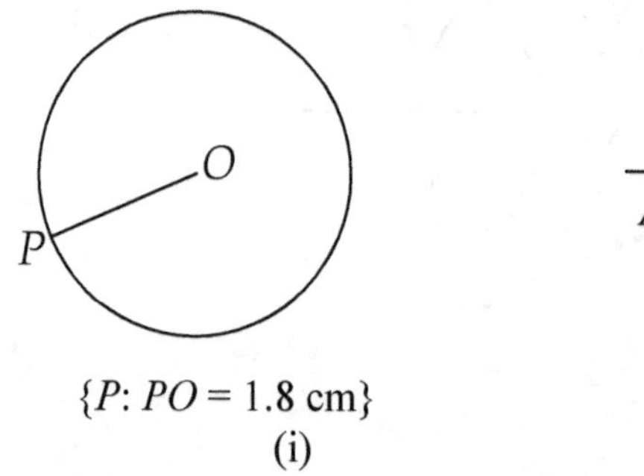

$\{P: PO = 1.8 \text{ cm}\}$
(i)

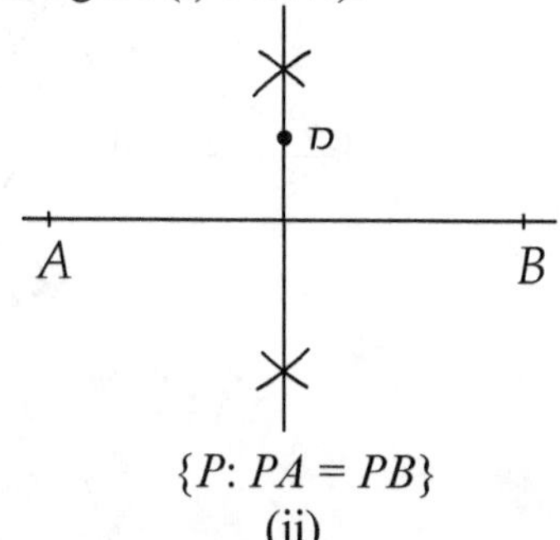

$\{P: PA = PB\}$
(ii)

(2) The locus of a point which is moving such that its distance from two fixed points is always equal is the mediator or perpendicular bisector of the line segment joining these two points. (See figure (ii) above).

(3) The locus of a point which is equidistant from two intersecting lines is the bisector of the angle between the lines. (See figure (i) below).

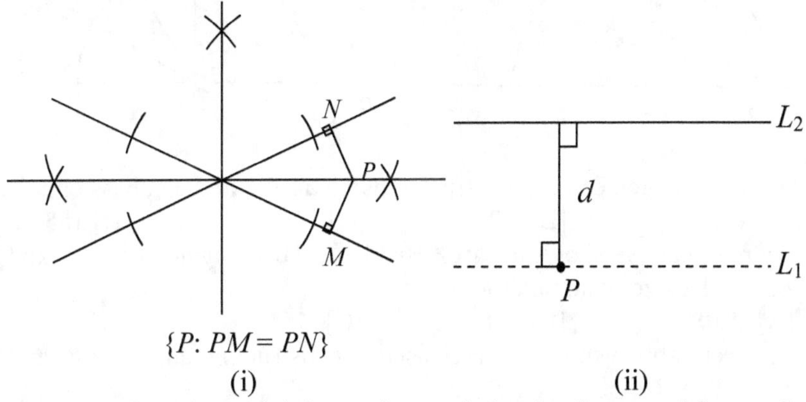

$\{P: PM = PN\}$

(i) (ii)

(4) The locus of a point which is moving such that its distance d from a fixed line L_1 is always constant is a line L_2 parallel to L_1. (See figure (ii) above).

MULTIPLE CHOICE EXERCISE 5:6

1. AB bisects PQ at point N. The statement that is true of N is:
   [A] N is the midpoint of AB.
   [B] N is the midpoint of AB and the midpoint of PQ.
   [C] N is the midpoint of PQ.
   [D] N divides PQ in the ratio 2:1.
2. Point P is the midpoint of AB. Complete the statement: $PB = 7$ cm, AB is equal to:
   [A] 7 cm [B] 14 cm [C] 3.5 cm [D] none of the above
3. The diagram(s) among the following that demonstrate(s) the correct way of constructing an angle of 60° is:

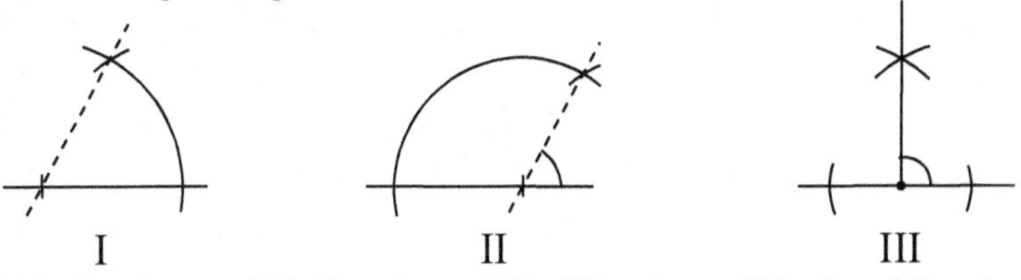

   [A] I only [B] II only [C] III only [D] I and II only
4. In the construction in figure (i) below, the size of angle BAC is:

[A] 60° [B] 90° [C] 120° [D] 150°

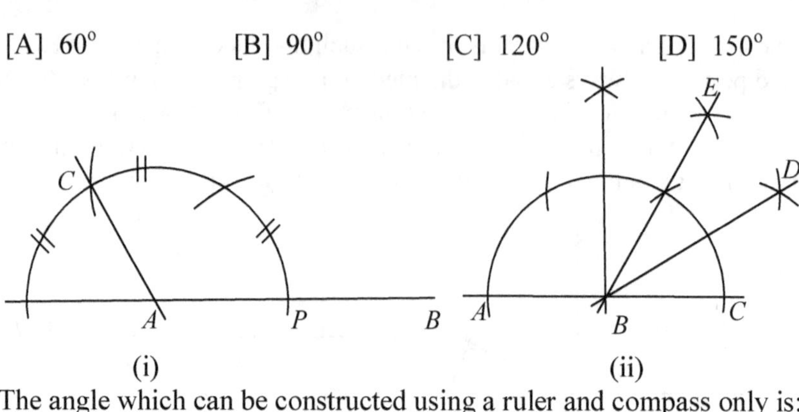

(i) (ii)

5. The angle which can be constructed using a ruler and compass only is:
   [A] 135° [B] 125° [C] 115° [D] 155°
6. Figure (ii) above shows the arcs used in constructing angles *DBC* and *EBC*. The size of angle *DBC* is:
   [A] 120° [B] 30° [C] 45° [D] 60°
7. Figure (ii) above shows the arcs used in constructing angles *DBC* and *EBC*. The size of angle *EBC* is:
   [A] 120° [B] 30° [C] 45° [D] 60°
8. Figure (ii) above shows the arcs used in constructing angles *DBC* and *EBC*. The size of angle *ABE* is:
   [A] 120° [B] 30° [C] 45° [D] 60°
9. The diagram among the following which shows the construction of $\angle ABC = 75^\circ$ using a ruler and a pair of compass only is:
   [A] I only [B] II only [C] III only [D] I and II only

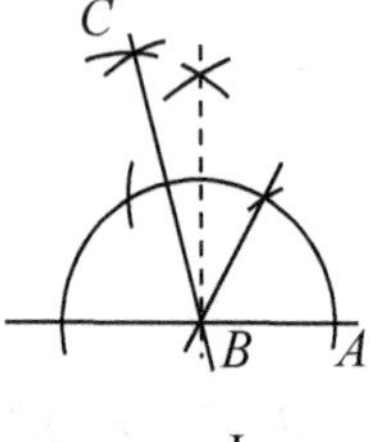

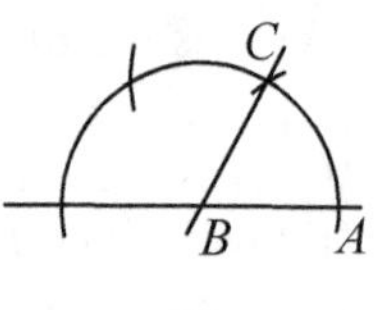

I II III

10. The name of the instrument in figure (i) below is:

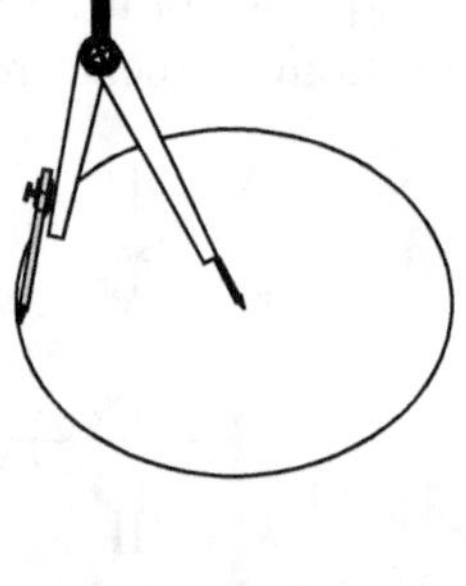

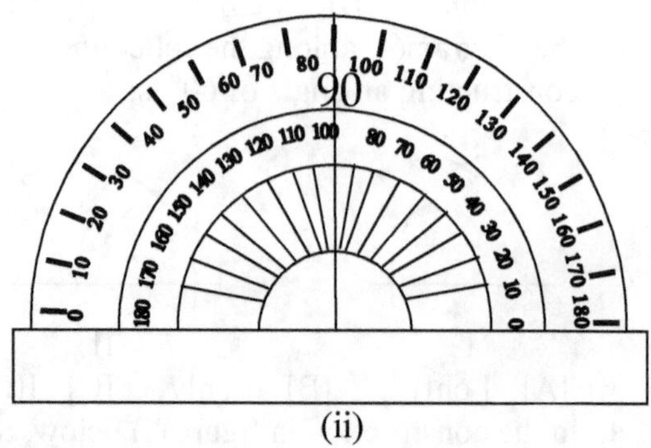

(i) (ii)

[A] a protractor [B] a pair of dividers
[C] a pair of compass [D] a set square

11. In order to construct an in-circle one needs at least:
    [A] a pair of compass, pencil and ruler
    [B] a protractor, pencil and ruler
    [C] a set square, pencil and ruler
    [D] a protractor, pencil and a set square
12. The name of the instrument in figure (ii) above is:
    [A] a set square [B] a pair of compasses
    [C] a pair of dividers [D] a protractor
13. The locus of a point which is moving such that its distance *d* from a fixed line *l* is always constant is:
    [A] a mediator of *l* [B] a line parallel to *l* [C] a parabola [D] a circle
14. The locus of a point which is moving such that its distance from a fixed point is always constant (the same) is:
    [A] a mediator.
    [B] a line parallel to the point.
    [C] the mediator constant from the point.
    [D] a circle
15. The locus of a point which is moving such that its distance from two fixed points is always equal is:
    [A] a perpendicular bisector *l*.
    [B] a line parallel to *l*.
    [C] the angle bisector from the two points.
    [D] a circle.
16. The locus of a point which is equidistant from two intersecting lines is:
    [A] the perpendicular bisector of the intersecting lines.
    [B] a line parallel to the intersecting lines.
    [C] a circle.
    [D] the angle bisector between the intersecting lines
17. The diagram among the following that shows the construction of a perpendicular bisector is:

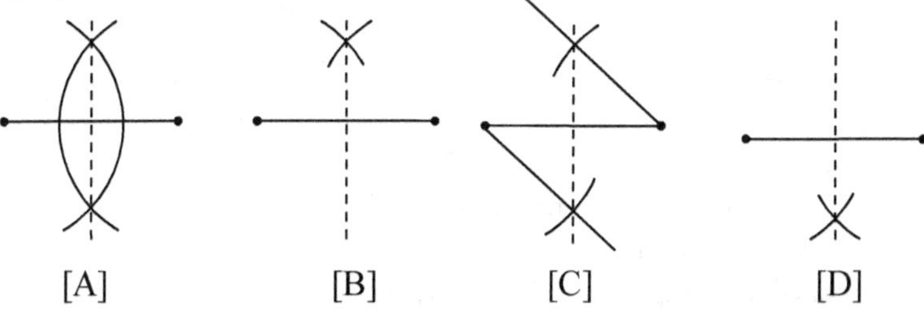

18. In the figure below, the correct construction of the bisector of the angle *BAC* is:

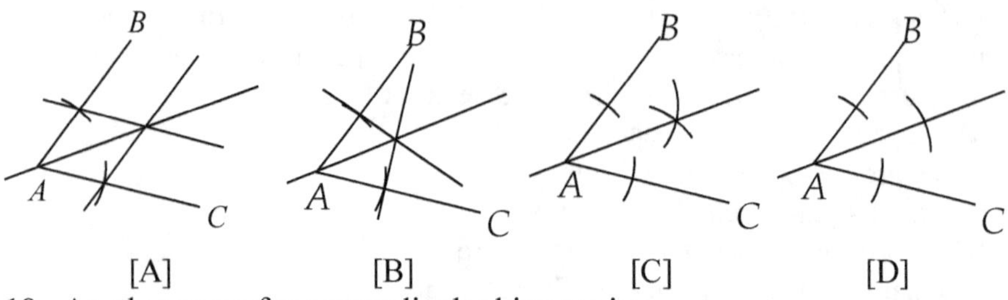

19. Another name for perpendicular bisector is:
    [A] Altitude [B] Midpoint [C] Angle bisector [D] Mediator
20. In figure (i) below, *AM* is said to be:
    [A] The median [B] The mediator
    [C] The altitude [D] The angle bisector

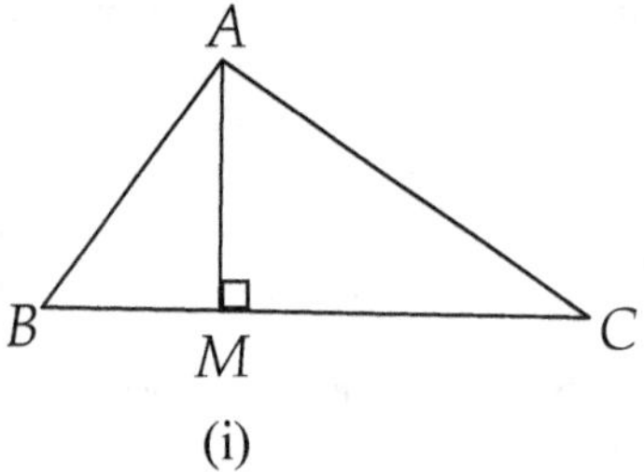

(i)

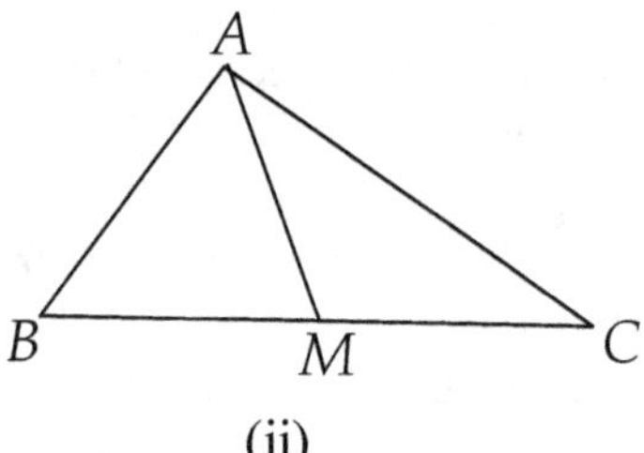

(ii)

21. In figure (ii) above, given that *M* is the midpoint of *BC*, *AM* is called:
    [A] The median [B] The mediator
    [C] The altitude [D] The angle bisector
22. In figure (i) below, given that M is the midpoint of *BC*, *NM* is called:
    [A] The median [B] The mediator
    [C] The altitude [D] The angle bisector

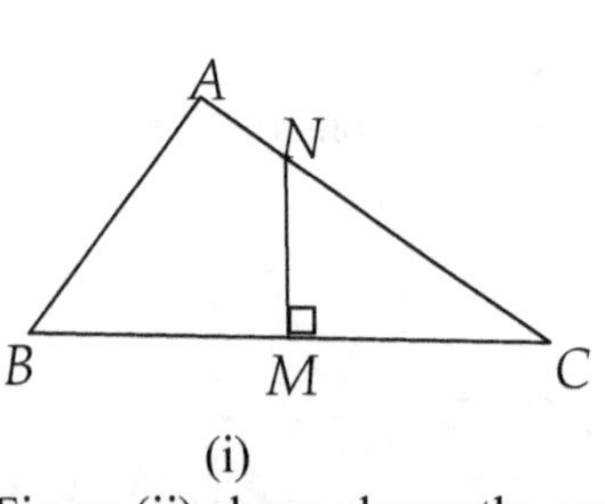

(i)

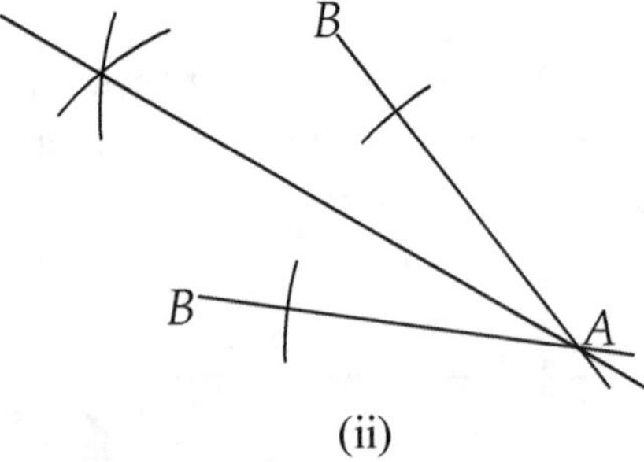

(ii)

23. Figure (ii) above shows the construction of:
    [A] A congruent segment [B] A congruent angle
    [C] A perpendicular bisector [D] An angle bisector
24. The figure below shows the construction of:
    [A] A mediator [B] A perpendicular to *AB*
    [C] An angle bisector [D] Intersecting lines

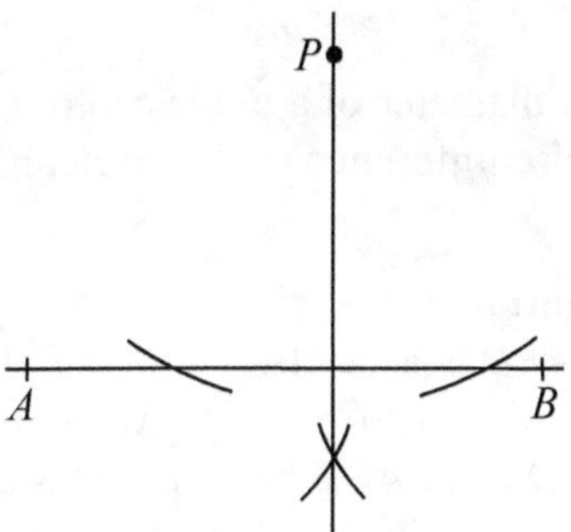

25. Among the following, the correct construction of a perpendicular bisector which is equal in length to the line segment $[XY]$ is:

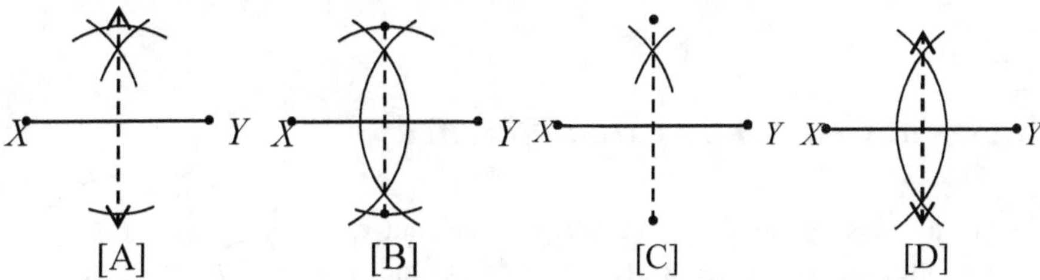

[A] [B] [C] [D]

26. Given that YN is the bisector of $\angle XYZ$ and $\lambda(\angle XYZ) = 88^\circ$ then, $\lambda(\angle XYN)$ is equal to:
    [A] 176° [B] 88° [C] 44° [D] 22°
27. Given that YN is the trisector of $\angle XYZ$ and $\lambda(\angle XYZ) = 72^\circ$ then, $\lambda(\angle XYN)$ is equal to:
    [A] 24° [B] 48° [C] 72° [D] 144°
28. The angle bisector of $\angle ABC$ is BD. If $\angle ABC = 18^\circ$, $\angle ABD$ is equal to:
    [A] 30° [B] 36° [C] 18° [D] 9°

5.7 CIRCLE GEOMETRY

(1) The perpendicular bisector of a chord passes through the centre (figure (i) below).

(2) Equal chords of a circle are equidistant from the centre (figure (ii) below).

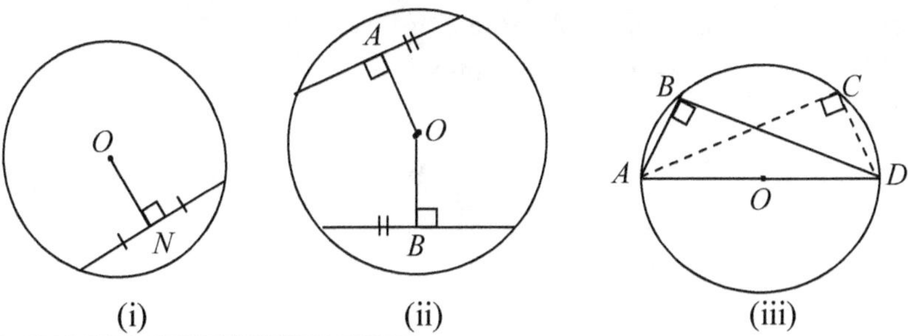

(i) (ii) (iii)

(3) The angle in a semi-circle is 90° (figure (iii) above)

Example

In the following figure, AOD is a diameter of a circle centre O and radius 10 cm; P is a variable point on the circumference of the circle. Find the length of AP when ΔAPD is isosceles.

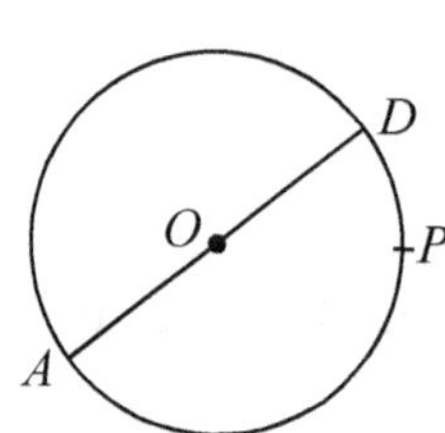

Solution

ΔAPD is isosceles $\Rightarrow AP = PD$

$\angle APD = 90^\circ$ [Angle in a semi-circle]

$\angle DAP = ADP = 45^\circ$ [APD is an isosceles Δ]

$$AP = AD\sin 45^\circ = 20\left(\frac{\sqrt{2}}{2}\right) = 10\sqrt{2} \text{ cm}$$

Angles at Centre and Circumference

(4) The angle subtended by an arc PQ at the centre of a circle is twice the angle subtended at the circumference by the same arc.

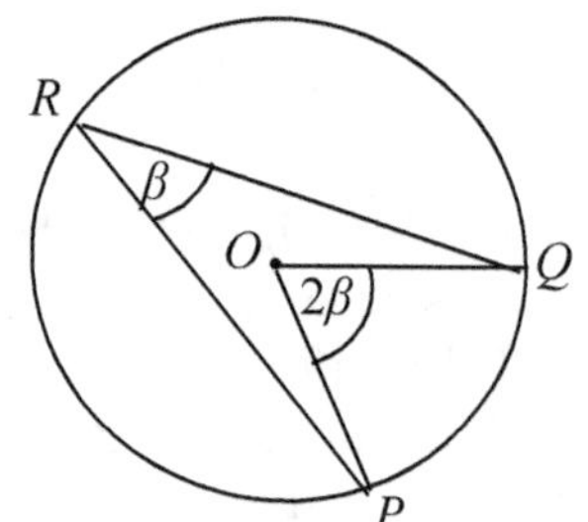

Example

In the figure on the right, A, B and C are points on the circle whose centre is O and $O\hat{C}A = 25^\circ$. Calculate

(i) Angle AOC (ii) Angle BAC

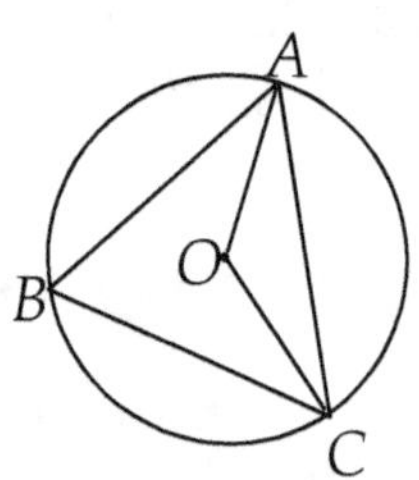

Solution

(i) $OA = OC$ [radii of same circle] $\Rightarrow \Delta AOC$ is isosceles

Hence, $\angle OAC = \angle OCA = 25^\circ$ [base $\angle$s of isosceles Δ]

$\angle OAC + \angle OCA + \angle AOC = 180^\circ$ [angles in a Δ]

$\angle AOC = 180^\circ - 2(25^\circ) = 130^\circ$

(ii) $\angle CBA = \frac{1}{2}\angle AOC$ [$\angle$at centre is twice $\angle$ at circumference]

$$\Rightarrow \angle CBA = \frac{1}{2}(130^\circ) = 65^\circ$$

Angles in the Same Segment

(5) The angles subtended in the same segment by the same arc are equal.
In figure (i) below the $\angle PSQ$ and $\angle PRQ$, are both angles subtended by the arc PQ, in the same segment. Hence, they are equal.

(6) The angles subtended by a chord in opposite segments are supplementary.
In figure (ii) below, AC subtends $\angle ABC$ and $\angle ADC$ in opposite segments so $\angle ABC$ and $\angle ADC$ are supplementary.

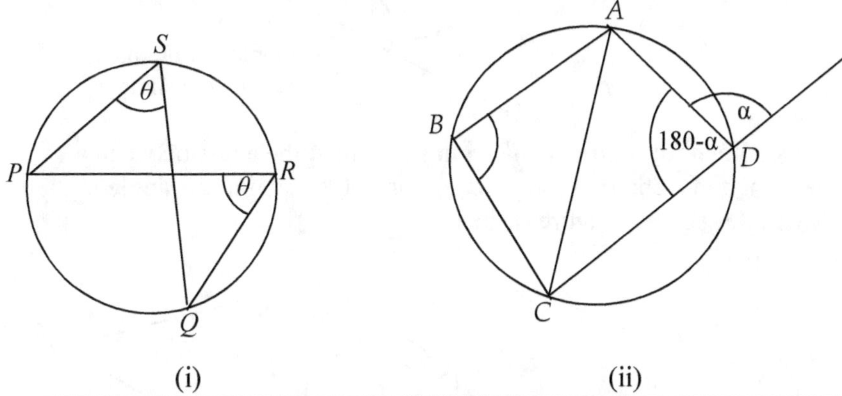

(7) Opposite angles of a cyclic quadrilateral are supplementary i.e. their sum is 180°. In figure (ii) above, $ABCD$ is a cyclic quadrilateral so $\angle ABC$ and $\angle ADC$ are supplementary.

(8) The exterior angle of a cyclic quadrilateral is equal to the opposite interior angle. In figure (ii) above, $\angle ABC = \alpha$ since $\angle ABC$ and $\angle ADC$ are supplementary and $\angle ADC$ and α are angles on a straight line.

To test whether or not a quadrilateral is a cyclic quadrilateral, it suffices to show that its opposite angles are supplementary as in theorem (7).

(9) A tangent drawn to a circle is perpendicular to the radius of the circle at the point of contact (figure (i) below).

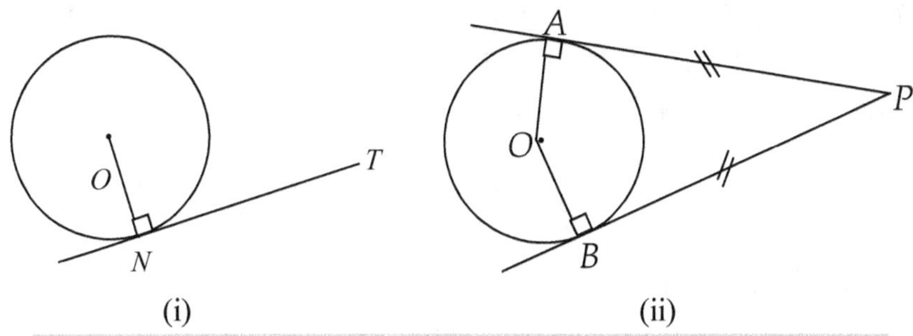

(10) Tangents to a circle from the same external point are equal in length (figure (ii) above).

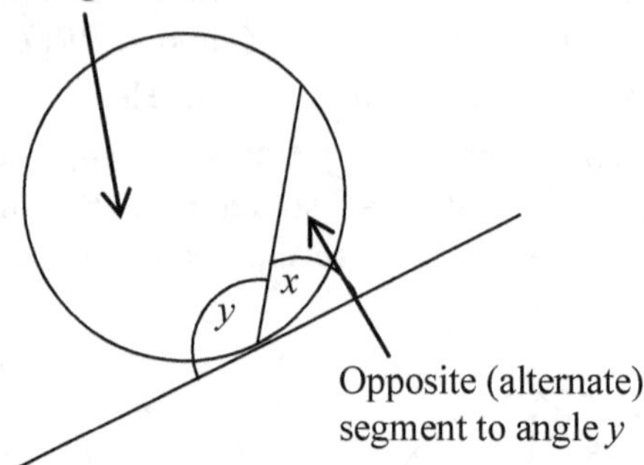

(11) The **alternate segment theorem** states that; the angle between a tangent and a chord at the point of contact is equal to the angle in the alternate segment (figure (i) below).

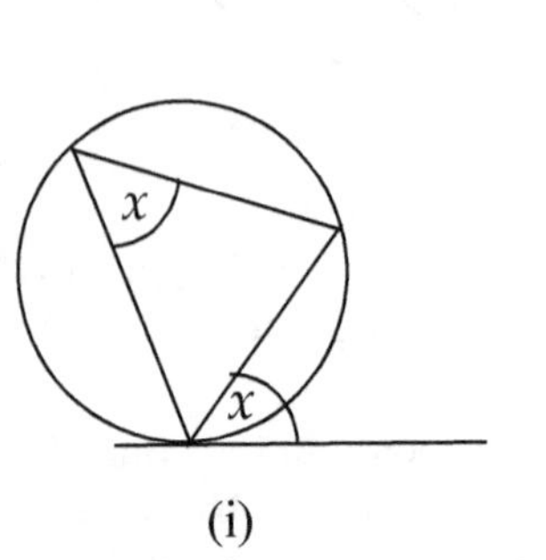

(i)

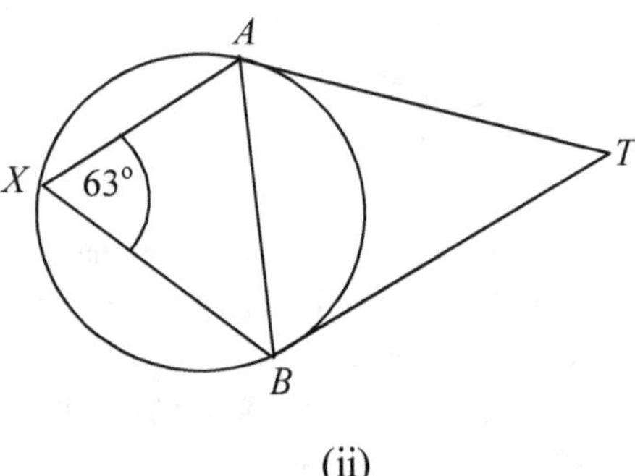

(ii)

Example

In figure (ii) above, TA and TB are tangents drawn to a circle from a point T. X is a point on the major arc of the circle. If $\angle AXB = 63^\circ$ calculate $\angle ATB$.

Solution

$AT = TB$ (tangents from same external point)

$\angle TAB = \angle TBA$ (ΔABT is isosceles)

Also $\angle TAB = \angle AXB = 63^\circ$ ($\angle$ in alt segment)

$\angle TAB + \angle TBA + \angle ATB = 180^\circ$ ($\angle$s in a Δ)

$\Rightarrow\ 63^\circ + 63^\circ + \angle ATB = 180^\circ$

$\Rightarrow \angle ATB = 180^\circ - 126^\circ = 54^\circ$

Intersecting chord theorem

In both cases, in figure (i) and (ii) below, $PA \cdot PB = PC \cdot PD$

If in (ii) PC is a tangent then, $PA \cdot PB = PC^2$

Note that in both cases all distances are reckoned from P.

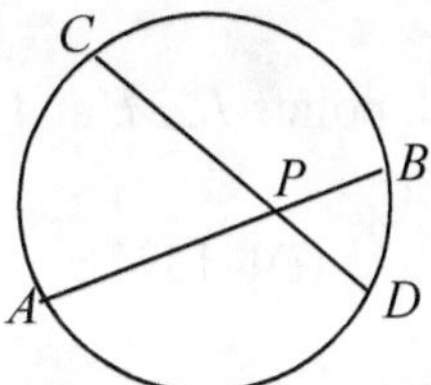

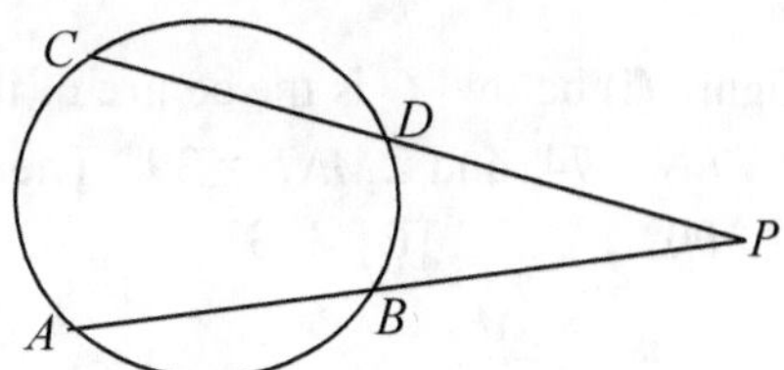

(i) : Internal intersection (ii) External intersection

Example

Find the value of x in the figure below.

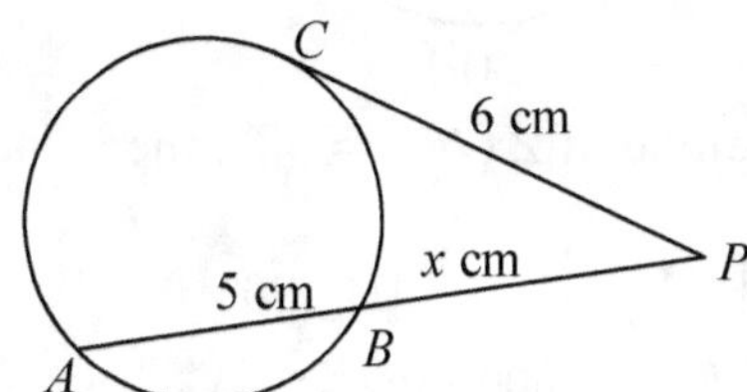

Solution

By the intersecting chord theorem

$x(x+5) = 6^2 \Rightarrow x^2 + 5x - 36 = 0$

$\Rightarrow (x+9)(x-4) = 0$

$\Rightarrow x = -9$ or $x = 4$.

Since $x \geq 0$, $x = 4$ cm.

Example

In the figure below, $AN = 7.2$ cm, $ND = 6$ cm, $BN = 9$ cm and $NC = x$ cm. Find x.

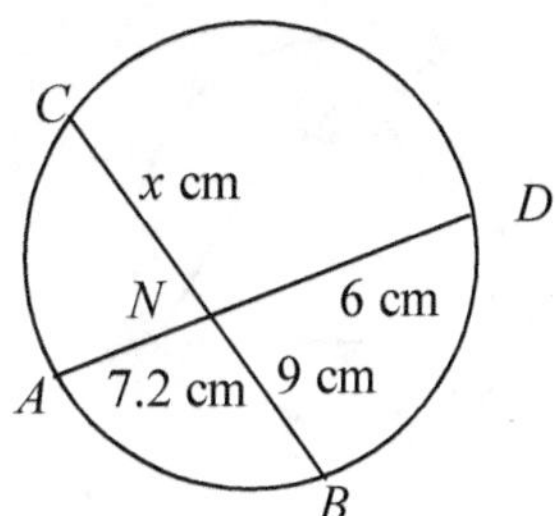

Solution

Using the intersecting chord theorem

$9x = 6(7.2) \Rightarrow x = \frac{6(7.2)}{9} = 4.8$ cm.

MULTIPLE CHOICE EXERCISE 5:7

1. In figure (i) below, O is the centre of the circle through points L, M, and N. If $\angle MLN = 74^\circ$ and $\angle MNL = 39^\circ$. The value of $\angle LON$ is:

   [A] 100° [B] 113° [C] 126° [D] 134°

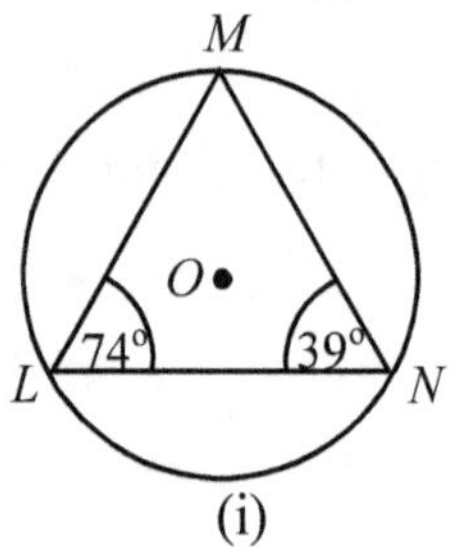

(i)

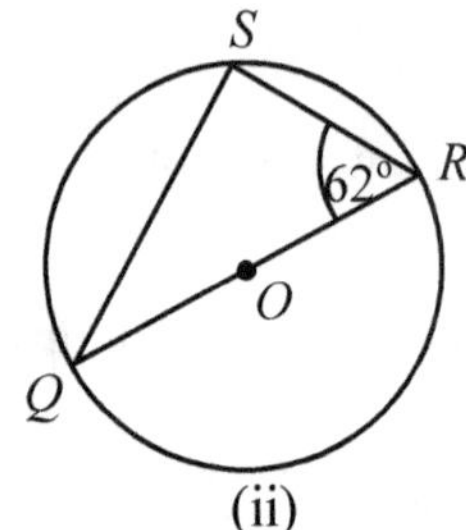

(ii)

2. In figure (ii) above, O is the centre of the circle. If $\angle QRS = 62^\circ$, the value of $\angle SQR$ is:

   [A] 14° [B] 28° [C] 31° [D] 90°

3. In figure (i) below, $PQRS$ is a circle. $\angle SPR = p^\circ$ and $\angle SQR = 2x^\circ$. The value of x in terms of p is:

   [A] $x = 2p$ [B] $x = p - 2$ [C] $x = p^2$ [D] $x = \frac{p}{2}$

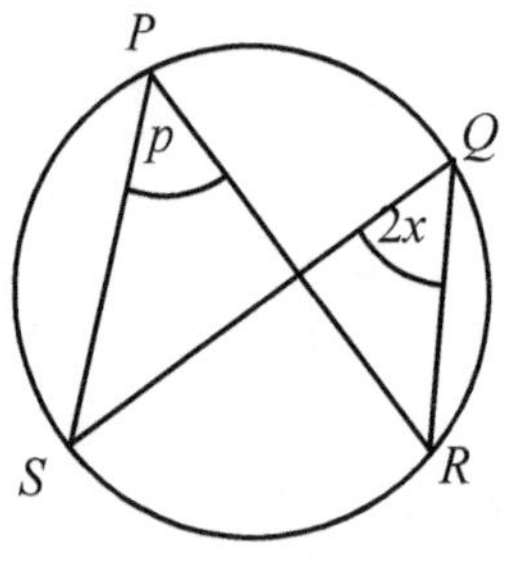

(i)

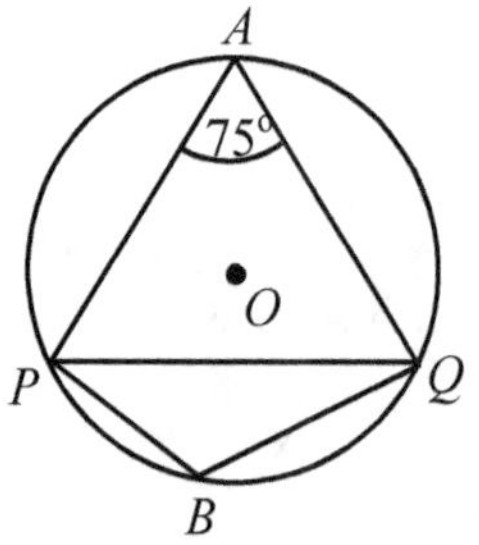

(ii)

4. In figure (ii) above, O is the centre of the circle. If $\angle PAQ = 75^\circ$, the value of $\angle PBQ$ is:

   [A] 51° [B] 75° [C] 105° [D] 150°

5. In figure (i) below, O is the centre of the circle QRT and PT is the tangent to the circle at T. The angle x is:

   [A] 40° [B] 35° [C] 25° [D] 20°

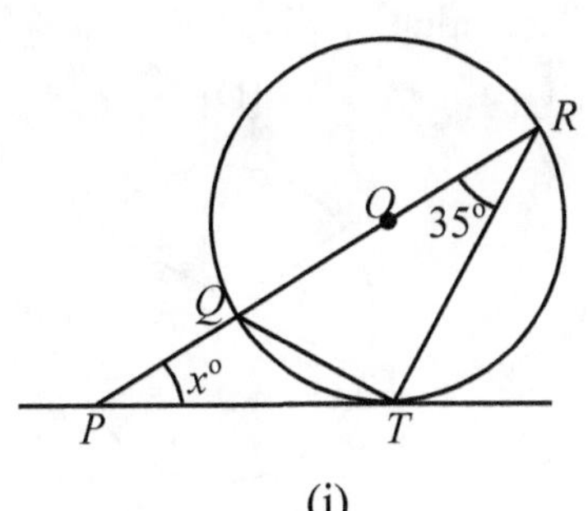

(i)

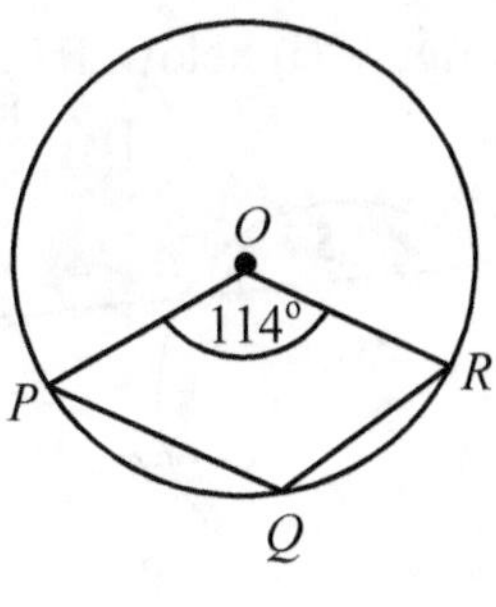

(ii)

6. In figure (ii) above, O is the centre of the circle. Given that $\angle POR = 114^{\circ}$ the value of $\angle PQR$ is:
   [A] 123° [B] 118.5° [C] 117° [D] 114°
7. In figure (i) below, O is the centre of the circle. If $\angle POQ = 39^{\circ}$ and $\angle PRQ = 5x^{\circ}$ the value of x is:
   [A] 4 [B] 8 [C] 16 [D] 20

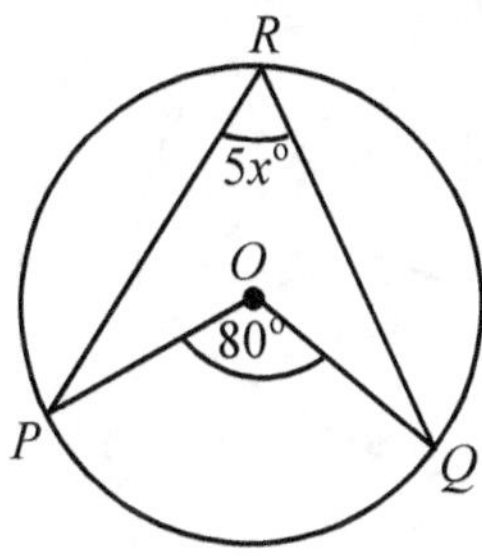

(i)

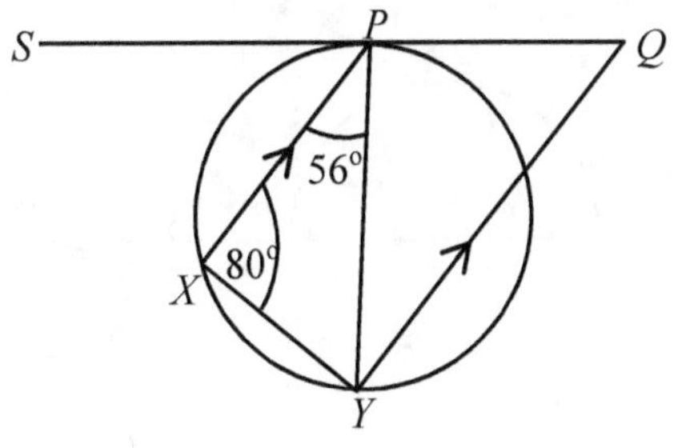

(ii)

8. In figure (ii) above, SQ is the tangent to the circle at P. $XP \| YQ$, $\angle XPY =$ 56° and $\angle PXY = 80°$. The value of angle PQY is:
   [A] 34° [B] 36° [C] 44° [D] 46°

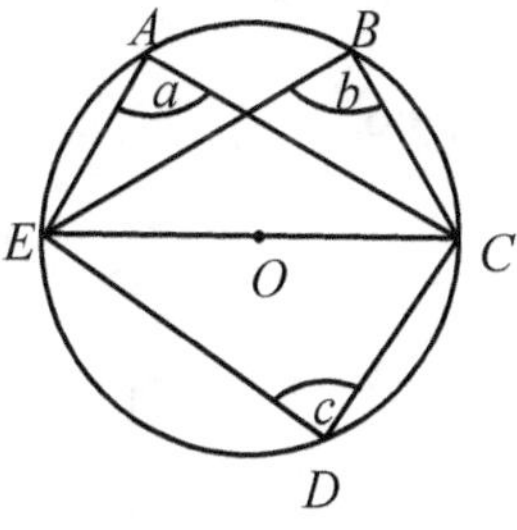

(i)

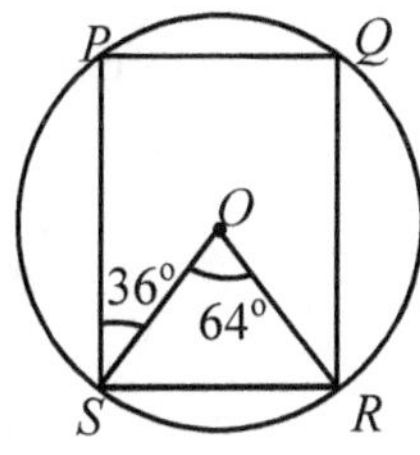

(ii)

9. In figure (i) above, O is the centre of the circle. It is true to say that:
   [A] $a = b$ [B] $b + c = 100$ [C] $a + b = c$ [D] $a = b$ and $b + c = 100$
10. In figure (ii) above, O is the centre of the circle, $\angle SOR = 64°$ and $\angle PSO = 36°$. The value of $\angle PQR$ is:
   [A] 100° [B] 96° [C] 94° [D] 86°

11. In figure (i) below, *ABCD* is a circle. The value of x is:

[A] $\frac{20}{9}$ [B] $\frac{36}{5}$ [C] 3 [D] $\frac{45}{4}$

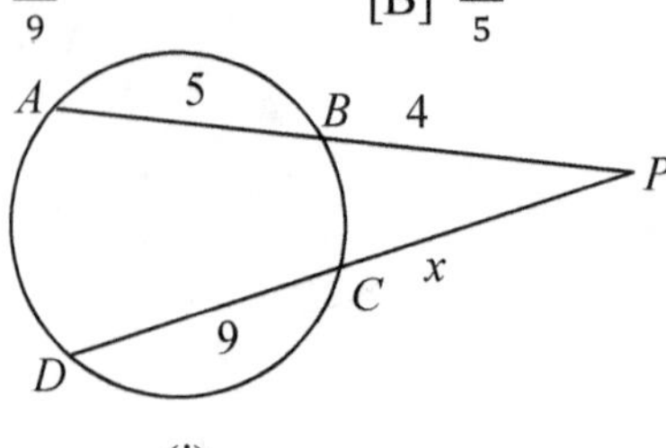

(i)

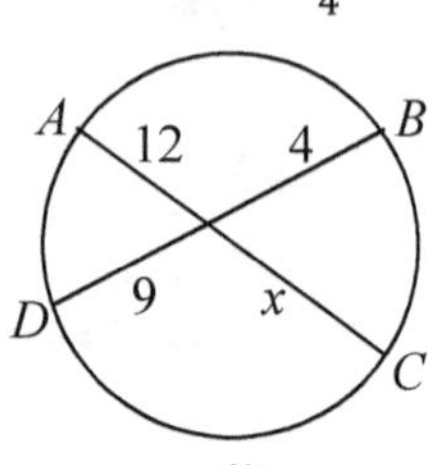

(ii)

12. In figure (ii) above, *ABCD* is a circle. The value of x is:

[A] $\frac{48}{9}$ [B] 3 [C] 36 [D] $\frac{16}{3}$

13. Given that in figure (i) below, $PQ = 6$ cm, $TR = 5$ cm and $RQ = 7$ cm. The radius of the circle with centre O is:

[A] 8 cm [B] 6 cm [C] 2 cm [D] 4 cm

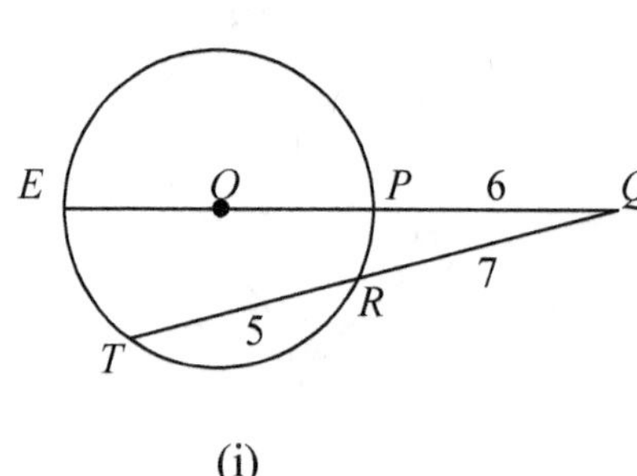

(i)

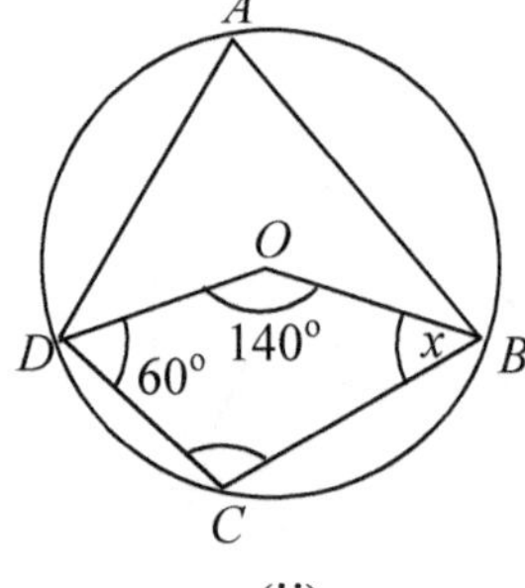

(ii)

14. O is the centre of the circle in figure (ii) above. The size of the angle marked x is:

[A] 120° [B] 40° [C] 50° [D] 70°

15. In figure (i) below, O is the centre of the circle. The value of $y - x$ is:

[A] 164° [B] 114° [C] 66° [D] 16°

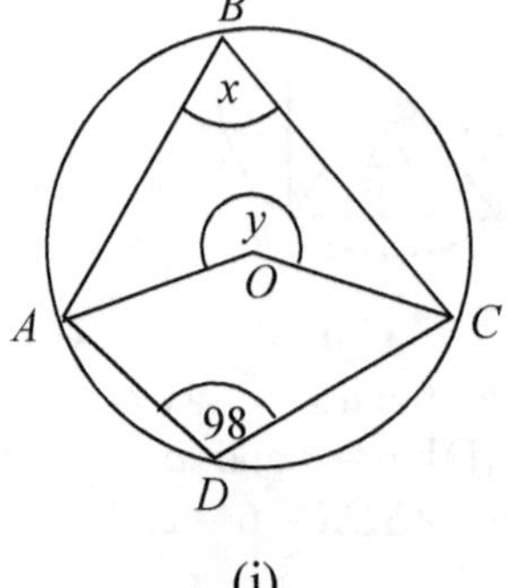

(i)

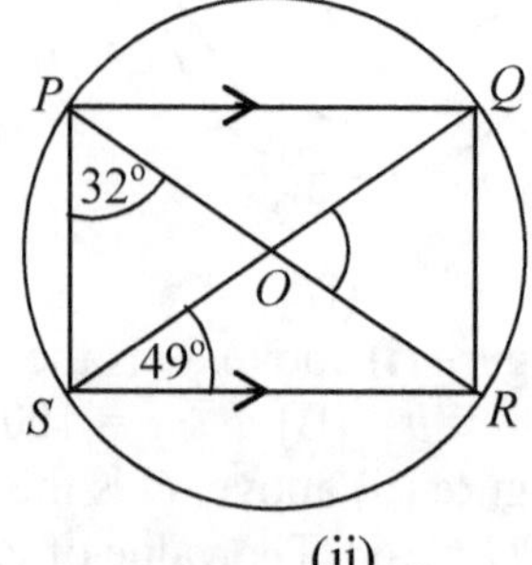

(ii)

16. In figure (ii) above, $\angle RPS = 32^\circ$ and $\angle QSR = 49^\circ$. The value of $\angle QOR$ is:
[A] 64° [B] 82° [C] 98° [D] 116°

17. In the figure below, O is the centre of the circle. $\angle BAO = 30^\circ$ and $\angle BCO = 20^\circ$. The value of reflex angle AOC is:
[A] 330° [B] 300° [C] 270° [D] 260°

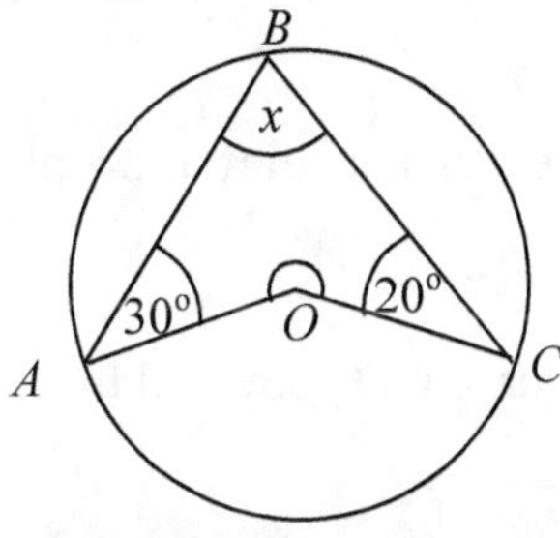

18. In figure (i) below, X, Y and Z are points on a circle centre O. WZ is a tangent to the circle at the point Z and $\angle XYZ = 22^\circ$. The value of θ is:
[A] 112° [B] 68° [C] 46° [D] 22°

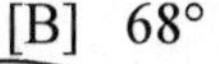

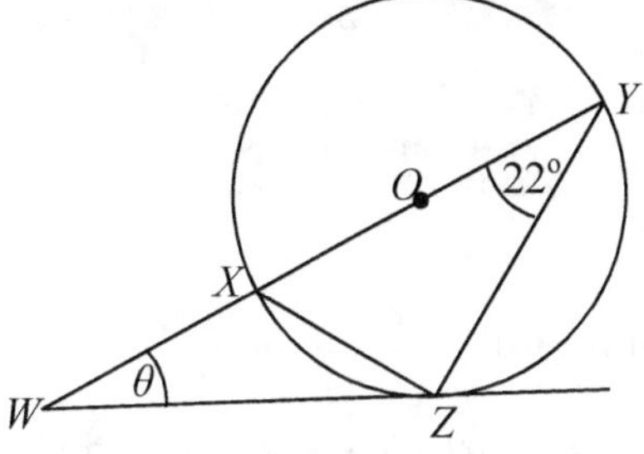

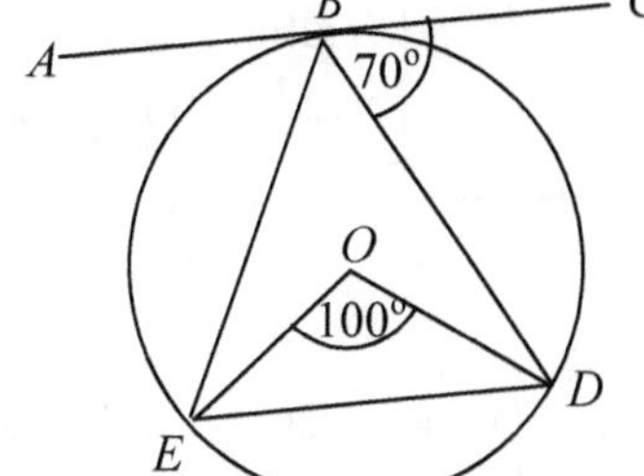

(i) (ii)

19. In figure (ii) above, O is the centre, $\angle DOE = 100^\circ$ and $\angle CBD = 70^\circ$. The value of $\angle BEO$ is:
[A] 20° [B] 30° [C] 40° [D] 60°

CHAPTER 6

MENSURATION

6.1 MENSURATION OF PLANE FIGURES

Perimeter

Perimeter is the length of the distance all round a plane figure.

Area

Area is the the number of square units covered by a plane figure.

The Rectangle

Length (l)

Width or breadth (w)

$$P = 2(l + w) \Rightarrow l = \frac{P}{2} - w \text{ and } w = \frac{P}{2} - l$$

$$A = lw \Rightarrow l = \frac{A}{w} \text{ and } w = \frac{A}{l}$$

Example

A rectangular floor has sides 4 m by 6 m. Calculate
(a)the perimeter
(b) the area of the floor

Solution

(a) $P = 2(l + w) = 2(4 + 6) = 20$ cm
(b) $A = lw = 4(6) = 24$ cm^2

The Square

l

l

$$P = 4l \Rightarrow l = \frac{P}{4}$$

$$A = l^2 \Rightarrow l = \sqrt{A}$$

Example

A square has side 11 cm. Find
(a) the perimeter (b) the area.

Solution

(a) $P = 4l = 4(11) = 44$ cm
(b) $A = l^2 = (11)^2 = 121$ cm^2

Area of a Parallelogram

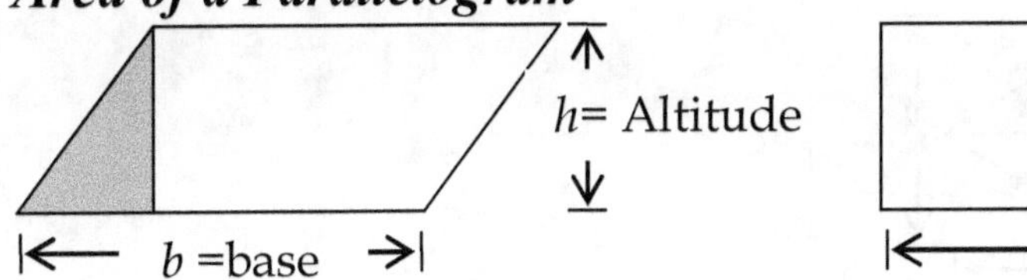

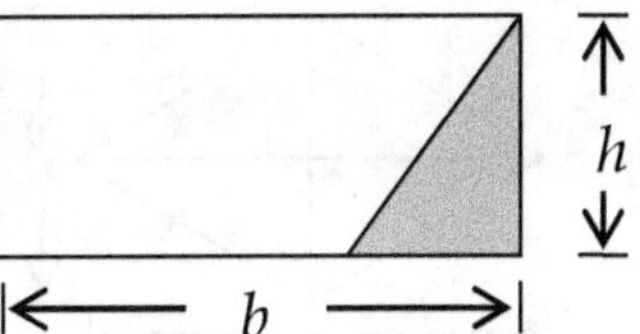

Area of parallelogram, $A = bh \Rightarrow b = \frac{A}{h}$ and $h = \frac{A}{b}$

Example
A parallelogram has an altitude of 13 cm and a base 15 cm, calculate its area.

Solution
$A = bh = (13 \text{ cm})(15 \text{ cm}) = 195 \text{ cm}^2$

The Triangle

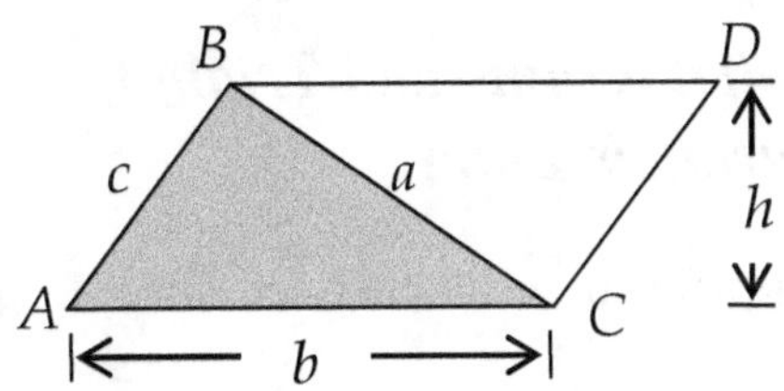

Area of triangle , $A = \frac{1}{2}bh$

Hero's Formula

$A = \sqrt{p(p-a)(p-b)(p-c)}$

Where $p = \frac{1}{2}(a+b+c)$ and the sides are of length a, b and c.

Example
A triangular lawn has a base of 4 m and a height of 8 m. What is its area?

Solution
$$A = \frac{1}{2}bh = \frac{1}{2}(4)(8) = 16 \text{ m}^2$$

Example
Calculate the area of a triangle whose sides are 5 cm, 7 cm and 10 cm.

Solution
$p = \frac{1}{2}(a+b+c) = \frac{1}{2}(5+7+10) = 11 \text{ cm}$

$A = \sqrt{p(p-a)(p-b)(p-c)}$

$\Rightarrow A = \sqrt{11(11-5)(11-7)(11-10)} = \sqrt{11(6)(4)(1)} = 16.25 \text{ cm}^2$

The Rhombus

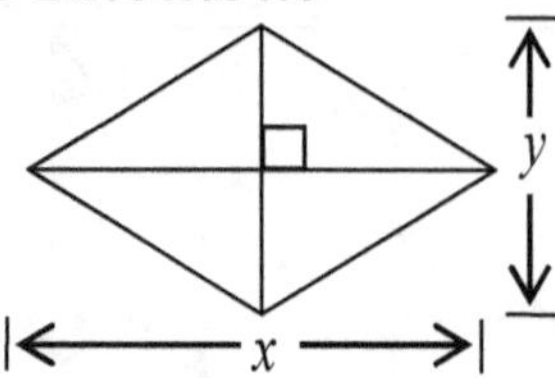

Area of rhombus = Half the product of the diagonals $\Rightarrow A = \frac{1}{2}xy$

Example

Find the area and perimeter of a rhombus whose diagonals are 12 cm by 16 cm.

Solution

$A = \frac{1}{2}xy = \frac{1}{2}(12)(16) = 96 \text{ cm}^2$

This rhombus is made up of 4 congruent triangles with arms 6 cm and 8 cm.

Hypotenuse of each triangle = $\sqrt{6^2 + 8^2} = 10 \text{ cm}$

$\Rightarrow \text{Perimeter} = 4 \times 10 \text{ cm} = 40 \text{ cm}$

Areas of Common Base Parallelograms and Triangles between Two Parallels Lines

(i) The areas of triangles, which are on the same base and between same parallels, are equal.

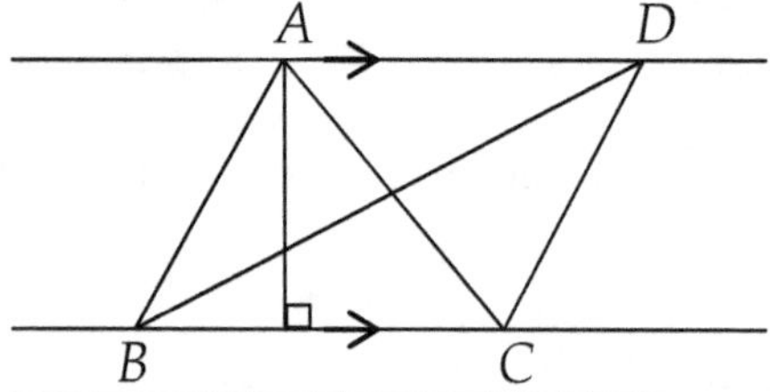

Area of ΔABC = Area of ΔBCD

(ii) The areas of parallelograms, which are on the same base and between same parallels, are equal.

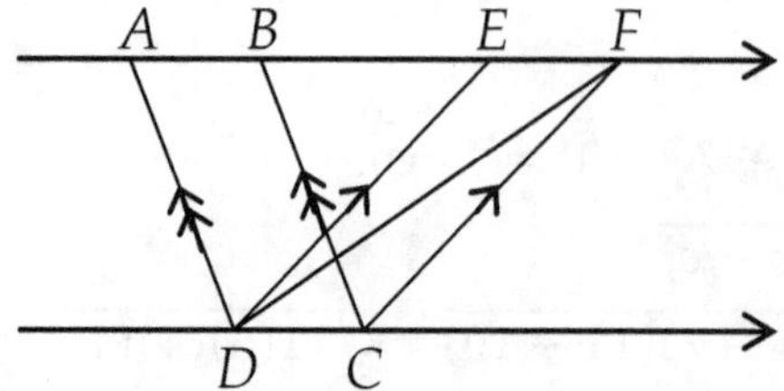

Area of parallelogram $ABCD$ = Area of parallelogram $CDEF$

(iii) The area of any parallelogram, which is on the same base and between same parallels as a triangle, is twice the area of the triangle.

Area of parallelogram $ABCD = 2 \times$ Area of triangle CDF

Example

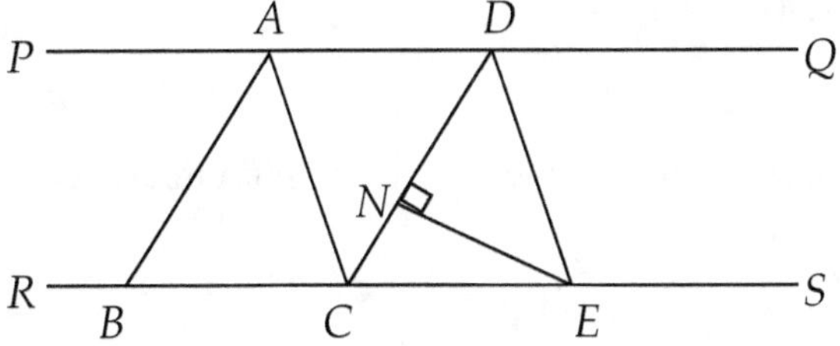

In the figure above, $PQ \| RS$, $AB \| DC$ and $AC \| DE$. Given that $CD = 8$ cm and $EN = 6$ cm, calculate the area of the parallelogram $ABCD$.

Solution

Area, A of $ABCD$ = Area of $ACED$ [common base between || lines]

$= 2 \times$ Area of triangle CDE

$$\Rightarrow A = 2 \times \frac{1}{2}(DC)(EN) = 2 \times \frac{1}{2}(8)(6) = 48 \text{ cm}^2$$

Trapezium

A trapezium is a quadrilateral with one pair of parallel sides.

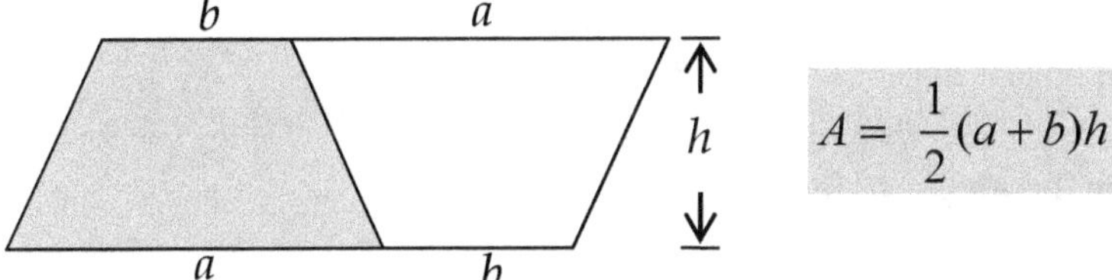

$$\text{Area of trapezium} = \frac{1}{2} \times \text{sum of parallel sides} \times \text{height}$$

Example

Find the area of a trapezium whose bases are 6 cm and 4 cm with the distance between parallel sides 5 cm.

Solution

$$A = \frac{1}{2}(a+b)h = \frac{1}{2}(6+4)5 = 25\text{cm}^2$$

Circumference and area of a Circle

$$C = 2\pi r \Rightarrow r = \frac{C}{2\pi} \text{ and } A = \pi r^2 \Rightarrow r = \sqrt{\frac{A}{\pi}}$$

Example

Taking π as $\frac{22}{7}$, find the radius and area of a circle whose circumference is 11 cm.

Solution

$$r = \frac{C}{2\pi} = \frac{11}{2(3.142)} = 1.75 \text{ cm}, \quad A = \pi r^2 = 3.142(1.75)^2 = 9.63 \text{ cm}^2$$

Arc length and Area of a Sector

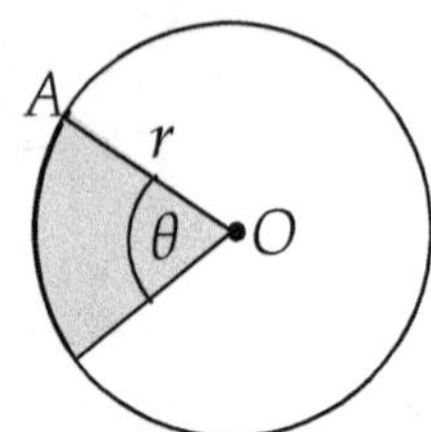

Arc length $l = \frac{\theta}{360} \times 2\pi r$.

Perimeter of sector $p = 2r + l$

Area of sector $S = \frac{\theta}{360} \times \pi r^2$

Example

An arc of a circle of radius 21 cm subtends an angle of 120° at the centre of a circle. Calculate the length of the arc. Take $\pi = \frac{22}{7}$.

Solution

$$l = \frac{\theta}{360} \times 2\pi r = \frac{120}{360} \times 2\left(\frac{22}{7}\right)(21) = 44 \text{ cm}.$$

Example

An arc of length 48 cm subtends an angle of 55° at the centre of a circle. Find the radius of the circle and the area of the sector. Take $\pi = \frac{22}{7}$.

Solution

$$l = \frac{\theta}{360} \times 2\pi r \Rightarrow r = \frac{360l}{2\pi\theta} = \frac{360(48)}{2\left(\frac{22}{7}\right)(55)} = 50 \text{ cm.}$$

$$S = \frac{\theta}{360} \times \pi r^2 = \frac{55}{360} \times \left(\frac{22}{7}\right)(50)^2 = 1200 \text{ cm}^2$$

Area of a Segment

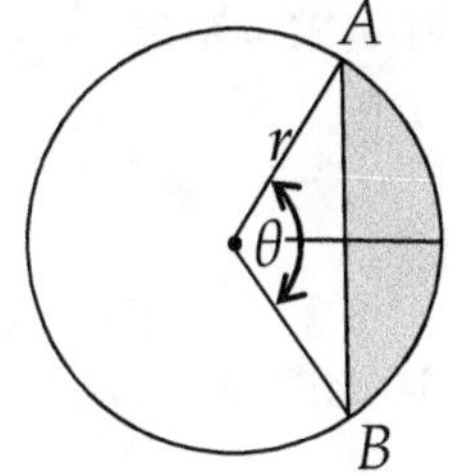

Area of Segment AB = Area of sector–Area of triangle

$$A_s = \frac{\theta}{360} \times \pi r^2 - \frac{1}{2} r^2 \sin\theta$$

Example

A chord subtends an angle of 135° at the centre of a circle of radius 21 cm. Calculate the area of the minor segment of the circle to the nearest square centimetre. $\left(\text{Take } \pi = \frac{22}{7}\right)$

Solution

$$A_s = \frac{\theta}{360} \times \pi r^2 - \frac{1}{2} r^2 \sin\theta$$

$$= \frac{135}{360} \times \frac{22}{7}(21)^2 - \frac{1}{2}(21)^2(\sin 135) = 364 \text{ cm}^2.$$

Composite Plane Figures

We can break composite figures into recognizable figures.

Example

The diagonals of a kite are of length 60 cm and 36 cm. Calculate its area.

Solution

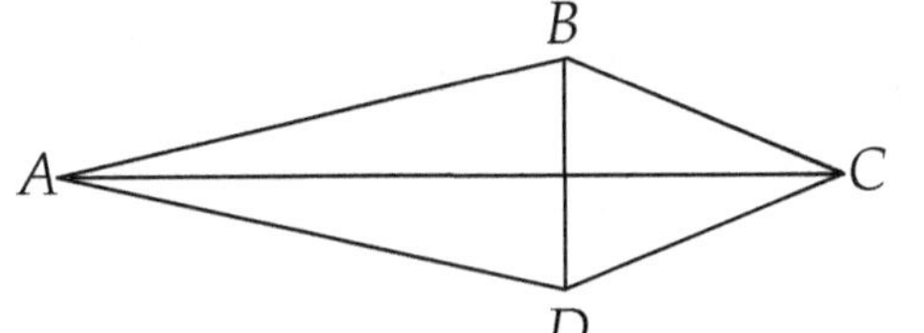

$$\text{Area of kite} = \frac{1}{2}(AC)(BD)$$
$$= \frac{1}{2}(60)(36)$$
$$= 1080 \text{ cm}^2$$

MULTIPLE CHOICE EXERCISE 6:1

In this exercise where necessary, take $\pi = \frac{22}{7}$.

1. If the perimeter of a square is 36 cm then the area of the square in square centimetres is:
   [A] 81 cm^2 [B] 36 cm^2 [C] 9 cm^2 [D] 36^2 cm^2
2. The area of a square is x^2 cm^2. Its perimeter is:
   [A] x^4 cm [B] $4x^2$ cm [C] $4x$ cm [D] $2x$ cm
3. The perimeter of the following rectangle is 36 m, the area of the rectangle is:
   [A] 64 m^2 [B] 65 m^2 [C] 84 m^2 [D] 124 m^2

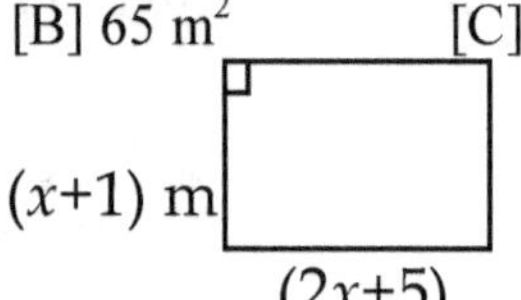

4. One side of a rectangular field is 8 m and the diagonal is 10 m. The area of the rectangle is:
   [A] 80 m^2 [B] 36 m^2 [C] 40 m^2 [D] 48 m^2
5. The length of a rectangle is twice the width. If the length is 8 cm, the perimeter in centimetres is:
   [A] 24 cm [B] 32 cm [C] 48 cm [D] 12 cm
6. The area of a rectangle with width 4 m and diagonal 8 m is:
   [A] $8\sqrt{3}$ m^2 [B] $12\sqrt{3}$ m^2 [C] $16\sqrt{3}$ m^2 [D] 48 m^2
7. A rectangular photograph 15 cm by 9 cm is pasted on a rectangular card. If a margin of 2.5 cm is left round the photograph, the perimeter of the card is:
   [A] 58 cm [B] 68 cm [C] 98 cm [D] 228 cm
8. In a trapezium, the lengths of the parallel sides are 4 cm and 6 cm and the perpendicular distance between these sides is 3 cm. The area of the trapezium is:
   [A] 36 cm^2 [B] 18 cm^2 [C] 30 cm^2 [D] 15 cm^2
9. The area in square units of the trapezium (a) below is:
   [A] 24 un^2 [B] 40 un^2 [C] 32 un^2 [D] 30 un^2

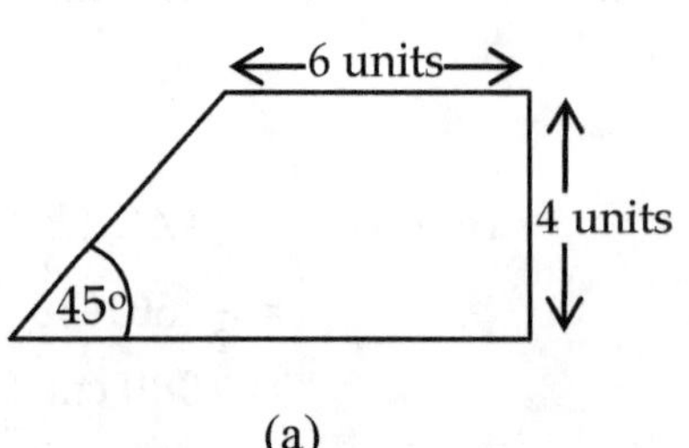

(a)

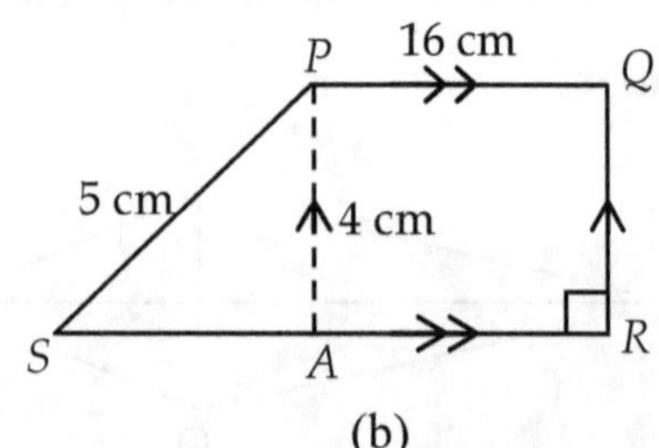

(b)

10. The perimeter of the trapezium $PQRS$ trapezium (b) above is:
[A] 24 cm [B] 44 cm [C] 36 cm [D] 70 cm
11. The area of the following parallelogram is:
[A] 2736 cm^2 [B] 936 cm^2 [C] 1368 cm^2 [D] 1872 cm^2

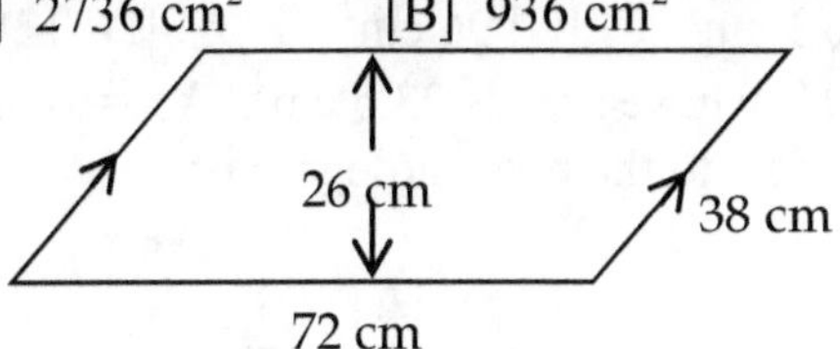

12. The lengths of the parallel sides of a trapezium are 5 cm and 7 cm. If its area is 120 cm^2, the perpendicular distance between the parallel sides is:
[A] 5.0 cm [B] 6.9 cm [C] 20.0 cm [D] 10.0 cm
13. The following is a trapezium $PQRS$ in which $|PS| = 9$ cm, $|QR| = 15$ cm, $|PR| = 2\sqrt{3}$ cm, $\angle PQR = 90°$ and $\angle QRS = 30°$. The area of the trapezium is:
[A] $24\sqrt{3}$ cm^2 [B] $36\sqrt{3}$ cm^2 [C] $42\sqrt{3}$ cm^2 [D] $72\sqrt{3}$ cm^2

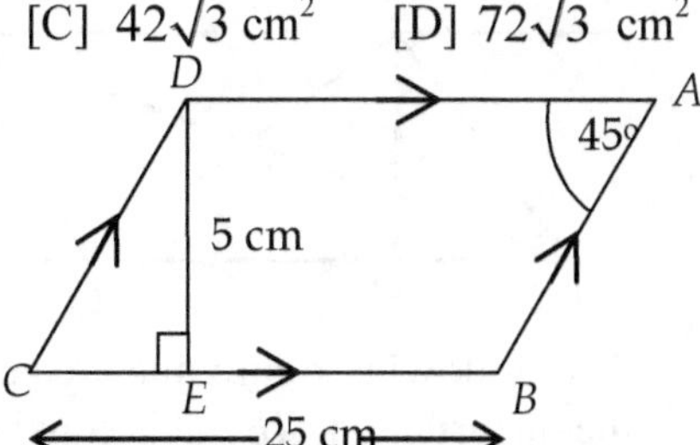

14. In the above triangle $ABCD$, DE is perpendicular to BC, $DE = 5$ cm, $BC = 25$ cm and angle $ADB = 45°$. The size of angle ABC is:
[A] 45° [B] 90° [C] 135° [D] 145°
15. In the above triangle $ABCD$, DE is perpendicular to BC, $DE = 5$ cm, $BC = 25$ cm and angle $ADB = 45°$. The length of EC is:
[A] 25 cm [B] 20 cm [C] 15 cm [D] 5 cm
16. In the above triangle $ABCD$, DE is perpendicular to BC, $DE = 5$ cm, $BC = 25$ cm and angle $ADB = 45°$. The area of $ABCD$ is:
[A] 625 cm^2 [B] 250 cm^2 [C] 125cm^2 [D] $62\frac{1}{2}$ cm^2
17. The area of the following triangle PQR is:

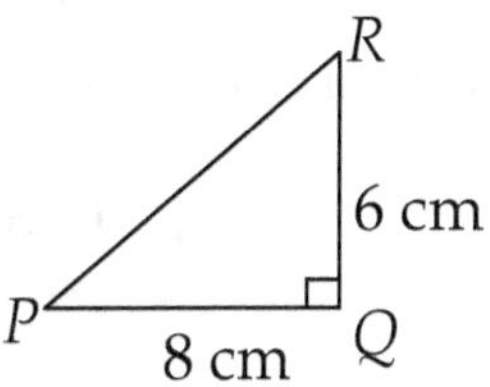

[A] 24 cm^2 [B] 12 cm^2
[C] 10 cm^2 [D] 48 cm^2

18. The sides of a triangle are 5 cm, 4 cm and 3 cm long. The area is:
    [A] 30 cm^2 [B] 15 cm^2 [C] 12 cm^2 [D] 6 cm^2
19. The area of an equilateral triangle of side 16 cm is:
    [A] $64\sqrt{3}$ cm^2 [B] $32\sqrt{3}$ cm^2 [C] 96 cm^2 [D] 128 cm^2
20. The following is a triangle *XYZ* whose area is 23.5 cm^2. $XY = 10$ cm and $YZ = 8$ cm. The value of θ, correct to the nearest degree is:

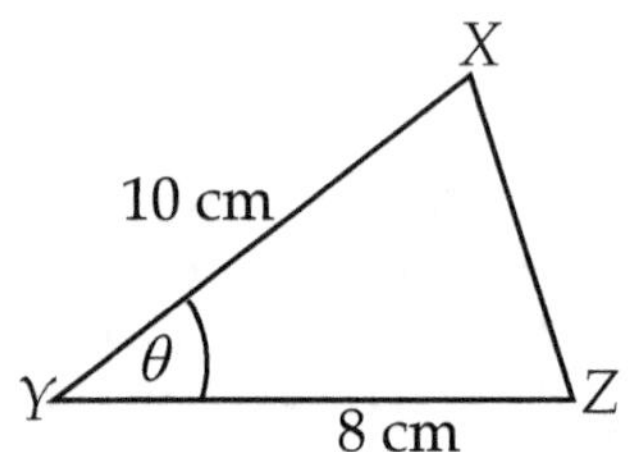

[A] 34°
[B] 35°
[C] 36°
[D] 37°

21. In following figure, $PS \| RQ$, $|RQ| = 6.4$ cm and perpendicular $PH = 3.2$ cm. The area of triangle *SQR* is:

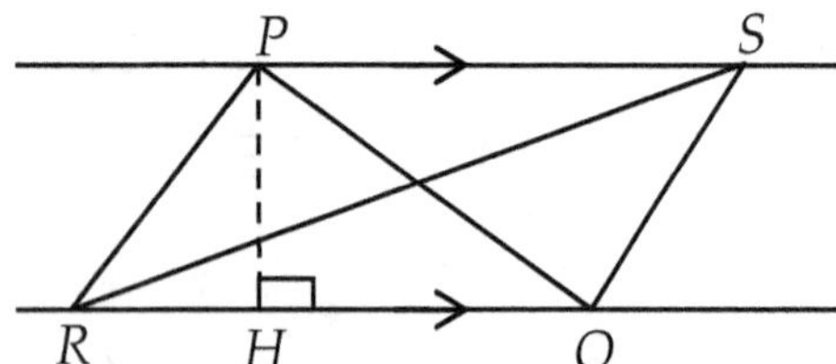

[A] 10.24 cm^2
[B] 9.60 cm^2
[C] 5.12 cm^2
[D] 20.48 cm^2

22. If in following figure, the area of the triangle $DCF = 24$ cm^2, the area of the quadrilateral *ABCD* is equal to:

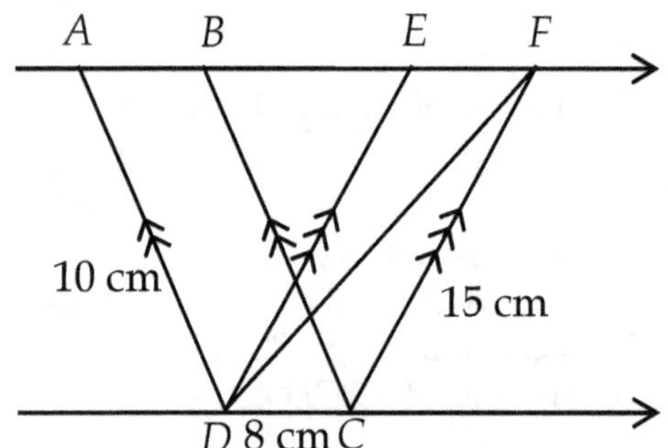

[A] 24 cm^2
[B] 48 cm^2
[C] 80 cm^2
[D] 96 cm^2

23. The diagonals *AC* and *BD* of a rhombus *ABCD* are 16 cm and 12 cm respectively. The area of the rhombus must be:
    [A] 36 cm^2 [B] 48 cm^2 [C] 60 cm^2 [D] 96 cm^2
24. The area of a rhombus is 24 cm^2 and one of its diagonals is 8 cm. The side of the rhombus is:
    [A] 4.3 cm [B] 5 cm [C] 6 cm [D 10 cm
25. If *O* is the centre of the following circle, the area of the shaded part is:

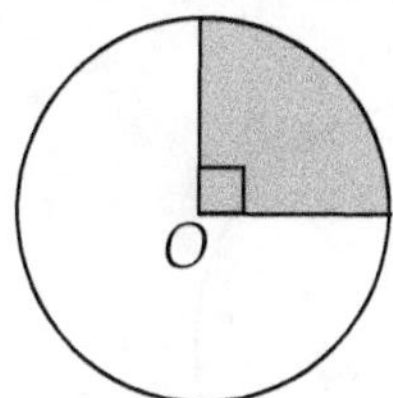

[A] $\frac{3\pi r^2}{8}$ [B] $\frac{\pi r^2}{2}$

[C] $\frac{\pi r^2}{4}$ [D] $\frac{\pi r}{2}$

26. The area of a circular field is 154 m^2. The perimeter of the field is:
[A] 44 m [B] 49 m [C] 88 m [D] 176 m
27. The area of a circle is 38.5 cm^2, its diameter is?
[A] 22 m [B] 14 m [C] 7 m [D] 6 m
28. The diameter of a circular field whose area is 616 cm^2 is:
[A] 98.00 m [B] 28.00 m [C] 49.00 m [D] 24.82 m
29. In the following diagram, *PXR* and *PYO* are two semi-circles with diameters 14 cm and 7 cm respectively. The area of the enclosed region *PXROY* correct to the nearest whole number is:
[A] 96 cm^2 [B] 116 cm^2 [C] 154 cm^2 [D] 192 cm^2
30. In the following diagram, *PXR* and *PYO* are two semi-circles with diameters 14 cm and 7 cm respectively. The perimeter of the region is:

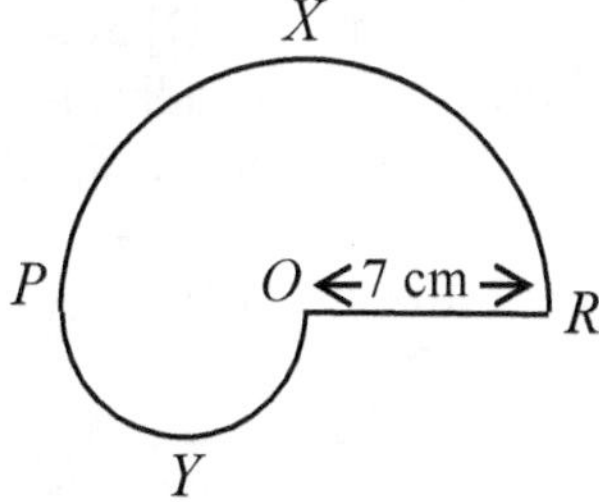

[A] 2 cm [B] 33 cm
[C] 40 cm [D] 66 cm

31. An arc of a circle of radius 7 cm is 14 cm long. The angle the arc subtends at the centre of the circle is:
[A] 44° [B] 51.43° [C] 98° [D] 114.55°
32. Correct to three significant figures the length of an arc which subtends an angle of 70° at the centre of a circle of radius 4 cm is:
[A] 2.44 cm [B] 4.89 cm [C] 9.78 cm [D] 25.1 cm
33. An arc of length 22 cm subtends an angle θ at the centre of a circle. If the radius of the circle is 15 cm the value of θ will be:
[A] 84° [B] 70° [C] 96° [D] 156°
34. *O* is the centre of circle (a) below with radius 10 cm and *ABC* = 30°. The length of the arc *AC*, correct to one decimal place is:
[A] 5.2 cm [B] 13.2 cm [C] 10.5 cm [D] 20.6 cm

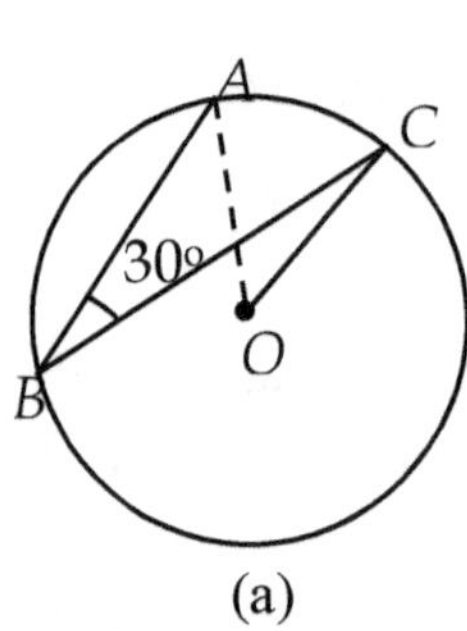

(a)

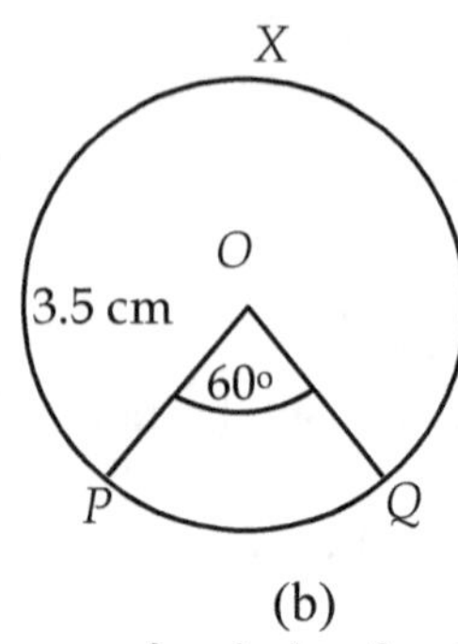

(b)

35. To 2 significant figures, the length of the arc of a circle of radius 3.5 cm which subtends an angle of 75° at the centre of the circle is:
[A] 2.3 cm [B] 4.6 cm [C] 8 cm [D] 16 cm

36. The length of the major arc *PXQ* in (b) above is:
[A] $18\frac{1}{3}$ cm [B] 11 cm [C] $9\frac{1}{6}$ cm [D] $7\frac{1}{3}$ cm

37. An arc of length 44 cm subtends an angle of 200° at the centre of a circle. The radius of the circle is:
[A] 3.9 cm [B] 12.6 cm [C] 25.2 cm [D] 38.4 cm

38. In figure (a) below, *O* is the centre of the circle *PRQ*. The radius is 3.5 cm and $\angle POQ = 50^\circ$. The length of the arc *PQ*, correct to one decimal place is:
[A] 157.1 cm [B] 37.7 cm [C] 11.0 cm [D] 3.1 cm

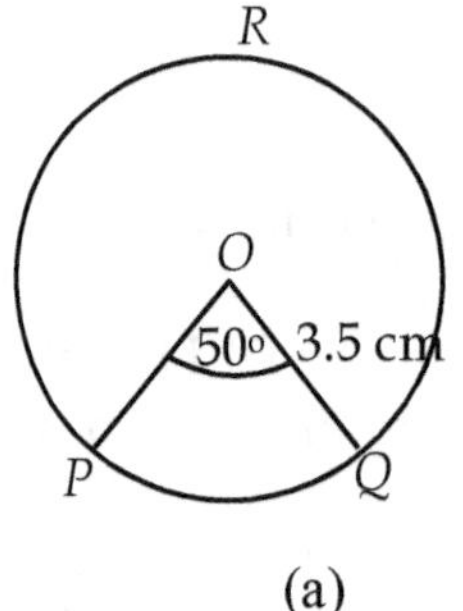

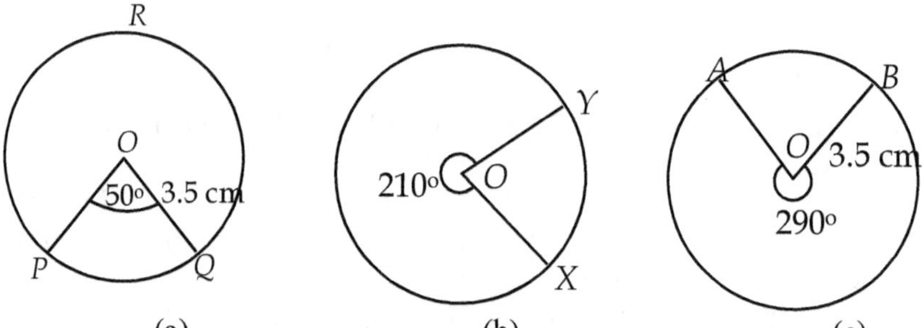

(a)

(b)

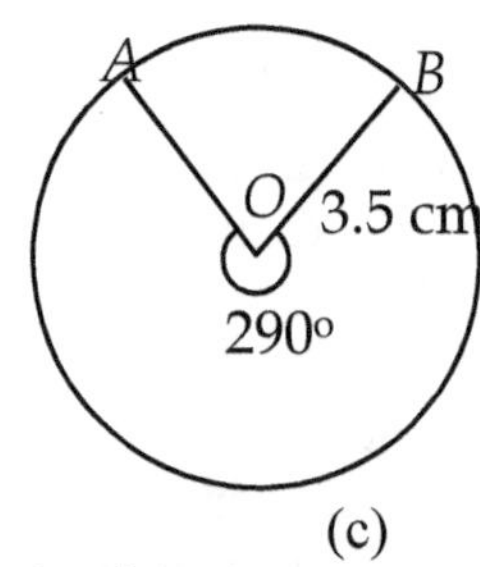

(c)

39. An arc of a circle 50 cm long subtends an angle of 75° at the centre of the circle. Correct to 3 significant figures the radius of the circle is:
[A] 8.74 cm [B] 38.2 cm [C] 61.2 cm [D] 76.4 cm

40. In figure (b) above, *O* is the centre of the circle. Reflex angle $XOY = 210^\circ$ and the length of the minor arc is 5.5 cm. The length of the major arc correct to the nearest metre is:
[A] 8 cm [B] 9 cm [C] 10 cm [D] 13 cm

41. In figure (c) above, *AOB* is a sector of the circle with centre *O* and radius 3.5 cm. If the reflex $\angle AOB = 290^\circ$ the length of the minor arc *AB* is:
[A] 17.72 cm [B] 14.97 cm [C] 4.28 cm [D] 2.14 cm

42. A rope of length 18 m is used to form a sector of circle of radius 3.5 m, on a school playing field. The size of the angle of the sector correct to the nearest degree is:
[A] 40° [B] 90° [C] 270° [D] 180°
43. The angle of a sector of a circle of radius 10.5 cm is 120°. The perimeter of the sector is:
[A] 22 cm [B] 33.5 cm [C] 43 cm [D] 66 cm
44. The angle of a sector of a circle is 108°. If the radius of the circle is $3\frac{1}{2}$ cm, the perimeter of the sector is:
[A] $6\frac{4}{5}$ cm [B] $7\frac{1}{10}$ cm [C] $10\frac{2}{3}$ cm [D] $13\frac{3}{5}$ cm
45. The angle of a sector of a circle of radius 35 cm is 28°. The perimeter of the sector is:
[A] $\frac{154}{9}$ cm [B] $\frac{784}{9}$ cm [C] $\frac{469}{9}$ cm [D] $\frac{286}{9}$ cm
46. A sector of a circle of radius 14 cm subtends an angle of 135° at the centre of the circle. The perimeter of the sector is:
[A] 47 cm [B] 61 cm [C] 88 cm [D] 231 cm
47. The angle of a sector of a circle of diameter 8 cm is 135°. The area of the sector is:
[A] $9\frac{3}{7}$ cm^2 [B] $12\frac{4}{7}$ cm^2 [C] $18\frac{6}{7}$ cm^2 [D] $25\frac{1}{7}$ cm^2
48. A sector of a circle of radius 7 cm has an area of 44 cm^2. The angle of the sector correct to the nearest degree is:
[A] 103° [B] 26° [C] 6° [D] 206°
49. A sector of a circle of radius 9 cm subtends an angle of 120° at the centre of the circle. The area of the sector to the nearest cm^2 is:
[A] 75 cm^2 [B] 84 cm^2 [C] 85 cm^2 [D] 86 cm^2
50. A circle has radius x cm. The area of a sector of the circle with angle 135°, in terms of x and π is:
[A] $\frac{\pi x^2}{8}$ [B] $\frac{3\pi x^2}{8}$ [C] $\frac{5\pi x^2}{8}$ [D] $\frac{\pi x}{8}$
51. The area of the minor sector *POQ* in (a) below is:
[A] $148\frac{1}{2}$ cm^2 [B] $32\frac{1}{12}$ cm^2 [C] $6\frac{5}{12}$ cm^2 [D] $1\frac{5}{6}$ cm^2

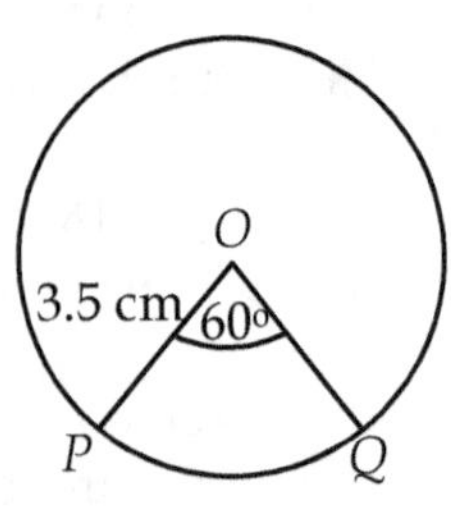

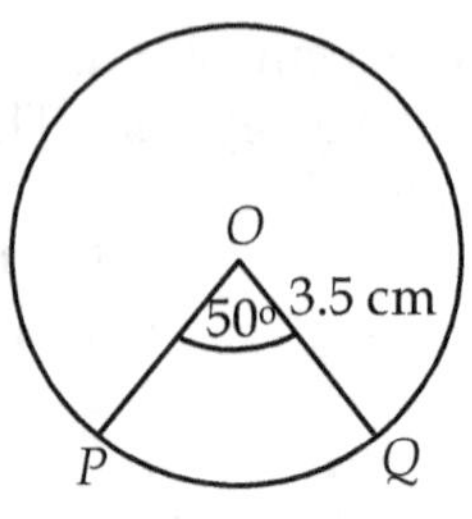

(a) (b)

52. The length of an arc of a circle of radius 5 cm is 4 cm. The area of the sector is:
[A] 2 cm^2 [B] 8 cm^2 [C] 10 cm^2 [D] 20 cm^2
53. Correct to three significant figures, the area of the minor sector, *OPQ* in (b) above is approximately equal to:
[A] 3.41 cm^2 [B] 157 cm^2 [C] 10.9 cm^2 [D] 5.35 cm^2
54. The angle subtended by a chord at the centre of a circle of radius 6 cm is 120°. The length of the chord should be:
[A] 6 cm [B] 12 cm [C] $6\sqrt{3}$ cm [D] $2\sqrt{3}$ cm
55. Given that, a chord of a circle 8 cm long subtends an angle of 120° at the centre. The radius of the circle is:
[A] 6.93 cm [B] 5.00 cm [C] 4.62 cm [D] 3.82 cm
56. In figure (a) below, *PQR* is a circle centre *O*. $|OQ| = |OR| = 7$ cm. The area of the shaded portion is:
[A] 11 cm^2 [B] 22 cm^2 [C] 38.5 cm^2 [D] 77 cm^2

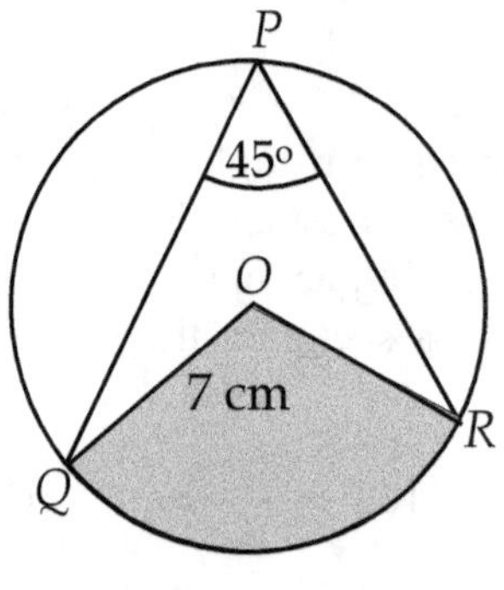

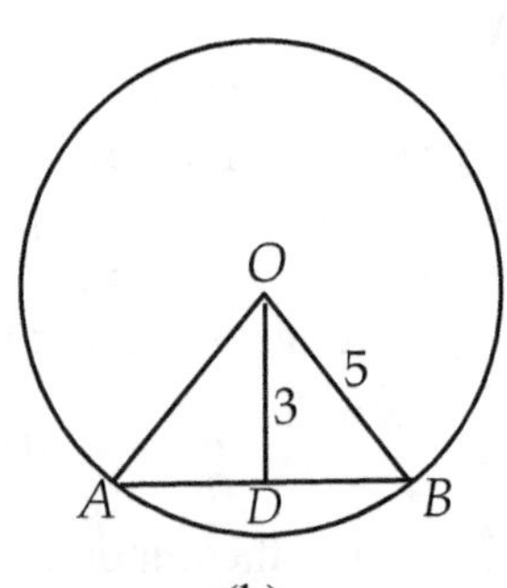

(a) (b)

57. In figure (b) above, *O* is the centre of the circle. Given that *OA* = 5 cm, *OD* = 3 cm and $\angle AOD = \angle BOD$, the length of the chord *AB* is:
[A] 8 cm [B] 5 cm [C] 3 cm [D] 15 cm
58. Figure (i) below, shows the shaded segment of a circle of radius 7 cm. The area of the triangle *OXY* is cm^2. The area of the segment is:

[A] $\frac{5}{12}$ cm [B] $\frac{7}{12}$ cm [C] $1\frac{1}{6}$ cm [D] $2\frac{1}{3}$ cm

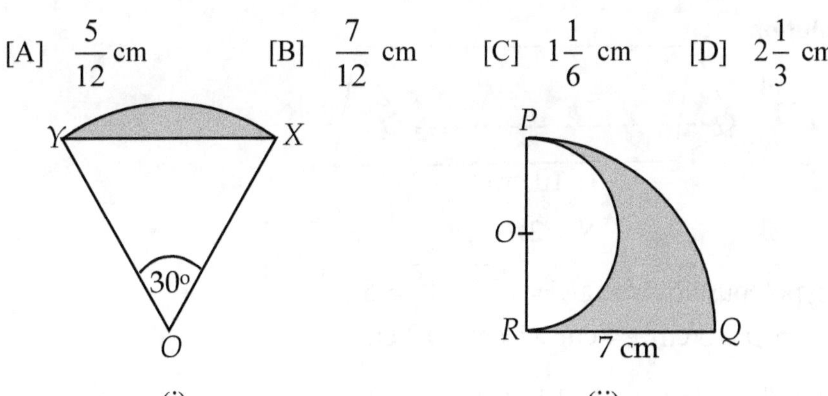

(i) (ii)

59. In figure (ii) above, *PQRO* is one quarter of a circle with centre *O*. $|RQ| = |PR| = 7$ cm. Correct to two decimal places, the area of the shaded portion is:
 [A] 57.70 cm^2 [B] 38.50 cm^2 [C] 27.00 cm^2 [D] 19.25 cm^2
60. The area of a square is equal to that of a triangle of base 9 cm and altitude 32 cm. The length of the side of the square must be:
 [A] 6 cm [B] 6.2 cm [C] 12 cm [D] 2.2 cm
61. The length of the side of a square, which is equal in area to a rectangle measuring 45 cm by 5 cm, is:
 [A] 25 cm [B] 23 cm [C] 16 cm [D] 15 cm
62. A square has a diagonal of 10 cm. the length of a side of the square is:
 [A] $\sqrt{10}$ cm [B] $\sqrt{50}$ cm [C] 10cm [D] 5cm

6.2 MENSURATION OF SOLID FIGURES

Surface Area and Volume of Prisms

The surface area *S* of the prism is given by $S = 2A + pl$

The volume *V* of the prism is given by $V = Al$

Where *A* and *p* are the area and perimeter of the cross-section, and *l* is the length of the prism.

Example

A right-angled triangular prism has arms 4 cm and 3 cm and length 10 cm. Calculate (a) its surface area. (b) its volume.

Solution

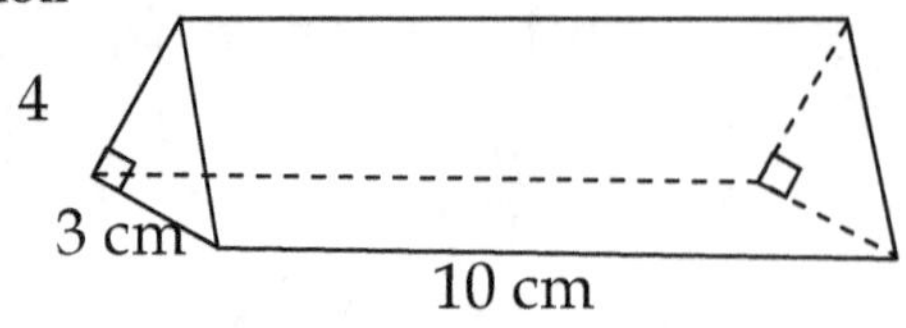

$$S = 2A + pl$$

Hypotenuse of triangle $= \sqrt{3^2 + 4^2} = 5$ cm

$\Rightarrow p = 3 \text{ cm} + 4 \text{ cm} + 5 \text{ cm} = 12 \text{ cm}$

(a) $A = \frac{1}{2}bh = \frac{1}{2}(4)(3) = 6 \text{ cm}^2$

$\therefore \text{S} = 2(6 \text{ cm}^2) + (12 \text{ cm})(10 \text{ cm}) = 132 \text{ cm}^2$

(b) $V = Al$, $A = 6 \text{ cm}^2$, $l = 10$ cm

$\Rightarrow V = 6 \text{ cm}^2 (10 \text{ cm}) = 60 \text{ cm}^3$

Example

The sides of a rectangular block (cuboid) are 5 cm, 11 cm and 40 cm. Calculate (a) its surface area (b) its volume.

Solution

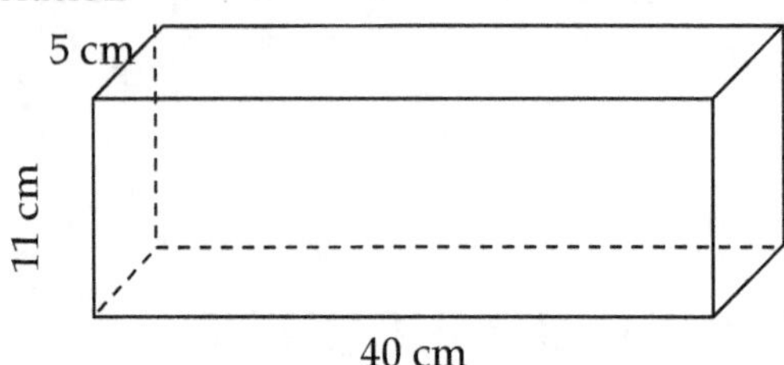

(a) $S = 2A + pl \Rightarrow S = 2(11)(5) + 2(11 + 5)40 = 1390\ cm^2$.

(b) $A = 11(5) = 55\ cm^2 \Rightarrow V = Al = (55\ cm^2)(40\ cm) = 2200\ cm^2$.

Surface Area and Volume of Cubes

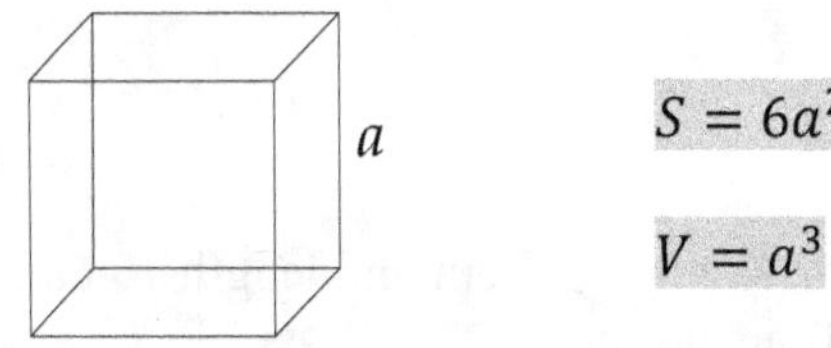

$S = 6a^2$

$V = a^3$

Example

A cube has side 9 cm. Find (a) its surface area. (b) its volume.

Solution

(a) $S = 6a^2 = 6(9)^2 = 486 \text{ cm}^2$ (b) $V = a^3 = (9)^3 = 729 \text{ cm}^3$

Surface Area and Volume of Cylinders

$$S = 2\pi r^2 + 2\pi rl \text{ or } 2\pi r(r + l)$$

$$V = \pi r^2 l$$

Example

A right-circular solid cylinder has a height of 30 cm and a radius of 7 cm. Calculate (a) its surface area (b) its volume

Solution

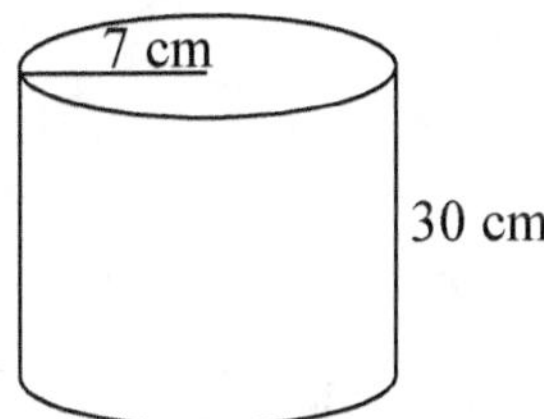

(a) $S = 2\pi r(r + l)$

$$\Rightarrow S = 2\left(\frac{22}{7}\right)(7)(7 + 30) = 1628 \text{ cm}^2$$

(b) $V = \pi r^2 l$

$$\Rightarrow V = \left(\frac{22}{7}\right)(7)^2(30) = 4620\ cm^3$$

Volume of Cones and Pyramids

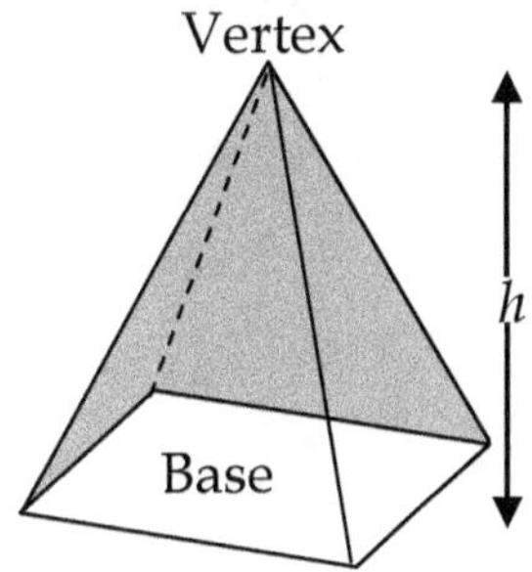

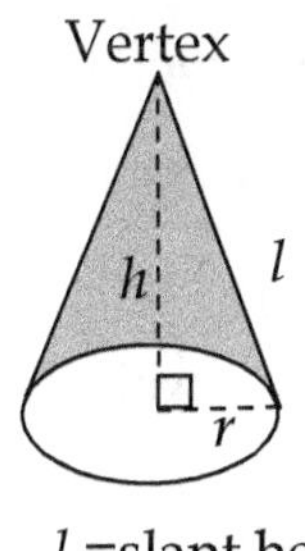

l =slant height

$$V = \frac{1}{3}\text{base area} \times \text{height}$$

$$V = \frac{1}{3}Ah$$

Example
The base of a pyramid is a square of side 5 cm. Find its volume if its height 12 cm

Solution
$V = \frac{1}{3}Ah = \frac{1}{3}(5\text{ cm})^2(12\text{ cm}) = 100\text{ cm}^3$

Example
Calculate the volume of a cone whose height is 12 cm and whose base radius is 7 cm.

Solution
$V = \frac{1}{3}Ah = \frac{1}{3}\pi r^2 h = \frac{1}{3}\left(\frac{22}{7}\right)(7\text{ cm})^2(12\text{ cm}) = 616\text{ cm}^3$

Surface Area of a Cone

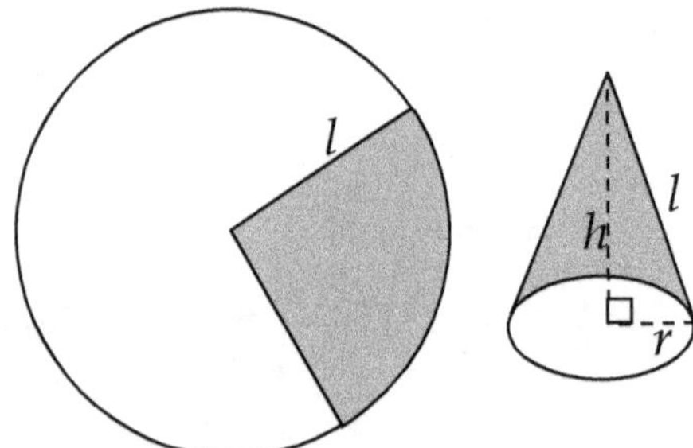

A cone can be made, by folding a sector, which subtends an angle θ at the center of a circle.

The radius r of the base of a cone, which is made out of a sector of radius l that subtends an angle θ at the center, is given by $r = \frac{\theta l}{360}$.

The lateral surface area of a cone is given by $A_l = \pi r l$

The total surface area of a solid cone $S = \pi r^2 + \pi r l$ or $S = \pi r(r + l)$

Example
Calculate the surface area of a solid cone with base radius 3.5 cm and slant height 5.5 cm.

Solution
$S = \pi r(r + l) = \frac{22}{7}(3.5)(3.5 + 5.5) = 99\text{ cm}^2$

Surface Area of a Pyramid

S = Area of base + Sum of area of all the triangular faces

Example

The faces of a pyramid are made of 4 isosceles triangles each with a base 8 cm and height 12 cm. Calculate the surface area of the pyramid.

Solution

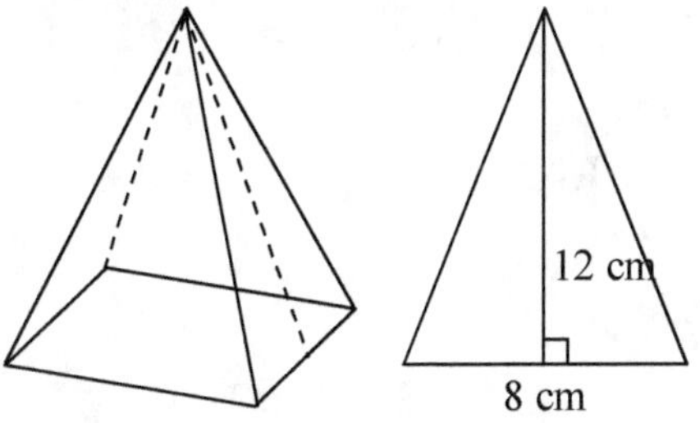

Area of each triangle $= \frac{1}{2}bh$

$= \frac{1}{2}(8)(12)$

$= 48\ \text{cm}^2$

Area of the 4 faces $= 4 \times 48\ \text{cm}^2$

$= 192\ \text{cm}^2$

Area of base $= (8\ \text{cm})^2 = 64\ \text{cm}^2$

Surface area $= 64\ \text{cm}^2 + 192\ \text{cm}^2 = 256\ \text{cm}^2$

Surface Area and Volume of a Sphere

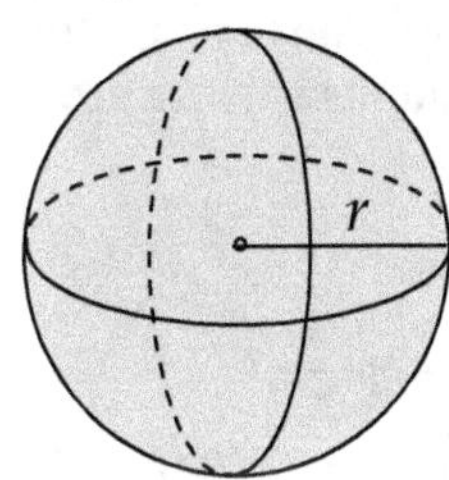

Area of sphere, $S = 4 \times \text{Area of circle} = 4\pi r^2$

The volume V of the sphere is given by, $V = \dfrac{4}{3}\pi r^3$

Example

A sphere has a radius of 14 cm. Calculate the (a) surface area (b) volume.

Solution

(a) $S = 4\pi r^2 = 4\left(\frac{22}{7}\right)(14)^2 = 2464\ \text{cm}^2$

(b) $V = \frac{4}{3}\pi r^3 = \frac{4}{3}\left(\frac{22}{7}\right)(14)^3 = 11498\ \text{cm}^3$

Composite Solid Figures

Break the composite figure into recognisable figures.

Volume and Surface Area of a Frustum

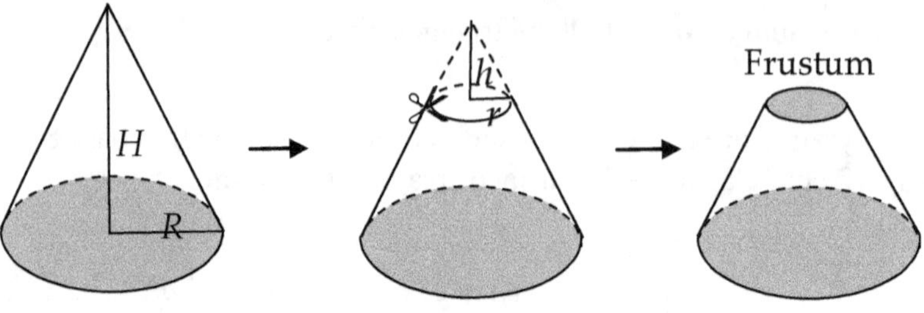

Volume of frustum $V = \frac{1}{3}\pi h\left(R^2 + rR + r^2\right)$.

Lateral surface area of frustum $S_l = \pi(R + r)l$.

Total surface area of a frustum open at the larger end $S = \pi\left(r^2 + (R + r)l\right)$.

Total surface area of a solid frustum $S = \pi\left(R^2 + r^2 + (R + r)l\right)$.

Example

A bucket is in the form of a frustum. The diameters of the two ends are 14 cm and 21 cm. If the slant height of the bucket is 30 cm, calculate its surface area when it is (a) open (b) closed

Solution

(a) $S = \pi(r^2 + (R + r)l) = \frac{22}{7}(7^2 + (10.5 + 7)30) = 1804\ \text{cm}^2$

(b) $S = \pi(R^2 + r^2 + (R + r)l) = \frac{22}{7}(10.5^2 + 7^2 + (10.5 + 7)30) = 2150.5\ \text{cm}^2$

Volume and Surface Area of a Hemisphere

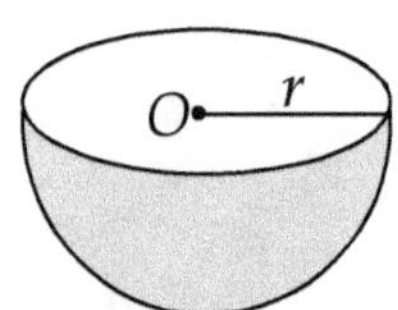

$V = \frac{2}{3}\pi r^3$

Surface area of a hemisphere open at the flat end $S = 2\pi r^2$

Total surface area of hemisphere $S = \pi r^2 + 2\pi r^2 = 3\pi r^2$

Example

A hemispherical bowl with a flat lid has a radius of 7 cm. Calculate the

(a) Volume of the bowl

(b) The surface area of the bowl when (i) Open. (ii) Closed

Solution

(a) Volume of bowl $V = \frac{2}{3}\pi r^3 = \frac{2}{3}\left(\frac{22}{7}\right)(7)^3 = 718.7 \text{ cm}^3$

(b) (i) $S_o = 2\pi r^2 = 2\left(\frac{22}{7}\right)(7)^2 = 308 \text{ cm}^3$

(ii) $S = 3\pi r^2 = 3\left(\frac{22}{7}\right)(7)^2 = 462 \text{ cm}^3$

THE EARTH AS A SPHERE

Longitudes and Latitudes

Longitudes and latitudes act like the coordinate plane.
Longitudes are measured from 0° to 90° N or S of the equator.
Latitudes are measured from 0° to 180° E or W of the Greenwich meridian.

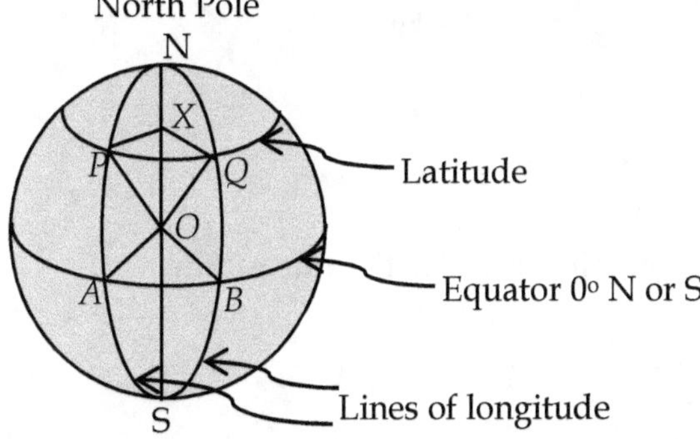

Points along the latitudes are measured from the Greenwich meridian, E or W of the Greenwich meridian.
Points along the longitudes are measured from the equator, N or S of the equator.

Great Circles

A great circle is a circle drawn round a sphere in such a way that its plane passes through the centre of the sphere. The equator and all longitudes are the great circles of the earth. The distance D along a great circle on the earths's surface is given by $D = \frac{\theta}{360} \times 2\pi R$

Where $R \approx 6370$ km, is the radius of the earth, and θ is the angular distance subtended by the two points at the centre of the earth.

Small Circles

A small circle is a circle drawn round a sphere in such a way that its plane does not passes through the centre of the sphere. Apart from the equator, all other latitudes are small circles.

Example

Calculate in km the shortest distance, measured over the earth's surface between the points $P(40°N, 20°W)$ and $Q(20°S, 20°W)$.

Solution

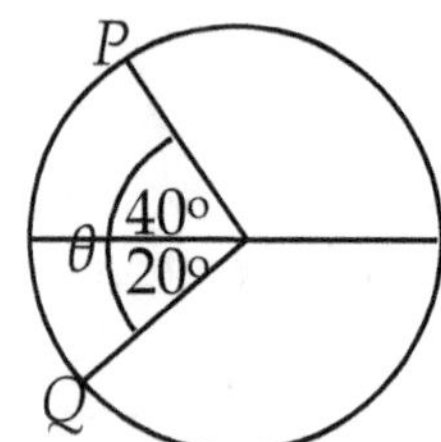

$$\text{Arc length } PQ = \frac{\theta}{360} \times 2\pi R$$

$$= \frac{60}{360} \times 2\left(\frac{22}{7}\right)(6370)$$

$$= 6673.3 \text{ km}$$

VOLUME OF FLOW

The volume of liquid flowing per unit time is called the **rate of flow**.

$$\text{Rate of flow} = \frac{\text{Volume of fluid}}{\text{Time taken}} \text{ or } R = \frac{V}{t}$$

Example

A circular tank of diameter 2.8 m and height 4 m is to be filled by a pipe through which water is flowing at the rate of 8 m^3/min. How long does it take to fill the tank?

Solution

$R = \frac{V}{t} \Rightarrow t = \frac{V}{R}$. But $V = \pi r^2 h$ and $r = \frac{2.8}{2} = 1.4, h = 4, R = 8$

$$\Rightarrow t = \frac{\pi r^2 h}{R} = \frac{\frac{22}{7}(1.4)^2(4)}{8} = 3.08 \text{ minutes.}$$

MULTIPLE CHOICE EXERCISE 6:2

In this exercises, where necessary, take $\pi = \frac{22}{7}$.

1. The shape of each side of a cuboid is:
   [A] A triangle [B] A trapezium [C] A circle [D] A rectangle
2. The number of vertices in a cuboid is:
   [A] 4 [B] 6 [C] 8 [D] 12
3. The number of faces in a cuboid is:
   [A] 4 [B] 6 [C] 8 [D] 12
4. The number of edges in a cuboid is:
   [A] 12 [B] 8 [C] 6 [D] 4
5. The total surface area of a cube of edge 3 cm is:
   [A] 27 cm^2 [B] 27 cm^3 [C]54 cm^2 [D] 36 cm^2
6. The sides of two cubes are in the ratio 2:5. The ratio of their volumes is:
   [A] 4:5 [B] 8:15 [C] 6:125 [D] 8:125
7. A rectangular tank 2.25 m long and 1.6 m wide contains 2800 litres of water. Correct to the nearest cm, the depth of water in the tank is:
   [A] 76 cm [B] 78 cm [C] 770 cm [D] 780 cm
8. A cylindrical container closed at both ends, has a radius of 7 cm and a height 5 cm. The total surface area of the container is:
   [A] 154 cm^2 [B] 220 cm^2 [C] 528 cm^2 [D] 770 cm^2
9. A cylindrical container closed at both ends, has a radius of 7 cm and a height 5 cm. The volume of the container is:
   [A] 154 cm^3 [B] 220 cm^3 [C] 528 cm^3 [D] 770 cm^3
10. The curved surface area of a cylindrical tin is 704 cm^2. The height when the radius is 8 cm is:
    [A] 3.5 cm [B] 7 cm [C] 14 cm [D] 28 cm
11. Correct to 1 decimal place the volume of a cylinder of height 8 cm and base radius 3 cm is:
    [A] 300.0 cm^3 [B] 250.0 cm^3 [C] 226.2 cm^3 [D] 150.9 cm^3
12. Water flows into a cylindrical container at the rate of 5π cm^3 per second. If the radius of the container is 3 cm, the level of the water in the container at the end of 9 seconds will be:
    [A] 2 cm [B] 5 cm [C] 8 cm [D] 15 cm
13. The following figure shows a rectangular sheet of thin metal from which a cylinder, 10 cm high, is to be made with no overlap. The radius of this cylinder is:

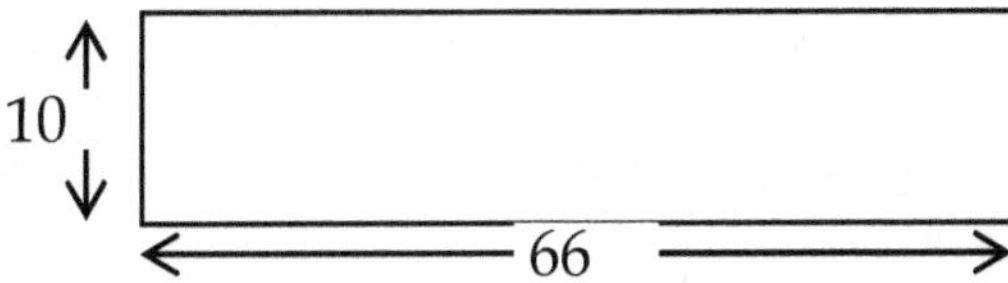

    [A] 3.3 cm [B] 6.6 cm [C] 10.5 cm [D] 21 cm
14. A solid cylinder of radius 7 cm is 10 cm long. Its total surface area is:

[A] 70π cm^2 [B] $18\,\pi$ cm^2 [C] $210\,\pi$ cm^2 [D] $238\,\pi$ cm^2

15. The volume of a cylinder of radius 14 cm is 210 cm^3. The curved surface area of the cylinder is:
    [A] 30 cm^2 [B] 15 cm^2 [C] 616 cm^2 [D] 1262 cm^2
16. The internal and external radii of a cylindrical bronze pipe are 1.5 cm and 2 cm respectively. If the pipe is 10 cm long, the volume of the bronze used is:
    [A] $5\frac{1}{2}$ cm^3 [B] 55 cm^3 [C] $196\frac{2}{5}$ cm^3 [D] 550 cm^3
17. A cone is made from a sector of a circle of radius 14 cm and angle 90°. The area of the curved surface of the cone is:
    [A] 22 cm^2 [B] 88 cm^2 [C] 77 cm^2 [D] 154 cm^2
18. The volume of a cone of radius 3.5 cm and vertical height 12 cm is:
    [A] 15.5 cm^3 [B] 21.0 cm^3 [C] 154.0 cm^3 [D] 42.0 cm^3
19. A sector is cut off from a circle of radius 8.2 cm to form a cone. If the radius of the resulting cone is 3.5 cm, then, the curved surface area of the cone is:
    [A] 12.83 cm^2 [B] 22.0 cm^2 [C] 67.2 cm^2 [D] 90.2 cm^2
20. The angle of a sector of a circle of radius 8 cm is 240°. This sector is bent to form a cone. The radius of the base of the cone must be:
    [A] $\frac{16}{3}$ cm [B] $\frac{15}{3}$ cm [C] $\frac{16}{5}$ cm [D] $\frac{8}{3}$ cm
21. The volume of a cone of height 9 cm is 1848 cm^3. Its radius is:
    [A] 7 cm [B] 14 cm [C] 28 cm [D] 98 cm
22. The total surface area of a cone whose height is 12 cm and whose base radius is 5 cm is:
    [A] $240\frac{2}{7}$ cm^2 [B] $235\frac{5}{7}$ cm^2 [C] $282\frac{6}{7}$ cm^2 [D] $251\frac{3}{7}$ cm^2
23. A cone is 14 cm deep and the base radius is $4\frac{1}{2}$ cm.
    The volume of water which is exactly half the volume of the cone is:
    [A] 49.5 cm^3 [B] 99 cm^3 [C] 148.5 cm^3 [D] 297 cm^3
24. The total surface area of a solid circular cone with base radius 3 cm and slant height 4 cm is:
    [A] 66 cm^2 [B] $\frac{753}{7}$ cm^2 [C] $\frac{782}{7}$ cm^2 [D] 88 cm^2
25. A 210° sector of a circle of radius 21 cm is bent to form a cone. The base radius of the cone is:
    [A] 3.5 cm [B] 7 cm [C] 10.5 cm [D] 12.25 cm
26. The total surface area of a solid cone of slant height 15 cm and base radius 8 cm in terms of π is:
    [A] 120π cm^2 [B] 184π cm^2 [C] $200\,\pi$ cm^2 [D] 320π cm^2
27. The curved surface area of a cone of radius 3 cm and slant height 7 cm is:
    [A] 22 cm^2 [B] 44 cm^2 [C] 66 cm^2 [D] 132 cm^2
28. The height of a right circular cone is 4 cm. The radius of the base is 3 cm. Its curved surface area is:
    [A] 9π cm^2 [B] $15\,\pi$ cm^2 [C] $16\,\pi$ cm^2 [D] $20\,\pi$ cm^2
29. The base diameter of a cone is 14 cm, and its volume is 462 cm^3. Its height is:

[A] 3.5 cm [B] 5 cm [C] 7 cm [D] 9 cm

30. The total surface area of a solid right circular cone of base radius r cm and height r cm is:
[A] $2\pi r^2$ cm^2 [B] $4\pi r^2$ cm^2 [C] $\frac{7}{3}\pi r^2$ cm^2 [D] $\frac{4}{3}\pi r^2$ cm^2

31. The surface area of a sphere of radius 7 cm is:
[A] 86 cm^2 [B] 154 cm^2 [C] 616 cm^2 [D] 143 cm^2

32. Two solid spheres have volumes 250 cm^3 and 128 cm^3 respectively. The ratio of their radii is certainly:
[A] 5:4 [B] 25:16 [C] 2:1 [D] 4:3

33. A hollow sphere has a volume of k cm^3 and a surface area of k cm^2. The diameter of the sphere is:
[A] 3 cm [B] 12 cm [C] 9 cm [D] 6 cm

34. A sphere has a surface area of 4312 cm^2. The radius of the sphere in cm correct to one decimal place is:
[A] 18.0 [B] 18.5 [C] 19.0 [D] 19.5

35. The cross-section of a prism is a right angled triangle 3 cm by 4 cm by 5 cm. The height of the prism is 8 cm. Its volume is:
[A] 48 cm^3 [B] 60 cm^3 [C] 96 cm^3 [D] 120 cm^3

36-37 The following figure shows a triangular prism of length 7 cm. The right angled triangle PQR is a cross section of the prism $|PR| = 5$ cm and $|RQ| = 3$ cm. Use the information to answer questions 36 to 37.

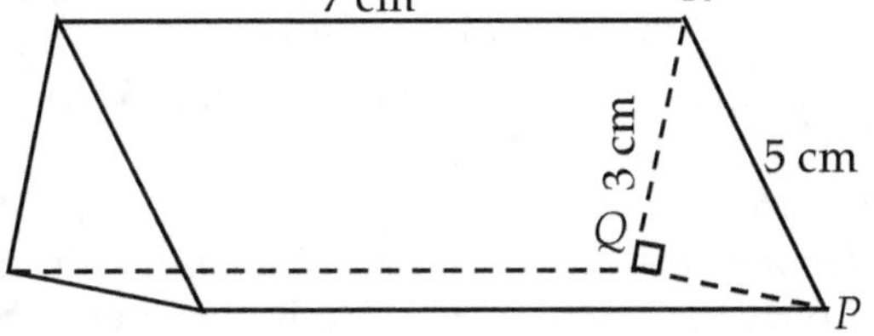

36. The area of the cross-section is:
[A] 4 cm^2 [B] 6 cm^2 [C] 15 cm^2 [D] 20 cm^2

37. The volume of the prism is:
[A] 28 cm^3 [B] 42 cm^3 [C] 70 cm^3 [D] 84 cm^3

38. The height of a pyramid on a square base is 15 cm. Given that the volume is 80 cm^3, the length of the side of the base in cm is:
[A] 3.3 [B] 5.3 [C] 4.0 [D] 8.0

39. The height of a pyramid on a square base is 15 cm. If the volume is 80 cm^3, the area of the square base is:
[A] 16 cm^2 [B] 9.6 cm^2 [C] 8 cm^2 [D] 25 cm^2

40. A right pyramid is on a square base of side 4 cm. The slanting side of the pyramid is $2\sqrt{3}$ cm. The volume of the pyramid is:
[A] $\frac{10}{3}$ [B] $\frac{16}{3}$ [C] $\frac{32}{3}$ [D] $\frac{64}{3}$

41. A pyramid on a square base of side 10 cm has a height of 15 cm, its volume must be:

[A] 150 cm^3 [B] 500 cm^3 [C] 1500 cm^3 [D] 5000 cm^3

42. The base of a pyramid is a 12 cm by 12 cm. If its height is 20 cm, the volume of the pyramid in cm^3 is:
[A] 960 [B] 80 [C] 1440 [D] 1600
43. The position of two countries P and Q are 15° N, 12° E and 65° N, 12° E respectively. Their difference in latitude is:
[A] 100° [B] 80° [C] 50° [D] 24°
44. Cotonou and Niamey are on the same line of longitude and Niamey is 7° north of Cotonou. If the radius of the earth is 6400 km, the distance of Niamey north of Cotonou along the line of longitude correct to the nearest kilometre is:
[A] 391 km [B] 503 km [C] 782 km [D] 1006 km
45. P and Q are two places on the same circle of latitude 79° S. P is on longitude 68° E, while Q is on longitude 22° W. The angular distance between P and Q is:
[A] 12° [B] 45° [C] 48° [D] 90°
46. Two ships on the equator are on longitude 45° W and 45° E respectively. Their distance apart along the equator, correct to 2 significant figures is:
[A] 3,200 km [B] 10,000 km [C] 6,400 km [D] 5,000 km
47. Two points P and Q are on longitude 67° W. Their latitudes differ by 90°. Taking the radius of the earth as 6400 km, their distance apart in terms of π is:
[A] 6400π km [B] $\frac{6400}{\pi}$ km [C] 3200π km [D] $\frac{3200}{\pi}$ km
48. Two places are 2816 km apart on the same line of longitude. The angular difference between their latitudes is: (Take $R = 6{,}400$ km)
[A] 25.2° [B] 26.1° [C] 51.3° [D] 63.9°
49. Abijan is 4° west of Accra and on the same circle of latitude. If the radius of this circle of latitude is 6370 km, the distance of Abijan west of Accra, correct to the nearest km is:
[A] 222 km [B] 445 km [C] 890 km [D] 5005 km
50. The following figure shows a cone with the dimensions of its frustum indicated. The height of the cone is:
[A] 12 cm [B] 15 cm [C] 18 cm [D] 24 cm

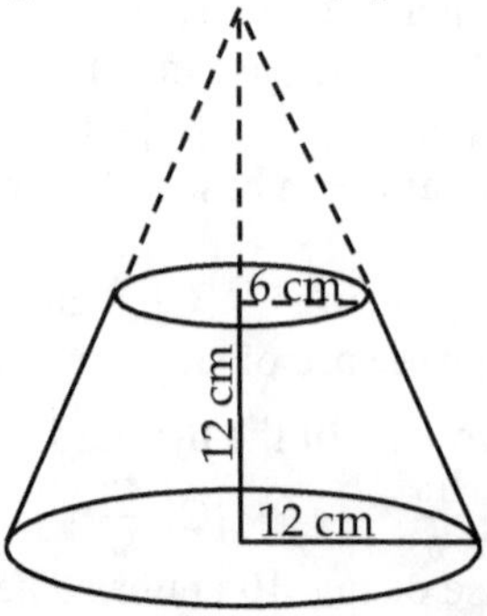

6 cm
12 cm
12 cm

CHAPTER 7

COORDINATE GEOMETRY AND GRAPHS

7.1 COORDINATE GEOMETRY

Line Segments

Let $A(x_1, y_1)$ and $B(x_2, y_2)$ be any two points.

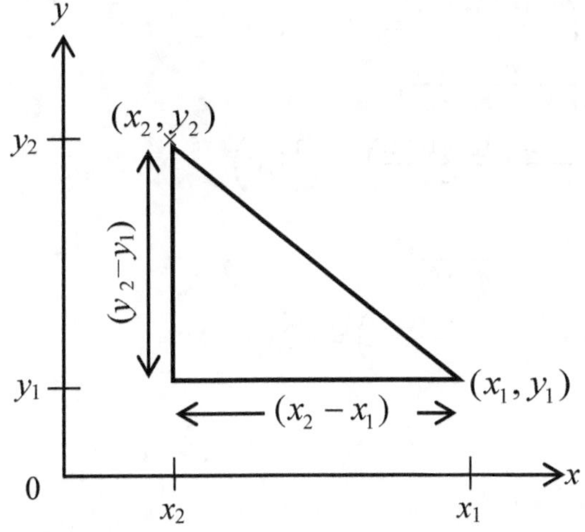

1. Distance between A and B is $AB = \sqrt{(x_2 - x_1)^2 + (y_2 - y_1)^2}$

2. Midpoint of the line segment AB is $M(x, y) = \left(\frac{x_1+x_2}{2}, \frac{y_1+y_2}{2}\right)$

3. The point $P(x, y)$ which divides AB internally in the ratio $m:n$, is given by
$$P(x, y) = \left(\frac{mx_2 + nx_1}{m+n}, \frac{my_2 + ny_1}{m+n}\right)$$

4. The point $P(x, y)$ which divides AB externally in the ratio $m:n$, is given by
$$P(x, y) = \left(\frac{mx_2 - nx_1}{m-n}, \frac{my_2 - ny_1}{m-n}\right)$$

A closer look at the formulae in ③ and ④ reveal that, for external division, we can simply substitute the ratio $m: -n$ in the formula in ③.

5. Then gradient or steepness of the line joining A and B is given by
$$m = \frac{y_2 - y_1}{x_2 - x_1}$$

Example

Let Q and R be the points $Q(8,6)$ and $R(16,12)$. Calculate

(a) The distance between Q and R. (b) The mid-point between Q and R.

(c) The coordinates of the point which divides the line joining Q and R in the ratio 2:1 (i) internally (ii) externally.

(d) The steepness of the line joining the points Q and R.

Solution

(a) $QR = \sqrt{(x_2 - x_1)^2 + (y_2 - y_1)^2} = \sqrt{(16-8)^2 + (12-6)^2} = 10$ units

(b) $M(x, y) = \left(\frac{x_1+x_2}{2}, \frac{y_1+y_2}{2}\right) = \left(\frac{8+16}{2}, \frac{6+12}{2}\right) = (12,9).$

(c) (i) Internally $(x, y) = \left(\frac{mx_2+nx_1}{m+n}, \frac{my_2+ny_1}{m+n}\right)$

$$= \left(\frac{2(8)+1(16)}{2+1}, \frac{2(6)+1(12)}{2+1}\right) = \left(\frac{32}{3}, 8\right).$$

(ii) Externally $(x, y) = \left(\frac{mx_2-nx_1}{m-n}, \frac{my_2-ny_1}{m-n}\right)$

$$= \left(\frac{2(8)-1(16)}{2-1}, \frac{2(6)-1(12)}{2-1}\right) = (0,0).$$

(d) $m = \frac{y_2-y_1}{x_2-x_1} = \frac{12-6}{16-8} = \frac{3}{4}$

The Equation of a Straight Line

The following table gives the various forms of the equations of a straight line, their gradients and the intercept with each axis.

Form	Gradient	Intercept	
		x-axis	y-axis
Gradient/ Intercept Form $y = mx + c$	m	$-\frac{c}{m}$	c
General Form $ax + by + k = 0$	$-\frac{a}{b}$	$-\frac{k}{a}$	$-\frac{k}{b}$
Double/Intercept Form $\frac{x}{a} + \frac{y}{b} = 1$	$-\frac{b}{a}$	a	b
Gradient/ one point form $y - y_1 = m(x - x_1)$	m	$\frac{mx_1 - y_1}{m}$	$y_1 - mx_1$

Parallel and Perpendicular Lines

Consider the lines l_1 and l_2 whose gradients are m_1 and m_2 respectively.

(a) $l_1 \parallel l_2 \Leftrightarrow m_1 = m_2$

(b) $l_1 \perp l_2 \Leftrightarrow m_1 m_2 = -1$

Example

Find the equation of the straight line, which passes through the point (−2, 5) and is

(a) Perpendicular to the line $y = 2x + 3$. (b) Parallel to the line $y = 3x + 2$.

Solution

(a) Let the gradient of the line $y = 2x + 3$ be $m_1 = 2$ and that of the perpendicular line be m_2. Then $m_1 m_2 = -1 \Rightarrow 2m_2 = -1 \Rightarrow m_2 = -\frac{1}{2}$.
Therefore, the equation of the perpendicular line is $y - y_1 = m_2(x - x_1)$.
Substitute $(-2,5)$ and $m_2 = -\frac{1}{2}$ into $y - y_1 = m_2(x - x_1)$,
$y - 5 = -\frac{1}{2}(x - (-2)) \Rightarrow y = -\frac{1}{2}x + 4.$

(b) Lines are parallel so $m_1 = m_2 = 3$.
Therefore, the equation of the parallel line is $y - y_1 = m_2(x - x_1)$.
Substitute (1, 3) and $m_2 = 3$ into the equation,
$y - 3 = 3(x - 1) \Rightarrow y = 3x$.

MULTIPLE CHOICE EXERCISE 7:1

1. A triangle with vertices at the points with coordinates (–4,4), (4,4), (1,–1) is:
   [A] Right–angled [B] Equilateral [C] Isosceles [D] Scalene
2. The statement which is true about the points $P(-1, -4)$, $Q(6, -5)$, $R(-1,5)$ and $S(3,2)$ is:
   [A] *R* and *Q* are in the second and third quadrants respectively.
   [B] *P* and *S* are in the fourth and third quadrants respectively.
   [C] *S* and *R* are in the first and second quadrants respectively.
   [D] *P* and *Q* are in the second and fourth quadrants respectively.
3. The coordinates of the midpoint of $P(-4,5)$ and $Q(2,1)$ are:
   [A] (–1,3) [B](–1,–3) [C] (1,–3) [D] (1,3)
4. A straight line passes through the points (5,3) and (8,4). The line:

[A] is parallel to the x-axis [B] slopes from left to right
[C] is parallel to the x-axis [D] slopes from right to left

5. Among the following, the straight line which slopes from left to right is the line which passes through the points:
   [A] (–1,5) and (2,0) [B] (3,–5) and (3,3)
   [C] (–2,1) and (1,1) [D] (0,0) and (3,6)
6. Among the following, the straight line which is parallel to the x-axis is the line that passes through the points:
   [A] (–1,5) and (2,0) [B] (3,–5) and(3,3)
   [C] (–2,1) and (1,1) [D] (0,0) and (3,6)
7. Among the following, the straight line which slopes from right to left is the line which passes through the points:
   [A] (–2,5) and (–2,3) [B] (3,–5) and (8,–5)
   [C] (–2,1) and (1,1) [D] (0,0) and (3,6)
8. The point which lies on the line $y = 2x–5$ is:
   [A] (1,3) [B] (2,5) [C] (3,–1) [D] (3,1)
9. The gradient of the line $2x+3y = 12$ is:
   [A] $-\frac{2}{3}$ [B] $-\frac{3}{2}$ [C] $\frac{2}{3}$ [D] $\frac{3}{2}$
10. The intercept of the line $2x+3y = 12$ with the x-axis is:
    [A] 2 [B] 3 [C] –6 [D] 6
11. The intercept of the line $2x+3y = 12$ with the y–axis is:
    [A] 2 [B] 3 [C] 4 [D] –4
12. The area of the triangle OAB, where O is the origin and A and B are the points where the line cuts the x- and y-axes respectively is:
    [A] 12 un^2 [B] 24 un^2 [C] 6un^2 [D] 18 un^2
13. The gradient of the line $2x+y = 8$ is:
    [A] 4 [B] 2 [C] –2 [D] –8
14. Given that the lines $2x+y = 8$ and $6y–mx = 3$ are parallel. The value of m is:
    [A] 2 [B] –2 [C] $\frac{1}{2}$ [D] $-\frac{1}{2}$
15. Given that the lines $2x+y = 8$ and $6y–mx = 3$ are perpendicular. The value of m is:
    [A] 2 [B] –2 [C] $\frac{1}{2}$ [D] $-\frac{1}{2}$
16. In the form $y = mx + c$, the equation of the line $2x–3y+5 = 0$ is:
    [A] $y=-\frac{2}{3}x+\frac{5}{3}$ [B] $y=\frac{2}{3}x-\frac{5}{3}$ [C] $y=\frac{2}{3}x+\frac{5}{3}$ [D] $y=-\frac{2}{3}x-\frac{5}{3}$
17. The gradient of the line $2x–3y+5 = 0$ is:
    [A] $-\frac{5}{3}$ [B] $\frac{5}{3}$ [C] $-\frac{2}{3}$ [D] $\frac{2}{3}$

18. The intercept of the line $2x–3y + 5 = 0$ with the x–axis is:
[A] $-\frac{5}{2}$ [B] $\frac{5}{2}$ [C] $-\frac{5}{3}$ [D] $\frac{5}{3}$
19. The coordinates of the midpoint of (3,2) and (–1,0) are:
[A] (–1,–1) [B] (1,1) [C] (–1,1) [D] (1,–1)
20. The gradient of the straight line which passes through the points (–1, 0) and (0,–2) is:
[A] –2 [B] 2 [C] $-\frac{1}{2}$ [D] $\frac{1}{2}$
21. The distance between the points (3,2) and (0,–2) is:
[A] 3 [B] $\sqrt{3}$ [C] 5 [D] $\sqrt{5}$
22. The coordinate of the point of intersection of the lines $x–y = 3$ and $x+2y = 6$ is:
[A] (4,1) [B] (6,1) [C] $\left(\frac{9}{4},\frac{3}{4}\right)$ [D] (1,4)
23. Given the straight lines $L_1: 2x = 3y + 5$, $L_2: 2y = 4x + 3$, $L_3: 2x + 4y = 5$, $L_4: 3x - 2y = 3$. The lines which are parallel are:
[A] L_1 and L_4 [B] L_2 and L_4 [C] L_2 and L_3 [D] L_1 and L_4
24. Given the straight lines $L_1: 2x = 3y + 5$, $L_2: 2y = 4x + 3$, $L_3: 2x + 4y = 5$, $L_4: 3x - 2y = 3$. The lines are perpendicular are:
[A] L_1 and L_4 [B] L_2 and L_4 [C] L_2 and L_3 [D] L_1 and L_4
25. Given the straight lines $L_1: 2x = 3y + 5$, $L_2: 2y = 4x + 3$, $L_3: 2x + 4y = 5$, $L_4: 3x - 2y = 3$. The line which passes through the point (–1,–3) is:
[A] L_1 [B] L_2 [C] L_3 [D] L_4
26. P is the midpoint of the line segment joining the points A(2,–3) and B(4,5). C is the point (5,9). The equation of the line PC is:
[A] $y = -4x + 11$ [B] $y = 4x - 11$ [C] $y = 4x + 11$ [D] $y = -4x - 11$
27. Given the straight lines $L_1: y - 2x = 5$, $L_2: y = 2x + 3$, $L_3: 4y = -2x - 6$, $L_4: 2y = x - 5$. The lines which are parallel are:
[A] L_1 and L_2 [B] L_2 and L_3 [C] L_1 and L_4 [D] L_3 and L_4
28. Given the straight lines $L_1: y - 2x = 5$, $L_2: y = 2x + 3$, $L_3: 4y = -2x - 6$, $L_4: 2y = x - 5$. One of the pair of perpendicular lines are:
[A] L_1 and L_2 [B] L_2 and L_3 [C] L_1 and L_4 [D] L_3 and L_4
29. Given that, $L_1: y = 2x + 3$, $L_2: 2x + y = 5$, $L_3: 4y = 2x + 3$, $L_4: x - \frac{y}{2} = 1$.
The lines which are parallel are:
[A] L_1 and L_4 [B] L_2 and L_4 [C] L_2 and L_3 [D] L_1 and L_4
30. Given that, $L_1: y = 2x + 3$, $L_2: 2x + y = 5$, $L_3: 4y = 2x + 3$,

$L_4: x - \frac{y}{2} = 1.$

The lines are perpendicular are:

[A] L_1 and L_4 [B] L_2 and L_4 [C] L_2 and L_3 [D] L_1 and L_2

31. The point of intersection of the lines $y-2x = 5$ and $3y = 4x+9$ is:
    [A] (3,–1) [B] (3,1) [C] (–3,1) [D] (–3,–1)
32. The line $3x-5y = 15$ cuts the y-axis at:
    [A] (0,5) [B] (0,–3) [C] (5,0) [D] (–3,0)
33. The line $3x-5y = 15$ cuts the x-axis at:
    [A] (0,5) [B] (0,–3) [C] (5,0) [D] (–3,0)
34. The gradient of the line $3x-5y = 15$ is
    [A] $-\frac{5}{3}$ [B] $\frac{5}{3}$ [C] $-\frac{3}{5}$ [D] $\frac{3}{5}$
35. The value of m for which the lines $y = 2x-13$ and $y - mx = -3$ are parallel is:
    [A] –2 [B] 2 [C] $\frac{1}{2}$ [D] $-\frac{1}{2}$
36. The value of m for which the lines $y = 2x-13$ and $y - mx = -3$ are perpendicular is:
    [A] –2 [B] 2 [C] $\frac{1}{2}$ [D] $-\frac{1}{2}$
37. Given that the lines $y = 2x - 13$ and $y - mx = -3$ are perpendicular. Their point of intersection is:
    [A] (–4,–5) [B] (4,5) [C] (–4,5) [D] (4,–5)
38. The gradient of the lines of the line $x + 2y = 2$ is:
    [A] $-\frac{1}{2}$ [B] $\frac{1}{2}$ [C] 2 [D] –2
39. The intercept of the line $x + 2y = 2$ with the x-axis is:
    [A] (0,–1) [B] (0,1) [C] (2,0) [D] (–2,0)
40. The intercept of the line $x+2y = 2$ with the y-axis is:
    [A] (0,–1) [B] (0,1) [C] (2,0) [D] (–2,0)
41. The area of the triangle between the line $x+2y = 2$ and the coordinate axes is:
    [A] 1 un^2 [B] 2 un^2 [C] 5 un^2 [D] $\sqrt{5}$ un^2
42. $P(3,0)$ and $Q(5,2)$ are two points on a straight line. The equation of the line PQ is:
    [A] $x + y = 3$ [B] $y = x-3$ [C] $y = x+3$ [D] $y = -x-3$
43. Given the lines L_1: $y = 2x-4$, L_2: $2y+x-6 = 0$, L_3: $y = \frac{1}{3}x + 7$, L_4: $3y = x-5$. The two perpendicular lines are:
    [A] L_1 and L_3 [B] L_2 and L_4 [C] L_3 and L_4 [D] L_1 and L_2
44. Given the lines L_1: $y = 2x-4$, L_2: $2y+x-6 = 0$, L_3: $y = \frac{1}{3}x + 7$,

L_4: $3y = x-5$.
The two parallel lines are:
[A] L_1 and L_3 [B] L_2 and L_4 [C] L_3 and L_4 [D] L_1 and L_2

45. Given that the points $P(2,2)$, $Q(3,5)$ and $R(4,a)$ are collinear, the coordinates of R are:
[A] (4,–8) [B] (3,–7) [C] (4,7) [D] (4,8)

46. The y intercept on the line $6x+3y-7 = 0$ is:
[A] $\frac{7}{6}$ [B] -7 [C] $\frac{7}{3}$ [D] -2

47. The x intercept on the line $6x+3y-7 = 0$ is:
[A] $\frac{7}{6}$ [B] -7 [C] $\frac{7}{3}$ [D] -2

48. The gradient of the line $6x+3y-7 = 0$ is:
[A] $\frac{7}{6}$ [B] -7 [C] $\frac{7}{3}$ [D] -2

49. The straight line parallel to the y-axis passes through the points:
[A] (–1,5) and (2,0) [B] (3,–5) and (3,3)
[C] (–2,1) and (1,1) [D] (0,0) and (3,6)

50. P is the midpoint of the line segment joining the point $A(2,-3)$ and $B(4,5)$. C is the point (5,9). The gradient of the line PC is:
[A] –4 [B] 4 [C] 11 [D] –11

51. P is the midpoint of the line segment joining the point $A(2,-3)$ and $B(4,5)$. C is the point (5,9). The intercept of the line PC with the y-axis is:
[A] 11 [B] -11 [C] $\frac{11}{4}$ [D] $-\frac{11}{4}$

52. P is the midpoint of the line segment joining the point $A(2,-3)$ and $B(4,5)$. C is the point (5,9). The intercept of the line PC with the x-axis is:
[A] 11 [B] -11 [C] $\frac{11}{4}$ [D] $-\frac{11}{4}$

53. In the following figure, the intercept of the line l_1 with the y-axis is:
[A] $-\frac{5}{4}$ [B] $-\frac{4}{5}$ [C] -4 [D] -5

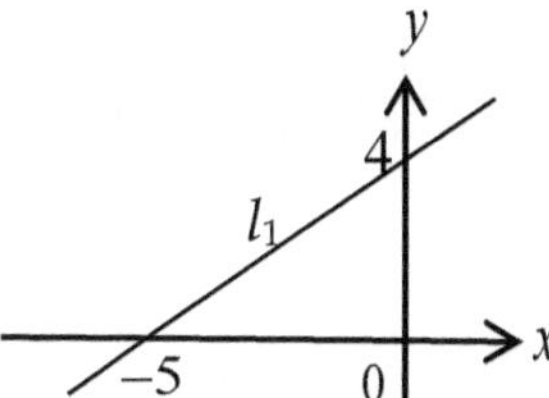

54. In the figure above, the intercept of the line l_1 with the x-axis is:
[A] $-\frac{5}{4}$ [B] $-\frac{4}{5}$ [C] -4 [D] -5

55. In the figure above, the gradient of the line l_1 is:

[A] $-\frac{5}{4}$ [B] $-\frac{4}{5}$ [C] -4 [D] -5

56. In the figure above, the equation of the line l_1 is:
[A] $5y = 4x - 20$ [B] $4y = 5x - 20$
[C] $5y = -4x + 20$ [D] $4y = -5x + 20$

57. In the following figure, the intercept of the line l_2 with the y-axis is:
[A] $-\frac{3}{5}$ [B] $\frac{3}{5}$ [C] 6 [D] 10

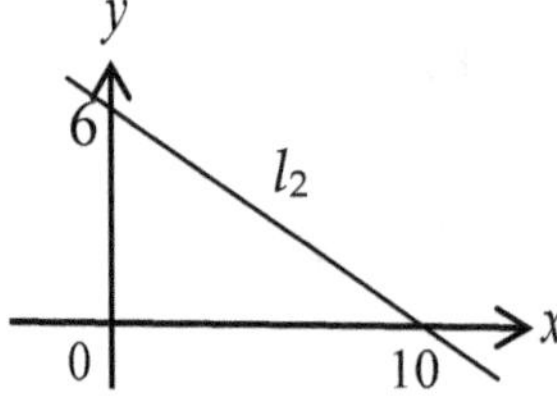

58. In the figure above, the intercept of the line l_2 with the x-axis is:
[A] $-\frac{3}{5}$ [B] $\frac{3}{5}$ [C] 6 [D] 10

59. In the figure above, the gradient of the line l_2 is:
[A] $-\frac{3}{5}$ [B] $\frac{3}{5}$ [C] 6 [D] 10

60. In the figure above, the equation of the line l_2 is:
[A] $5y = 3x + 30$ [B] $3y = 5x - 30$
[C] $5y = 3x - 30$ [D] $3y = -5x + 30$

7.2 GRAPHS OF FUNCTIONS

Graphing Straight Lines

Example

The equation of a straight line is given as $y - 2 = -2(x - 3)$. Make a table of values and hence draw the graph of this equation.

Solution

$$y - 2 = -2(x - 3) \Rightarrow y = 8 - 2x$$

x	1	3	5
y	6	2	–2

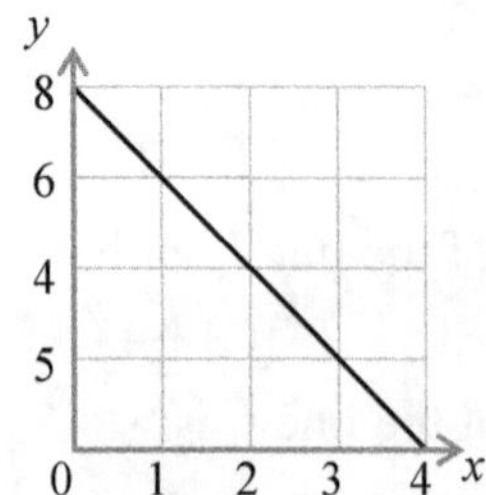

Simultaneous Linear Equations (Graphical Method)

The coordinates of the point of intersection of two lines gives the solution of the equations.

Example
Solve the simultaneous equations $y = 2x$ and $y + x = 3$, using the graphical method.

Solution

$y = 2x$

x	0	1	2
y	0	2	4

$y + x = 3$

x	0	2	4
y	3	1	−1

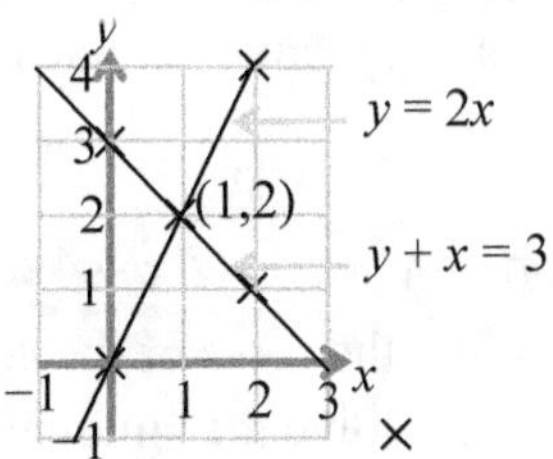

The two lines meet at the point (1,2) so the solution of the simultaneous equations is $x = 1, y = 2$.

Graphs of Quadratic Functions

Example
Given that $f(x) = 4 - 3x - x^2$, make up a table of values of x against $y = f(x)$, for integral values of x in the range $-5 \le x \le 2, x \in \mathbb{R}$. Hence draw the graph of $y = f(x)$.

Solution

x	$4-3x-x^2$	$y=f(x)$
−5	$4+15-25$	−6
−4	$4+12-16$	0
−3	$4+9-9$	4
−2	$4+6-4$	6
−1	$4+3-1$	6
0	$4+0-0$	4
1	$4-3-1$	0
2	$4-6-4$	−6

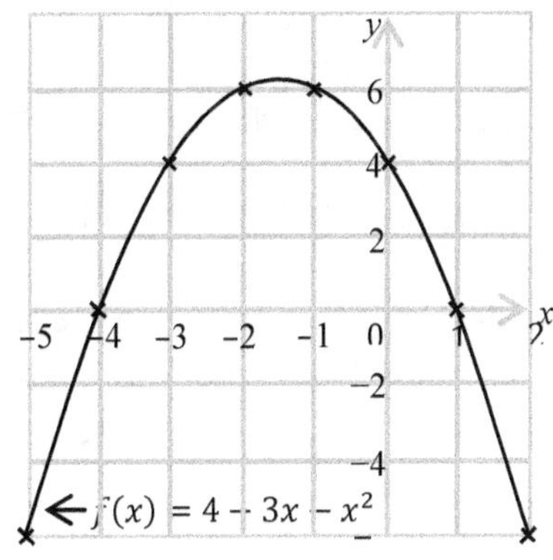

Graphical Solutions of Quadratic Equations

Draw the graph of $y = ax^2 + bx + c$, and find the roots of the equation $ax^2 + bx + c = 0$ from the graph. The roots are the values of x for which the line $y = 0$ (the x-axis) intersects the graph.

Simultaneous Equations - One Linear, One Quadratic - Graphical Method

Simultaneous equations-one linear, one quadratic can be solved graphically by determining the point of intersection of the curve and the straight line in the same way as simultaneous linear equations are solved graphically.

Example

Plot the graph of $y = x^2 - 5x + 4$ for integral values of x from 0 to +5. Use your graph to find the roots of the equation $x^2 - 5x + 4 = 0$. Also use your graphical method Solve the simultaneous equations $y = 2x - 6$ and $y = x^2 - 5x + 4$.

Solution

x	$x^2 - 5x + 4$	y
0	$0 - 0 + 4$	4
1	$1 - 5 + 4$	0
2	$4 - 10 + 4$	-2
3	$9 - 15 + 4$	-2
4	$16 - 20 + 4$	0
5	$25 - 25 + 4$	4

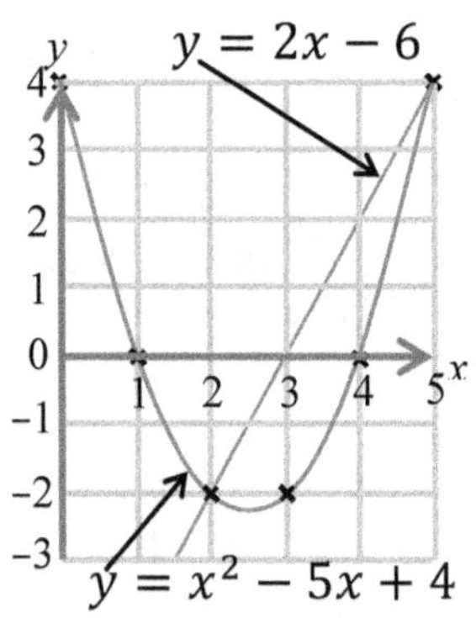

From the graph, the roots of the equation are $x = 1$ or $x = 4$.

By graphing $y = 2x - 6$ on the same axes we see that the graphs intersect at (2,–2) and (5,4). Hence the solutions of the simultaneous equations are $x = 2$, when $y = -2$ and $x = 5$, when $y = 4$.

Solving Quadratic Equations from the Graph of any Other Quadratic Function

It is possible to solve quadratic equations using the graph of any other quadratic function. To do this, simultaneously eliminate the term in x^2 from the quadratic equation and quadratic function then graph the resulting straight line on the same Cartesian plane as the graph of the quadratic function.

Example

The function f is defined on the set $\mathbb{R}$ of real numbers as $f(x) = x^2 + x - 12$.

(a) Using a scale of 1 cm to represent 2 units on the x-axis, and 1 cm to represent 4 units on the y-axis draw the graph of $y = f(x)$ for $\{x: -6 \leq x \leq 5\}$.

(b) By drawing a suitable straight line on your graph, solve the equation $x^2 - 14 = 0$.

(c) Also solve from your graph the equation $2x^2 + 3x - 2 = 0$.

Solution

The table of values for the function is

x	$x^2 + x - 12$	$f(x)$
–6	36 – 6 – 12	18
–5	25 – 5 – 12	8
–4	16 – 4 – 12	0
–3	9 – 3 – 12	–6
–2	4 – 2 – 12	–10
–1	1 – 1 – 12	–12
0	0 + 0 – 12	–12
1	1 + 1 – 12	–10
2	4 + 2 – 12	–6
3	9 + 3 – 12	0
4	16 + 4 – 12	8
5	25 + 5 – 12	18

The graph is shown below.

(b) $f(x) = x^2 + x - 12$.............①

$0 = x^2 - 14$ ... ... ②

① – ②: $f(x) = x + 2$ ... ... ③

The required straight line is $y = x + 2$ and its table of values is

x	0	–2	2
y	2	0	4

From the graph shown below, the solution of $x^2 - 14 = 0$ is $x \approx -3.8$ or $x \approx 3.8$.

(c) We The can solve the equation $2x^2 + 3x - 2 = 0$ in a similar manner.

$f(x) = x^2 + x - 12$.......................①

$0 = 2x^2 + 3x - 2$ ... ... ③

① $-\frac{1}{2}\times$ ③: $f(x) = -\frac{1}{2}x - 11$

The required straight line is $f(x) = -\frac{1}{2}x - 11$ whose table of values is

x	–6	2	6
y	–8	–12	–14

The line is shown on the graph below and from the graph, the solution of $2x^2 + 3x - 2 = 0$ is $x \approx -2$ and $x \approx 0.5$.

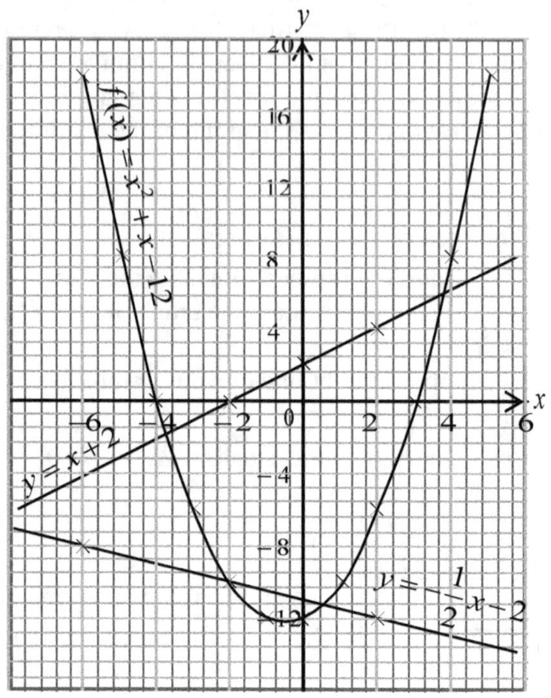

Turning Points

The **turning point** is the point where the curve changes direction.

Depending on the signs of Δ and *a*, the graph of the quadratic function $y = ax^2 + bx + c$ usually takes one of the forms in the following table.

Δ and nature of roots	$a > 0$ minimum value	$a < 0$ maximum value
Real and distinct roots $\Delta > 0$	y x	y x
Real and equal roots $\Delta = 0$	y x	y x
Roots are not real (Imaginary or Complex roots) $\Delta < 0$	y x	y x

Note!

1. A graph of quadratic function is symmetrical about an axis through the turning point.
2. The intercept with the y-axis is c.

Example

Find the turning point on the quadratic curve $y = 4 - 3x - x^2$ and the point where the curve cuts the y-axis.

Solution

The intercept with the x-axis occur when $y = 0$ i. e. $4 - 3x - x^2 = 0$.

$(4 + x)(1 - x) = 0 \Rightarrow x = -4$ or $x = 1$

But midpoint $= \dfrac{x_1 + x^2}{2} = \dfrac{-4+1}{2} = -\dfrac{3}{2}$

Therefore, the graph is symmetrical about the line$x = -\frac{3}{2}$.

Substituting in $y = 4 - 3x - x^2$, we have $y = 4 - 3\left(-\frac{3}{2}\right) - \left(-\frac{3}{2}\right)^2 = \frac{25}{4}$.

Therefore, the turning point is $\left(-\dfrac{3}{2}, \dfrac{25}{4}\right)$.

The intercept with the y-axis occurs when $x = 0$. $\Rightarrow y = 4$ is the intercept with the y-axis.

Curves and Tangents

For a straight line to be a tangent to a curve at a point, the line must touch the curve at one and only one point. The implication of this is that equal roots will result when the equation of the curve and that of the straight line are solved simultaneously.

Example

Determine which of the following lines, is a tangent to the curve $y = x^2$. (a) $y = x + 2$ (b) $y = 6x - 9$

Solution

(b) $x^2 = x + 2 \Rightarrow x^2 - x - 2 = 0 \Rightarrow (x-2)(x+1) = 0 \Rightarrow x = 2$ or $x = -1$

Therefore, the line cuts the curve at two points. Hence $y = x + 2$ is not a tangent to the curve.

(b) $x^2 = 6x - 9 \Rightarrow x^2 - 6x + 9 = 0 \Rightarrow (x-3)^2 = 0 \Rightarrow x = 3$.

Therefore, the line touches the curve at the point where $x = 3$. Hence, the line $y = 6x - 9$ is a tangent to the curve.

The Gradient of a Curve at a Point

The gradient of a straight line is constant, but the gradient of a curve varies from point to point. The gradient of a curve at a particular point is the same as the gradient of the tangent drawn to the curve at that point.

To find the gradient of a curve $y = f(x)$at a point $P(x, y)$,

(a) Draw the graph of the function.

(b) Draw the tangent to the curve at that point.

(c) Find the gradient of this tangent by choosing two points on the curve.

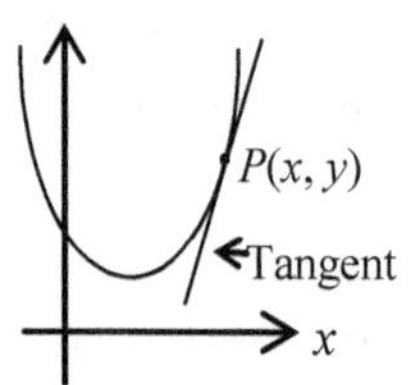

Example

Find the gradient of the curve $y = x^2$ at the point $P(2,4)$.

Solution

The table of values and the graph of $y = x^2$ are as shown below.

x	−3	−2	−1	0	1	2	3
y	9	4	1	0	1	4	9

The graph is shown below.

Gradient m at P $= \dfrac{y_2 - y_1}{x_2 - x_1} = \dfrac{8-0}{3-1} = 4$

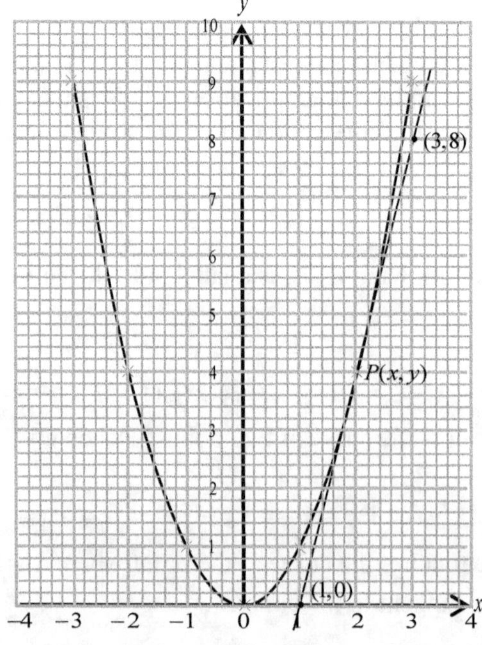

Gradient as a Rate of Change

The gradient of a function at any point is a measure of the rate of change of some quantity with respect to another at that point. For instance, the gradient of a distance/time graph of a moving particle at any point is the velocity of the particle at that point. Similarly, the gradient of a velocity/time graph of a moving particle at any point is the acceleration of the particle at that point.

Example

When a stone is thrown vertically upwards, its distance *d* m after *t* seconds, is given by the equation $s = 6t - t^2$.

(a) Using a scale of 1 cm to represent 1 s on the *t*-axis and 1 cm to represent 1 m on the *s*-axis, draw the graph of $s = 6t - t^2$ for values of *t* from 0 to 6 seconds.

From your graph, determine

(b) The maximum height attained by the stone.

(c) The time when the stone returns to its original position.

(d) The gradient of the curve when $t = 8$ seconds, stating it units and what the gradient stand for.

Solution

(a) The table of values and the graph are as below.

t (s)	0	1	2	3	4	5	6
s (m)	0	5	8	9	8	5	0

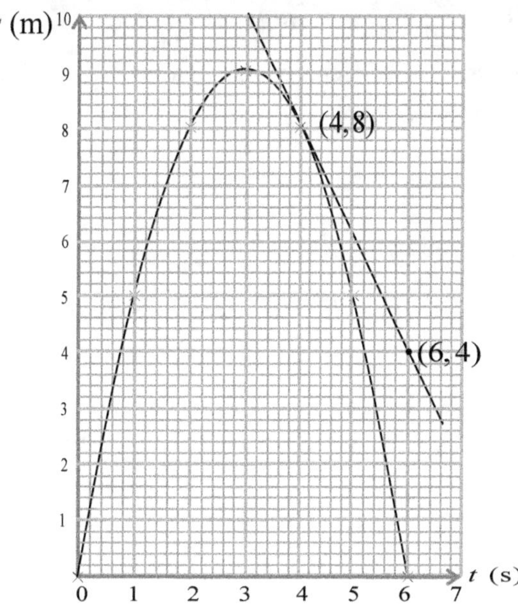

(b) Maximum height attained is 9 m.
(c) The stone returns to its original position at the sixth second.
(d) The gradient at $P(4,8) = \dfrac{s_2 - s_1}{t_2 - t_1} = \dfrac{8-4}{4-6} = -2.$

The units of the gradient are m/s. This gives the velocity of the stone when $t = 8$ seconds, so the units of the gradient are correct.

Elementary Differentiation

Differentiation of some Simple Functions

If c and n are real constants:

1. The derivative of $y = c$ is $\dfrac{dy}{dx} = 0.$
2. The derivative of $y = x^n$ is $\dfrac{dy}{dx} = nx^{n-1}.$
3. The derivative of $y = ax^n$ is $\dfrac{dy}{dx} = anx^{n-1}.$

Since $\dfrac{1}{x^n} = x^{-n}$,

4. The derivative of $y = \dfrac{a}{x^n}$ is $\dfrac{dy}{dx} = -anx^{-n-1}$
5. The derivative of a sum (or difference) is equal to the sum (or difference) of the derivatives of the terms.

Example

Find $\dfrac{dy}{dx}$ in each of the following.

(a) $y = x^3$ (b) $y = x$ (c) $y = 6$ (d) $y = x^5 + 3x^2$ (e) $y = x^2 - 4x + 2$

Solution

(a) $y = x^3 \Rightarrow \dfrac{dy}{dx} = 3x^2$ (b) $y = x \Rightarrow \dfrac{dy}{dx} = 1$ (c) $y = 6 \Rightarrow \dfrac{dy}{dx} = 0$

(d) $y = x^5 + 3x^2 \Rightarrow \dfrac{dy}{dx} = 5x^4 + 3(2)x = 5x^4 + 6x$ (e) $y = x^2 - 4x + 2 \Rightarrow \dfrac{dy}{dx} = 2x - 4$

Example

Find the gradient of the curve $y = 3x^2 - 11x + 9$ at the point where $x = 4$.

Solution

$\frac{dy}{dx} = 6x - 11 \Rightarrow$ when $x = 4, \frac{dy}{dx} = 6(4) - 11 = 13$

Therefore the gradient at $x = 4$ is 13.

Maximum and Minimum Values

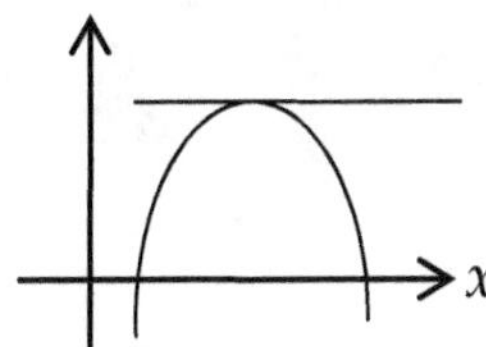

At the maximum point $\frac{dy}{dx} = 0$ and $\frac{d^2y}{dx^2} < 0$.

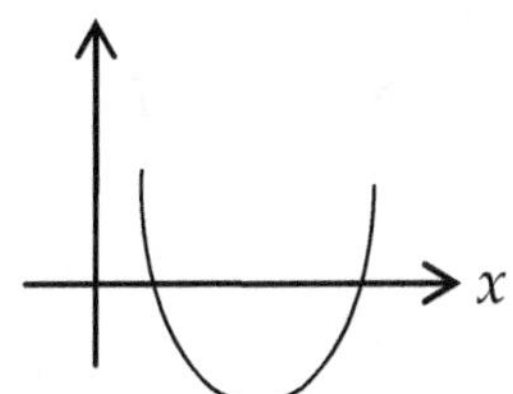

At a minimum value $\frac{dy}{dx} = 0$ and $\frac{d^2y}{dx^2} > 0$.

Velocity and Acceleration

$$v = \frac{ds}{dt} \Rightarrow a = \frac{dv}{dt} \text{ or } a = \frac{d^2s}{dt^2}.$$

Certain values of velocity and acceleration have implications as follows,

1. When $v = 0$, the particle is at rest.
2. When $v < 0$ the particle is moving in the opposite direction to that in which the displacement is measured.
3. When $v > 0$ the particle is moving in the same direction to that in which the displacement is measured.
4. When $a = 0$, the velocity of the particle is constant (steady motion).
5. When $a > 0$ the particle is accelerating or speeding up.
6. When $a < 0$ the particle is retarding or slowing down.

Example

A particle moves in a straight line so that its distance from a fixed point O after t seconds is s metres, where $s = \frac{1}{3}t^3 - \frac{3}{2}t^2 + 2t$. Find the two times when the particle is at rest. Find the acceleration of the particle at these times and interpret the result.

Solution

$s = \frac{1}{3}t^3 - \frac{3}{2}t^2 + 2t \Rightarrow \frac{ds}{dt} = t^2 - 3t + 2$

At rest $\frac{ds}{dt} = 0 \Rightarrow t^2 - 3t + 2 = 0$

$\Rightarrow (t-1)(t-2) = 0 \Rightarrow t = 1$ or $t = 2$

Therefore, the particle is at rest when $t = 1$ s and when $t = 2$ s.

$\frac{ds}{dt} = t^2 - 3t + 2 \Rightarrow \frac{d^2s}{dt^2} = 2t - 3$

When t = 1, $\Rightarrow \frac{d^2s}{dt^2} = 2(1) - 3 = -1$ ms^{-2} . $\Rightarrow$The particle is slowing down.

When t = 2, $\Rightarrow \frac{d^2s}{dt^2} = 2(2) - 3 = 1$ ms^{-2} .$\Rightarrow$ The particle is speeding up.

Speed time Graphs

The figure below, shows a speed time graph for a body that starts from rest (i.e. with zero velocity) at time $t = 0$ s and accelerates uniformly until it attains a velocity of 10 m/s at A. The body then moves with uniform velocity of 10 m/s for 5 s from A to B. At B, the velocity of the body decreases steadily until the body comes to rest again at $t = 12$ s .

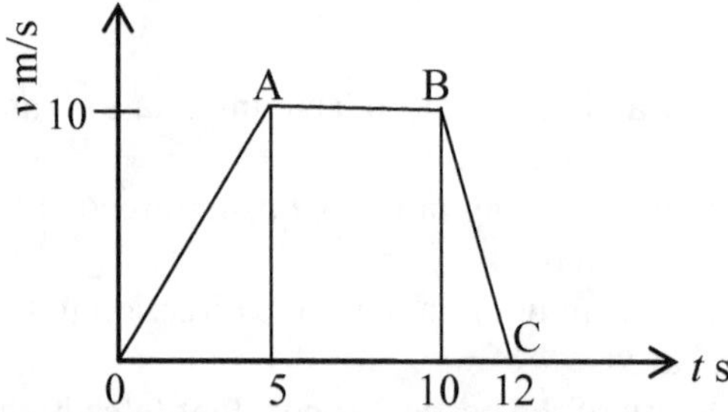

A graph such as this gives two pieces of information.

(a) The total distance covered by the body is the area under the graph and is equal to the area of trapezium.

(b) The gradient $\frac{ds}{dt}$ of the graph in each section of its motion gives the acceleration of the particle in that section.

Example

From the figure above, calculate

(i) The total distance covered by the body.

(ii) The acceleration of the particle in each section.

Solution

(i) Total distance = Area of trapezium $= \frac{1}{2}(OC + AB)h$

$$\Rightarrow \text{Total distance} = \frac{1}{2}(12+5)10 = 85 \text{ m}$$

(ii) From O to A, acceleration $= \frac{10}{5} = 2 \text{ m/s}^2$

From A to B, acceleration $= 0 \text{ m/s}^2$

From B to C, acceleration $= -\frac{10}{2} = -5 \text{ m/s}^2$.

Distance Time Graphs

Example

A boy left a town at 8 a.m. trekking to his village 8 km away at a speed of 5 km/h. After trekking 2 km he meets a car coming from the village at a steady speed. Arriving in town at 8:30 a.m. the driver loads passengers for 15 minutes then immediately returns to the village at the same speed. Draw a graph for these journeys and use it to determine

(a) The time at which the car left the village.

(b) The time when the car overtakes the man.

(c) The distance covered by the man by the time the car overtakes him.

Solution

From the graph shown below,

(a) The car left the village at 8:45 a.m.

(b) The car overtakes the man at 8:48 a.m.

(c) The distance covered by the man 4.6 km.

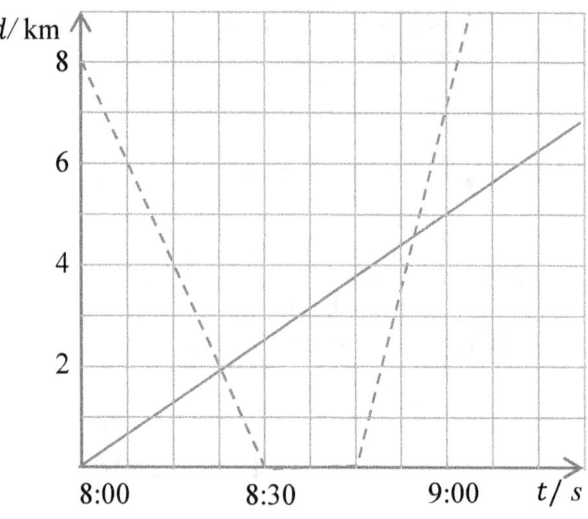

MULTIPLE CHOICE EXERCISE 7:2

1. The root of the equation represented by the graph in Figure 34:19 is:
   [A] 4 [B] 7 [C] –4 [D] –7

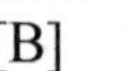

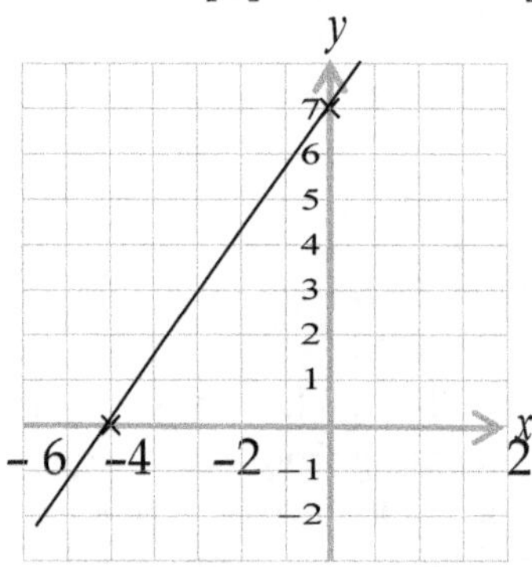

2. The graph of $f(x) = x^2 + x - 6$ is most likely:

[A]

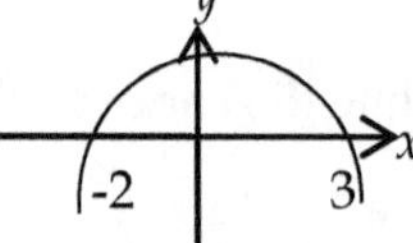

[B]

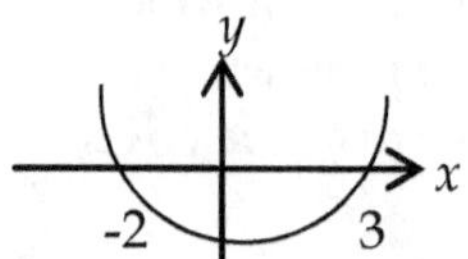

[C]

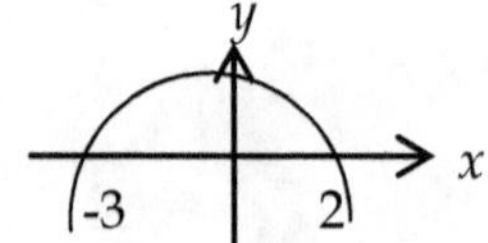

[D]

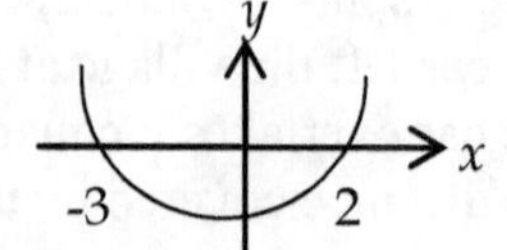

3. The graph of $y = 4 - 3x - x^2$ is most certainly:

[A]

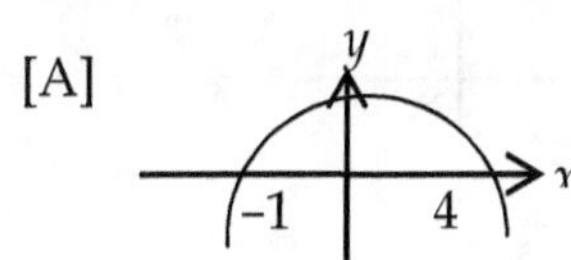

[B]

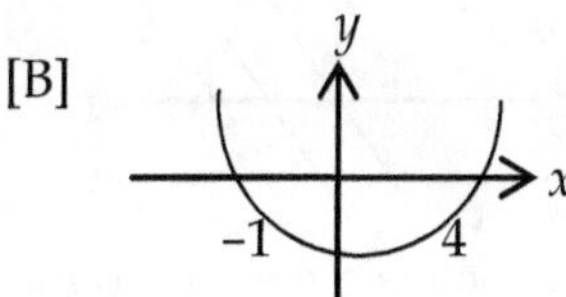

[C]

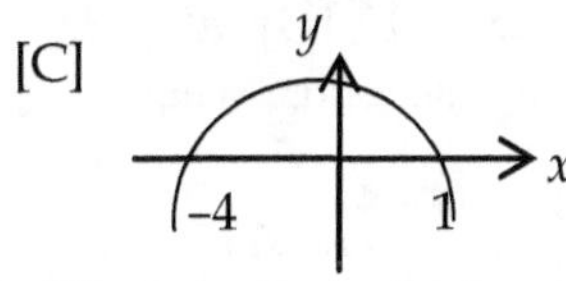

[D]

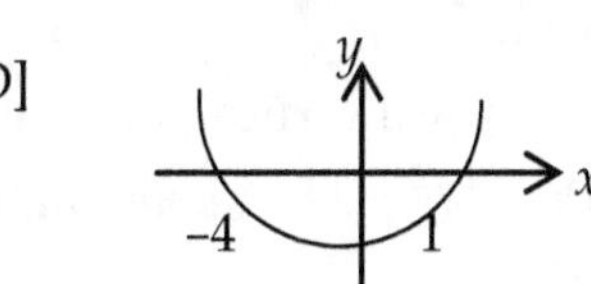

4. The equation which represents the sketch graph below is:

[A] $y = 8 - 2x + x^2$ [B] $y = 8 + 2x + x^2$

[C] $y = 8 - 2x - x^2$

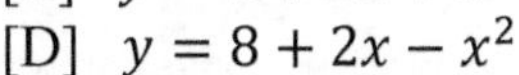
[D] $y = 8 + 2x - x^2$

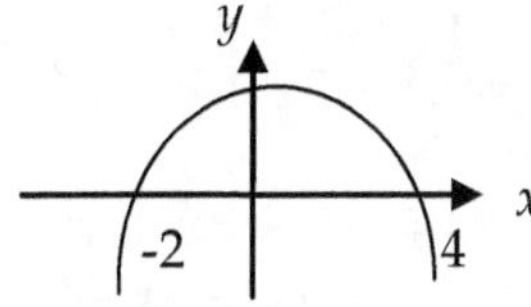

5. The equation which represents the sketch graph (i) below is:

[A] $y = x^2 - 2x - 3$ [B] $y = x^2 + 2x - 3$

[C] $y = x^2 + 2x + 3$

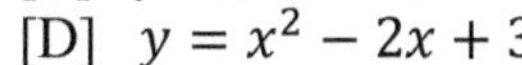
[D] $y = x^2 - 2x + 3$

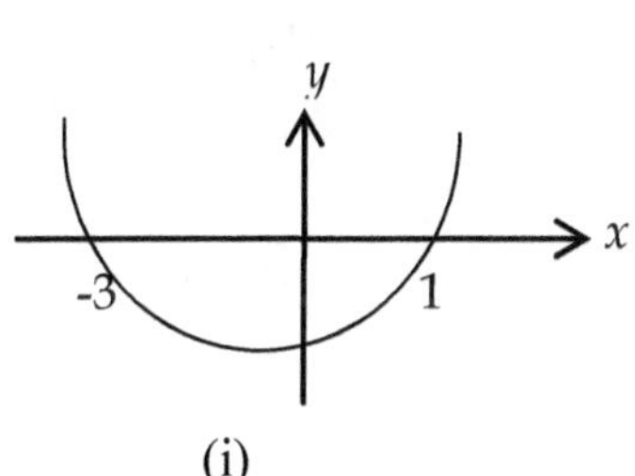

(i)

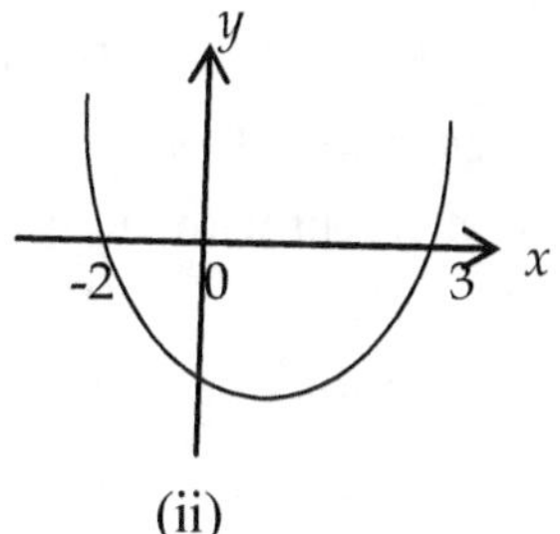

(ii)

6. Graph (ii) above is the graph of the function:

[A] $y = x^2 - x - 6$ [B] $y = x^2 + x - 6$

[C] $y = x^2 + x + 6$ [D] $y = x^2 - x + 6$

7. The graph representing inconsistent lines is:

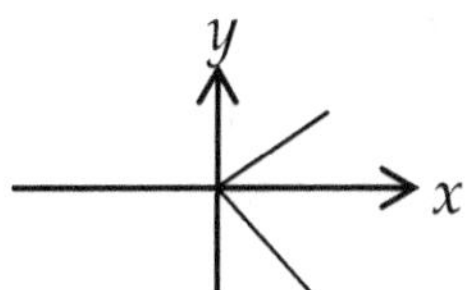

[B]

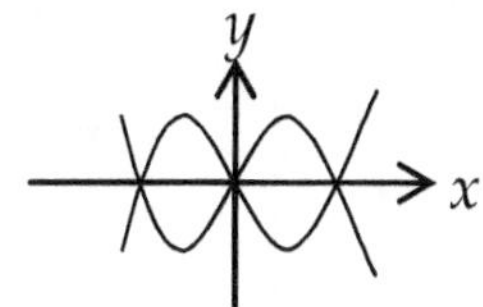

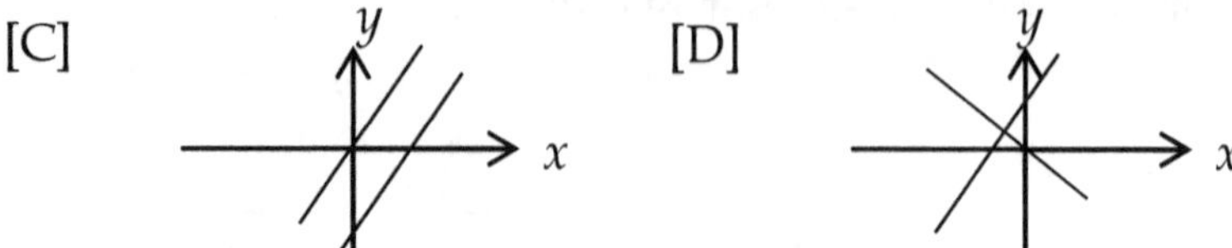

8. The line which is a tangent to the curve $y = x^2$ is:
   [A] $y = 2x - 3$ [B] $y = 2x - 2$ [C] $y = 2x - 1$ [D] $y = 2x$
9. A stone thrown vertically upward moves s metres in t seconds, where $s = 80t - 5t^2$. The maximum height it reaches is:
   [A] 640 m [B] 320 m [C] 75 m [D] 40 m
10. The maximum value of $x^2 - 4x + 5$ is:
    [A] –5 [B] 0 [C] –1 [D] 1
11. The function whose curve has a maximum point is:
    [A] $y = 8 - 2x - x^2$ [B] $y = 8 + 2x + x^2$
    [C] $y = x^2 + 2x - 8$ [D] $y = x^2 - 2x + 8$
12. The function whose curve has a minimum point is:
    [A] $f(x) = 12 + 8x - x^2$ [B] $f(x) = 12 - 8x + x^2$
    [C] $f(x) = 12 - 8x - x^2$ [D] $f(x) = -12 + 8x - x^2$
13. The graph of the curve $y = 2x^2 - 5x - 1$ and a straight line PQ were drawn to solve the equation $y = 2x^2 - 5x - 1$. The equation of the straight line PQ is:
    [A] $y = 1$ [B] $y = 0$ [C] $y = 3$ [D] $y = -3$
14. The equation of a curve is given by $y = 2x^2 - x - 1$.
    The intercept with the y-axis is:
    [A] $(-1, 0)$ [B] $(1, 0)$ [C] $(0, -1)$ [D] $(0, 1)$
15. The equation of a curve is given by $y = 2x^2 - x - 1$.
    The intercepts with the x-axis is:
    [A] $(0, 1)$ and $\left(0, -\frac{1}{2}\right)$ [B] $(-1, 0)$ and $\left(-\frac{1}{2}, 0\right)$
    [C] $(-1, 0)$ and $\left(\frac{1}{2}, 0\right)$ [D] $(1, 0)$ and $\left(-\frac{1}{2}, 0\right)$
16. A particle P moves so that its distance, s metres, from a fixed point O, at time t seconds, $t \geq 0$ is given by $s = 3 + 8t - \frac{1}{12}t^2$. The velocity V of the particle when $t = 4$ is:
    [A] $\frac{77}{3}$ m/s [B] $\frac{22}{3}$ m/s [C] $\frac{68}{3}$ m/s [D] $\frac{72}{3}$ m/s

17. A particle P moves so that its distance, s metres, from a fixed point O, at time t seconds, $t \geq 0$ is given by $3 + 8t - \frac{1}{12}t^2$. The value of t when P finally comes to rest is:
[A] $t = 24$ s [B] $t = 32$ s [C] $t = 48$ s [D] $t = 64$ s
18. A train of length 100 m is traveling at a speed of 40 km per hour. In minutes, the length of time taken for the train to pass completely over a bridge of length 0.7 km is:
[A] 1.2 [B] 0.02 [C] 12 [D] 0.2
19. A particle P moves so that its distance, s metres, from a fixed point O, at time t seconds is given by $s = 20 + 24t \quad t^2$. The distance traveled during the first 3 seconds of motion is:
[A] 20 m [B] 72 m [C] 83 m [D] 92 m
20. A particle P moves so that its distance, s metres, from a fixed point O, at time t seconds is given by $s = 20 + 24t - t^2$. The initial velocity of the particle is:
[A] 20 m/s [B] 24 m/s [C] 22 m/s [D] 45 m/s
21. A particle P moves so that its distance, s metres, from a fixed point O, at time t seconds is given by $s = 20 + 24t - t^2$. The velocity of the particle when $t = 3$ s is:
[A] 20 m/s [B] 45 m/s [C] 18 m/s [D] 21 m/s
22. The distance s in metres of a particle from a fixed point O is given by $s = t^2 = \frac{13}{2}t^2 + 14t + 5$ where t is the time taken in seconds. The velocity of the particle when $t = 1$ second is:
[A] 5 m/s [B] –1 m/s [C] 14 m/s [D] 4 m/s
23. Figure 34:27 not drawn to scale shows the speed time graph of a cyclist. The length of time in hours for which the cyclist rode at uniform speed is:
[A] 3 [B] 4 [C] 6 [D] 18

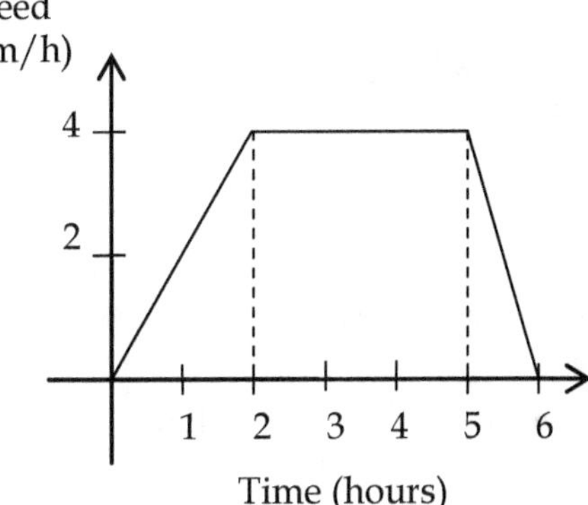

24. The table of values for $y = x - 6$ is:

[A]

x	–5	–8	–7
y	1	–14	–13

[B]

x	–5	–8	–7
y	–11	–2	–13

[C]

x	–5	–8	–7
y	–11	–14	–13

[D]

x	–5	–8	–7
y	1	–2	–1

25. The equation which corresponds to the table of values is:
[A] $y = 4 + 5x$ [B] $y = 3 + 6x$ [C] $y = 5 + 4x$ [D] $y = 6 + 3x$

Input (x)	1	2	3	4	5
Output (y)	9	12	15	18	21

26. The graph of a quadratic function is given as
[A] $\Delta > 0, a > 0$ [B] $\Delta < 0, a < 0$ [C] $\Delta > 0, a < 0$ [D] $\Delta < 0, a > 0$

7.3 VARIATION (PROPORTIONS)

DIRECT VARIATION

If y is **directly proportional** to x we write $y \propto x \Leftrightarrow y = kx \Leftrightarrow k = \frac{y}{x}$.

If y is **directly proportional** to x^2 we write $y \propto x^2 \Leftrightarrow y = kx^2 \Leftrightarrow k = \frac{y}{x^2}$.

Example

Given that y is directly proportional to x and $y = 64$ when $x = 8$. Find the value of y when $x = 20$.

Solution

$$y \propto x \Rightarrow k = \frac{y}{x} \Rightarrow k = \frac{64}{8} = 8$$

When $x = 20$, $y = kx \Rightarrow y = 8(20) = 160$

INVERSE VARIATION

If y is **inversely proportional** to x we write $y \propto \frac{1}{x} \Leftrightarrow y = \frac{k}{x} \Leftrightarrow k = xy$.

If y is **inversely proportional** to x^2 we write $y \propto \frac{1}{x^2} \Leftrightarrow y = \frac{k}{x^2} \Leftrightarrow k = x^2y$.

Example

Given that y is inversely proportional to x and $y = 4$ when $x = 12$, find the value of y when $x = 3$.

Solution

$y \propto \frac{1}{x} \Leftrightarrow k = yx \Rightarrow$ when $y = 4$ and $x = 12$, $k = 4(12) = 48$.

When $x = 3$, $y = \frac{k}{x} = \frac{48}{3} \Rightarrow y = 16$.

JOINT OR COMBINED VARIATION

Join or **combined variation** is a variation in which one quantity varies as two or more quantities vary. For instance $V = \pi r^2 h$.

Example

The pressure of a fixed mass of gas varies directly as the temperature and inversely as the volume of the gas. Express this statement symbolically.

Solution

Combining $P \propto T$ and $P \propto \frac{1}{V}$ gives $P \propto \frac{T}{V}$.

The constant of proportionality is m, since the mass, m of the gas is fixed;

Therefore $P = \frac{mT}{V}$.

Partial Variation

A **partial variation** is a variation which is expressed as two or more terms e.g. $S = \pi r^2 + \pi rl$.

MULTIPLE CHOICE EXERCISE 7:3

1. Given that n varies directly as m and if $n = 8$ when $m = 20$. The value of m when $n = 7$ is:

   [A] 13 [B] 15 [C] $17\frac{1}{2}$ [D] $18\frac{1}{2}$

2. Given that $(x+3)$ varies directly as y and $x = 3$ when $y = 12$, the value of x when $y = 8$ is:

[A] 1 [B] $\frac{1}{2}$ [C] $-\frac{1}{2}$ [D] -1

3. Given that y is directly proportional to x^2 and $y = 5$ when $x = 2$, then when $x = 6$, $y =$:
   [A] 18 [B] 21 [C] 27 [D] 45
4. P varies inversely as the square of W. When $W = 4$, $P = 9$, then the value of P when $W = 9$ is:
   [A] $\frac{4}{3}$ [B] 6 [C] 4 [D] $\frac{16}{9}$
5. Given that y varies inversely as x^2 then x varies:
   [A] inversely as y^2 [B] inversely as $\sqrt{y}$.
   [C] directly as y^2 [D] directly as $\sqrt{y}$
6. Given that x varies inversely as y and $x = \dfrac{2}{3}$ when $y = 9$, the value of y when $x = \dfrac{3}{4}$ is:
   [A] $\frac{1}{18}$ [B] $\frac{81}{8}$ [C] $\frac{9}{2}$ [D] 8
7. Given that $y \propto \dfrac{1}{\sqrt{x}}$ and $x = 16$ when $y = 2$ when $y = 24$, x will be:
   [A] $\frac{1}{9}$ [B] $\frac{1}{6}$ [C] $\frac{1}{3}$ [D] $\frac{2}{3}$
8. Given that x is inversely proportional to m^2 and $x = 3$ when $m = 9$, the value of x when $m = 3$ is:
   [A] 3 [B] 6 [C] 9 [D] 27
9. Given that $p \propto \dfrac{1}{\sqrt{r}}$ and $p = 3$ when $r = 16$ the value of r when $p = \dfrac{3}{2}$ is:
   [A] 48 [B] 72 [C] 64 [D] 324
10. Given that R is inversely proportional to S and $R = 15$ when $S = 12$. The value of S when $R = 60$ is:
    [A] $\frac{1}{4}$ [B] 3 [C] 4 [D] 5
11. m varies directly as n and inversely as the square of p; Given that $m = 3$, when $n = 2$ and $P = 1$. The value of m in terms of n and p is:
    [A] $m = \frac{2n}{3p}$ [B] $m = \frac{3n}{2p}$ [C] $m = \frac{2n}{3p^2}$ [D] $m = \frac{3n}{2p^2}$
12. Given that p varies directly as q while q varies inversely as r. The statement which is true:
    [A] r varies directly as p. [B] p varies inversely as r.
    [C] p varies directly as r. [D] q varies inversely as p.
13. Given that 20 men take 6 days to clear a field. The time it would take 12 men working at the same rate to clear a similar field is:
    [A] 40 days [B] 2 days [C] $3\frac{1}{2}$ days [D] 10 days

14. K varies directly as N and inversely as the square of L. Given that $L = 1$ when $N = 3$ and $K = 2$. The value of K in terms of N and L is:

[A] $K = \frac{2N}{3L^2}$ [B] $K = \frac{3N}{2L^2}$ [C] $K = \frac{2L^2}{3N}$ [D] $K = \frac{3L^2}{2N}$

15. Given that $x \propto y$ and $y \propto \frac{1}{z^2}$. The way x varies with z is:

[A] $x \propto \frac{1}{z}$ [B] $x \propto \frac{1}{\sqrt{z}}$ [C] $x \propto \frac{1}{z^2}$ [D] $x \propto \frac{1}{\sqrt{z}}$

16. $x \propto y$ and that $x = 28$ when $y = 4$. The formula connecting x and y is:

[A] $x = 2y$ [B] $x = 4y$ [C] $x = 7y$ [D] $x = 14y$

17. $x \propto y$ and when $x = 4$, $y = 20$. The value of x when $y = 5$ is:

[A] 4 [B] 3 [C] 2 [D] 1

18. $m \propto \frac{1}{n}$ and $m = 3$ when $n = 2$. The law connecting m and n is:

[A] $m = 6n$ [B] $m = 3n$ [C] $m = \frac{6}{n}$ [D] $m = \frac{3}{n}$

19. Given that $x \propto \frac{1}{y}$ and that $x = 9$ when $x = 4$. The formula which connects x and y is:

[A] $x = \frac{36}{y}$ [B] $x = \frac{13}{y}$ [C] $x = \frac{9}{y}$ [D] $x = \frac{5}{y}$

20. L varies jointly as M and N. When $M = 2$, $N = 3$ and $L = 9$. The law connecting L, M and N is:

[A] $M = \frac{2}{3}LN$ [B] $L = MN$ [C] $L = \frac{2}{3}MN$ [D] $M = \frac{3}{2}LN$

21. R varies directly as t and inversely as m. K is the constant of proportionality. The relationship between R, t and m is:

[A] $R = \frac{Km}{t}$ [B] $R = \frac{Kt}{m}$ [C] $R = K + \frac{m}{t}$ [D] $R = t + \frac{K}{m}$

22. The energy E of a moving body varies partly as the square of the height, H of the body above sea level and partly as the square root of its velocity, V. Given that a and b are constants, the equation representing the above expression is:

[A] $E = aH^2 + b\sqrt{V}$ [B] $E = a\sqrt{H} + bV^2$

[C] $E = \frac{a}{H^2} + \frac{b}{\sqrt{V}}$ [D] $E = \frac{a}{\sqrt{H}} + \frac{b}{V^2}$

23. The equation which represents the relation in the following table is:

[A] $y = -3x + 8$ [B] $y = \frac{1}{3}x + 8$ [C] $y = -\frac{1}{3}x + 8$ [D] $y = 3x + 8$

x	−3	−2	−1	0	1
y	1	−2	−5	−8	−11

24. The equation which represents the relation in the following table is:

[A] $y = 8x + 9$ [B] $y = \frac{1}{8}x + 9$ [C] $y = -8x + 9$ [D] $y = -\frac{1}{8}x + 9$

x	−2	−1	0	1	2
y	−7	1	9	17	25

25. The equation which best describes the relation between x and y in the following table is:

[A] $y = x^2 + 5$ [B] $y =- x^2 - 5$ [C] $y = x^2 - 5$ [D] $y =- x^2 + 5$

x	-1	0	1	2	3
y	4	5	4	1	-4

7.4 GRAPHICAL INEQUALITIES

Vertical and Horizontal Boundary Lines

The line $x = a$ is vertical and the line $y = b$ is horizontal.

Example

Represent the following inequations on separate Cartesian planes.

(a) $x > 2$ (b) $x < 0$ (c) $y \leq 3$ (d) $y > 0$

Solution

(a)

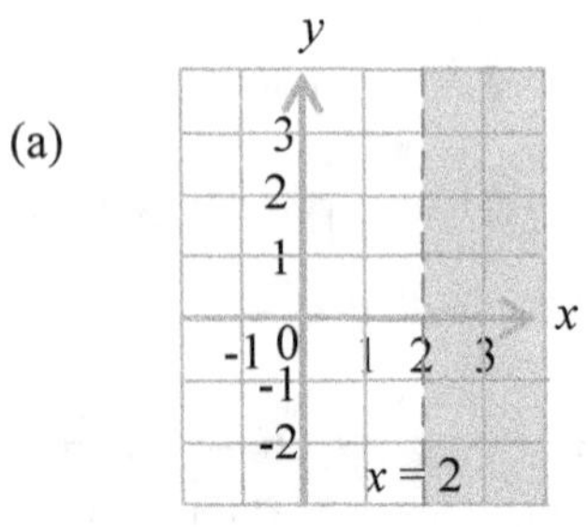

(b)

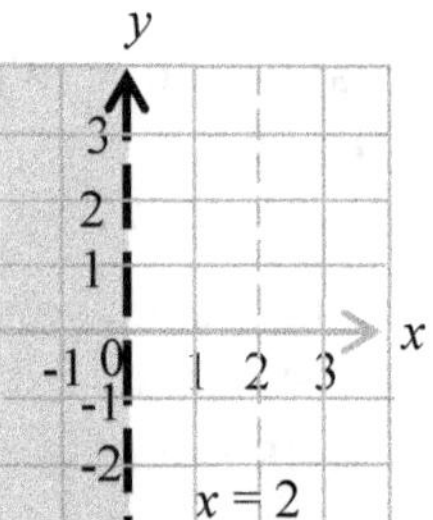

(c)

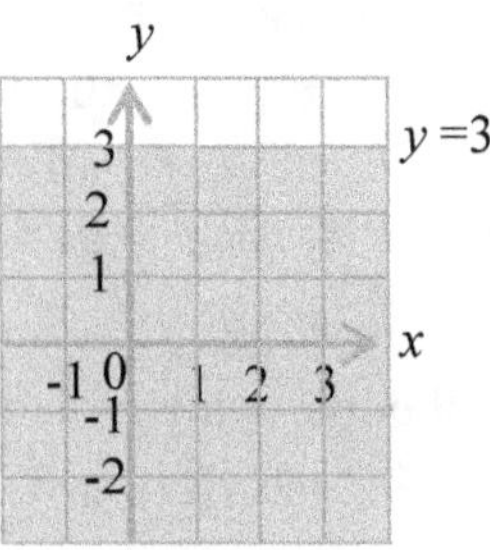

(d)

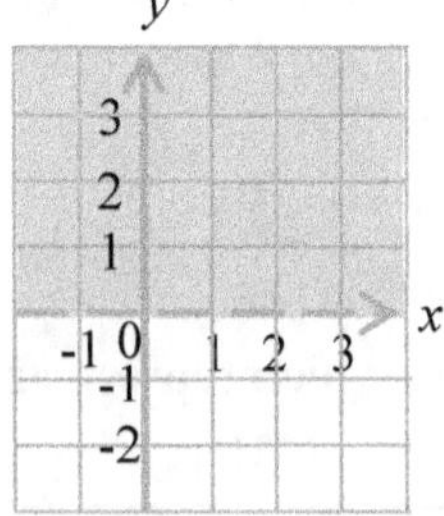

Simultaneous Linear Inequations

Graphically, the region described by the intersection of two or more inequations give the solution of the inequations.

Example

Shade the region in which the inequations $x + y < 7$, $x - y < 3$ and $y \geq -4$ are all satisfied.

Solution

For the line $y + x = 7$,

x	0	2	7
y	7	5	0

The graph of the above inequalities is shown below.

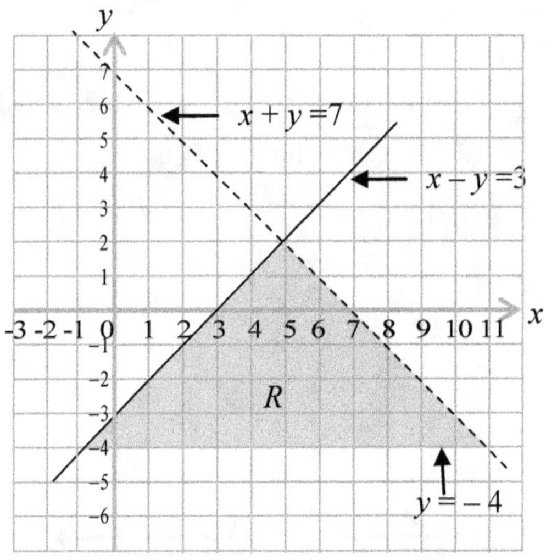

$R = \{(x, y) : x - y \geq 3\} \cap \{(x, y) : x + y < 7\} \cap \{(x, y) : y \geq -4\}$

MULTIPLE CHOICE EXERCISE 7:4

1. In the following graph, the shaded portion shows the boundary of the half plane defined by the inequality:

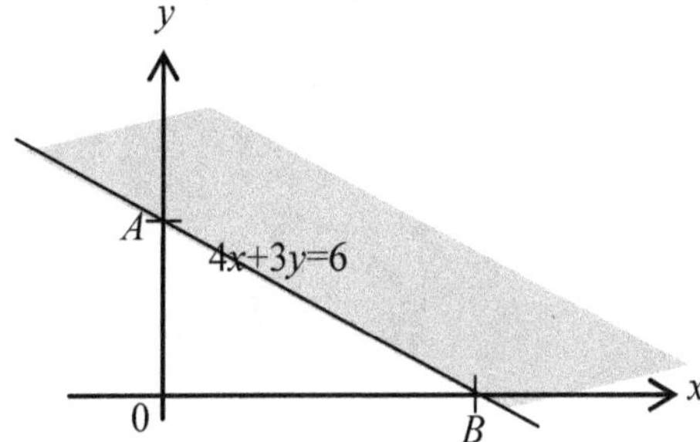

[A] $4x + 3y > 6$

[B] $4x + 3y < 6$

[C] $4x + 3y \geq 6$

[D] $4x + 3y \leq 6$

2. In the figure above, the co-ordinates of point B is:

[A] $\left(0,1\frac{1}{2}\right)$ [B] $(0,2)$ [C] $(2,0)$ [D] $\left(1\frac{1}{2},0\right)$

3. The inequality illustrated in the sketch graph below is:

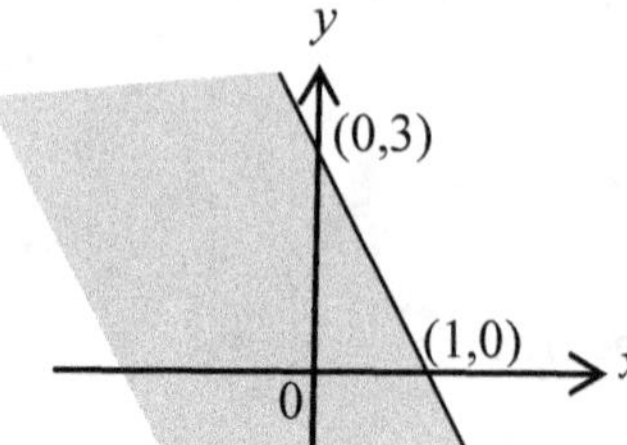

[A] $y > -2x + 3$

[B] $y \geq -3x + 3$

[C] $y \geq 3x + 2$

[D] $y > -2x + 3$

4. In The following graph, the region P, Q, R or T which satisfies the inequalities $0 < y < 1$, $y < x + 2$ and $x < 0$ is:

[A] P [B] Q [C] R [D] S

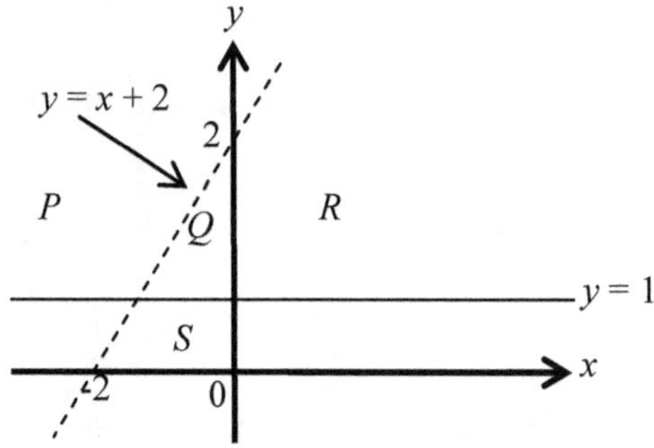

5. In the following graph, the region defined by triangle OPQ can be represented by the inequalities:

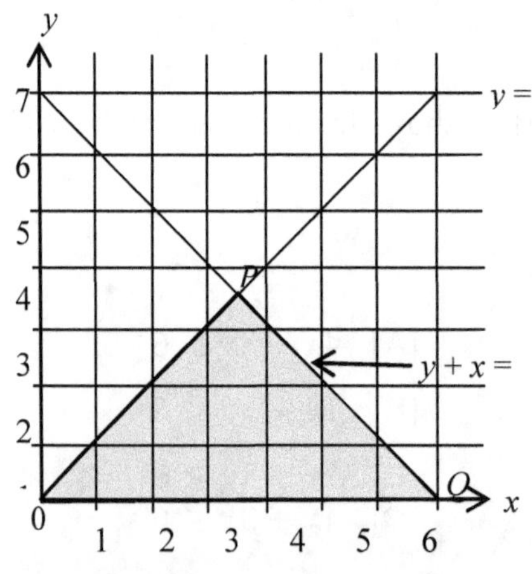

[A] $y \geq x, y \leq 0,\ x + y \geq 7$

[B] $y \leq x, y \geq 0,\ x + y \leq 7$

[C] $y \geq x, y \geq 0,\ x + y \leq 7$

[D] $y \leq x, y \leq 0,\ x + y \leq 7$

6. In the following figure, the equations of the lines *AC*, *AB*, and *BC* are:

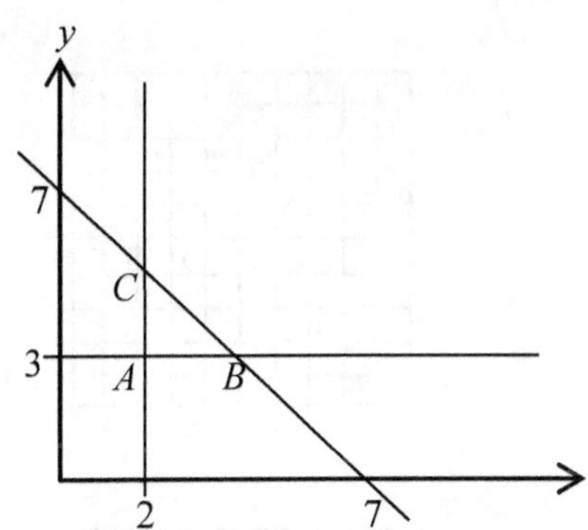

[A] $x = 2, y = 3,\ x + y = 7$

[B] $x = 3, y = 2,\ x + y = 7$

[C] $x = -2, y = 3,\ x + y = 7$

[D] $x = 2, y = -3,\ x + y = 7$

7. In the figure above, the three inequalities which define the triangle *ABC* are:

[A] $x \geq 2, y \leq 3,\ x + y \leq 7$ [B] $x \geq 3, y \geq 2,\ x + y \leq 7$

[C] $x \geq 2, y \geq 3,\ x + y \leq 7$ [D] $x \geq 2, y \geq -3,\ x + y \geq 7$

8. The shaded region in the following graph is best described by:

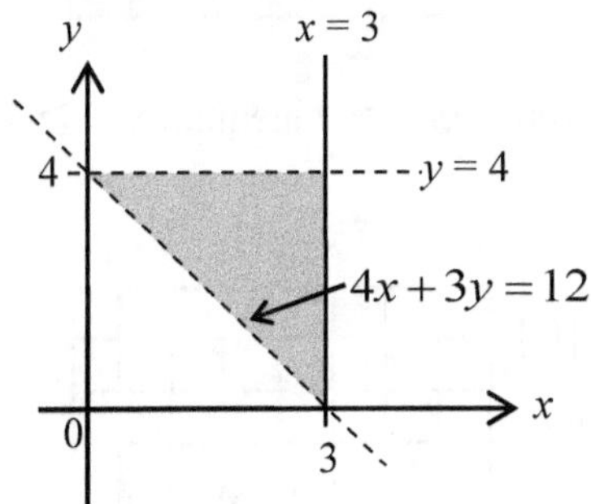

[A] $x \leq 3, y \leq 4$ and $4x + 3y \leq 12$

[B] $x \leq 3, y < 4$ and $4x + 3y < 12$

[C] $x \geq 3, y \geq 4$ and $4x + 3y > 12$

[D] $x \leq 3, y < 4$ and $4x + 3y > 12$

9. The graph which represents the inequality $x + 1 \leq 0$ is:

[A]

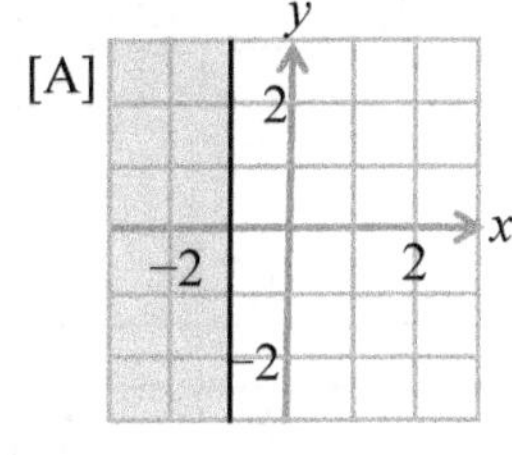

[B]

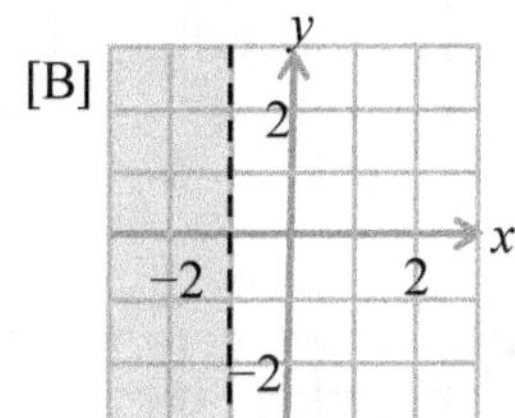

[C]

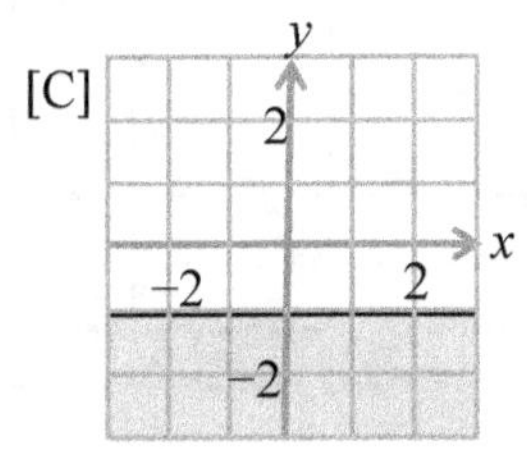

[D]

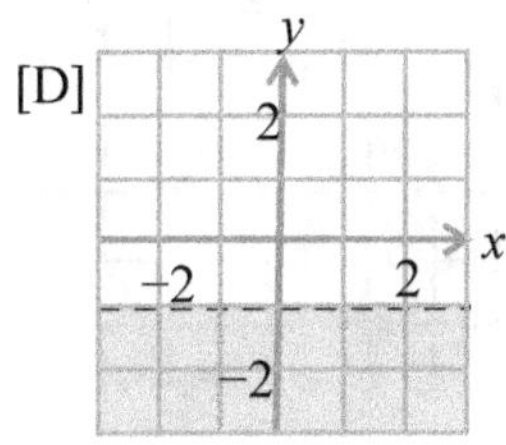

10. The inequalities $y > x + 2$ and $y \geq 1 + 2x$ are represented by the graph:

[A]

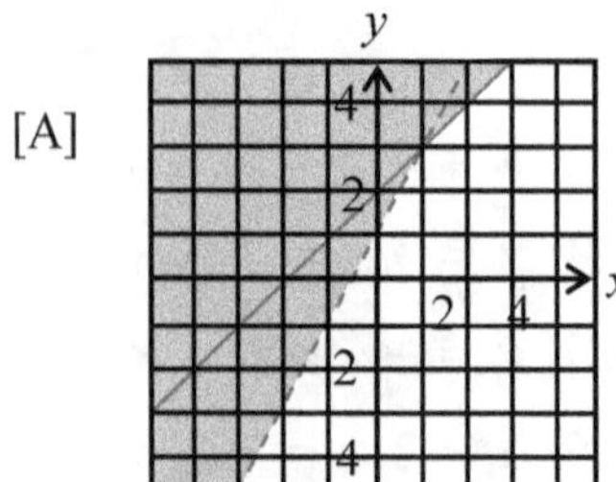

[B]

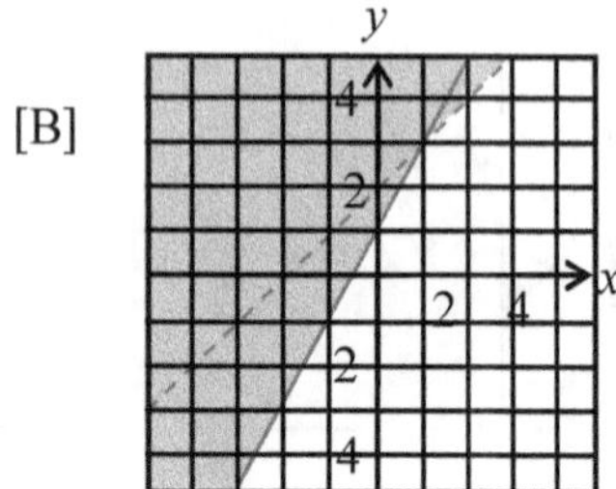

[C]

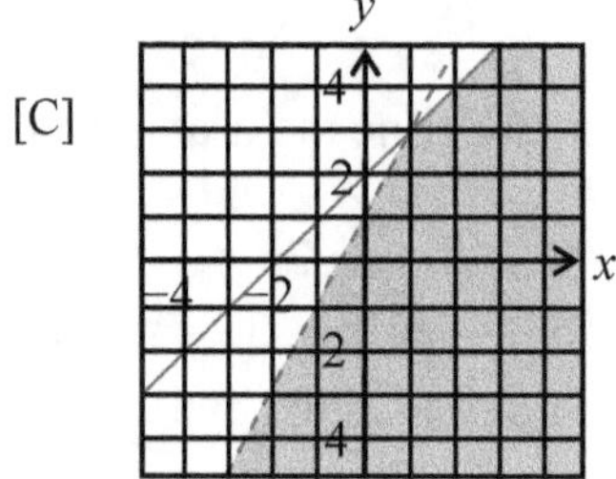

[D]

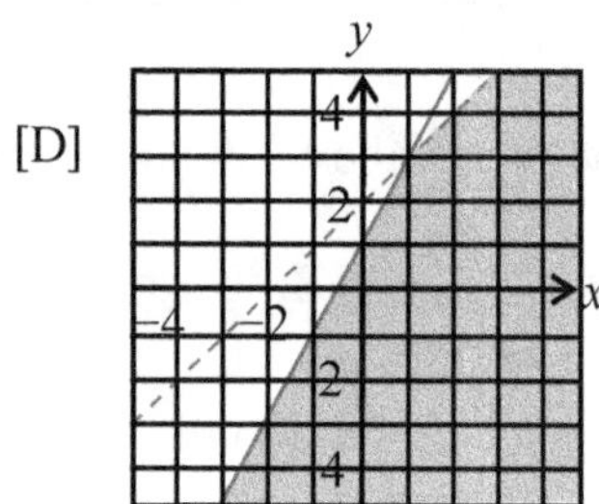

11. The graph among the following which represents the inequality $y \geq x - 1$ is:

[C]

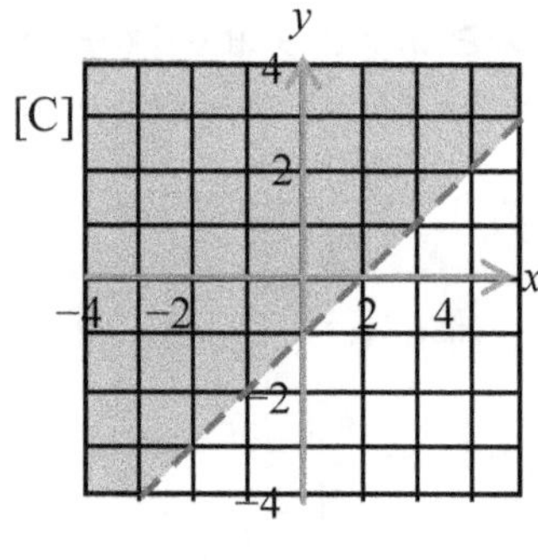

[D]

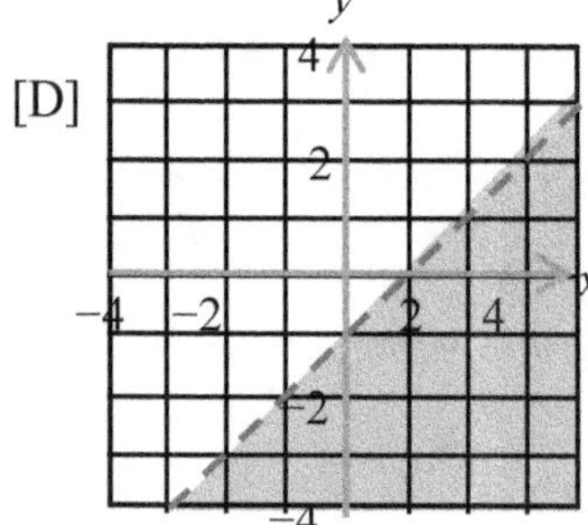

[A]

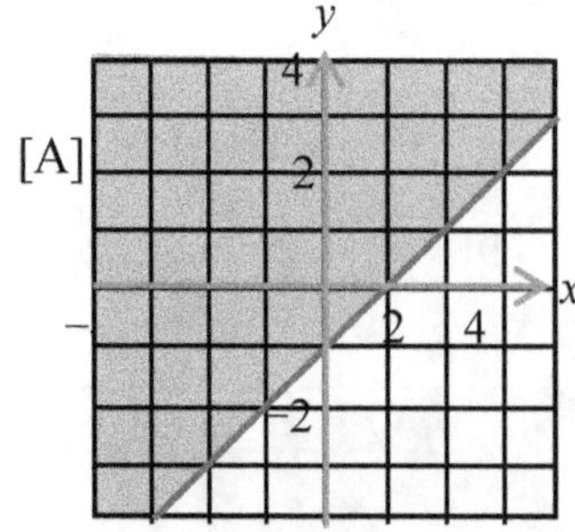

[B]

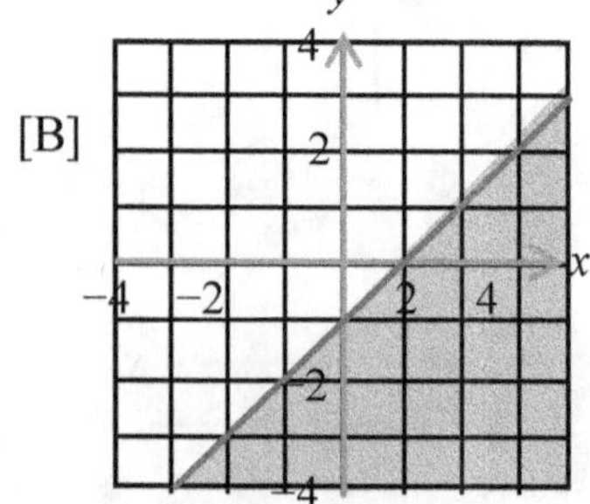

12. The linear inequality which represents the graph is:

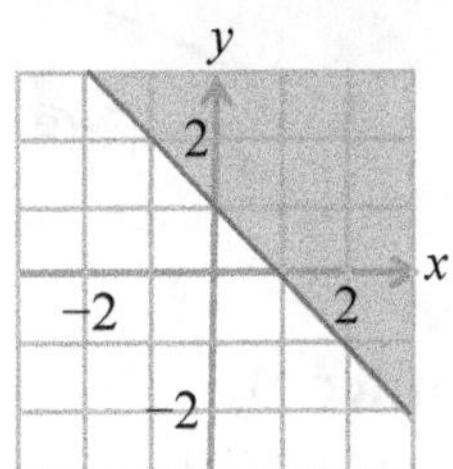

[A] $y \le -x + 1$

[B] $y \ge -x + 1$

[C] $y \le x + 1$

[D] $y \ge x + 1$

7.5 NETWORKS

Network Terminology

A **network** is a collection of points called **vertices (**or **nodes)** and lines, called **edges** (or **arcs** or **links)**, connecting these points.
A **region** is the area bounded by the vertices and edges.

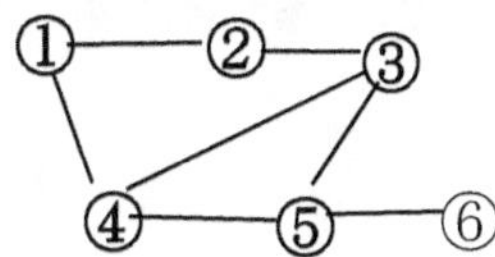

The network on the left has 6 vertices, 7 edges and 3 regions. The area outside the network diagram is counted as one region.

Euler's formula states that in any network, $R + V - E = 2$.

Example

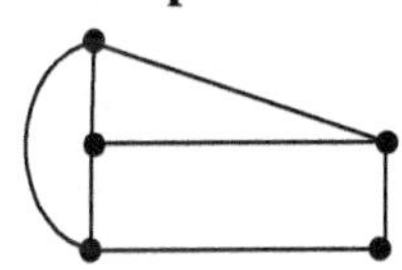

State the number of vertices, regions, and edges in the network. Hence verify that $R + V - E = 2$.

Solution
The network has 5 vertices, 4 regions, and 7 edges.
$R + V - E = 5 + 4 - 7 = 2$ as required.

Ordered Lists and Unordered Lists

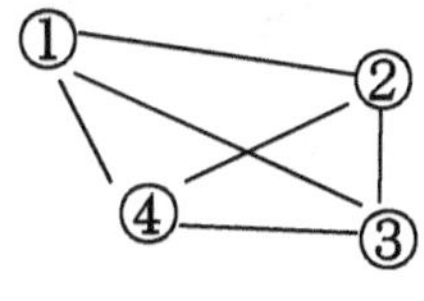

An **unordered list** usually enclosed in braces is a list in which the order in which the elements are written is immaterial. For instance the network in the figure on the left can be represented by the unordered list $\{1, 2, 3, 4\}$.

An **ordered list** usually enclosed in parentheses is a list in which the order in which the elements are written is important. For instance the network in the figure on the right can be represented by the ordered list (1, 2, 4, 3) which indicates that the flow is from 1 to 2 to 4 and to 3.

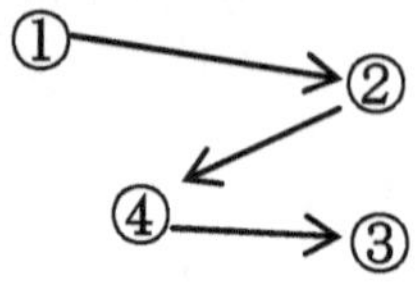

The Degree or Order of a Vertex

The number of edges meeting at a vertex is called the **order** of the vertex. If the order of a vertex is *even*, the vertex is called an **even vertex** otherwise it is an **odd vertex**.

Example

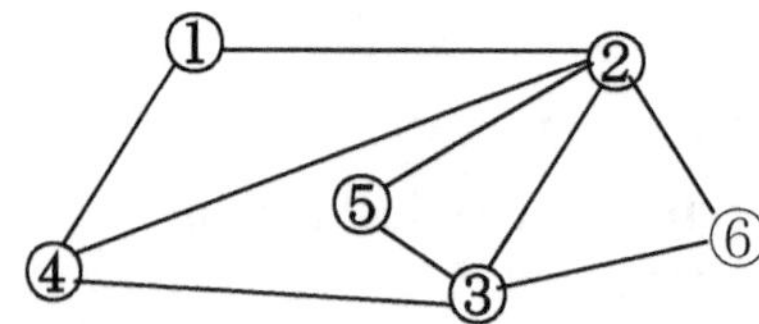

Use the diagram on the left to list the set:

(a) E of even vertices.

(b) O of odd vertices.

Solution

(a) $E = \{1, 3, 5, 6\}$ (b) $O = \{2, 4\}$.

Traversable Networks

A **traversable network** is one that can be traced exactly once beginning at some point without retracing any arc.

Conditions for a Network to be Traversable

For a network to be traversable,

1. All its vertices should be even or
2. The network should have exactly two odd vertices.

Example

In the following diagrams (a) to (g) determine which of the networks are traversable, giving reasons for your answer.

(a) 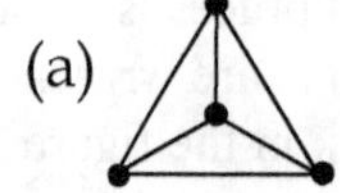 (b) 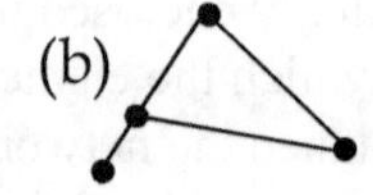 (c) 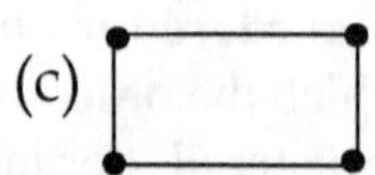 (d) 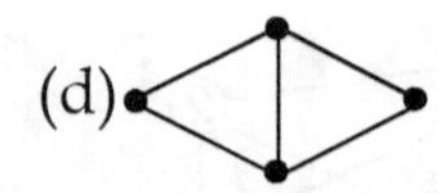

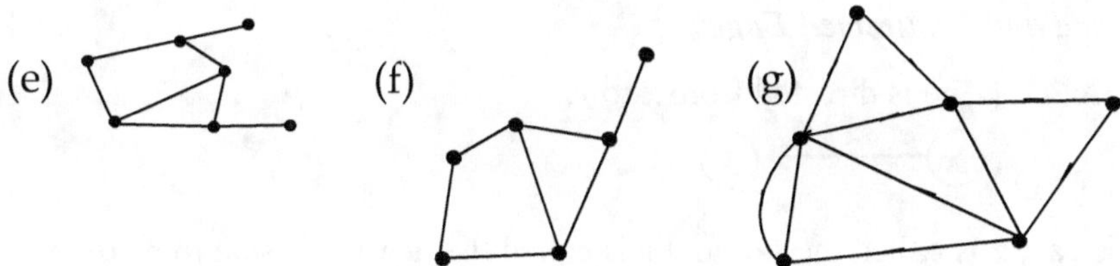

Solution

(a), (e) and (f) are not traversable because they each have more than 2 odd vertices.

(c) is traversable because all the vertices are even.

(b), (d) and (g) are traversable because each has exactly 2 odd vertices.

Network Graphs

A network graph is an ordered pair $G = (V, E)$ comprising the set V of vertices together with the set E of edges, such that each element of E is a paired subset of V.

Example

Given the diagram below and the graph $G = (V, E)$. List the elements of V and E.

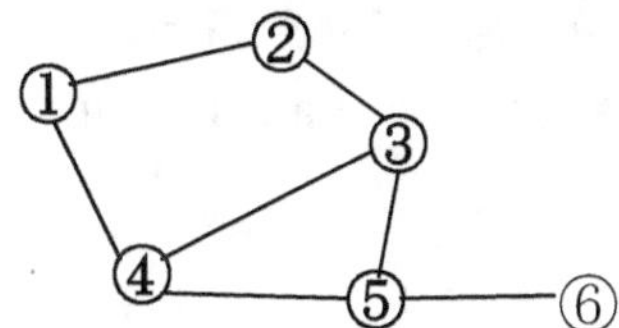

Solution

$V = \{1, 2, 3, 4, 5, 6\}$

$E = \{\{1,2\},\{1,4\},\{2,3\},\{3,4\},\},\{3,5\},\{4,5\},\{5,6\}\}$

Example

A network is defined by $G = (V, E)$ where $V = \{a, b, c, d, e\}$ and $E = \{\{a, b\}, \{a, c\}, \{b, c\}, \{b, d\}, \{c, d\}, \{a, e\}\}$.

(a) Sketch the network diagrammatically.

(b) State the number of regions, vertices and edges of the network.

Solution

(a)

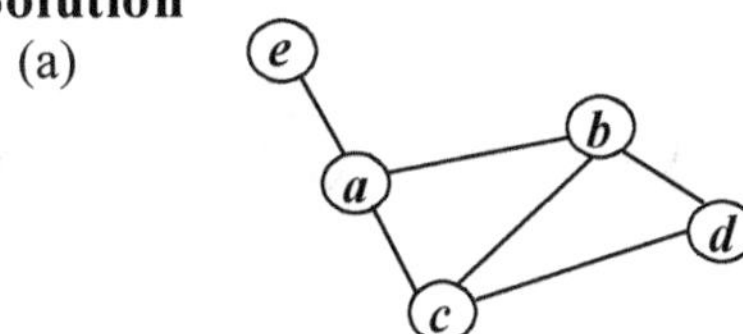

(b) The network has 3 regions, 5 vertices and 6 edges.

Directed and Undirected Edges

The arc $a = (x, y)$ is directed from x to y.

In this case x is called the **tail** and y is called the **head**. y is said to be the successor (**direct successor**) of x and x is said to be the predecessor (**direct predecessor**) of y.
The arc (y, x) is called arc (x, y) **inverted.**

An arc $a = \{x, y\} = \{y, x\}$ is conventionally considered to be directed in either ways. Both x and y are heads and tails. Diagrammatically, this is represented as

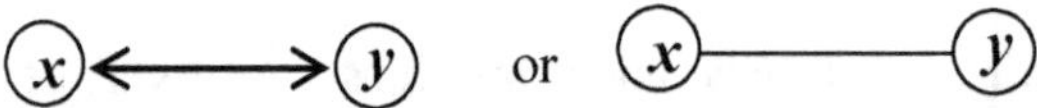

Properties of Network Graphs

(i) **Adjacent or Coincident Edges** are edges with a common vertex.

In the figure on the right, the edges AX and BX are adjacent or coincident because they have a common vertex X.

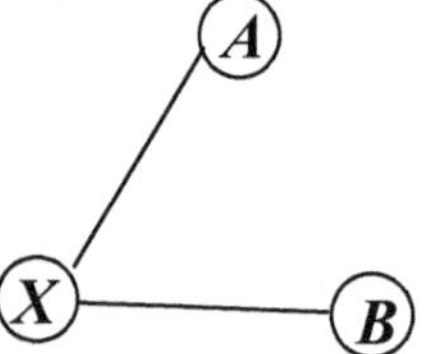

(ii) **Consecutive Arrows** are two edges which are such that the head of one arrow is at the tail of another arrow

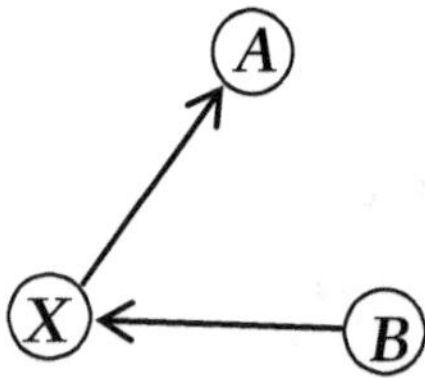

In the figure on the left, the arrows BX and XA are consecutive because the head of BX is at the tail of XA.

(iii) **Adjacent Vertices** (a) below are two vertices which share a common edge.

(a) (b)

(iv) **Consecutive Vertices** (b) above are two adjacent vertices which share a directed edge.

(v) **Incident edge and Vertex** is simply an edge and a vertex on the edge.

Types of Network Graphs

1. A **simple graph** is a graph with no loops or multiple edges.

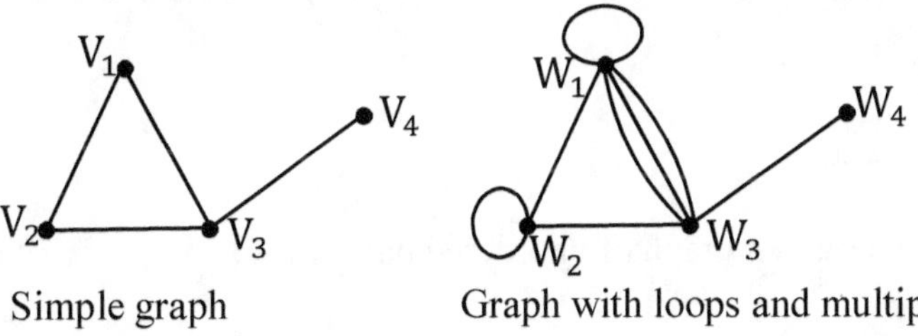

Simple graph (i) Graph with loops and multiple edges (ii)

Graph (i) is a simple graph. It has no loops and no multiple edges.
Graph (ii) is not a simple graph. It has two loops at W_1 and W_2 and has multiple edges at W_1 and W_3.

2. A **directed graph** sometimes referred to simply as a **digraph** is an ordered pair $D = (V, A)$ such that V is the set of vertices and A is the set of ordered edges.

Example

The graph of a network structure is $D = (V, A)$, where $V = \{1, 2, 3\}$ and $A = \{(1, 3), (2, 1), (3, 1), (3, 2)\}$. Represent this network diagrammatically.

Solution

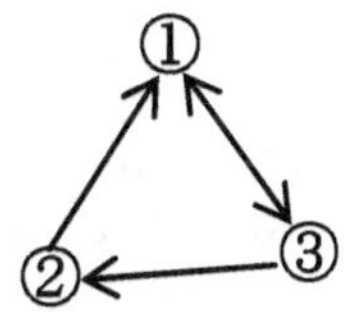

This is an example of a digraph because each edge in a directed graph has at least an arrow at one end.

Example

Represent the network in the following diagram symbolically.

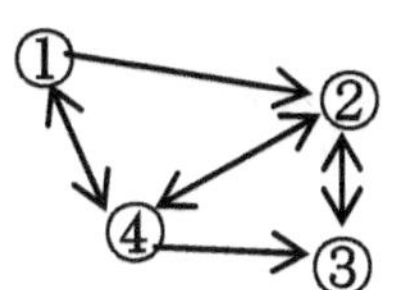

Solution

The symbolic representation of the graph is
$D = (V, A)$ where $V = \{1, 2, 3, 4\}$ and
$A = \{(1, 2), (1,4), (2,3), (2,4), (3,2), (4, 1), (4, 2), (4, 3)\}$.

Example

Draw the network structure for the graph $D = (V, A)$ where $V = \{1, 2, 3, 4, 5, 6\}$ and $A = \{(1,2),(2,1),(3,1),(3,2),(3,4), (3,5),(4,1), (4,5),(5,3),(5,6)\}$.

Solution

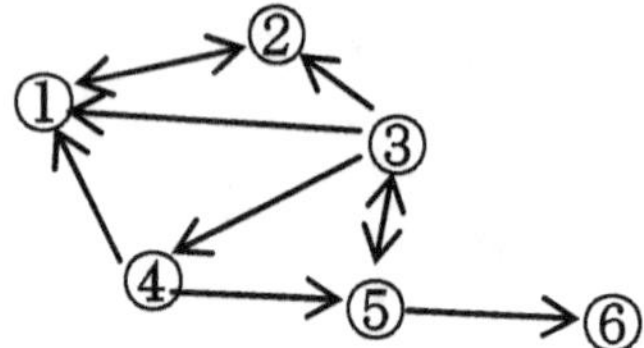

3. An **undirected graph** is a graph of unordered pairs $\{a, b\}$.

Example

Draw the network structure for the graph $G = (V, E)$ where $V = \{1, 2, 3\}$ and $E = \{\{1, 2\}, \{1, 3\}, \{2, 3\}\}$.

Solution

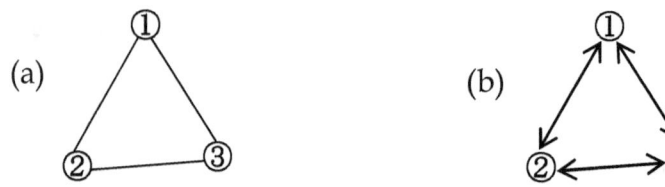

Figure (i) above, is an example of an undirected graph.
Notice that though the graphs in figure (i) and (ii) are equivalent, it is needless putting the arrows as in (ii).

4. ***Mixed Graph***

A **mixed graph** is a graph $G = (V, E, A)$ in which some edges are directed and some are undirected. V is the set of vertices, E is the set of unordered edges and A is the set of ordered edges.

Example

Represent the network below symbolically.

Solution

Symbolically, the graph of this network is $G = (V, E, A)$, where $V = \{1, 2, 3, 4, 5\}$, $E = \{\{1, 5\}, \{2, 3\}, \{3, 4\}\}$ and $A = \{(1, 2), (4, 1), (4, 5)\}$.
This is an example of a mixed graph.

5. ***Complete Graph***

A **complete graph** is a simple graph in which each pair of vertices is connected by an edge. The complete graph with n vertices denoted by K_n, is regular of degree (number of edges at each vertex) $n-1$ and has $\frac{1}{2}n(n-1)$ edges.

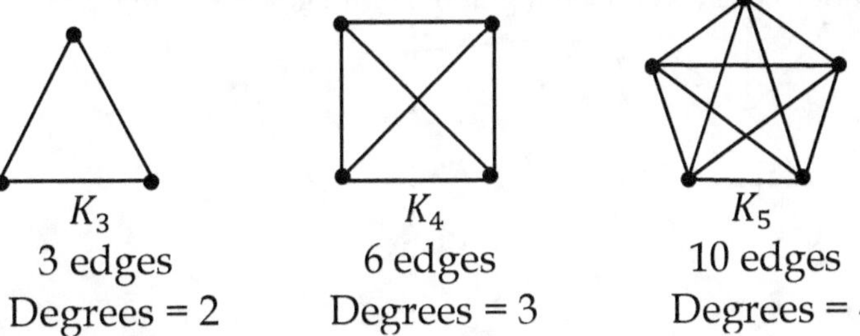

6. ***Weighted Graph***

   A **weighted graph** is one for which a number (weight) is assigned to each edge.

Example

Given that the numbers in the following network show distances in km between towns in a municipality. Find the shortest route from A to G, using the distances shown on the network.

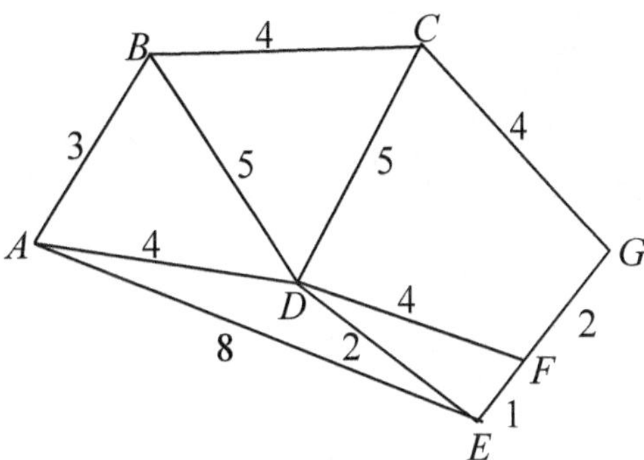

Solution

By adding the distances, the shortest distance is 9 km and is the road $ADEFG$.

7. ***Null Graph***

   A **null graph** on n vertices denoted by Nn is a graph whose edge set is empty.

① ② ③ ④ ⑤

We can represent the graph in the figure on the left symbolically by $G = (V, E)$ where $V = \{1, 2, 3, 4, 5\}$ and $E = \{\ \}$.

Network Trees

A **network tree** can be defined in any of the following ways:

(i) A network tree is a connected graph with no cycles.

(ii) A network tree is a graph in which any two vertices are connected by exactly one path.

(iii) A network tree is a connected graph in which $n(E) = n(V) - 1$.

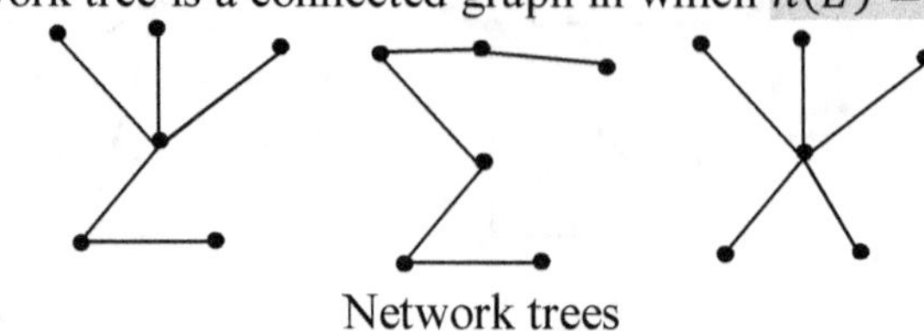

Network trees

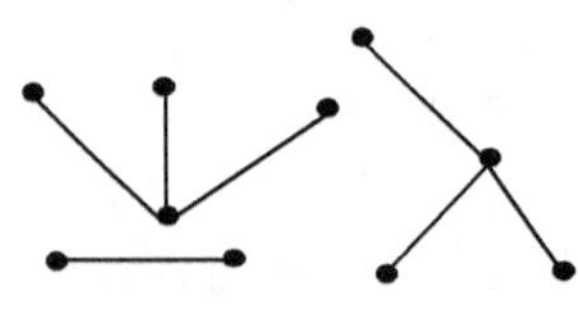

A **forest** is a collection of trees.

MULTIPLE CHOICE EXERCISE 7:5

2. In the following the network which is traversable is:

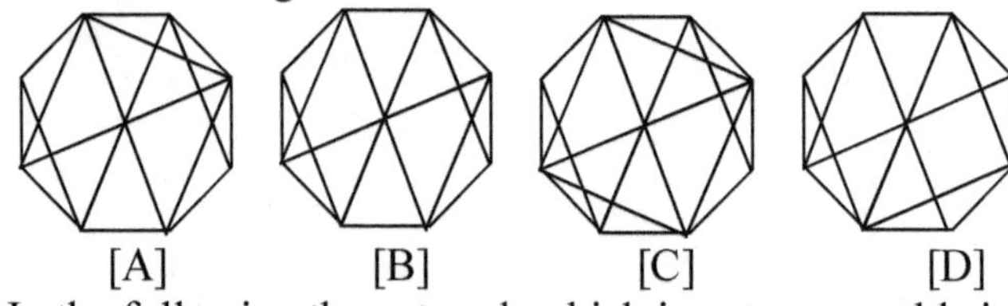

[A] [B] [C] [D]

3. In the following the network which is not traversable is:

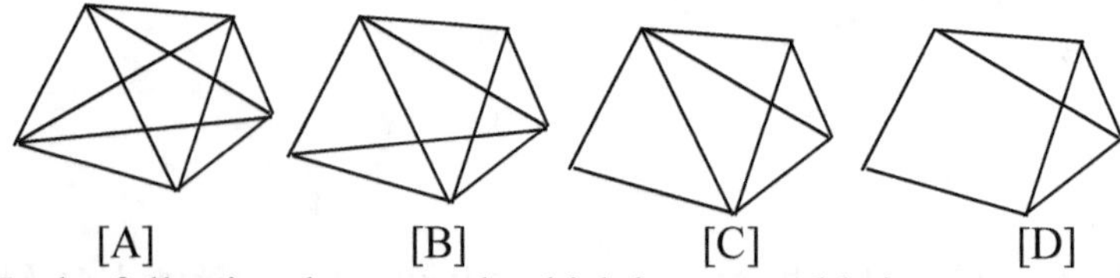

[A] [B] [C] [D]

4. In the following the network which is traversable is:

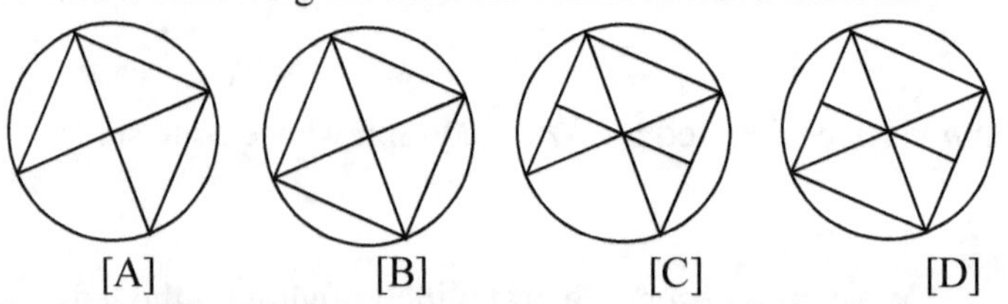

[A] [B] [C] [D]

5. One of the possible routes to traverse the following network is:

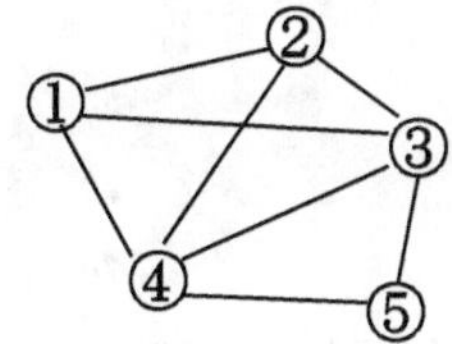

[A] (4, 2, 3, 5, 4, 3, 1, 4, 2)
[B] (3, 2, 3, 5, 4, 3, 1, 4, 2)
[C] (1, 2, 3, 5, 4, 3, 1, 4, 2)
[D] (5, 2, 3, 1, 4, 3, 1, 4, 2)

6. The number of vertices, regions and arcs the following network are respectively:

[A] 6, 9, 5 [B] 6, 5, 9
[C] 5, 9, 9 [D] 5, 6, 9

7. The smallest number of diagonals that can be drawn on the faces of a cube to make the network between its vertices and edges traversable is:
[A] 1 [B] 2 [C] 3 [D] 4
8. The vertices and edges of polyhedra are three-dimensional networks. Among the following regular polyhedra, the one which is traversable is:

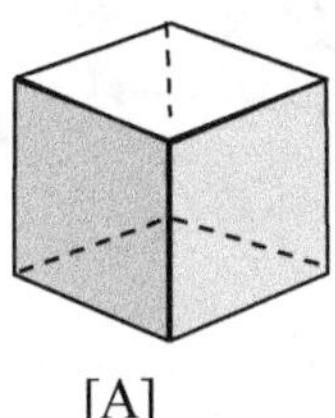
[A]

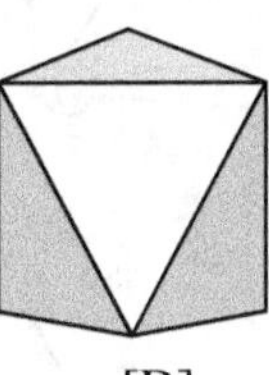
[B]

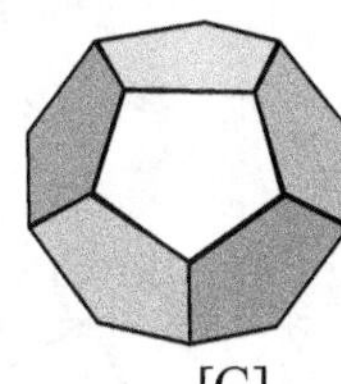
[C]

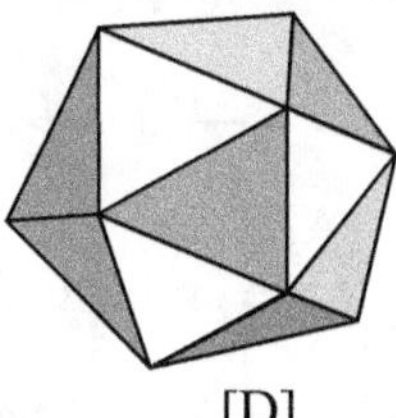
[D]

9. The following floor plan of a house forms a network which is:

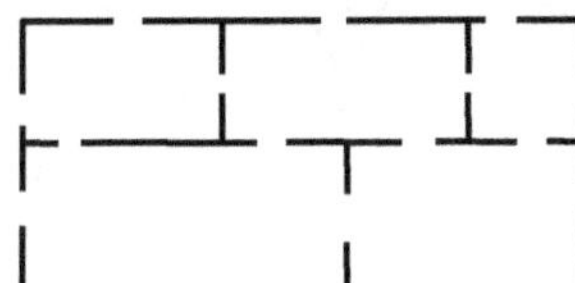

[A] Traversable, with 16 nodes and 6 arcs.
[B] Traversable, with 6 nodes and 16 arcs.
[C] Not traversable, with 16 nodes and 6 arcs.
[D] Not traversable, with 6 nodes and 16 arcs

10. A house is said to be traversable if one can enter all the rooms without passing through any door more than once. This is possible if:
[A] the network for the floor plan has no even vertices.
[B] the network for the floor plan has no even edges.
[C] the network for the floor plan has no odd vertices.
[D] the network for the floor plan has no odd edges.
11. The statement which is certainly true of networks is:
[A] A network with exactly two odd vertices is not traversable.
[B] A network with no odd vertices is traversable.
[C] A network with more than two odd vertices is traversable.
[D] A network with more than two even vertices is traversable.
12. The statement which is certainly true of networks is that the starting point in a traversable network with:
[A] all even vertices is also the ending point.

[B] all even vertices cannot be the ending point.
[C] two odd vertices is also the ending point.
[D] no odd vertices cannot be the ending point.

13. Given that $V = \{a, b, c, d, e\}$ and
$E = \{\{a, b\}, \{a, c\}, \{b, c\}, \{b, d\}, \{c, d\}, \{a, e\}\}$.
The number of regions in the network defined by $G = (V, E)$ is:
[A] 3 [B] 4 [C] 5 [D] 6

14. Given that $V = \{1, 2, 3, 4, 5\}$ and $E = \{\{1, 2\}, \{1, 3\}, \{2, 3\}, \{2, 4\}, \{3, 4\}, \{1, 5\}\}$. The number of regions, nodes and arcs in the network defined by $G = (V, E)$ are respectively:
[A] 3, 5 and 6 [B] 3, 6 and 5 [C] 5, 3 and 6 [D] 6, 3 and 5

15. In the following, the graph which represents the network $G = (V, E)$, $V = \{1,2,3\}$ and $E = \{\{1,2\},\{1,3\},\{2,3\}\}$ is:

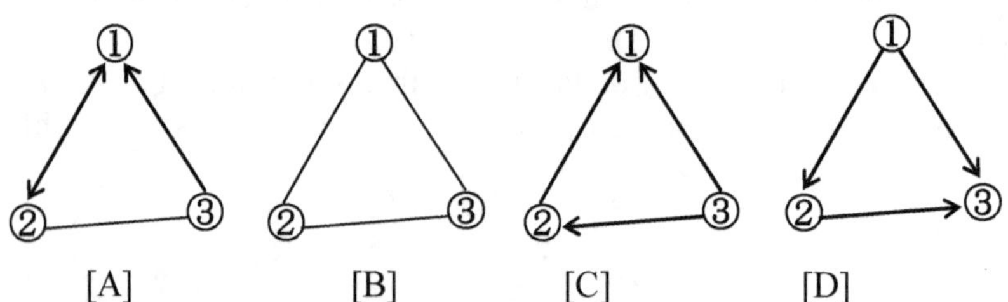

[A] [B] [C] [D]

16. A complete graph among the following graphs is:

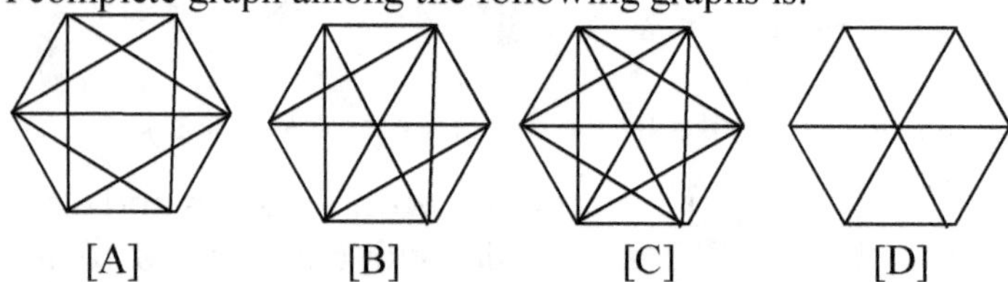
[A] [B] [C] [D]

17. In the following graph of network, the set of edges is:

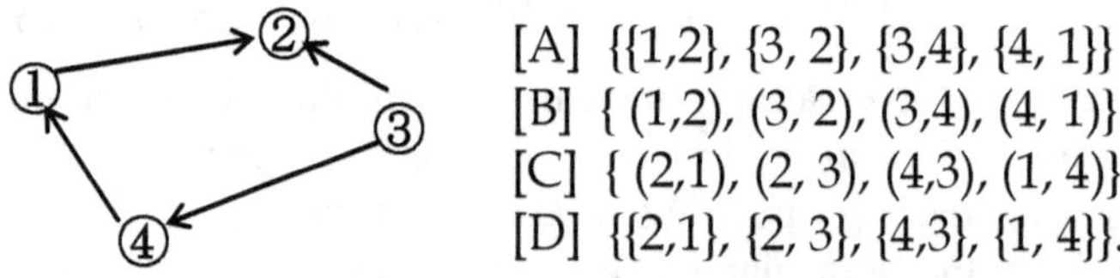

[A] {{1,2}, {3, 2}, {3,4}, {4, 1}}
[B] { (1,2), (3, 2), (3,4), (4, 1)}.
[C] { (2,1), (2, 3), (4,3), (1, 4)}.
[D] {{2,1}, {2, 3}, {4,3}, {1, 4}}.

18. The network graph which is a tree among the following is:

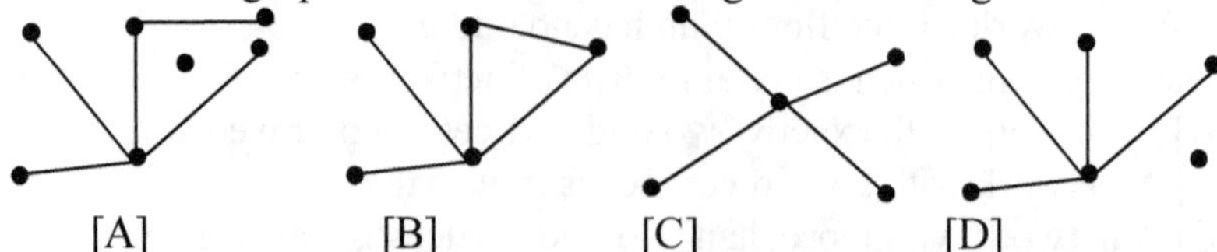
[A] [B] [C] [D]

19. The number of regions in a network with 4 vertices and 6 edges is:
[A] 3 [B] 4 [C] 5 [D] 6

20. In the following net work, the set of odd vertices is:

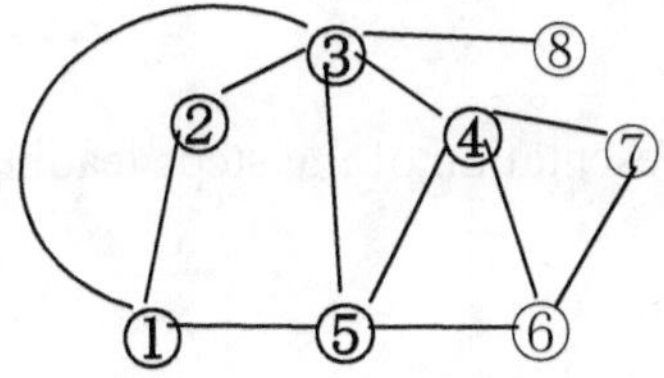

[A] {1, 2, 7, 8} [B] {1, 3, 5, 7}
[C] {3, 4, 5, 6} [D] {1, 3, 6, 8}

21. In the above graph of network, the set of even vertices is:
[A] {1, 2, 7, 8} [B] {2, 4, 5, 7} [C] {3, 4, 5, 6} [D] {2, 4, 6, 8}
22. In the following graph of network, the edge {1,4} is incident to node:

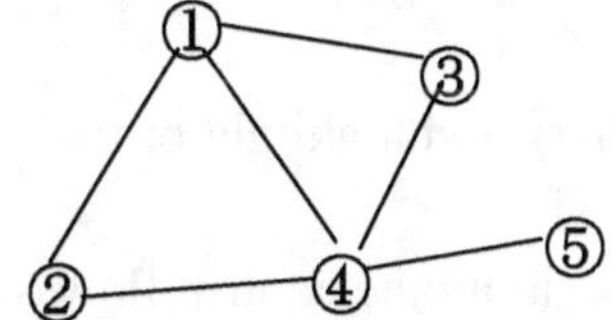

[A] 2 [B] 1
[C] 3 [D] 5

23. In the following graph of network, node 3 and 4 are said to be:

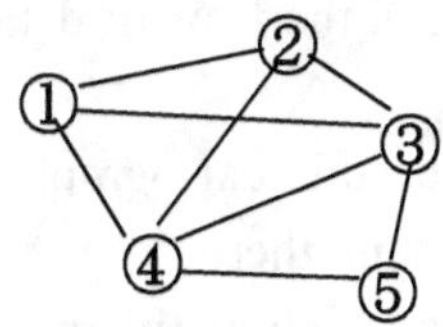

[A] consecutive [B] coincident
[C] adjacent [D] incident

24. In the above graph of network, edge {3,4} is said to be:
[A] consecutive to node 3 and 4. [B] coincident to node 3 and 4.
[C] adjacent to node 3 and 4. [D] incident to node 3 and 4.
25. In the above graph of network, edge {3,4} and node 3 are said to be:
[A] consecutive [B] inverted [C] adjacent [D] incident
26. In the network diagram below, the edges {1,2} and {1,3} are said to be:

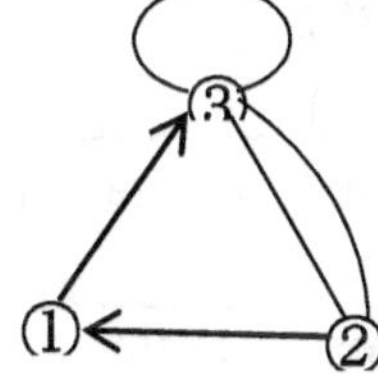

[A] consecutive [B] inverted
[C] adjacent [D] incident

27. In the network diagram above, the edges {1,2} and {2,3} are said to be:
[A] consecutive [B] inverted [C] adjacent [D] incident
28. The network which is most likely to be a tree is the network which has:
[A] 5 edges and 4 nodes [B] 4 edges and 5 nodes
[C] 6 edges and 4 nodes [D] 4 edges and 6 nodes
29. The number of arcs in the following network is:
[A] 2 [B] 3 [C] 4 [D] 5

7.6 FLOW DIAGRAMS

A flow diagram is a schematic or graphic representation of the steps leading to the solution of a given problem.

TYPES OF FLOW DIAGRAMS

1. **A functional flow block diagram** is a sketch showing the various parts of a system and their functions in relation to one another.
2. **A process flow diagram** is a sketch showing the graphical representation of a process.
3. **An alluvial diagram** is a graphical summary and highlight of the important structural changes in networks.
4. **A control flow diagram** is a diagram describing the control flow of a business process, a program or any other process.
5. **A data flow diagram** is a graphical representation of the flow of data through an information system.
6. **A flow map** is a mixture of maps and flow charts used in cartography, to show the movement of objects from one location to another.
7. **A signal flow graph** is a graph used in mathematics to show the relations among the variables of a set of linear algebraic relations.
8. **A flowchart** is a graphical representation of the step-by-step solution to a given problem.

BASIC FLOW DIAGRAM SYMBOLS

	Symbol	Function
1	(START)	Used at the beginning of a flow diagram to indicate the starting point of the program.
2	(STOP)	Used at the end of a flow diagram to indicate the ending point of the program.
3	▭	The **instruction block** is used to issue intermediate instructions.
4	◇	The **decision block** is used to ask precise questions with “Yes” or “No” answers. A decision block usually has two exits, one for “Yes” and the other for “No”.
5	↓	Arrows connecting the instruction and decision boxes are used to show the direction in which the instructions have to be followed.

Example

Draw a flow chart which can be used to find the area of a trapezium.

Solution

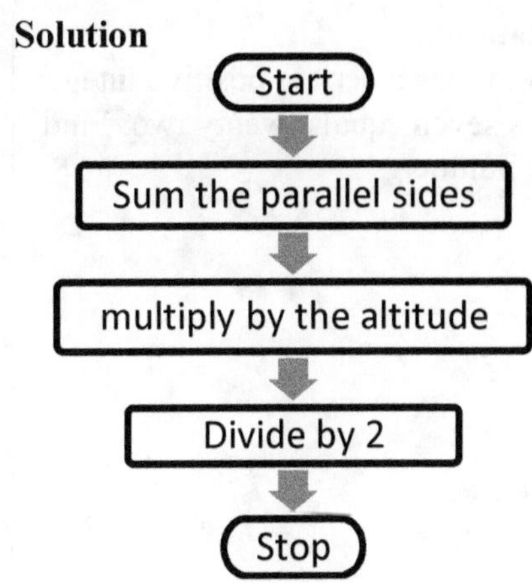

Example

Use a flow Chart to evaluate $9 - 6 + 4 \times 6 \div 3$.

Solution

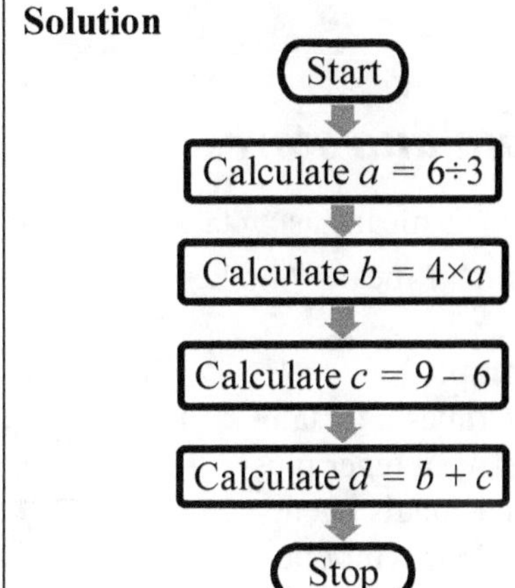

Example

Draw a flow diagram which you would use to draw a diagram with a pencil.

Solution

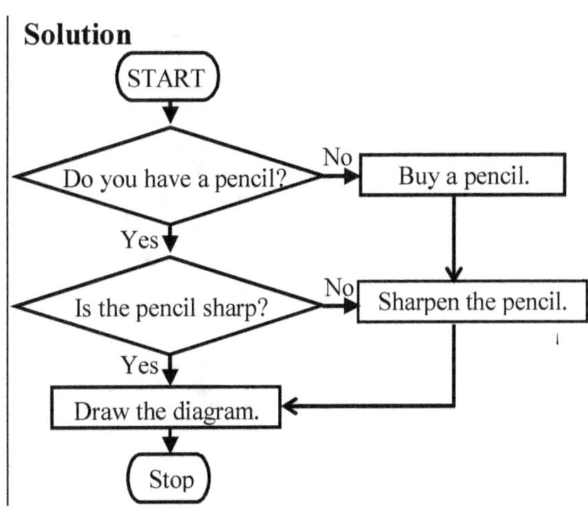

Example
Five times a certain positive integer plus seven equals twenty two. Find the number.

Solution
An inverse operation method may be used in solving the problem as follows.

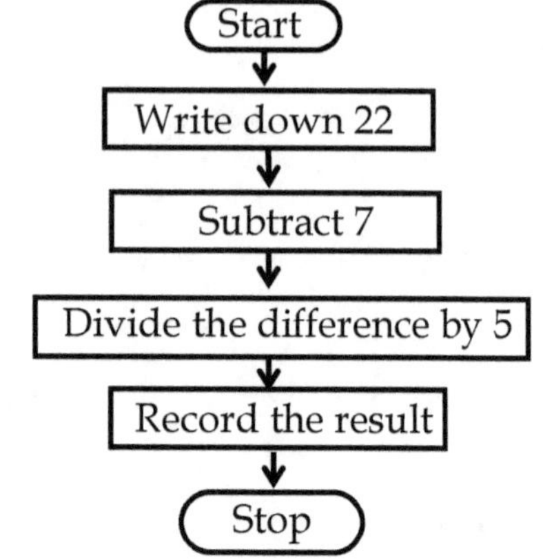

Data and stores

n: = 1 is means as "n takes the value 1".

n: = n +1 means "n takes the new value n +1".

Example
Five times a certain positive integer plus seven equals twenty two. Use symbolic language to find the number.

Solution

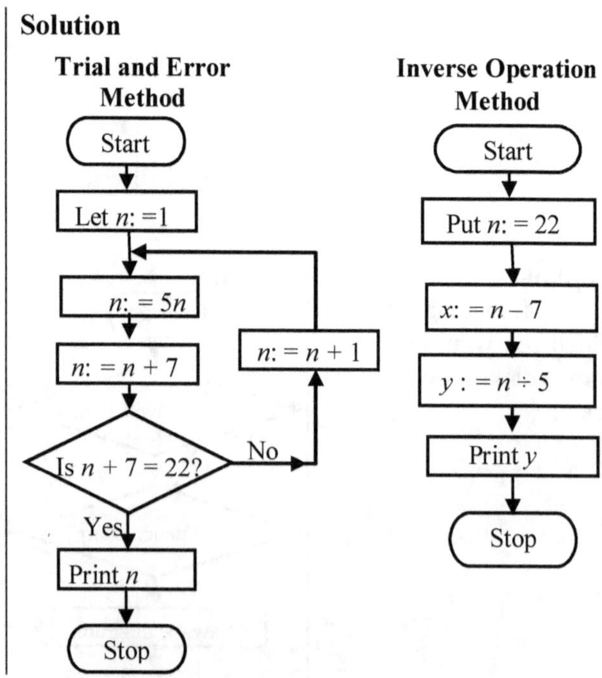

MULTIPLE CHOICE EXERCISE 7:6

2. On a flow chart, the shape used for a decision block is:

[A] [B] [C] [D]

3. The shape ▭ is used on a flow chart as:
   [A] a decision block   [B] an instruction block
   [C] a question block   [D] a description block
4. The type of flow diagram which highlights and summarizes the significant structural changes in networks is:
   [A] a data flow diagram   [B] a process flow diagram
   [C] an alluvial diagram   [D] a signal flow diagram
5. Given the line segment $XY = 8$ cm. The construction described by the flow chart below is:
   [A] the inscribed circles C_1 and C_2 [B] the locus of C_1 and C_2
   [C] the intersection of C_1 and C_2 [D] the mediator of XY

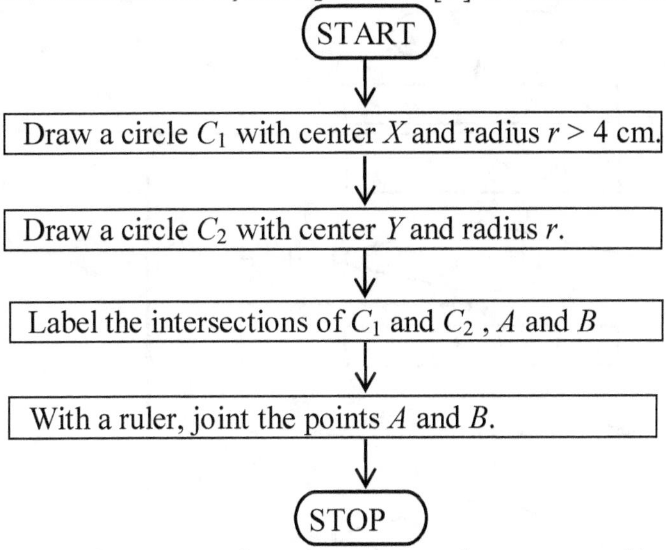

START
↓
Draw a circle C_1 with center X and radius $r > 4$ cm.
↓
Draw a circle C_2 with center Y and radius r.
↓
Label the intersections of C_1 and C_2 , A and B
↓
With a ruler, joint the points A and B.
↓
STOP

6. Given that $a, b, r \in \mathbb{R}$. A flow chart to show the sequence of instructions performed to make b the subject of the formula $r = \sqrt{a^2 + b^2}$ contains the following instructions.
   I: Find the square root of both sides   II: Square both sides
   III: Subtract a^2 from both sides.
   The correct order of instructions is:
   [A] I, II, III   [B] II, III, I   [C] III, I, I   [D] II, I, III
7. The sequence of instructions performed to make r the subject of the

formula $P = \frac{mv^2}{r}$ is:

[A] Multiply both sides by r, Divide both sides by P
[B] Multiply both sides by r, Divide both sides by r
[C] Multiply both sides by r, Multiply both sides by P
[D] Divide both sides by r, Multiply both sides by P

8. In a flow chart to show the sequence of instructions performed to make C the subject of the formula $F = \frac{9}{5}C + 32$, the first two instructions are; multiply both sides by 5, divide both sides by 9 respectively. The third and final instruction should be:

[A] Multiply both sides by $\frac{160}{9}$ [B] Divide both sides by $\frac{160}{9}$

[C] Add $\frac{160}{9}$ to both sides [D] Subtract $\frac{160}{9}$ from both sides

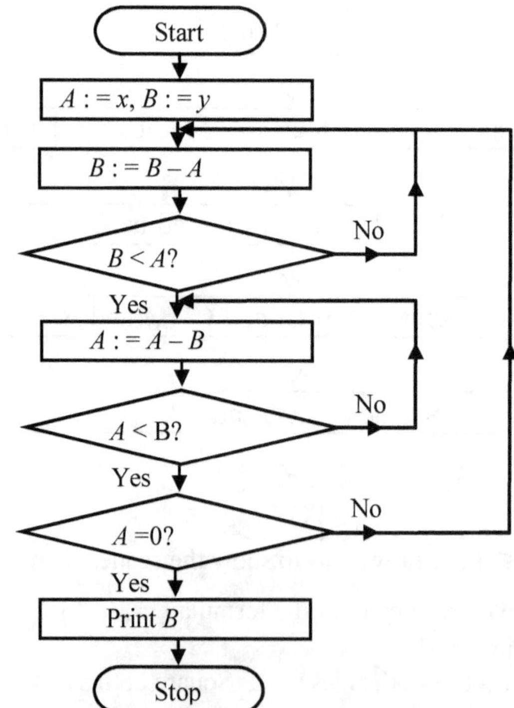

9. In the flow diagram below, the number of loops is:
[A] 9 [B] 4 [C] 2 [D] 3

10. In the flow diagram below, the number of stores is:
[A] 9 [B] 4 [C] 2 [D] 3

11. The instruction boxes for the flow chart which shows the sequence of operations to change the subject of the formula $A = \pi r^2$ chronologically contains contain the instructions:
    [A] Find the square root of both sides, divide both sides by π ..
    [B] Divide both sides by π, multiply both sides by π.
    [C] Divide both sides by π, find the square root of both sides.
    [D] Find the square root of both sides, multiply both sides by π.
12. The flow chart below shows the probability $P(X)$ of obtaining a sum which is a multiple of 6 when two dice are rolled. According to the flow chart if the sum is 9 then:
    [A] the dice should be rolled again.
    [B] subtracts 3 from 9 to continue.
    [C] adds 3 to 9 to continue.
    [D] the probability is $\frac{3}{4}$.

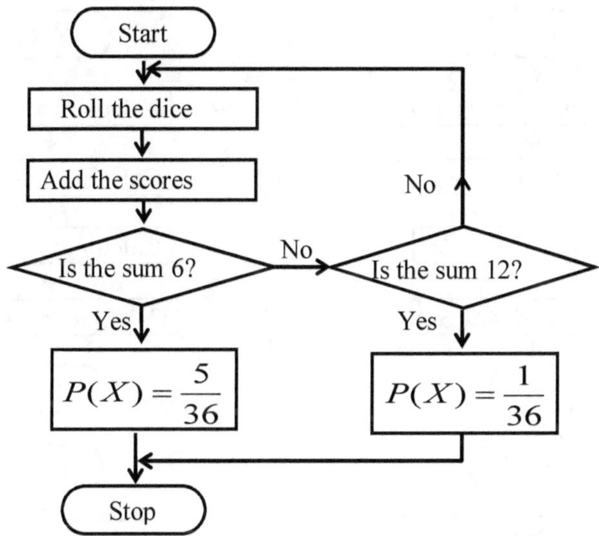

13. The expression which is being evaluated by the flow chart in figure (i) below is:
    [A] $3x^2 + 2x - 5$ [B] $2x^2 + 3x - 5$
    [C] $3a^2 + 2a - 5$ [D] $2a^2 + 3a - 5$
14. The expression which is being evaluated by the flow chart in figure (ii) below is:
    [A] $13 - 12 \div 3 + 2 \times 8$ [B] $(13 - 12) \div 3 + 2 \times 8$
    [C] $13 - 12 \div (3 + 2) \times 8$ [D] $(13 - 12) \div (3 + 2) \times 8$

START
Calculate $a = 2x$
Calculate $b = a + 3$
Calculate $c = bx$
Calculate $d = c - 5$
STOP

(i)

START
Calculate $a = 12 \div 3$
Calculate $b = 2 \times 8$
Calculate $c = 13 - a$
Calculate $d = c + b$
STOP

(ii)

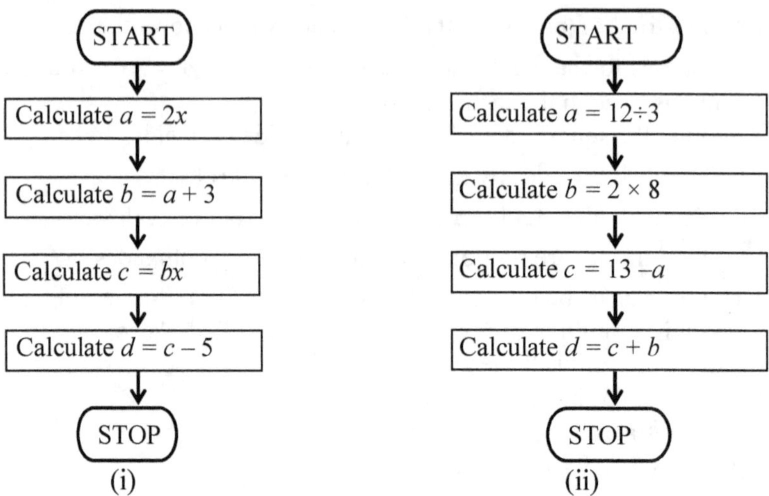

15. The flow chart which represents the evaluation of $9 - 6 + 4 \times 6 \div 3$ is:

[A]

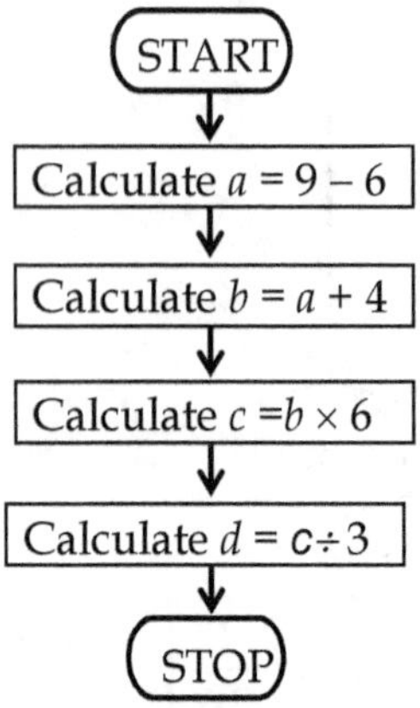

START
Calculate $a = 9 - 6$
Calculate $b = a + 4$
Calculate $c = b \times 6$
Calculate $d = c \div 3$
STOP

[B]

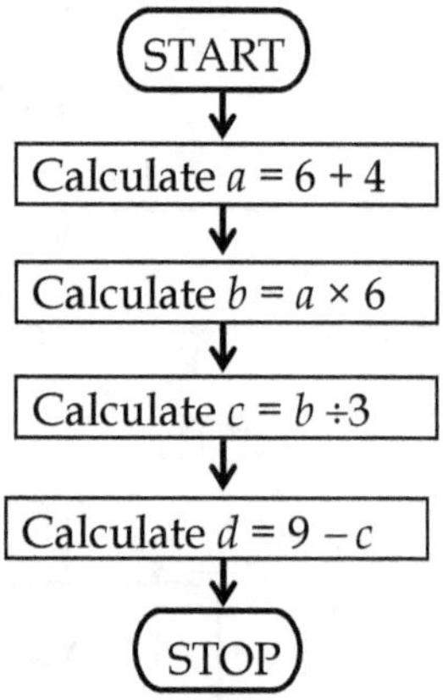

START
Calculate $a = 6 + 4$
Calculate $b = a \times 6$
Calculate $c = b \div 3$
Calculate $d = 9 - c$
STOP

[C]

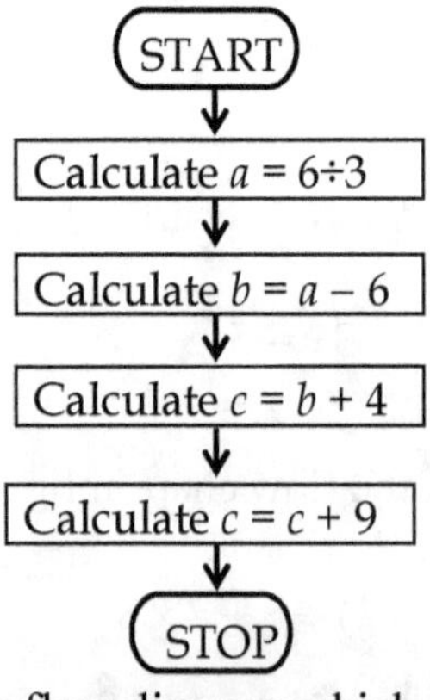

START
Calculate $a = 6 \div 3$
Calculate $b = a - 6$
Calculate $c = b + 4$
Calculate $c = c + 9$
STOP

[D]

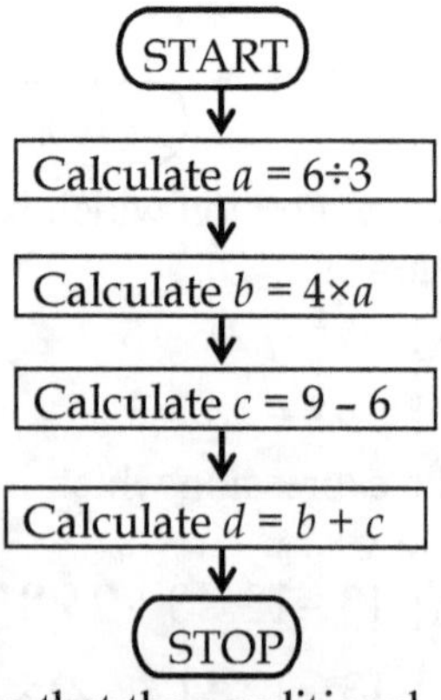

START
Calculate $a = 6 \div 3$
Calculate $b = 4 \times a$
Calculate $c = 9 - 6$
Calculate $d = b + c$
STOP

16. The flow diagram which can be used to prove that the conditional statement $p \longrightarrow q$ is true is:

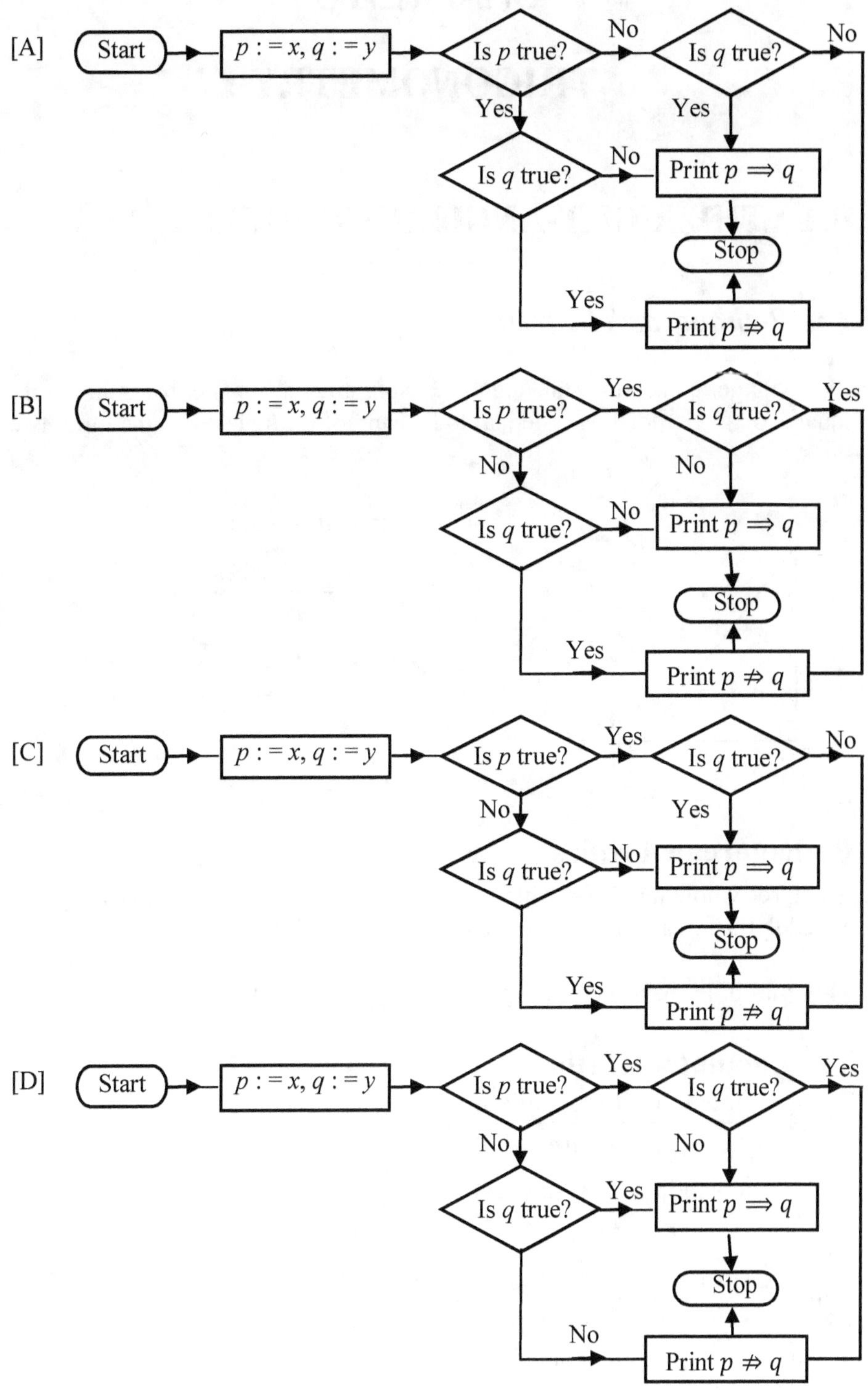
[A]
Start
$p := x, q := y$
Is p true?
No
Is q true?
No
Yes
Yes
Is q true?
No
Print $p \Rightarrow q$
Stop
Yes
Print $p \not\Rightarrow q$
[B]
Start
$p := x, q := y$
Is p true?
Yes
Is q true?
Yes
No
No
Is q true?
No
Print $p \Rightarrow q$
Stop
Yes
Print $p \not\Rightarrow q$
[C]
Start
$p := x, q := y$
Is p true?
Yes
Is q true?
No
No
Yes
Is q true?
No
Print $p \Rightarrow q$
Stop
Yes
Print $p \not\Rightarrow q$
[D]
Start
$p := x, q := y$
Is p true?
Yes
Is q true?
Yes
No
No
Is q true?
Yes
Print $p \Rightarrow q$
Stop
No
Print $p \not\Rightarrow q$

CHAPTER 8

TRIGONOMETRY

8.1 THE RIGHT-ANGLED TRIANGLE

The Pythagoras Theorem

The Pythagoras theorem states that in a right-angled triangle, the number of squares on the hypotenuse is equal to the sum of the squares on the other two sides.

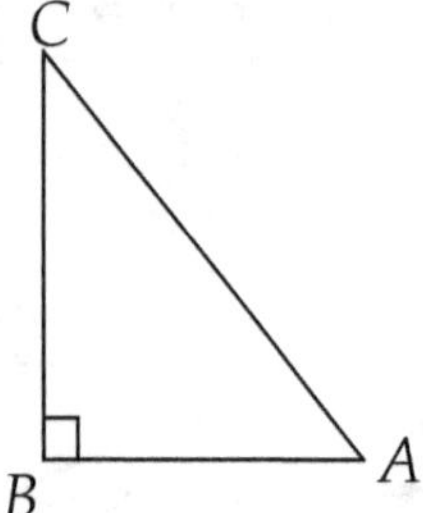

For the triangle on the left,

$$AC^2 = AB^2 + BC^2$$

$$\Rightarrow AC = \sqrt{AB^2 + BC^2}$$

$$\Rightarrow AB = \sqrt{AC^2 - BC^2}$$

$$\Rightarrow BC = \sqrt{AC^2 + AB^2}$$

Pythagorean Triples

Any three whole numbers, which can form sides of a right-angled triangle, are called Pythagorean triples. E.g. 3,4,5; 6,8,10; 5,12,13; 8,15,17; 7,24,25 etc

Multiples of Pythagorean triples are also Pythagorean triples.

Trigonometric Ratios

$$\sin\hat{A} = \frac{\text{side opposite to angle } A}{\text{hypotenuse}} = \frac{\text{opp}}{\text{hyp}} = \frac{\text{O}}{\text{H}}$$

$$\cos\hat{A} = \frac{\text{side adjacent to angle } A}{\text{hypotenuse}} = \frac{\text{adj}}{\text{hyp}} = \frac{\text{A}}{\text{H}}$$

$$\tan\hat{A} = \frac{\text{side opposite to angle } A}{\text{side adjacent to angle } A} = \frac{\text{opp}}{\text{adj}} = \frac{\text{O}}{\text{A}}$$

The trig ratios of angle B are similarly defined.

Note that!!
Adjacent to ∠*A* means 'next to ∠*A*' and opposite to ∠*A* means 'on the other side of ∠*A*'.
The mnemonic **RAT-SOH-CAH-TOA** may be a useful mental aid.

Example

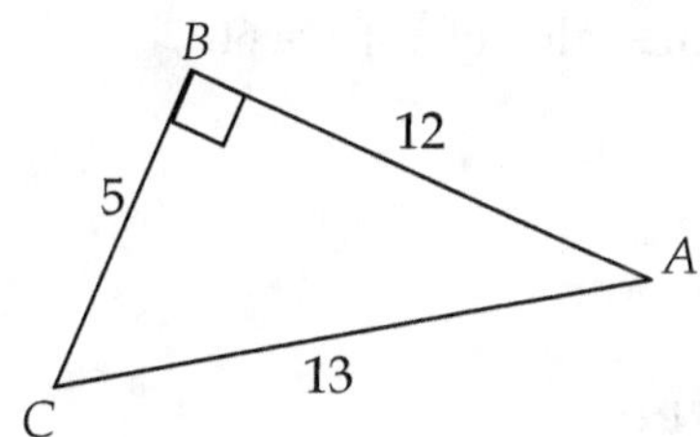

Use the figure on the left to write down as fractions the value of the given trigonometric ratios.

(a) $\sin A$ (b) $\cos C$ (c) $\sin C$
(d) $\cos A$ (e) $\tan A$ (f) $\tan C$

Solution

(a) $\sin A = \frac{\text{opp}}{\text{hyp}} = \frac{5}{13}$ (b) $\cos C = \frac{5}{13}$ (c) $\sin C = \frac{\text{opp}}{\text{hyp}} = \frac{12}{13}$

(d) $\cos A = \frac{12}{13}$ (e) $\tan A = \frac{\text{opp}}{\text{adj}} = \frac{5}{12}$ (f) $\tan C = \frac{\text{opp}}{\text{adj}} = \frac{12}{5}$

Inverse Trigonometric Ratios

The corresponding angle to a given trigonometric ratio is called the **arc trigonometric ratio** of the angle.

Finding other Trig Ratios Given Another

Example

Given that $\cos\theta = \frac{15}{17}$, find the values of $\sin\theta$ and $\tan\theta$ without using tables or calculators.

Solution

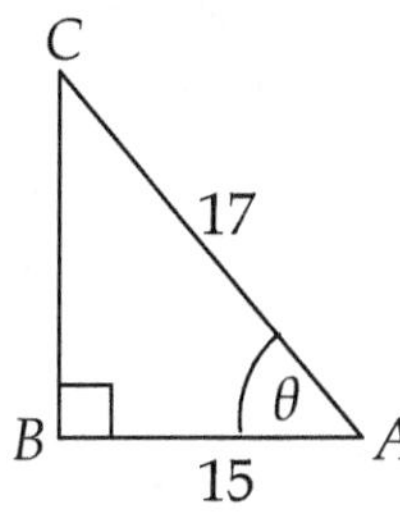

Since $\cos\theta = \dfrac{A}{H}$

Let $AB = 15$ units and $AC = 17$ units.
Then by the Pythagoras theorem,
$BC = \sqrt{AC^2 - AB^2} = \sqrt{17^2 - 15^2} = \sqrt{64} = 8$ units

$\Rightarrow \sin\theta = \dfrac{BC}{AC} = \dfrac{8}{17}$ and $\tan\theta = \dfrac{BC}{AB} = \dfrac{8}{15}$

Complementary Angles

Two angles whose sum is 90° are said to be complementary. Generally, *The sine of any acute angle is equal to the cosine of its complement and vice versa.* i.e. $\sin\theta = \cos(90° - \theta)$ and $\cos\theta = \sin(90° - \theta)$

Example
Two angles x and y are complementary. Find the value of x if $y = 60°$.

Solution
$x = 90° - y$ and $y = 60° \Rightarrow x = 90° - 60° = 30°$

Trigonometric Ratios of Special Angles

The following table is a summary of the trig ratios of special angles.

θ	$\sin\theta$	$\cos\theta$	$\tan\theta$	$\cot\theta$	$\operatorname{cosec}\theta$	$\sec\theta$
$0°$	0	1	0	∞	∞	1
$30°$	$\frac{1}{2}$	$\frac{\sqrt{3}}{2}$	$\frac{\sqrt{3}}{3}$	$\sqrt{3}$	2	$\frac{2\sqrt{3}}{3}$
$45°$	$\frac{\sqrt{2}}{2}$	$\frac{\sqrt{2}}{2}$	1	1	$\sqrt{2}$	$\sqrt{2}$
$60°$	$\frac{\sqrt{3}}{2}$	$\frac{1}{2}$	$\sqrt{3}$	$\frac{\sqrt{3}}{3}$	$\frac{2\sqrt{3}}{3}$	2
$90°$	1	0	∞	0	0	∞

The General Angle

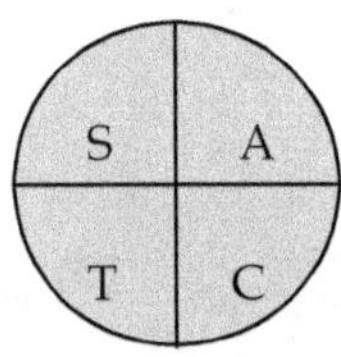

In the first quadrant ALL the three trig ratios are positive.
In the second quadrant only the SINE ratio is positive.
In the third quadrant only the TANGENT ratio is positive.
In the fourth quadrant only the COSINE ratio is positive.
The mnemonic **ACTS** may be useful as a mental aid.

The Meaning of Negative Angles

Conventionally angles measured anticlockwise are positive while angles measured clockwise are negative.

$$(-\theta) \equiv (360-\theta)$$
$$\Leftrightarrow \sin(-\theta) \equiv \sin(360-\theta) = -\sin\theta$$
$$\Leftrightarrow \cos(-\theta) \equiv \cos(360-\theta) = \cos\theta$$
$$\Leftrightarrow \tan(-\theta) \equiv \tan(360-\theta) = -\tan\theta$$

Example

Evaluate the following

(a) $\sin 330°$ (b) $\tan 315°$ (c) $\cos 240°$

Solution

(a) $\sin 330° = \sin(360-330)° = -\sin 30° = -\frac{1}{2}$

(b) $\tan 315° = -\tan(360-315)° = -\tan 45° = -1$

(c) $\cos 240° = \cos(360-240)° = \cos 120°$

$$\cos 120° = \cos(180-120)° = -\cos 60° = -\frac{1}{2}$$

MULTIPLE CHOICE EXERCISE 8:1

1. A triangle has sides 8 cm, 15 cm and 17 cm. Therefore, the best name for it is:
   [A] a equilateral triangle. [B] an obtuse triangle.
   [C] a right-angled triangle. [D] an isosceles triangle.
2. The pair of trigonometric ratios which are equal is:
   [A] sin 50° and cos 50° [B] sin 50° and tan 50°
   [C] sin 50° and tan 40° [D] sin 50° and cos 40°
3. The pair of trigonometric ratios such that one is the inverse of the other is:
   [A] $\sin\theta$ and $\operatorname{cosec}\theta$ [B] $\cos\theta$ and $\cot\theta$
   [C] $\sin\theta$ and $\sec\theta$ [D] $\cos\theta$ and $\operatorname{cosec}\theta$
4. Given the triangle, below. The incorrect relation is:
   [A] $r^2 = p^2 + q^2$ [B] $p^2 = r^2 - q^2$
   [C] $q^2 = r^2 - p^2$ [D] $q^2 = r^2 + p^2$

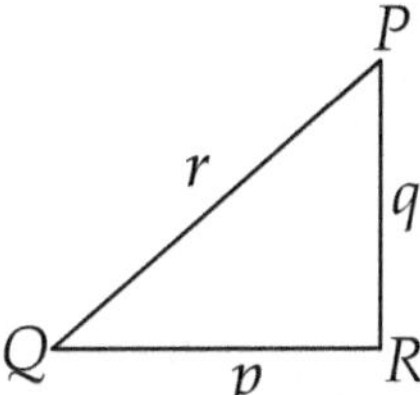

5. Given the triangle, above. The value of p is:
   [A] $\sqrt{q^2 - r^2}$ [B] $\sqrt{r^2 - q^2}$ [C] $\sqrt{q^2 + r^2}$ [D] $\sqrt{r^2 + q^2}$
6. A Pythagorean triple among the following is:

[A] 6, 9, 11 [B] 7, 9, 12 [C] 8, 15, 17 [D] 14, 17, 20

7. The triplet which does not represent the lengths of the sides of a right-angled triangle is:
   [A] 6, 8, 10 [B] 5, 12, 10 [C] 8, 15, 17 [D] 7, 23, 24
8. 36.4° is equal to:
   [A] $36°34'$ [B] $36°60'$ [C] $36°24'$ [D] $36°40'$
9. $41°27'$ in degrees as a decimal is:
   [A] $41.45°$ [B] $41.33°$ [C] $41.25°$ [D] $41.60°$
10. The ratio which is equal to $\sin 60^\circ$ is:
   [A] $\cos 30°$ [B] $\sin 30°$ [C] $\cos 60°$ [D] $\tan 30°$
11. The ratio which is equal to $\sin 30^\circ$ is:
   [A] 0.0500 [B] 0.5050 [C] 0.866 [D] 0.5000
12. The value of $\tan 45^\circ$ is:
   [A] 2.0000 [B] 0.5000 [C] 1.0000 [D] 1.5000
13. The cosine of $60^\circ \frac{1}{2}$. The value of $28–20\cos 60^\circ$ is:
   [A] 18 [B] 10 [C] 4 [D] 24
14. If $\sin\theta = \frac{3}{5}$, the value of $\tan\theta$ for $0<\theta<90^\circ$ is:
   [A] $\frac{4}{5}$ [B] $\frac{3}{4}$ [C] $\frac{5}{8}$ [D] $\frac{1}{2}$
15. If $\cos\theta = \frac{5}{13}$ the value of $\tan\theta$ for $0<\theta<90^\circ$ is:
   [A] $\frac{5}{12}$ [B] 5 [C] $\frac{13}{5}$ [D] $\frac{12}{5}$
16. Given that $\tan x = \frac{5}{12}$. The value of $\sin x + \cos x$ is:
   [A] $\frac{5}{13}$ [B] $\frac{7}{13}$ [C] $\frac{17}{13}$ [D] $\frac{5}{12}$
17. If $\tan x = 2\frac{1}{2}$ the value of $\sin x$ for $0^\circ<x<90^\circ$ is:
   [A] $\frac{\sqrt{29}}{5}$ [B] $\frac{5\sqrt{29}}{29}$ [C] $\frac{2\sqrt{29}}{29}$ [D] $\frac{\sqrt{29}}{2}$
18. Given that $\sin P = \frac{5}{13}$ where P is acute. The value of $\cot P– \tan P$ is:
   [A] $\frac{79}{156}$ [B] $\frac{95}{156}$ [C] $\frac{5}{13}$ [D] $\frac{13}{12}$
19. Given that $\sin\theta =– 0.5000$, where $0^\circ \leq \theta \leq 90^\circ$, θ is:
   [A] 30° [B] 120° [C] 150° [D] 210°
20. If $\sin x = \cos 50^\circ$ then, x equals:

[A] 40° [B] 45° [C] 50° [D] 90°

21. If $\sin (x+30)^{o} = \cos 40^{o}$, the value of x^{o} is:
[A] 15° [B] 20° [C] 60° [D] 90°

22. If $10 \tan 60^{o} = 20 \tan x$. Correct to the nearest degree x is equal to:
[A] 30° [B] 40° [C] 41° [D] 60°

23. If $\sin x = \cos 70^{o}$, x equals:
[A] 110° [B] 70° [C] 30° [D] 20°

24. If $\cos x = \sin 27.3^{o}$, the value of x where $0^{o} \le x \le 90^{o}$ is:
[A] 27.3° [B] 35.4° [C] 54.6° [D] 62.7°

25. Without using tables or calculators, the value of $\dfrac{\sin 20°}{\cos 70°} + \dfrac{\cos 25°}{\sin 65°}$ is:
[A] 2 [B] 1 [C] –2 [D] –1

26. If $\sin x = \dfrac{12}{13}$, where $0^{o} < x < 90^{o}$. The value of $1 - \cos^2 x$ is:
[A] $\dfrac{25}{169}$ [B] $\dfrac{64}{169}$ [C] $\dfrac{105}{169}$ [D] $\dfrac{144}{169}$

27. Using four figure tables, the sine of 70° is:
[A] 0.9390 [B] 0.9394 [C] 0.9397 [D] 0.9399

28. Using four figure tables $\cos 80^{o}$ is:
[A] 0.173 [B] 0.1736 [C] 0.1740 [D] 0.1744

29. Using four figure tables, the angle whose sine is 0.841 to one decimal place is:
[A] 57.2° [B] 56.8° [C] 32.8° [D] 32.2°

30. Given that $\cos x = 0.5321$, then x is equal to:
[A] 56.2° [B] 57.9° [C] 33.2° [D] 32.1°

31. If $\sin \theta = 0.6088$, two possible values of θ are:
[A] 37°30', 142°30' [B] 52°30', 127°30'
[C] 37°30', 127°30' [D] 52°30', 142°30'

32. Using Mathematical tables $\cos 40^{o} - \sin 30^{o}$ equals:
[A] – 0.2660 [B] 0.2660 [C] – 0.0266 [D] 0.0266

33. If $\cos 60° = \dfrac{1}{2}$, The angle whose cosine is equal to $\dfrac{1}{2}$ is:
[A] 120° [B] 150° [C] 210° [D] 300°

34. If $\cos x$ is negative and $\sin x$ is negative, it is true that:
[A] $0° < x < 90°$ [B] $90° < x < 180°$ [C] $180° < x < 270°$ [D] $270° < x < 360°$

35. If $\sin \theta = \dfrac{\sqrt{3}}{2}$ and $\cos \theta = -\dfrac{1}{2}$. The value of θ is:
[A] 30° [B] 60° [C] 90° [D] 120°

36. The value of $\tan 315^{o}$ is:

[A] -1 [B] $-\frac{\sqrt{2}}{2}$ [C] 0 [D] 1

37. The value of $\sin 210^\circ$ is:
    [A] $\frac{1}{2}$ [B] $-\frac{\sqrt{3}}{2}$ [C] $-\frac{1}{2}$ [D] $\frac{\sqrt{3}}{2}$
38. $\cos 75^\circ$ has the same value as:
    [A] $\cos 115^\circ$ [B] $\cos 255^\circ$ [C] $\cos 285^\circ$ [D] $-\cos 255^\circ$
39. If $\sin\theta = \cos\theta$ for $0^\circ \le \theta \le 360^\circ$. The possible values of θ are:
    [A] 45°, 225° [B] 135°, 315° [C] 45°, 315° [D] 135°, 225°
40. $\cos 57^\circ$ has the same value as:
    [A] $\sin 213^\circ$ [B] $\cos 303^\circ$ [C] $\cos 127^\circ$ [D] $\cos 137^\circ$
41. If $\sin\theta = -\frac{1}{2}$. The values of θ between 0° and 360° are:
    [A] 120°,240° [B] 120°,180° [C] 210°,330° [D] 210°,300°
42. $\cos 65^\circ$ has the same value as:
    [A] $\sin 65^\circ$ [B] $\cos 25^\circ$ [C] $\cos 205^\circ$ [D] $\cos 295^\circ$
43. In $\Delta PQR, \angle PQR$ is a right angle $|QR| = 2$ cm and $\angle PRQ = 60^\circ$. $|PR|$ is equal to:
    [A] $4\sqrt{3}$ cm [B] 4 cm [C] $2\sqrt{3}$ cm [D] 1 cm
44. If $\cos x = \frac{5}{8}$ for $0^\circ \le x \le 180^\circ$. The value of x is:
    [A] 141.3° [B] 128.7° [C] 51.3° [D] 48.7°
45. $25^\circ\,45'$ as a decimal is:
    [A] 25.75° [B] 25.55° [C] 25.45° [D] 25.15°
46. In surd form $\sin 45^\circ \cos 30^\circ + \cos 45^\circ \sin 30^\circ$ is equal to:
    [A] $\frac{\sqrt{2}}{2}$ [B] $\frac{\sqrt{3}}{2}$ [C] $\sqrt{2}$ [D] $\frac{\sqrt{6}+\sqrt{2}}{4}$

8.2 APPLICATIONS OF TRIGONOMETRY

SOLVING RIGHT-ANGLED TRIANGLES

Solving Right-Angled Triangles Given One Side and One Acute Angle

Example

In the figure below, find to 2 decimal points (a) AB (b) AC

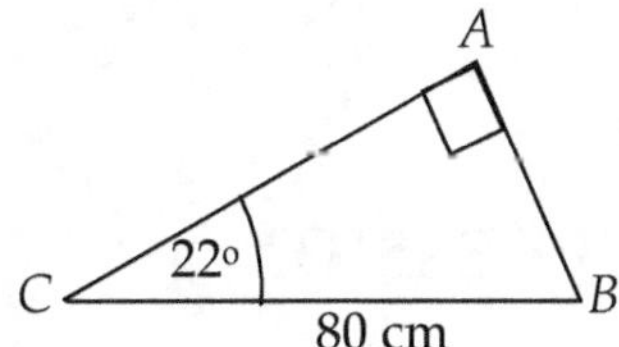

Solution

(a) $AB = 80 \sin 22° = 29.97$ cm (b) $AC = 80 \cos 22° = 74.17$ cm

Solving Right-Angled Triangles Given Two Sides

Use the Pythagoras theorem to find the third side and trig ratios to find the angles.

Example

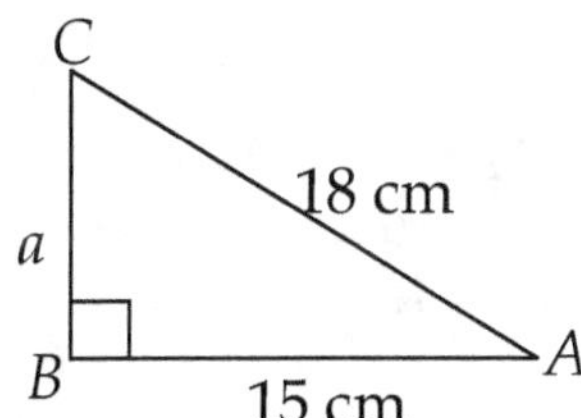

Given the triangle on the left, find
(i) a, leaving your answer in surd form.
(ii) Angle A (iii) Angle C

Solution

(i) Using the Pythagoras theorem; $a = \sqrt{18^2 - 15^2} = 3\sqrt{11}$.

(ii) $\hat{A} = \cos^{-1}\left(\frac{15}{18}\right) = 33.6°$ (iii) $\hat{C} = 90° - 33.6° = 56.4°$

Solving Isosceles Triangles With Given Sides

Draw the angle bisector of the angle between the equal sides to the opposite side, and use the Pythagoras theorem to solve the two congruent triangles.

Example 29:5

Find the angles of the triangle ABC with $AB = 9.6$ cm and $AC = BC = 8.2$ cm.

Solution

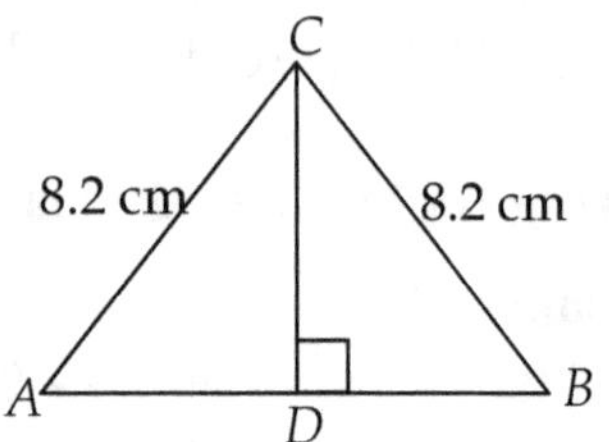

Since CD bisects $AB, AD = \frac{9.6}{2} = 4.8$ cm.

From $\Delta ACD, \hat{A} = \cos^{-1}\left(\frac{4.8}{8.2}\right) = 54.2°$

$\Rightarrow \hat{B} = 54.2°$

$\Rightarrow \angle ACB = 180° - 2(54.2°) = 71.6°$.

ANGLES OF ELEVATION AND DEPRESSION

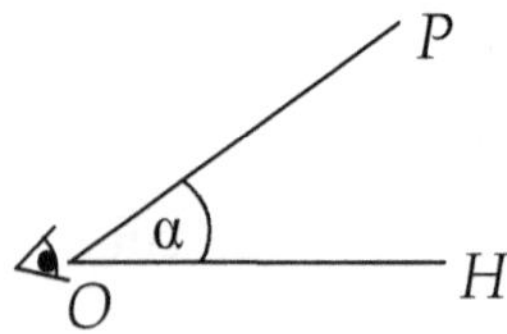

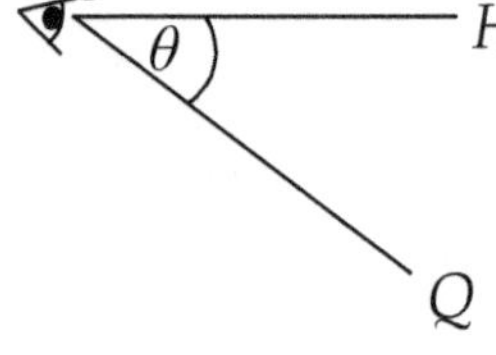

α is the angle of elevation. θ is the angle of depression.

Example

The angle of elevation of the top of a house is 30° to a man lying 11 m away from the house. Calculate the height of the house.

Solution

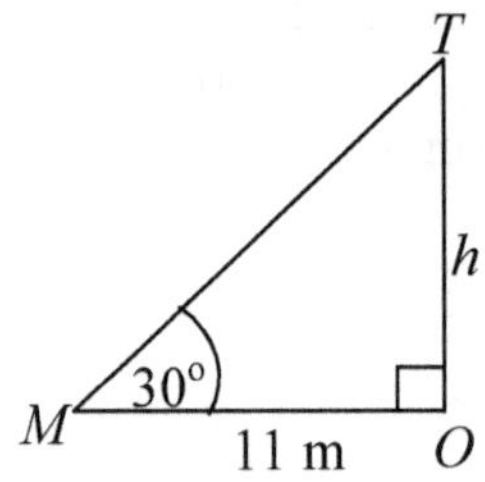

Let OT be the tree of height h and OM the distance of the man from the tree.

Then from the figure on the left

$$h = 11 \tan 30° = \frac{11\sqrt{3}}{3} \text{ m.}$$

Example

A road surveyor measures the difference in altitude of two points to be 2 m. If the distance between these two points is 12 m, calculate the angle of elevation of the higher point from the lower one. Hence, find the angle of depression of the lower point from the higher one.

Solution

Let P and Q be the lower and higher points respectively and the difference in altitude between the points be OQ as shown in the figure on the below.

Then the angle θ of elevation of Q from P is equal to the angle α of depression of P from Q [alternate angles between parallel lines].

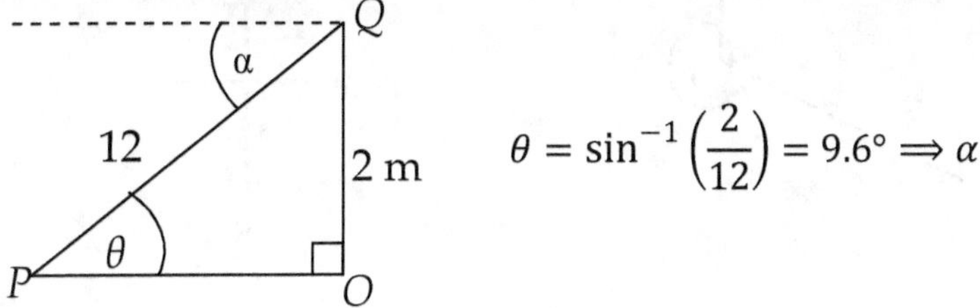

$$\theta = \sin^{-1}\left(\frac{2}{12}\right) = 9.6° \Rightarrow \alpha$$

Example

Two points A and B 15 m apart are in line and on the same level with a tree. If the angles of elevation of the top of the tree from the points A and B are 16° and 25° respectively, calculate the height of the tree and the distance of each of the points from the tree.

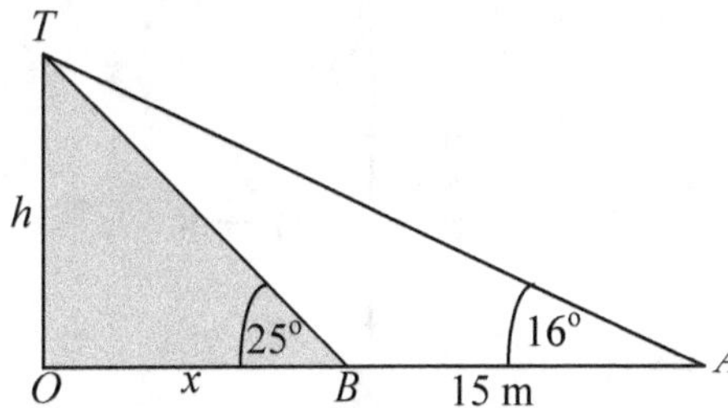

Solution

$h = x\tan 25^\circ$ ....①

$h = (x+15)\tan 16^\circ$ ....②

② − ①:

$(x+15)\tan 16^\circ - x\tan 25^\circ = 0$

$\Rightarrow x = \dfrac{15\tan 16^\circ}{\tan 25^\circ - \tan 16^\circ} = 23.95$

$\Rightarrow h = 23.95\tan 25^\circ = 11.17$ m.

⇒ Distance of B from the tree = 23.95 m
Distance of A = 23.95 +15 = 38.95 m.

BEARINGS IN TWO DIMENSIONS

Bearings deal with the angular direction of one point from another.

Cardinal Points or Compass Bearing

Compass bearings are measured from North or South to East or West. Thus N 35° E means an angle of 35° from the North towards the East.
S 42° W means an angle of 42° from the South towards the West.

Three Digit Bearings

Conventionally and for analytical purposes bearings are measured in the clockwise direction from the north which is considered to be 0°. Thus N 35° E is written as 035°, S 30° W is written as 210°. By this convention, all bearings are given to three digits. Thus, east will be represented by 090°, South West by 225°, North West by 315°, west by 270° etc.

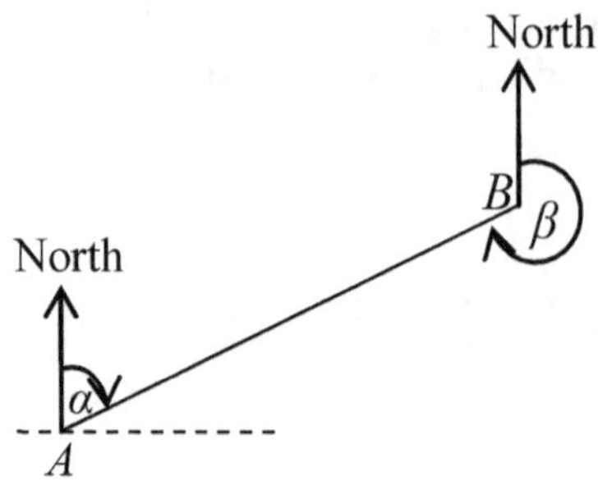

N
W E
S

Three Digit Bearings | The Cardinal Points

Example
Convert the following 3 digit bearings to compass bearings.
(a) 075° (b) 324° (c) 138° (d) 249°

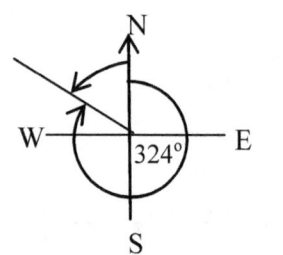

Solution

(a)

$075^\circ = \text{N } 75^\circ \text{ E}$

(b)

$324^\circ = 360^\circ - 324^\circ = \text{N } 36^\circ \text{ W}$

(c)

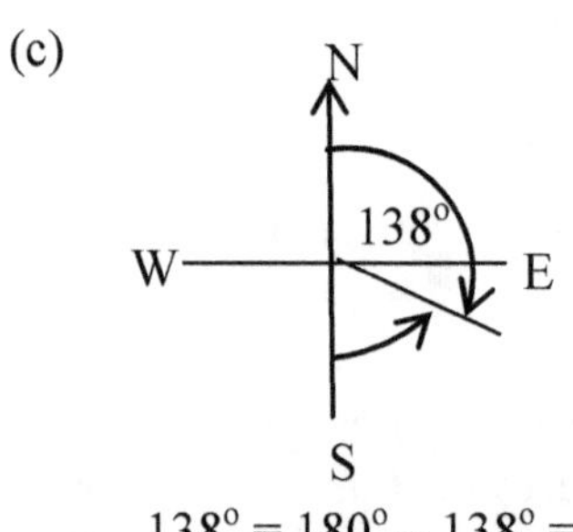

$138^\circ = 180^\circ - 138^\circ = \text{S } 42^\circ \text{ E}$

(d)

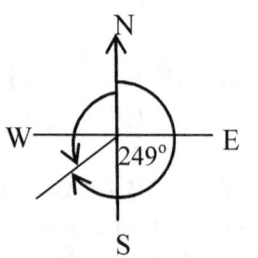

$249^\circ = 270^\circ - 249^\circ = \text{S } 21^\circ \text{ W}$

Example
Convert the following compass bearings to 3 digit bearings.
(a) N 85° E (b) S 28° W

Solution

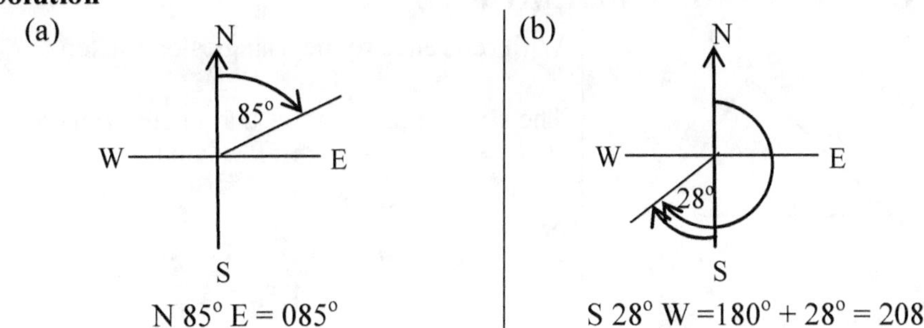

N 85° E = 085°

S 28° W = 180° + 28° = 208°

If the bearing of a point *B* from another point *A* is θ, then the bearing of the point *A* from *B* is $(180^\circ + \theta)$.

Example
A boy starts from a point *A* and moves on a bearing of 020° to a point *B*, which is 5 km away from *A*. He then changes his course to a bearing of 110° and moves to a point *C*, which is 12 km from *B*. Find the distance and bearing of the point *C* from the point *A*, giving your answer to one decimal place.

Solution

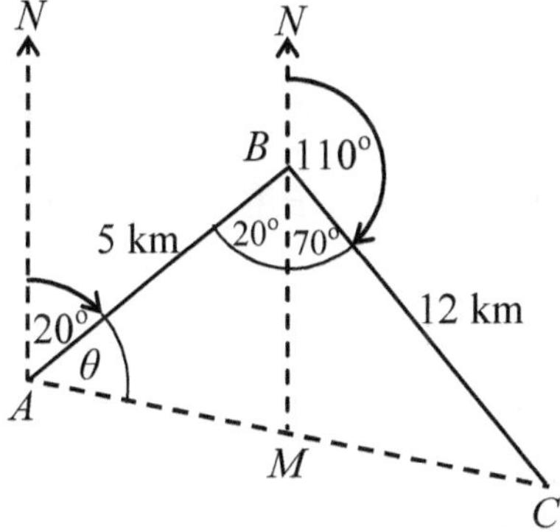

From the figure above, $\angle ABM = 20^\circ$ (alternate angles)
$\angle CBM = 180^\circ - 110^\circ = 70^\circ$ (angles on straight line)
$\angle ABC = 70^\circ + 20^\circ = 90^\circ$

Using the Pythagoras theorem,
$AC = \sqrt{AB^2 + BC^2} = \sqrt{5^2 + 12^2} = 13$ cm

$$\sin\theta = \frac{12}{13} \Rightarrow \theta = \sin^{-1}\left(\frac{12}{13}\right) = 67.38^\circ$$

Bearing of C from A = 20° + 67.38° ≈87.4°, and the distance of *C* from *A* is 13 m.

SINE AND COSINE FORMULAE

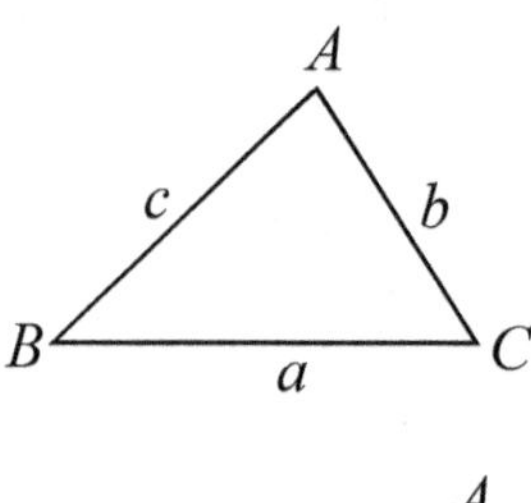

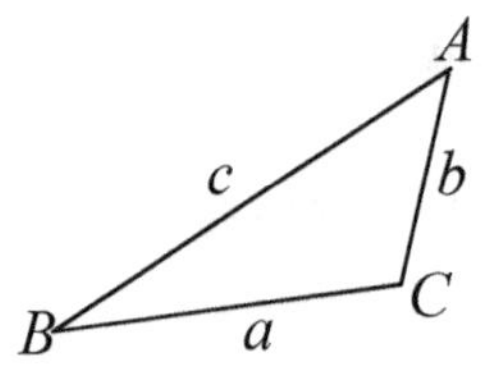

With reference to the triangles on the left,

The **sine formula** states that in any triangle

$$\frac{\sin A}{a} = \frac{\sin B}{b} = \frac{\sin C}{c}$$

or

$$\frac{a}{\sin A} = \frac{b}{\sin B} = \frac{c}{\sin C}$$

The **cosine formula** states that in any triangle

$$a^2 = b^2 + c^2 - 2bc\cos A$$
$$b^2 = a^2 + c^2 - 2ac\cos B$$
$$c^2 = a^2 + b^2 - 2ab\cos C$$

The sine rule is used when:

(i) Two angles and one side opposite one of the given angles are given.

(ii) Two sides and an angle opposite one of the given sides are given.

The cosine rule is used when:

(i) Two sides and the included angle are given.

(ii) All the three sides are given.

Example

In the following figure, calculate (a) AC (b) $\angle BAC$ (c) $\angle ACB$.

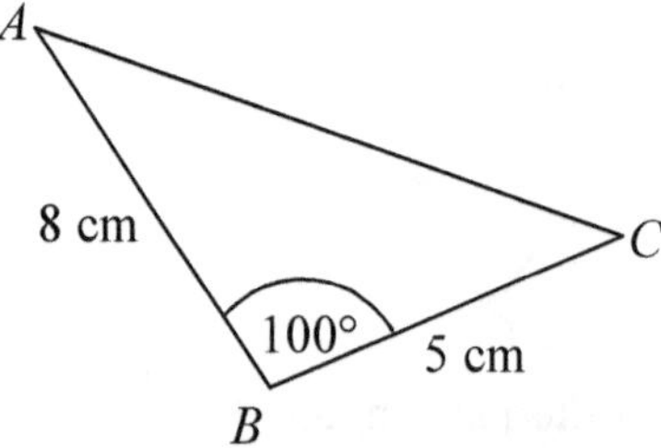

Solution

(a) By the cosine rule, $b^2 = a^2 + c^2 - 2ac\cos B$.

$\Rightarrow b^2 = 5^2 + 8^2 - 2(5)(8)\cos 100 = 102.89 \Rightarrow b = 10.14$ cm.

(b) By the sine rule, $\frac{\sin A}{5} = \frac{\sin 100}{10.14} \Rightarrow \sin A = \frac{5\sin 100}{10.14} = 0.4856$

$\Rightarrow \angle BAC = \sin^{-1} 0.4856 = 29.1°$

(c) $\Rightarrow \angle ACB = 180 - (100 + 29.1) = 50.9°$

MULTIPLE CHOICE EXERCISE 8:2

1. In figure (i) below, the length of the side marked x is:
   [A] 7sin 56° [B] 7tan 56° [C] 7tan 34° [D] 7cos 34°

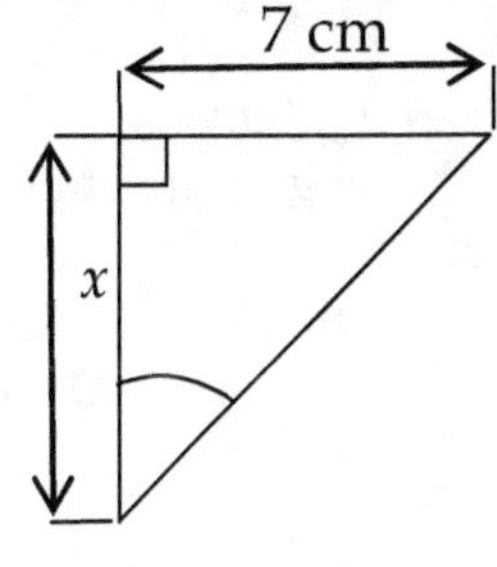

(i) (ii)

2. In figure (ii) above, $AC = 42$ cm, $AD = 12$ cm and $BD = 16$ cm. As a vulgar fraction tan A is:
   [A] $\frac{4}{5}$ [B] $\frac{3}{5}$ [C] $\frac{5}{6}$ [D] $\frac{4}{3}$
3. In figure (ii) above, $AC = 42$ cm, $AD = 12$ cm and $BD = 16$ cm. Sin A as a decimal fraction is:
   [A] 0.8 [B] 1.3 [C] 1.6 [D] 0.6
4. In figure (ii) above, $AC = 42$ cm, $AD = 12$ cm and $BD = 16$ cm. The size of C to the nearest degree is:
   [A] 27° [B] 32° [C] 28° [D] 58°
5. In figure (ii) above, $AC = 42$ cm, $AD = 12$ cm and $BD = 16$ cm. The perimeter of triangle BDC in cm is:
   [A] 70 [B] 80 [C] 90 [D] 92
6. The following is a trapezium $ABCD$. Angle D is a right angle and BE is perpendicular to DC. $AB = 12$ cm and $AD = 8$ cm. The value of $DC = 18$ cm. The value of tan C is:

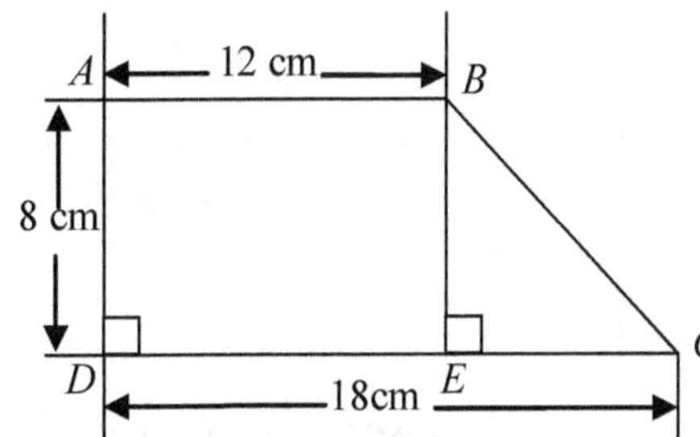

   [A] $\frac{4}{9}$ [B] $\frac{2}{3}$ [C] $\frac{4}{3}$ [D] $\frac{4}{5}$

7. A ladder 9 m long leans against a vertical wall, making an angle of 64° with the horizontal ground. To one decimal place, the distance of the foot of the ladder from the wall is:
   [A] 3.9 m [B] 5.8 m [C] 7.9 m [D] 8.1 m

8. When helicopter is 800 m above the ground, its angle of elevation from a point P on the ground is 30°. The distance of the helicopter from P by the line of sight is:
   [A] 400 m [B] 800 m [C] 1600 m [D] 1700 m
9. The angle of elevation of X from Y is 30°. If $XY = 40$ m the height of X above the level of Y is:
   [A] 10 m [B] 20 m [C] 40 m [D] 50 m
10. If the shadow of a pole 7 m high is equal to half its length, the angle of elevation of the sun; correct to the nearest degree is:
   [A] 63° [B] 0° [C] 60° [D] 26°
11. The angle of elevation of the top of a tree 39 m away from the point on the ground is 30°.
   The height of the tree is:
   [A] $39\sqrt{3}$ m [B] $13\sqrt{3}$ m [C] $\frac{13}{\sqrt{3}}$ m [D] $\frac{\sqrt{3}}{13}$ m
12. A ladder leans against the wall at an angle 60° to the wall. If the foot of the ladder is 5 m away from the wall, the length of the ladder is:
   [A] $\frac{5\sqrt{3}}{3}$ m [B] 5 m [C] $5\sqrt{3}$ m [D] $\frac{10\sqrt{3}}{3}$ m
13. The angle of elevation of the top of a tower from a point on the horizontal ground, 40 m away from the foot of the tower is 30°. The height of the tower is:
   [A] 20 m [B] $\frac{40\sqrt{3}}{3}$ m [C] $20\sqrt{3}$ m [D] $40\sqrt{3}$ m
14. At a point 500 m from the base of a water-tank, the angle of elevation of the top of the tank is 45°. The height of the tank is:
   [A] 500m [B] 353 m [C] 354 m [D] 250 m
15. A ladder 6 m long leans against a vertical wall, so that it makes an angle of 60° with the wall. The distance of the foot of the ladder from the wall is:
   [A] 3 m [B] 6 m [C] $2\sqrt{3}$ m [D] $3\sqrt{3}$ m
16. The angle of elevation of the top X of a vertical pole from a point P on a level ground is 60°. The distance from P to the foot of the pole is 55 m. Without using tables, the height of the pole is:
   [A] $\frac{50}{3}$ m [B] 50 m [C] $55\sqrt{3}$ m [D] 60 m
17. The angle of elevation of a point T on a tower from a point U on the horizontal ground is 30°. If $TU = 54$ m, the height of T above the ground is:
   [A] 108 m [B] 72 m [C] 31.2 m [D] 27 m
18. In the following figure, $\angle PQR$ is a right angle, $\angle PRQ = 60^\circ$ and

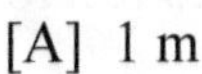 $|QR| = 2$ cm. $|PR|$ is equal to:

[A] 1 m [B] 4 m [C] $2\sqrt{3}$ m [D] $\frac{4}{3}$ m

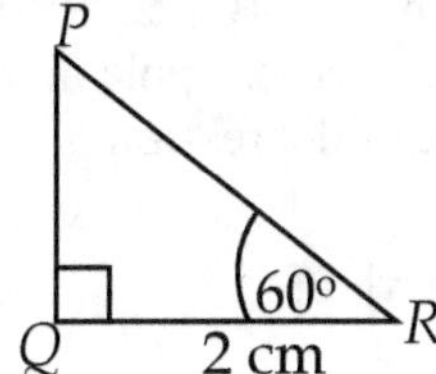

19. A ladder 5 m long rest against a wall so that it foot makes an angle of 30° with the horizontal. The distance of the foot of the ladder from the wall is:

[A] $\frac{5\sqrt{3}}{3}$ m [B] $2\frac{1}{2}$ m [C] $\frac{5\sqrt{3}}{2}$ m [D] $\frac{10\sqrt{3}}{3}$ m

20. A pole of length l leans against a vertical wall so that it makes an angle of 60° with the horizontal ground. If the top of the pole is 8 m above the ground, l must be:

[A] $16\sqrt{3}$ m [B] $\frac{\sqrt{3}}{16}$ m [C] 16 m [D] $\frac{16\sqrt{3}}{3}$ m

21. The angle of depression of a point on the ground from the top of a building is 20.3°. If the distance of the point from the foot of the building is 40 m, the height of the building, correct to one decimal place is:

[A] 13.9 m [B] 28.1 m [C] 27.8 m [D] 14.8 m

22. From the top of a cliff 20 m high, a boat can be sighted at sea 75 m from the foot of the cliff. The angle of depression of the boat from the top of the cliff is:

[A] 14.9° [B] 15.5° [C] 74.5° [D] 75.1°

23. From the top of a building 10 m high, the angle of depression of a stone lying on the ground is 69°. Correct to one decimal place, the distance of the stone from the foot of the building is:

[A] 3.6 m [B] 3.8 m [C] 6.0 m [D] 9.3 m

24. A cliff on the bank of a river is 300 metres high. If the angle of depression of a point on the opposite side of the river is 60°, the width of the river is:

[A] 100 m [B] $75\sqrt{3}$ m [C] $100\sqrt{3}$ m [D] $200\sqrt{3}$ m

25. From the top of a cliff, the angle of depression of a boat on the sea is 60°; if the top of the cliff is 25 m above the sea level, the horizontal distance from the bottom of the cliff to the boat is:

[A] $\frac{\sqrt{3}}{25}$ m [B] $25\sqrt{3}$ m [C] $\frac{25\sqrt{3}}{3}$ m [D] $\frac{25}{3}$ m

26. The angle of depression of a point Q from a vertical tower PR, 30m high, is 40°. If the foot P of the tower is on the same level ground as Q, correct to two decimal places, $|PQ|$ should be:

    [A] 35.75 m [B] 25.00 m [C] 22.98 m [D] 19228 m

27. In the following figure, AB is a vertical pole and BC is horizontal. If $|AC| = 10$ m and $|BC| = 5$ m. The angle of depression of C from A is:

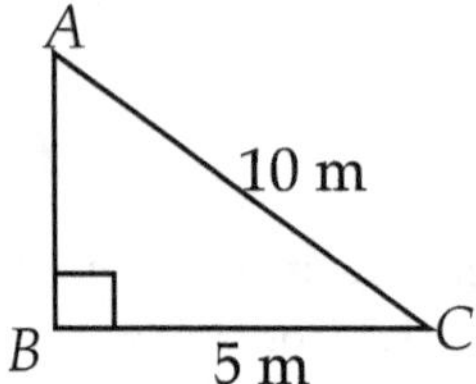

    [A] 63° [B] 60°
    [C] 45° [D] 30°

28. In the following figure, QRT is a straight line. If $\angle PTR = 90^\circ$, $\angle PRT = 60^\circ$, $\angle PQR = 30^\circ$ and $|PQ| = 6\sqrt{3}$ m. $|RT|$ is equal to:

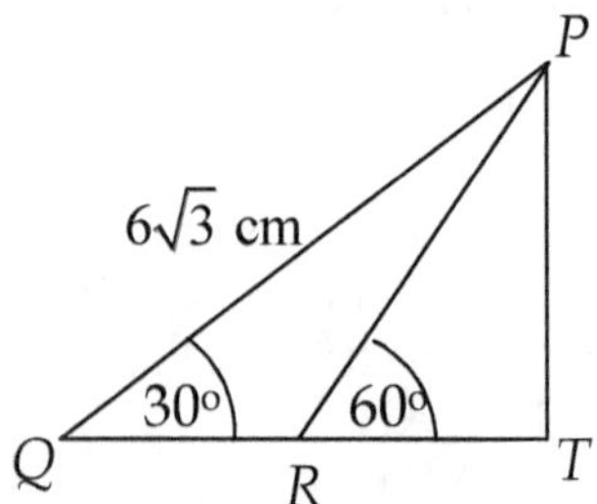

    [A] 0.3 m
    [B] $\frac{\sqrt{3}}{2}$ m
    [C] 3 m
    [D] $3\sqrt{3}$ m

29. The true bearing of 250° as a compass bearing is:

    [A] S 70° E [B] N 70° E [C] S 70° W [D] N 70° W

30. The bearing which is equivalent to S 50° W is:

    [A] 230° [B] 220° [C] 130° [D] 040°

31. The bearing S 40° E is the same as:

    [A] 040° [B] 050° [C] 130° [D] 140°

32. The bearing S 50° W is the same as:

    [A] 050° [B] 130° [C] 140° [D] 230°

33. Town P is 150 km from a town Q in the direction 050°. The bearing of Q from P is:

    [A] 050° [B] 150° [C] 230° [D] 310°

34. The bearing of P from Q is x, where $270^\circ < x < 360^\circ$. The bearing of Q from P is:

    [A] $(x-90)^\circ$ [B] $(x+180)^\circ$ [C] $(x-135)^\circ$ [D] $(x-180)^\circ$

35. The figure below shows the position of three ships A, B and C at sea. B is due north of C such that $|AB| = |BC|$ and the bearing of B from A is 040°. The bearing of A from C is:

    [A] 040° [B] 070° [C] 110° [D] 290°

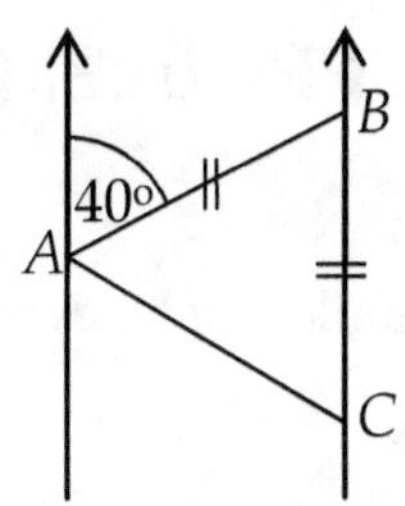

36. A tree is 8 km due south of a building. Ambe is standing 8 km west of the tree. The distance of Ambe from the building is:
    [A] $4\sqrt{2}$ km [B] 8 km [C] $8\sqrt{2}$ km [D] 16 km
37. A tree is 8 km due south of a building. Ambe is standing 8 km west of the tree. The bearing of Ambe from the building is:
    [A] 315° [B] 270° [C] 225° [D] 135°
38. Three observation posts P, Q and R are such that Q is due east of P and R is due north of Q, if $|PQ| = 5$ km and $|PR| = 10$ km, $|QR|$ equals:
    [A] 50 km [B] 9.5 km [C] 7.6 km [D] 8.7 km
39. The following figure shows that, the bearing of Q from P is:
    [A] 236° [B] 214° [C] 146° [D] 124°

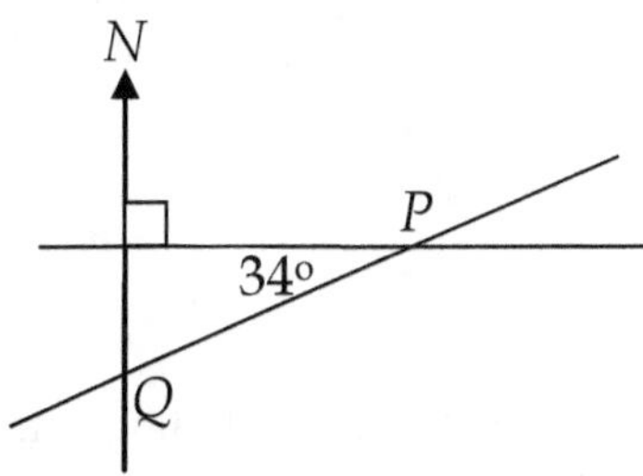

40. The bearing of a town Q from a town P is 215°. P is 80 km north of R while R is due east of Q. The distance between Q from R correct to the nearest km is:
    [A] 46 km [B] 56 km [C] 38 km [D] 98 km
41. The following figure shows that, the bearing of C from B is:
    [A] 060° [B] 090° [C] 120° [D] 240°

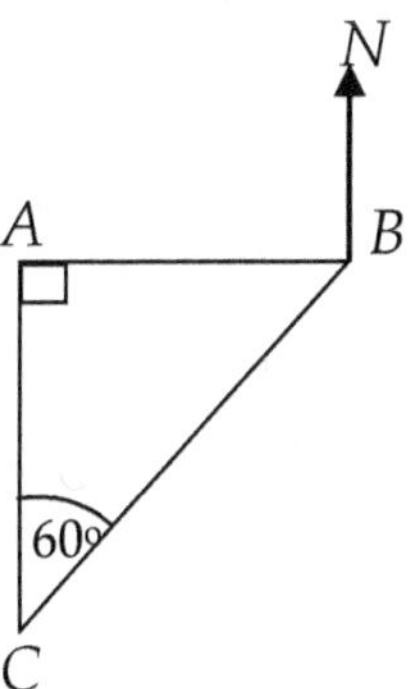

8.3 THREE DIMENSIONAL PROBLEMS

Lines and Planes

In drawing three-dimensional diagrams, the following rules may be followed

I. ***West-East lines*** are drawn parallel to the top and bottom edges of the page.

II. ***North-South lines*** are drawn inclined to the right at an acute angle to the west east line.

III. ***Vertical lines*** are drawn parallel to the left and right edges of the page.

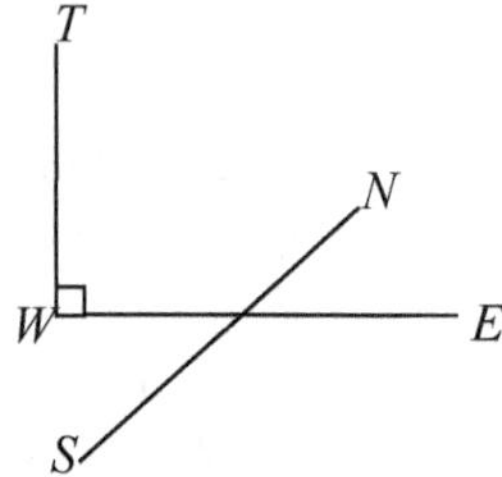

VII. Line Perpendicular to a Plane

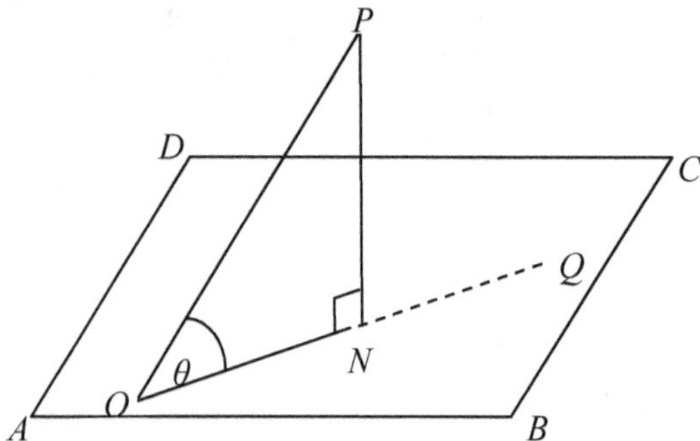

The figure above shows a line *OP* that intersects with the plane *ABCD* making an angle θ with the plane at *O*. The line *PN* is perpendicular to the plane as indicated. Any line such as *PN* that is perpendicular to a plane is necessarily perpendicular to any line on the plane.

Parallel lines do not meet. Perpendicular lines meet at 90°. **Skew lines** are lines that are neither parallel nor intersecting.

Lines of Greatest Slope

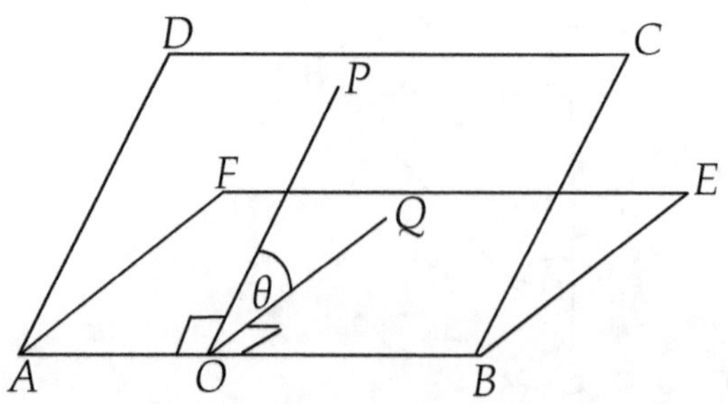

In the figure below, the lines *AD*, *OP*, *BC* or any other line parallel to these lines are called **lines of greatest slope**, because these are the only lines on the plane *ABCD* with maximum gradient.

Example

The following figure shows a cuboid. Given that $AB = 8$ cm, $BC = 6$ cm and $CG = 4$ cm, calculate:

(a) AH (b) BD (c) $\angle DBH$ (d) $\angle CBG$

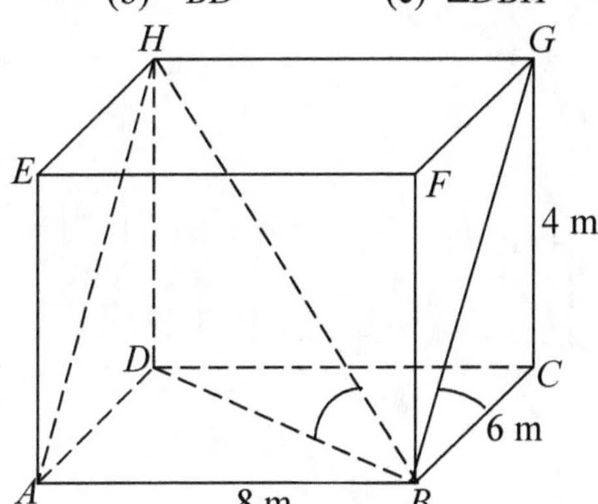

Solution

(a)

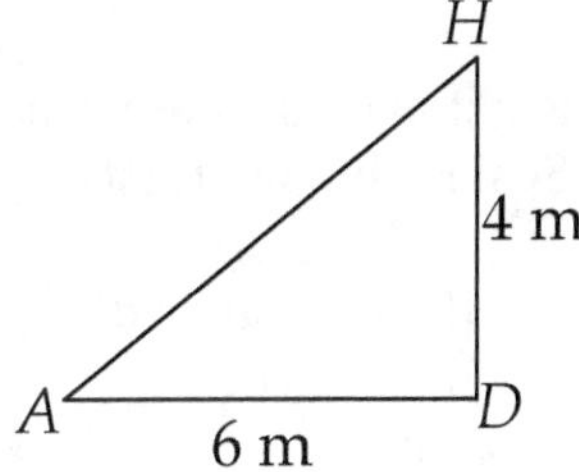

$$AH = \sqrt{AD^2 + DH^2}$$
$$= \sqrt{6^2 + 4^2}$$
$$=7.21 \text{ cm}$$

(b)

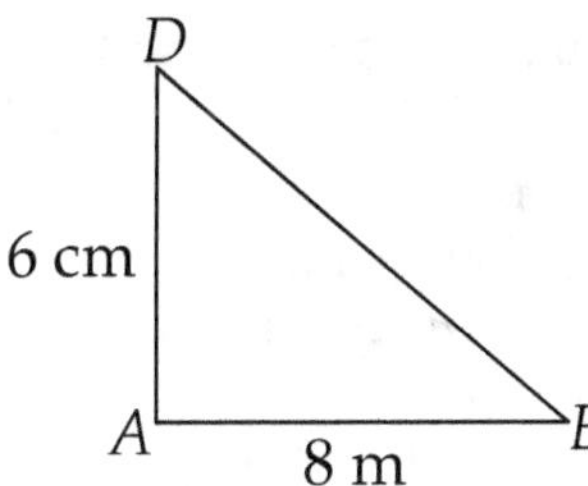

$$BD = \sqrt{AB^2 + AD^2}$$
$$= \sqrt{8^2 + 6^2}$$
$$=10 \text{ cm}$$

(c)

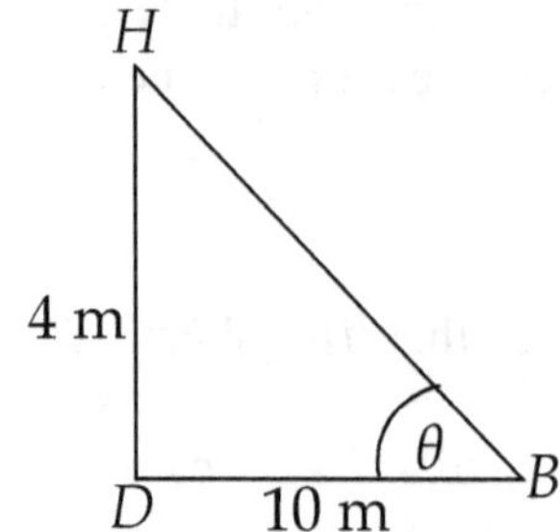

$$\theta = \tan^{-1}\left(\frac{4}{10}\right) = 21.8°$$

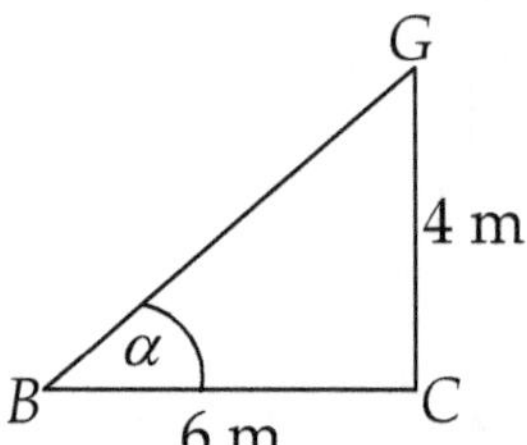

$$\alpha = \tan^{-1}\left(\frac{4}{6}\right)$$
$$= 33.7°$$

MULTIPLE CHOICE EXERCISE 8:3

1. In the cube in figure (i) below, ΔHDB is:
   [A] Isosceles but not equilateral [B] Right-angled and scalene
   [C] Equilateral [D] Scalene but not right-angled
2. In the cube in figure (i) below, ΔHAC is:
   [A] Isosceles but not equilateral [B] right-angled
   [C] Equilateral [D] scalene
3. In the cube in figure (i) below, angle HAC is equal to:
   [A] 30° [B] 45° [C] 90° [D] 60°
4. In the cube in figure (i) below, the segment which intersects with DC is:
   [A] EF [B] GF [C] BC [D] AB
5. In the cube in figure (i) below, EF and AE:
   [A] have point C in common [B] are parallel
   [C] will intersect [D] will never meet
6. In figure (i) below, the number of sets of parallel lines is:
   [A] 12 [B] 4 [C] 3 [D] 2

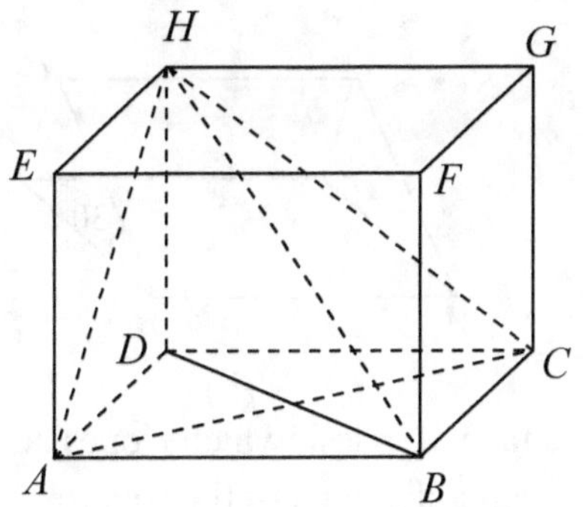

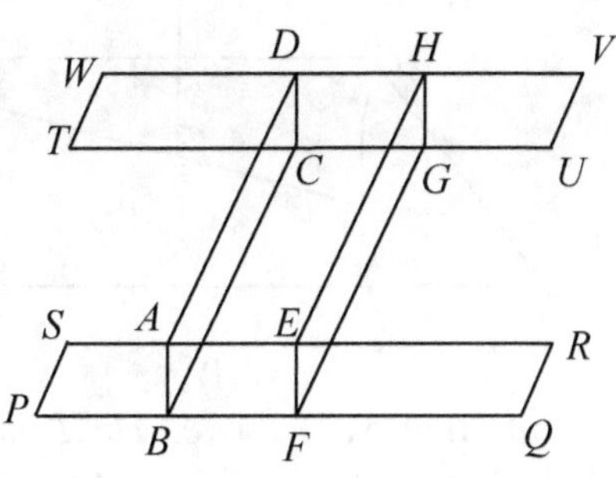

(i) (ii)

7. In figure (ii) above, the number of planes is:
   [A] 12 [B] 4 [C] 3 [D] 2
8. A fly crawls from X to Y by the shortest route on the surface of the cuboid in figure (i) below. The length of this journey is:
   [A] $2\sqrt{34}$ cm [B] $10\sqrt{2}$ cm [C] 20 cm [D] $\sqrt{58}$ cm

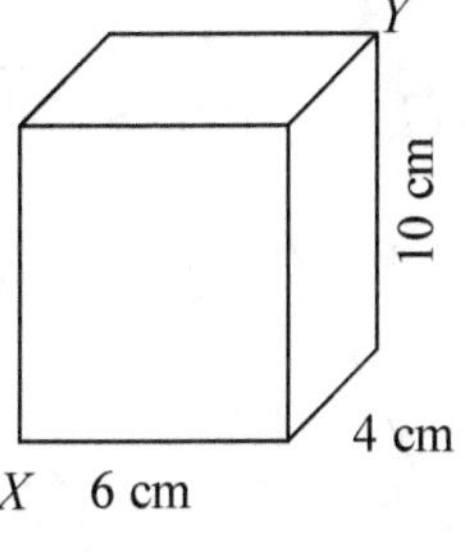

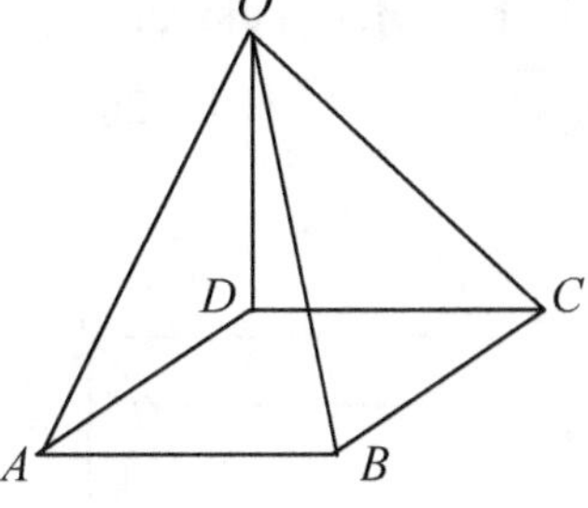

(i) (ii)

9. Figure (ii) above shows a pyramid on a square base. All the eight edges of the pyramid are equal. The value in degrees, of angle *OAC* is:
   [A] 30° [B] 75° [C] 60° [D] 45°
10. In figure (ii) above, a line segment parallel to *DC* is:
    [A] *AB* [B] *CO* [C] *CB* [D] *BO*
11. Figure 41:29, line segments that intersect *AD* are:
    [A] *AD, DO, CB, CO* [B] *DC, DO, AB, AO*
    [C] *BO, AO, DO, CD* [D] *BO, BC, DO, CD*
12. In Figure (ii) above, a line segments skew to *DC* are:
    [A] *CO, DO* [B] *AD,BC* [C] *BO, CO* [D] *BO, AO*
13. In figure (i) below, *PN* is a vertical pole standing on a horizontal plane *ABCD*. *XY* is a line on the plane. Therefore, the pole *PN* is:
    [A] Parallel to the line *XY*. [B] Perpendicular to the line *XY*.
    [C] Identical to the line *XY*. [D] Is in the same plane with the line *XY*.

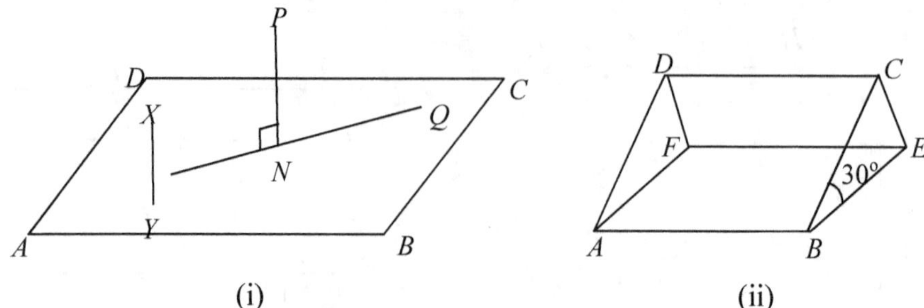

(i) (ii)

14. In figure (ii) above, $ABCD$ is the square face of a desk which slopes at 30° to the horizontal $ABEF$. The angle which a diagonal of the square makes with the horizontal is:
    [A] 30° [B] 10.2° [C] 20.7° [D] 60°
15. In figure (ii) above,, $ABCD$ is the square face of a desk which slopes at 30° to the horizontal $ABEF$. Given that the side of the square is 70 cm, the height of CE is:
    [A] 35 cm [B] 17.5 cm [C] 8.8 cm [D] 52.5 cm
16. In Figure (i) below, $AB = 20$ cm, $BC = 16$ cm and $AE = 12$ cm. The length of the diagonal AG in cm is:
    [A] $4\sqrt{41}$ [B] 20 [C] $20\sqrt{2}$ [D] $4\sqrt{34}$

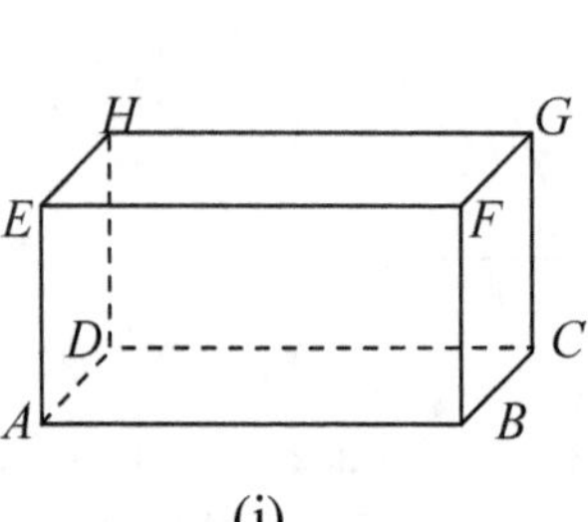

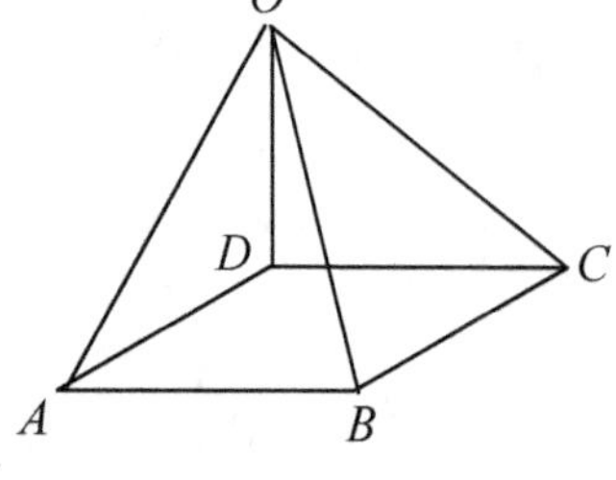

(i) (ii)

17. In Figure (i) above, $AB = 20$ cm, $BC = 16$ cm and $AE = 12$ cm. The angle which AB makes with BE is:
    [A] 31° [B] 53° [C] 37° [D] 59°
18. In Figure (i) above, $AB = 20$ cm, $BC = 16$ cm and $AE = 12$ cm. The angle which the plane $ADGF$ makes with the plane $ADHE$ is:
    [A] 31° [B] 53° [C] 37° [D] 59°
19. Figure (ii) above is a pyramid with a square base $ABCD$ of side 5 cm. $OA = OB = OC = OD = 8$ cm. The height of the pyramid to the nearest whole number is:
    [A] 7 cm [B] 13 cm [C] 8.8 cm [D] 8 cm

CHAPTER 9

MATRICES AND TRANSFORMATIONS

9.1 MATRICES

A matrix is a rectangular array of numbers in rows and columns.

The size or order of a matrix with r rows and c columns is specified as $r \times c$, read 'r by c'.

Example

State the size of the matrix $\begin{pmatrix} 2 & 4 & 7 \\ 3 & 6 & 5 \end{pmatrix}$.

Solution

2×3

Types of Matrices

1. **Square matrix**: number of rows = columns. E.g. $\begin{pmatrix} 2 & 1 \\ 4 & 3 \end{pmatrix}$.

   A square matrix has a **leading diagonal** and a **minor diagonal**.

   $$\begin{pmatrix} 0 & 1 & 2 \\ 8 & 3 & 7 \\ 5 & -4 & 6 \end{pmatrix} \begin{matrix} \leftarrow \text{Minor diagonal} \\ \\ \leftarrow \text{Leading diagonal} \end{matrix}$$

2. **Diagonal matrix:** a square matrix with all the elements zeros except those in the leading diagonal. E.g. $\begin{pmatrix} 1 & 0 & 0 \\ 0 & 3 & 0 \\ 0 & 0 & 6 \end{pmatrix}$.

3. **Unit** or **identity Matrix I:** a diagonal matrix with all the elements in the leading diagonal ones. E.g. $\begin{pmatrix} 1 & 0 \\ 0 & 1 \end{pmatrix}$.

4. **Rectangular matrix:** number of rows ≠number of columns. E.g. $\begin{pmatrix} 2 & 4 & 7 \\ 3 & 6 & 5 \end{pmatrix}$.

5. **Column matrix:** all the elements are in a single column. E.g. $\begin{pmatrix} -2 \\ 4 \end{pmatrix}$.

6. **Row matrix:** all the elements are in a single row. E.g. $\begin{pmatrix} -2 & 4 \end{pmatrix}$.
7. **Zero or null matrix:** all the elements are zeros. E.g. $\begin{pmatrix} 0 & 0 \\ 0 & 0 \end{pmatrix}$.

Equality of Matrices

Two matrices **A** and **B** are equal if and only if:

(i) They have the same size or order.

(ii) Their corresponding elements are equal.

Examples

Find x and y given that $\begin{pmatrix} 3x & y \\ 0 & 4x \end{pmatrix} = \begin{pmatrix} 6 & 2 \\ 0 & 8 \end{pmatrix}$.

Solution

Equating corresponding entries, $3x = 6 \Rightarrow x = 2$ and $y = 2$.

Addition and Subtraction of Matrices

If matrices are of the same size then they are compatible for matrix addition and subtraction and we can add or subtract corresponding elements.

Example

Let $\mathbf{A} = \begin{pmatrix} 1 & 3 \\ 1 & 2 \end{pmatrix}$ and $\mathbf{B} = \begin{pmatrix} 0 & -1 \\ 2 & 3 \end{pmatrix}$. Find (i) $\mathbf{A} + \mathbf{B}$ (ii) $\mathbf{A} - \mathbf{B}$

Solution

(i) $\mathbf{A} + \mathbf{B} = \begin{pmatrix} 1 & 3 \\ 1 & 2 \end{pmatrix} + \begin{pmatrix} 0 & -1 \\ 2 & 3 \end{pmatrix} = \begin{pmatrix} 1 & 2 \\ 3 & 5 \end{pmatrix}$

(ii) $\mathbf{A} - \mathbf{B} = \begin{pmatrix} 1 & 3 \\ 1 & 2 \end{pmatrix} - \begin{pmatrix} 0 & -1 \\ 2 & 3 \end{pmatrix} = \begin{pmatrix} 1 & 4 \\ -1 & -1 \end{pmatrix}$

Scalar Multiplication of Matrices

To multiply a matrix by a scalar, multiply each element by the scalar.

Example

Evaluate $\frac{1}{3}\begin{pmatrix} 6 & 9 \\ -15 & 12 \end{pmatrix}$

Solution

$$\frac{1}{3}\begin{pmatrix} 6 & 9 \\ -15 & 12 \end{pmatrix} = \begin{pmatrix} \frac{1}{3}\times 6 & \frac{1}{3}\times 9 \\ \frac{1}{3}\times -15 & \frac{1}{3}\times 12 \end{pmatrix} = \begin{pmatrix} 2 & 3 \\ -5 & 4 \end{pmatrix}$$

Multiplication of Matrices

If the number of columns in the first matrix is equal to the number of rows in the second matrix, we can multiply them and generally
(m by n) × (n by m) = m by p.

Example

Compute $\begin{pmatrix} 2 & 7 \\ 1 & 3 \end{pmatrix}\begin{pmatrix} 1 & 4 & -3 \\ 5 & 2 & 0 \end{pmatrix}$.

Solution

$$\begin{pmatrix} 2 & 7 \\ 1 & 3 \end{pmatrix}\begin{pmatrix} 1 & 4 & -3 \\ 5 & 2 & 0 \end{pmatrix} = \begin{pmatrix} 2\times 1+7\times 5 & 2\times 4+7\times 2 & 2\times(-3)+7\times 0 \\ 1\times 1+3\times 5 & 1\times 4+3\times 2 & 1\times(-3)+3\times 0 \end{pmatrix}$$

$$= \begin{pmatrix} 2+35 & 8+14 & -6+1 \\ 1+15 & 4+6 & -3+0 \end{pmatrix} = \begin{pmatrix} 37 & 22 & -6 \\ 16 & 10 & -3 \end{pmatrix}$$

Note that $\mathbf{AB} \neq \mathbf{BA}$.

The Transpose of a Matrix

The transpose $\mathbf{A}^{\mathrm{T}}$ of a matrix A is obtained by interchanging the rows and columns of the matrix.

E.g. the transpose of $\mathbf{B} = \begin{pmatrix} 1 & 3 \\ 2 & 4 \\ 7 & -9 \end{pmatrix}$ is $\mathbf{B}^{\mathrm{T}} = \begin{pmatrix} 1 & 2 & 7 \\ 3 & 4 & -9 \end{pmatrix}$.

The Determinant of a 2×2 *Matrix*

The **determinant** of a 2×2 matrix $\mathbf{A} = \begin{pmatrix} a & b \\ c & d \end{pmatrix}$, is denoted by det **A,** det$\begin{pmatrix} a & b \\ c & d \end{pmatrix}$, $|\mathbf{A}|$ or $\begin{vmatrix} a & b \\ c & d \end{vmatrix}$. Det $\mathbf{A} = ad - bc$.

Note! $\begin{pmatrix} a & b \\ c & d \end{pmatrix} \neq \begin{vmatrix} a & b \\ c & d \end{vmatrix}$

A **singular matrix** is a matrix whose determinant is equal to zero.

Example

Determine whether the matrices $\begin{pmatrix} 3 & 2 \\ 1 & 5 \end{pmatrix}$ and $\begin{pmatrix} 2 & -1 \\ 4 & -2 \end{pmatrix}$ are singular.

Solution

Det $\begin{pmatrix} 3 & 2 \\ 1 & 5 \end{pmatrix} = 3(5) - 1(2) = 13 \Rightarrow \begin{pmatrix} 3 & 2 \\ 1 & 5 \end{pmatrix}$ is not a singular matrix.

Det $\begin{pmatrix} 2 & -1 \\ 4 & -2 \end{pmatrix} = 2(-2) - 4(-1) = 0 \Rightarrow \begin{pmatrix} 2 & -1 \\ 4 & -2 \end{pmatrix}$ is a singular matrix.

The Adjoint of a 2×2 *Matrix*

The adjoint of a 2×2 matrix $\mathbf{A} = \begin{pmatrix} a & b \\ c & d \end{pmatrix}$ is denoted by Adj $\begin{pmatrix} a & b \\ c & d \end{pmatrix}$ or adj **A**.

$$\text{Adj}\begin{pmatrix} a & b \\ c & d \end{pmatrix} = \text{Adj}\mathbf{A} = \begin{pmatrix} d & -b \\ -c & a \end{pmatrix}$$

Thus $\text{Adj}\begin{pmatrix} 15 & 9 \\ 10 & 5 \end{pmatrix} = \begin{pmatrix} 5 & -9 \\ -10 & 15 \end{pmatrix}$

The Inverse of a 2×2 *Matrix*

$$\mathrm{A}^{-1} = \frac{1}{ad-bc}\begin{pmatrix} d & -b \\ -c & a \end{pmatrix} = \frac{1}{\text{Det }\mathbf{A}} \times \text{Adj }\mathbf{A}$$

$$\mathbf{A}^{-1}\mathbf{A} = \mathbf{A}\mathbf{A}^{-1} = \mathbf{I}$$

Example

Find the inverse of $\mathbf{A} = \begin{pmatrix} 4 & 1 \\ 6 & 2 \end{pmatrix}$.

Solution

$$\mathbf{A}^{-1} = \frac{1}{4(2)-1(6)}\begin{pmatrix} 2 & -1 \\ -6 & 4 \end{pmatrix} = \frac{1}{2}\begin{pmatrix} 2 & -1 \\ -6 & 4 \end{pmatrix} \Rightarrow \mathbf{A}^{-1} = \begin{pmatrix} 1 & -\frac{1}{2} \\ -3 & 2 \end{pmatrix}$$

Solving Simultaneous Linear Equations Using the Matrix Method

The concept is based on the fact that $\mathbf{A^{-1}A = I}$ and $\mathbf{IQ = Q}$.

Example

Using the matrix method, solve the equations $4x + y = 3, 6x + 2y = 5$.

Solution

$$\begin{pmatrix} 4 & 1 \\ 6 & 2 \end{pmatrix}\begin{pmatrix} x \\ y \end{pmatrix} = \begin{pmatrix} 3 \\ 5 \end{pmatrix} \ldots\ldots\ldots\ldots\ldots\ldots (1)$$

$$\begin{pmatrix} 4 & 1 \\ 6 & 2 \end{pmatrix}^{-1} = \begin{pmatrix} 1 & -\frac{1}{2} \\ -3 & 2 \end{pmatrix}$$

Pre-multiply both sides of equation (1) by $\begin{pmatrix} 4 & 1 \\ 6 & 2 \end{pmatrix}^{-1}$.

$$\begin{pmatrix} x \\ y \end{pmatrix} = \begin{pmatrix} 1 & -\frac{1}{2} \\ -3 & 2 \end{pmatrix}\begin{pmatrix} 3 \\ 5 \end{pmatrix} = \begin{pmatrix} \frac{1}{2} \\ 1 \end{pmatrix}$$

$$\Rightarrow x = \frac{1}{2} \text{ and } y = 1$$

Route and Incidence Matrices

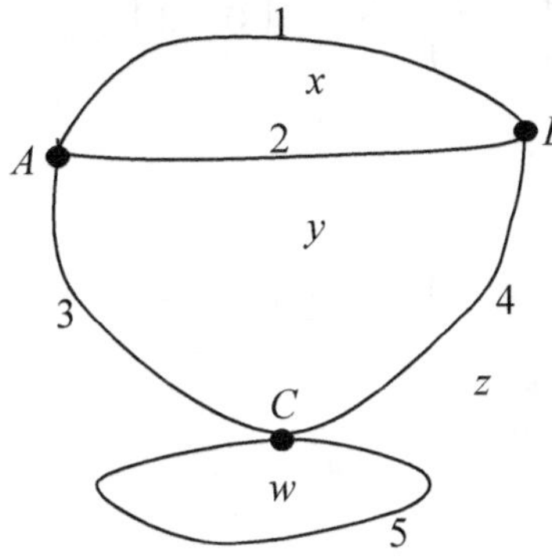

A **route matrix** is a matrix that represents the number of direct routes linking each node in a network to another.

A route matrix for the network diagram on the left is

$$\begin{matrix} & A & B & C \\ A & 0 & 2 & 1 \\ B & 2 & 0 & 1 \\ C & 1 & 1 & 1 \end{matrix}$$

An **incidence matrix** is a matrix that shows the relationship between each node and each region or each node and each route or each region and each route in a network. For the network diagram above, the following three matrices are incidence matrices

$$\begin{matrix} & w & x & y & z \\ A & 0 & 1 & 1 & 1 \\ B & 0 & 1 & 1 & 1 \\ C & 1 & 0 & 1 & 1 \end{matrix}, \quad \begin{matrix} & 1 & 2 & 3 & 4 & 5 \\ A & 1 & 1 & 1 & 0 & 0 \\ B & 1 & 1 & 0 & 1 & 0 \\ C & 0 & 0 & 1 & 1 & 1 \end{matrix}, \quad \begin{matrix} & w & x & y & z \\ 1 & 0 & 1 & 0 & 1 \\ 2 & 0 & 1 & 1 & 0 \\ 3 & 0 & 0 & 1 & 1 \\ 4 & 0 & 0 & 1 & 1 \\ 5 & 1 & 0 & 0 & 1 \end{matrix}$$

MULTIPLE CHOICE EXERCISE 9:1

1. The order of the matrices $\begin{pmatrix} 3 & 1 & 2 \\ 4 & 0 & 3 \end{pmatrix}$ and $\begin{pmatrix} 4 \\ 0 \\ 3 \end{pmatrix}$ are respectively:

   [A] 2×3 and 1×3 [B] 2×3 and 3×1
   [C] 3×2 and 1×3 [D] 3×2 and 3×1

2. **A** is a 2 by 3 matrix and **B** is a 4 by 2 matrix. The number of elements in **A** and **B** are respectively:
   [A] 4 and 6 [B] 6 and 4 [C] 8 and 6 [D] 6 and 8

3. An example of a 3 by 2 matrix is:

   [A] $\begin{pmatrix} 3 \\ 2 \end{pmatrix}$ [B] $\begin{pmatrix} 0 & 2 & 4 \\ 0 & 0 & 1 \end{pmatrix}$ [C] $\begin{pmatrix} 0 & 1 \\ 0 & -1 \\ 0 & 4 \end{pmatrix}$ [D] $\begin{pmatrix} 3 & 0 \\ 0 & 3 \end{pmatrix}$

4. Let $\mathbf{A} = \begin{pmatrix} 4 & 1 \\ -2 & 3 \end{pmatrix}$. The transpose of **A** is:

   [A] $\begin{pmatrix} 4 & -2 \\ 1 & 3 \end{pmatrix}$ [B] $\begin{pmatrix} 3 & -2 \\ 1 & 4 \end{pmatrix}$ [C] $\begin{pmatrix} 3 & 1 \\ -2 & 4 \end{pmatrix}$ [D] $\begin{pmatrix} 4 & 1 \\ -2 & 3 \end{pmatrix}$

5. Given that $\mathbf{P} = \begin{pmatrix} 3 & 2 \\ 4 & 5 \end{pmatrix}$ then Adj **P** is equal to:

   [A] $\begin{pmatrix} -3 & 4 \\ 2 & -5 \end{pmatrix}$ [B] $\begin{pmatrix} 5 & 4 \\ 2 & 3 \end{pmatrix}$ [C] $\begin{pmatrix} 5 & -2 \\ -4 & 3 \end{pmatrix}$ [D] $\begin{pmatrix} 5 & -4 \\ -2 & 3 \end{pmatrix}$

6. Given that **A** is a 3 by 2 matrix and **B** is a 2 by 4 matrix. The matrix product **AB** will have size:
   [A] 2 by 3 [B] 3 by 4 [C] 3 by 2 [D] 2 by 4

7. The values of m for which the matrix $\begin{pmatrix} 6-m & 2 \\ 25 & 1-m \end{pmatrix}$ is singular are:

   [A] -11 and 4 [B] -11 and -4 [C] 11 and 4 [D] 11 and -4

8. Given that $\mathbf{A} = \begin{pmatrix} 2 & -3 \\ -2 & 4 \end{pmatrix}$ and $\mathbf{B} = \begin{pmatrix} -2 & 1 \\ -5 & 3 \end{pmatrix}$. The matrix product **AB** is:

[A] $\begin{pmatrix} -11 & 7 \\ 16 & 10 \end{pmatrix}$ [B] $\begin{pmatrix} 11 & -7 \\ 16 & 10 \end{pmatrix}$ [C] $\begin{pmatrix} 11 & -7 \\ -16 & 10 \end{pmatrix}$ [D] $\begin{pmatrix} 11 & 7 \\ -16 & 10 \end{pmatrix}$

9. Given that $\mathbf{A} = \begin{pmatrix} 2 & -3 \\ -2 & 4 \end{pmatrix}$ and $\mathbf{B} = \begin{pmatrix} -2 & 1 \\ -5 & 3 \end{pmatrix}$. The determinant of the product **AB** is:

[A] 2 [B] 222 [C] −222 [D] −2

10. Given that $\mathbf{M} = \begin{pmatrix} 4 & 2 \\ 6 & 1 \end{pmatrix}$, $\mathbf{M}^{-1}$ is equal to:

[A] $\frac{1}{8}\begin{pmatrix} 1 & -2 \\ -6 & 4 \end{pmatrix}$ [B] $-\frac{1}{8}\begin{pmatrix} -1 & 2 \\ 6 & -4 \end{pmatrix}$

[C] $-\frac{1}{8}\begin{pmatrix} 1 & 2 \\ 6 & -4 \end{pmatrix}$ [D] $\frac{1}{8}\begin{pmatrix} 1 & -2 \\ -6 & 4 \end{pmatrix}$

11. Given that $\mathbf{A} = \begin{pmatrix} 0 & 2 \\ 4 & -2 \\ -1 & 1 \end{pmatrix}$ and $\mathbf{B} = \begin{pmatrix} -2 & 4 & -1 & 2 \\ 0 & 3 & 0 & 1 \end{pmatrix}$. The size of the matrix product **AB** is:

[A] 2×3 [B] 3×4 [C] 3×2 [D] 2×4

12. The singular matrix among the following matrices is:

[A] $\begin{pmatrix} -2 & 2 \\ 2 & 2 \end{pmatrix}$ [B] $\begin{pmatrix} -2 & -2 \\ 2 & 2 \end{pmatrix}$ [C] $\begin{pmatrix} -2 & -2 \\ 2 & -2 \end{pmatrix}$ [D] $\begin{pmatrix} -2 & -2 \\ -2 & 2 \end{pmatrix}$

13. Given that the matrix $\begin{pmatrix} a & 1 \\ 4 & a \end{pmatrix}$ is singular, the possible values of a are:

[A] 4 and −1 [B] 4 and 1 [C] −4 and 1 [D] −2 and 2

14. The inverse of the matrix $\begin{pmatrix} 5 & 2 \\ 7 & 3 \end{pmatrix}$ is:

[A] $\begin{pmatrix} -3 & -7 \\ -2 & -5 \end{pmatrix}$ [B] $\begin{pmatrix} -3 & 2 \\ 7 & -5 \end{pmatrix}$ [C] $\begin{pmatrix} 3 & -2 \\ -7 & 5 \end{pmatrix}$ [D] $\begin{pmatrix} 3 & -7 \\ -2 & 5 \end{pmatrix}$

15. The value of a for which the matrix $\begin{pmatrix} a & 3-a \\ 2 & 1 \end{pmatrix}$ is singular is:

[A] 2 [B] −2 [C] 3 [D] −3

16. Given that **PQ = I**, where $\mathbf{P} = \begin{pmatrix} 5 & -4 \\ -1 & 1 \end{pmatrix}$ and $\mathbf{P} = \begin{pmatrix} 1 & 0 \\ 0 & 1 \end{pmatrix}$. The matrix **Q** is:

[A] $\begin{pmatrix} 1 & 4 \\ -1 & 5 \end{pmatrix}$ [B] $\begin{pmatrix} 1 & -4 \\ -1 & 5 \end{pmatrix}$ [C] $\begin{pmatrix} 1 & -4 \\ 1 & 5 \end{pmatrix}$ [D] $\begin{pmatrix} 1 & 4 \\ 1 & 5 \end{pmatrix}$

17. The matrices $\mathbf{P}=\begin{pmatrix} m & 2 \\ 5 & 0 \end{pmatrix}$ and $\mathbf{Q}=\begin{pmatrix} 2 & -m \\ -2 & 1 \end{pmatrix}$ are such that $\mathbf{P}+\mathbf{Q}$ is singular. The value of m is:

[A] -1 [B] 1 [C] 4 [D] -4

18. The inverse of the 2×2 matrix $\begin{pmatrix} 6 & 10 \\ 2 & 4 \end{pmatrix}$ is:

[A] $\begin{pmatrix} 1 & -2\frac{1}{2} \\ -\frac{1}{2} & 1\frac{1}{2} \end{pmatrix}$ [B] $\begin{pmatrix} 4 & -10 \\ -2 & 6 \end{pmatrix}$ [C] $\begin{pmatrix} -2 & 6 \\ -10 & 4 \end{pmatrix}$ [D] $\begin{pmatrix} 1 & 2\frac{1}{2} \\ \frac{1}{2} & 1\frac{1}{2} \end{pmatrix}$

19. The transpose of the matrix $\begin{pmatrix} 7 & 3 \\ -1 & 5 \end{pmatrix}$ is:

[A] $\begin{pmatrix} 5 & -3 \\ 1 & 7 \end{pmatrix}$ [B] $\begin{pmatrix} 7 & -1 \\ 3 & 5 \end{pmatrix}$ [C] $\begin{pmatrix} -7 & -1 \\ 3 & -5 \end{pmatrix}$ [D] $\begin{pmatrix} -7 & 1 \\ -3 & -5 \end{pmatrix}$

20. Given that $\begin{pmatrix} 2 & 1 \\ 1 & -1 \end{pmatrix}\begin{pmatrix} 1 & -2 \\ 1 & 2 \end{pmatrix}=3\mathbf{A}$. The matrix $\mathbf{A}$ is:

[A] $\begin{pmatrix} 3 & -2 \\ 0 & -4 \end{pmatrix}$ [B] $\begin{pmatrix} 1 & \frac{2}{3} \\ 0 & \frac{4}{3} \end{pmatrix}$ [C] $\begin{pmatrix} 3 & \frac{2}{3} \\ 0 & \frac{4}{3} \end{pmatrix}$ [D] $\begin{pmatrix} 3 & -\frac{2}{3} \\ 0 & -\frac{4}{3} \end{pmatrix}$

21. The determinant of the matrix $\mathbf{A}=\begin{pmatrix} x & 7 \\ 4 & 2 \end{pmatrix}$ is 6. The value of x is:

[A] -17 [B] -34 [C] 17 [D] 34

22. Let $\mathbf{M}=\begin{pmatrix} 3 & 1 \\ 1 & 2 \end{pmatrix}$ and $\mathbf{N}=\begin{pmatrix} 1 & 4 \\ -3 & 2 \end{pmatrix}$. As a single matrix, $\mathbf{M}+3\mathbf{N}$ is equal to:

[A] $\begin{pmatrix} 10 & 7 \\ 1 & 4 \end{pmatrix}$ [B] $\begin{pmatrix} 10 & 7 \\ -1 & 4 \end{pmatrix}$ [C] $\begin{pmatrix} 6 & 13 \\ 8 & -8 \end{pmatrix}$ [D] $\begin{pmatrix} 6 & 13 \\ -8 & 8 \end{pmatrix}$

23. The inverse of the 2×2 matrix $\begin{pmatrix} 6 & 11 \\ 2 & 4 \end{pmatrix}$ is:

[A] $\begin{pmatrix} 2 & -5\frac{1}{2} \\ -1 & 3 \end{pmatrix}$ [B] $\begin{pmatrix} 4 & 2 \\ -11 & 6 \end{pmatrix}$ [C] $\begin{pmatrix} 3 & -5\frac{1}{2} \\ -1 & 2 \end{pmatrix}$ [D] $\begin{pmatrix} 0 & 0 \\ 0 & 0 \end{pmatrix}$

24. The adjoint of the matrix $\begin{pmatrix} 7 & 3 \\ -1 & 5 \end{pmatrix}$ is:

[A] $\begin{pmatrix} 5 & -3 \\ 1 & 7 \end{pmatrix}$ [B] $\begin{pmatrix} 7 & -1 \\ 3 & 5 \end{pmatrix}$ [C] $\begin{pmatrix} -7 & -1 \\ 3 & -5 \end{pmatrix}$ [D] $\begin{pmatrix} -7 & 1 \\ -3 & -5 \end{pmatrix}$

25. Given that $p = 2q - 1$ and $2p = 3q + 2$, then:

[A] $\begin{pmatrix} 1 & 2 \\ 2 & 3 \end{pmatrix}\begin{pmatrix} p \\ q \end{pmatrix} = \begin{pmatrix} -1 \\ 2 \end{pmatrix}$ [B] $\begin{pmatrix} 2 & -1 \\ 3 & 2 \end{pmatrix}\begin{pmatrix} p \\ q \end{pmatrix} = \begin{pmatrix} 1 \\ 2 \end{pmatrix}$

[C] $\begin{pmatrix} 1 & -1 \\ 2 & 2 \end{pmatrix}\begin{pmatrix} p \\ q \end{pmatrix} = \begin{pmatrix} 2 \\ 3 \end{pmatrix}$ [D] $\begin{pmatrix} 1 & -2 \\ 2 & -3 \end{pmatrix}\begin{pmatrix} p \\ q \end{pmatrix} = \begin{pmatrix} -1 \\ 2 \end{pmatrix}$

26. The incorrect statement among the following is:
[A] A square matrix is necessarily a diagonal matrix.
[B] A diagonal matrix is necessarily a square matrix.
[C] A unit matrix is necessarily a square matrix.
[D] A unit matrix is necessarily a diagonal matrix.

9.2 TRANSFORMATIONS AND SYMMETRY

TRANSFORMATION

A **transformation** is a change in position, size or shape of an object.

Translation, Reflection, Rotation

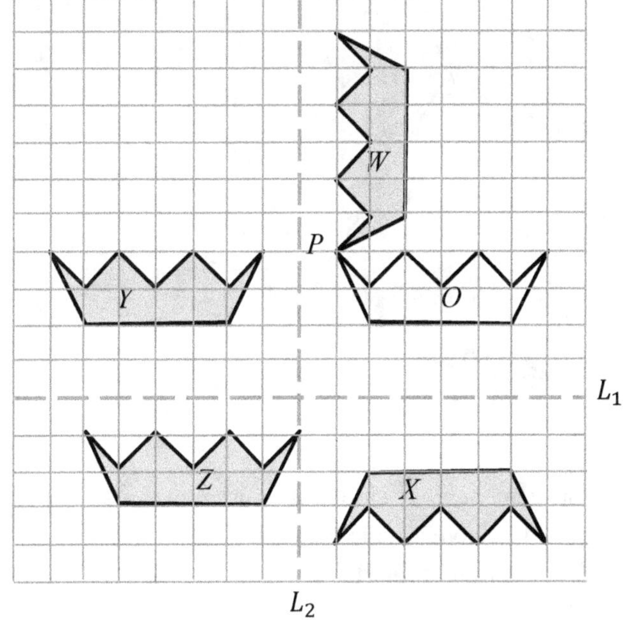

1. In the figure above, W is a rotation of the plane figure O through an angle of 90° in an anticlockwise sense about the point P as axis of rotation. Note that clockwise rotations are conventionally taken as negative.
2. X is a reflection of the plane figure O in the dotted line L_1 as mirror line. In a reflection, every point on the object is the same perpendicular distance from the mirror line as the corresponding point on the image.
3. Y is a reflection of the plane figure O in the dotted line L_2 as mirror line.
4. Z is a translation of the plane figure O, 7 units to the left and 5 units down the page. This translation can be described by the column vector $\begin{pmatrix} -7 \\ -5 \end{pmatrix}$ called a **translation vector**.

In a translation, every point moves the same distance in the same direction, without any rotation.

NOTE!!

In each of the above transformations, the shape and size of the figure remains unchanged (shape and size are invariant). A transformation in which shape and size are invariant is called an **isometry**.

Enlargements

In the figure on the below, $A'B'C'$ is an enlargement of ABC, with enlargement factor 3 and centre O.

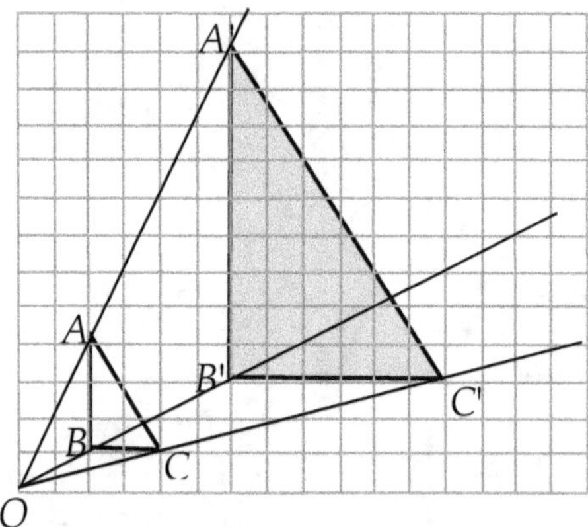

Remarks

1. If during an enlargement, each side of a figure increases k times, the enlargement is said to have a scale factor k.

   Image length $= k \times$ object length

   Image area $= k^2 \times$ object area

2. If the object is instead reduced k times, the scale factor will be $\frac{1}{k}$.
3. In an enlargement object and image are similar.

Example

An irregular figure of area 252 cm^2 is enlarged by scale factor $\frac{5}{6}$; find the area of its image.

Solution

Image area $= k^2 \times$ object area $= \left(\frac{5}{6}\right)^2 \times 252 = 175 \text{ cm}^2$

Shear

In the figure below $OAB'C'$ is a shear of the unit square *OABC* **parallel to the *x*-axis with shear factor** –2.

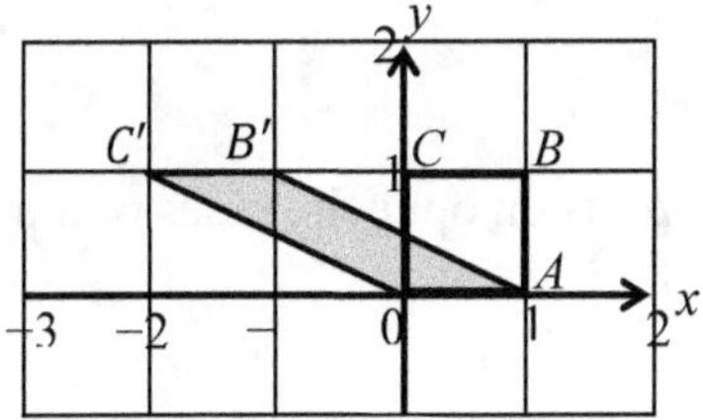

Example

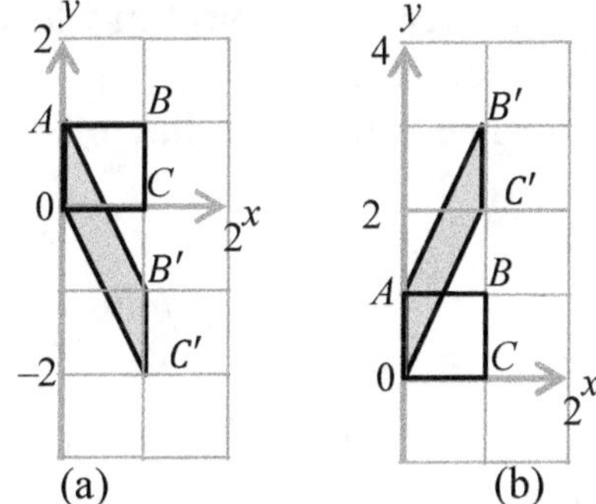

In the graphs (a) and (b) above, the unit square *OABC* is transformed into the parallelograms $OAB'C'$ and $OAB''C''$. Describe these transformations completely.

Solution

(a) A shear parallel to the *y*-axis with shear factor –2.
(b) A shear parallel to the *y*-axis with shear factor 2.

Stretch

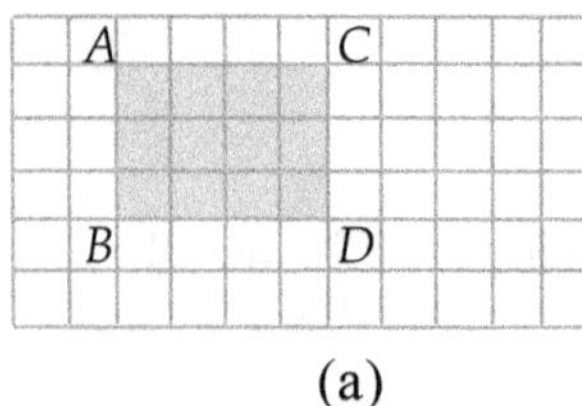

(a)

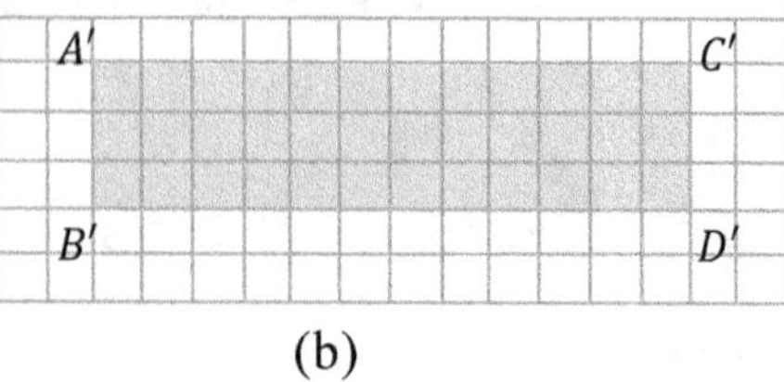

(b)

In the above graphs (b) is a **stretch** of (a) towards the right, **stretch factor** 4.
In a stretch,

1. Length of stretched side = k × Length of corresponding object side.
2. Image area = k × object area.
3. When $k < -1$, the stretch is in the negative direction and when $k > 1$, the stretch is in the positive direction.
4. When $-1 < k < 1$, the object is compressed in one direction only.

SYMMETRY

Symmetry is the correspondence of parts on opposite sides of a point, line or plane.

Types of Symmetry

(a) Reflective or Mirror Symmetry

Mirror symmetry is that property of an object in which one half of an object is similar to the other half.
In plane figures, the mirror line is termed a **line of symmetry**.

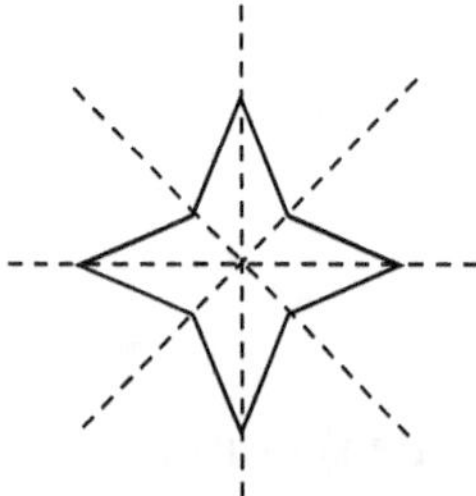

4 lines of symmetry

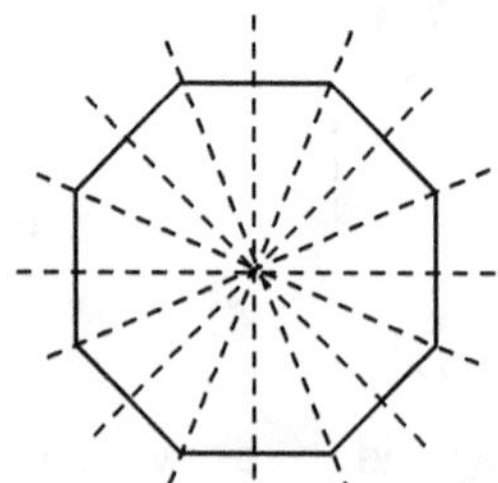

An octagon has 8 lines of symmetry

The following table shows more plane figures and their lines of symmetry.

	Plane Figure	Number of lines of symmetry
Isosceles trapezium		1
Square		4
Rhombus		2
Rectangle		2
Kite		1
Equilateral triangle		3
Isosceles triangle		1
Parallelogram		none

In solid figures, the mirror line is termed a **plane of symmetry**

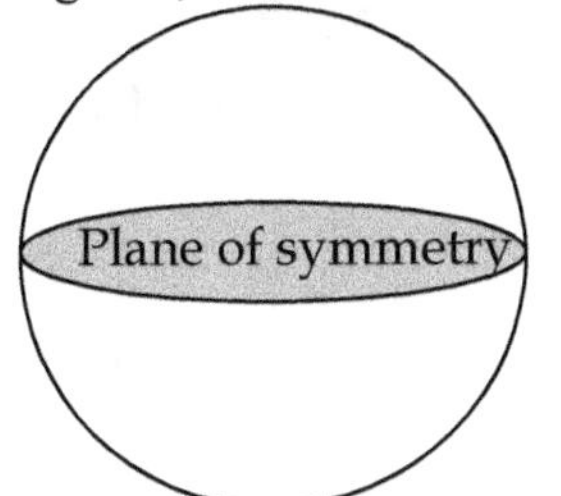

Infinite number of planes of symmetry

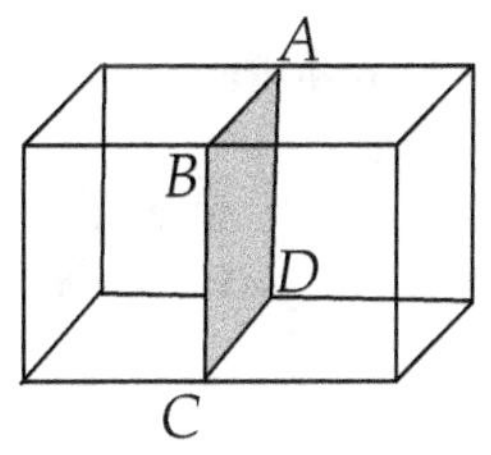

3 planes of symmetry

(b) Rotational or Radial symmetry

Radial symmetry or rotational symmetry is the proportional arrangement of similar parts of a body around a central axis or point.
A circle exhibits radial symmetry of infinite order about its centre.
An equilateral triangle exhibits radial symmetry of order 3.

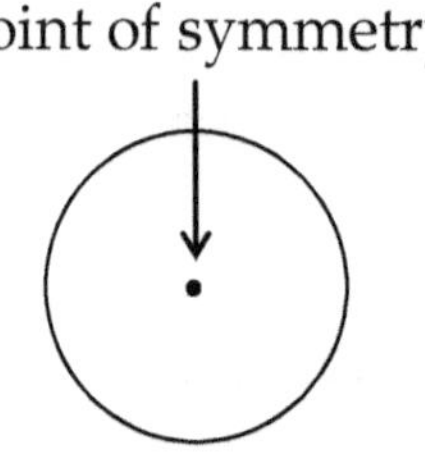

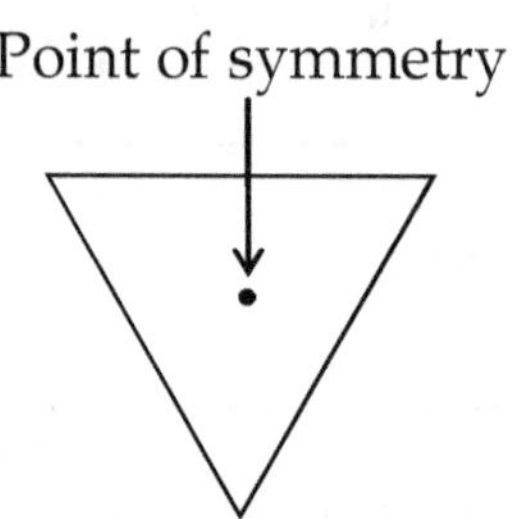

Any regular polygon with n sides has rotational symmetry of order .n

Regular Polygon	Order of rotational symmetry
Equilateral triangle	3
Square	4
Regular pentagon	5
Regular hexagon	6
n-gon	n

The figure on the left below shows that a parallelogram exhibits point symmetry of order 2 about the point O. A_2, B_2, C_2, and D_2 can respectively fit on A_1, B_1, C_1 and D_1.

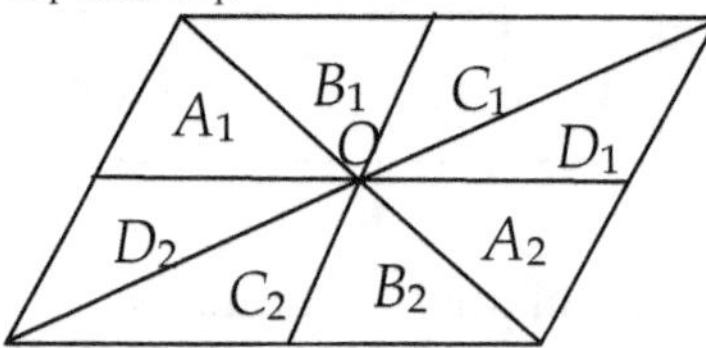

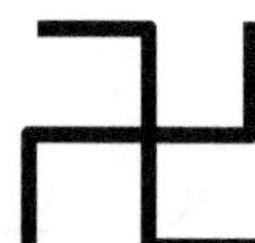

The order of rotational symmetry of swastika (the German army batch on the right above) is 4.
For solid figures, that exhibit radial or rotational symmetry, the centre of symmetry is called the **axis of symmetry**.

Exercise

The diagrams (a) to (g) show some solids with one axes of symmetry shown. Draw the diagrams and mark the remaining axes of symmetry if any.

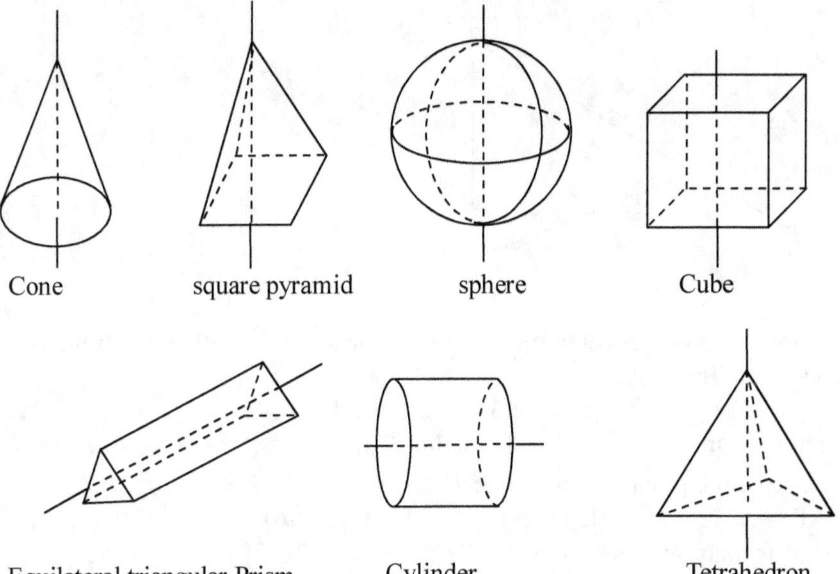

MULTIPLE CHOICE EXERCISE 9:2

1. In the following graph, triangle $A'B'C'$ is an enlargement of triangle ABC. The scale factor of the enlargement is:

   [A] $\frac{3}{2}$ [B] 2 [C] $\frac{5}{2}$ [D] 3

2. In the following diagram, not drawn to scale, triangle $OA'B'$ is an enlargement of triangle OAB. The scale factor of the enlargement above is:

   [A] $-\frac{1}{3}$ [B] -3 [C] $\frac{1}{3}$ [D] 3

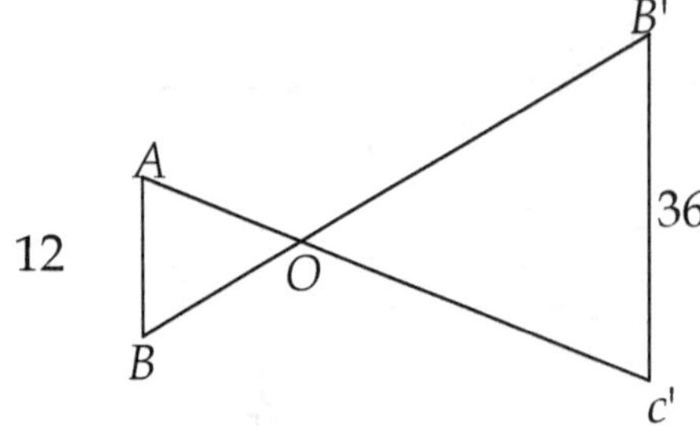

3. The point (4,3) is reflected in the x-axis followed by a reflection in the y-axis. The final image is:
   [A] (4,–3) [B] (–4,3) [C] (–4,–3) [D] (–3,4)
4. A certain transformation T is defined by $T:(x,y)\mapsto(2x+y,-x+y)$. The image of the point (–2, 6) under T is:
   [A] (–2,6) [B] (2,8) [C] (2,–6) [D] (2,–8)
5. A transformation T is defined by $T:(x,y)\mapsto(2x,2y)$.
   The image of (-3,4) under T is:
   [A] (–6,–8) [B] (6,–8) [C] (6,8) [D] (–6,8)
6. A transformation T is defined by $T:(x,y)\mapsto(2x,2y)$. The statement that correctly describes the transformation T is:
   [A] An enlargement scale factor 2 centre (0, 0)
   [B] An enlargement scale factor 2 centre (0, 2)
   [C] An enlargement scale factor 2 centre (2, 2)
   [D] A translation 2 units to the right.
7. The square with vertices W (0,0), X (1,0), Y (1,1), Z (0,1) is transformed into a square with vertices W (0,0), X (1,0), Y (3,1), Z (2,1). The statements that correctly describes the transformation T is:
   [A] A shear, shear factor 2 where points on the y-axis are invariant.
   [B] A stretch, stretch factor 2 where points on the y-axis are invariant.
   [C] A shear, shear factor 2 where points on the x-axis are invariant.
   [D] A stretch, stretch factor 2 where points on the x-axis are invariant.
8. A triangle whose vertices are A(1,2), B(3,4) and C(6,2) is transformed to a triangle whose vertices are A'(1,–2), B'(3,–4) and C'(6,–2). The statement which correctly describes the transformation is:
   [A] T is a reflection in the line $y=-1$ [B] T is a reflection in the line $y=x$
   [C] T is a reflection in the line $y=-x$ [D] T is a reflection in the line $y=0$
9. The order of rotational symmetry of a pyramid with a square base and axis through the vertex and centre of the square is:
   [A] 2 [B] 3 [C] 4 [D] 5
10. The graph among the following which shows the reflection of triangle ABC on the left in the line XY is:

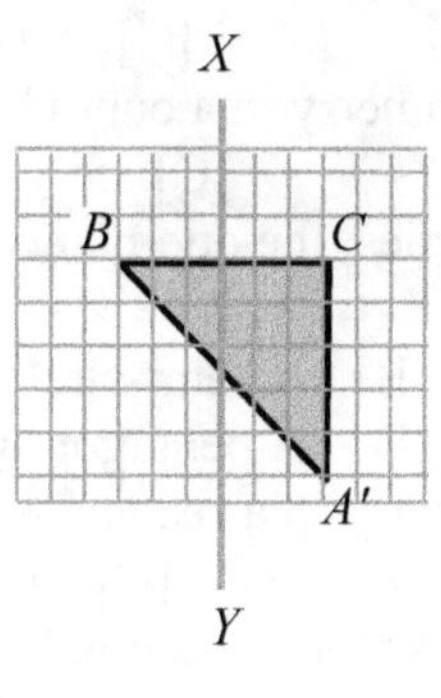

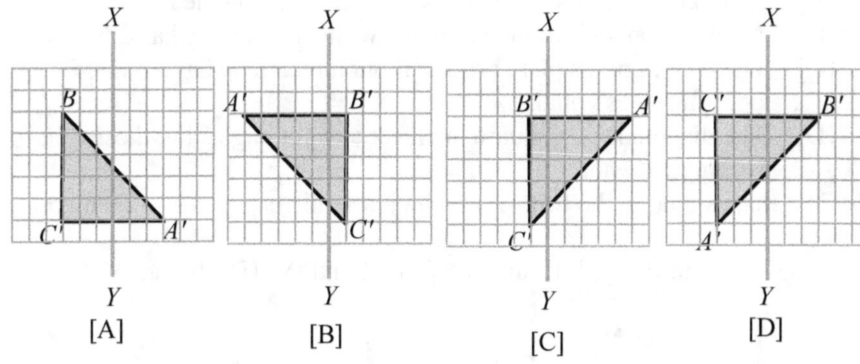

11. Among the following, the diagram which represents a shear, shear factor –3 parallel to the y-axis is:

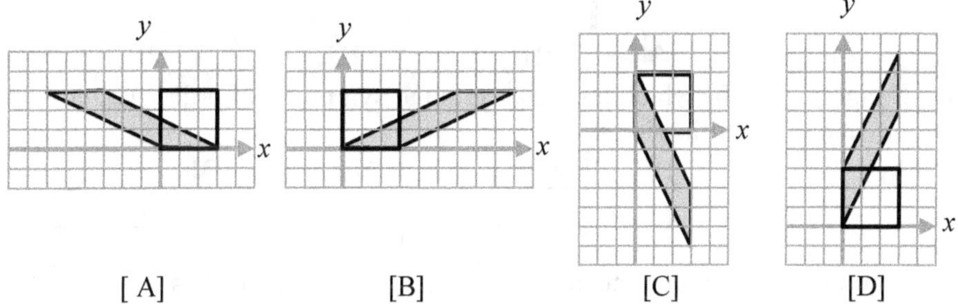

12. In the following transformation, the invariant line is:

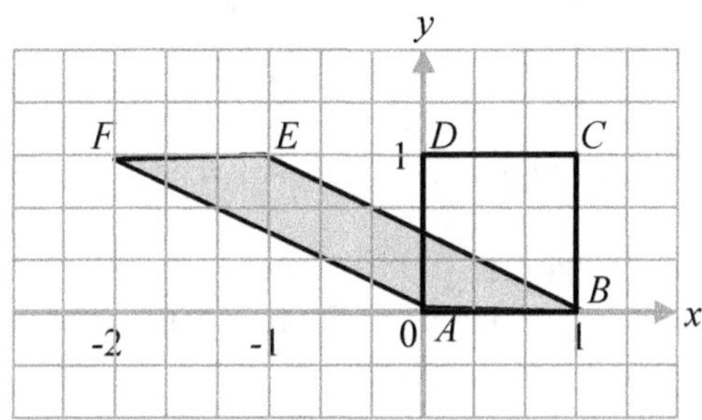

[A] AD [B] BC [C] CD [D] AB

13. The number of planes of symmetry in a cuboid is:
    [A] 3 [B] 4 [C] 5 [D] 6
14. Using their symmetric properties, the odd plane figure among the following is:
    [A] An isosceles triangle. [B] A semi-circle.
    [C] A rectangle. [D] A pentagon with four sides equal.
15. The number of lines of symmetry in a rectangle is:
    [A] 1 [B] 2 [C] 4 [D] 8
16. Symmetrically a square differs from a rectangle because:
    [A] A square has 2 lines of symmetry while a rectangle has 4.
    [B] A square has 4 lines of symmetry while a rectangle has 2.
    [C] The 4 sides of a square are equal but a rectangle has a pair of opposite sides equal.
    [D] The diagonals of a square intersect at right angles but those of a rectangle do not.
17. The shapes among the following which are mirror images of each other are:
    [A] I and II [B] III and V [C] I and IV [D] II and IV

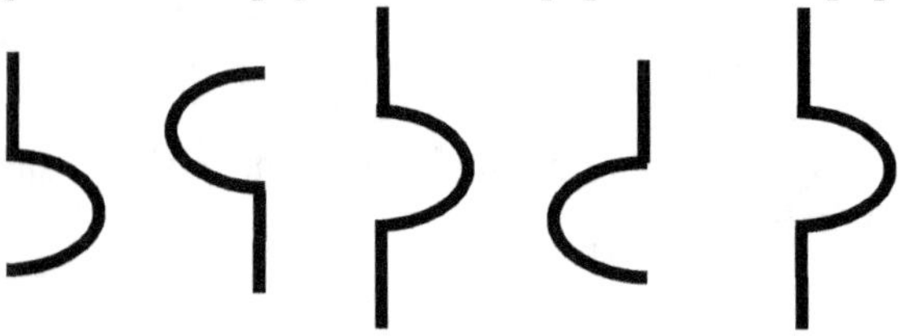

I. II. III. IV.

18. The net of a cube which possess rotational symmetry of order greater than one among the following is:
    [A] 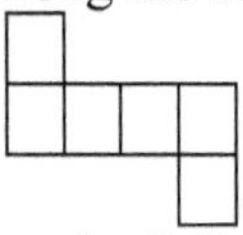 [B] 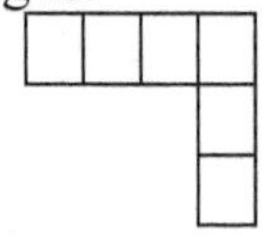 [C] 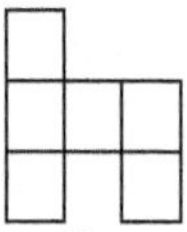 [D] 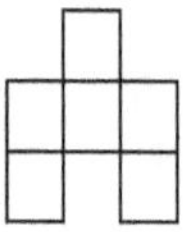
19. Among the following, the net which has no line symmetry is:

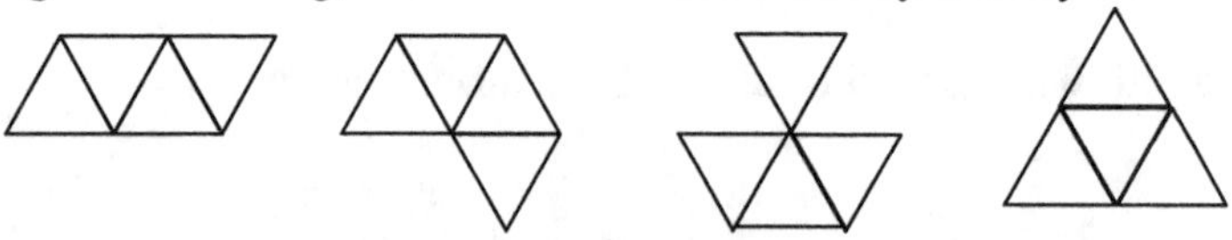

[A] [B] [C] [D]

20. Among the following, the quadrilateral which has exactly one line of symmetry is:
    [A] A kite [B] A rectangle [C] A Parallelogram [D] A Rhombus
21. The number of lines of symmetry in the following triangle is:
    [A] 2 [B] 3 [C] 4 [D] 5

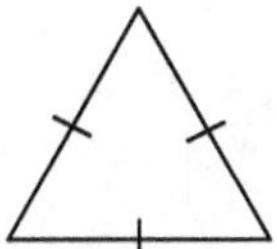

22. The figure which has two lines of symmetry is:
    [A] An isosceles triangle [B] A square
    [C] An equilateral triangle [D] A rhombus
23. The figure which has 9 planes of symmetry is:
    [A] A regular nonagon [B] A cube
    [C] A regular octagon [D] A regular hexagon
24. As drawn the diagram among the following that is not symmetrical about a horizontal axis is:

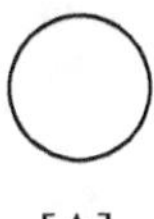

[A]

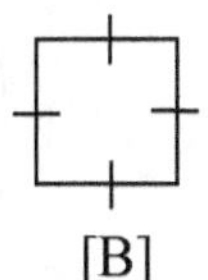

[B]

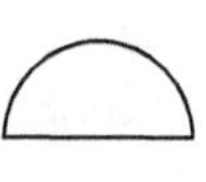

[C]

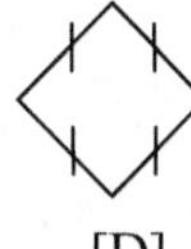

[D]

25. A plane figure may have:
    [A] A line of symmetry and a plane of symmetry.
    [B] A line of symmetry and a point of symmetry.
    [C] A point of symmetry and a plane of symmetry.
    [D] A axis of symmetry and a plane of symmetry.
26. A solid figure may have:
    [A] A line of symmetry and a plane of symmetry.
    [B] A line of symmetry and a point of symmetry.
    [C] A point of symmetry and a plane of symmetry.
    [D] A axis of symmetry and a plane of symmetry.

9.3 TRANSFORMATIONS WITH MATRICES

Transformation Matrix (Matrix Operator)

We can use a matrix operator $\mathbf{A} = \begin{pmatrix} a & b \\ c & d \end{pmatrix}$ to transform a point $P(x, y)$ to a new point $P'(X, Y)$ as follows $\mathbf{AP} = \begin{pmatrix} a & b \\ c & d \end{pmatrix}\begin{pmatrix} x \\ y \end{pmatrix} = \begin{pmatrix} ax + by \\ cx + dy \end{pmatrix} = \begin{pmatrix} X \\ Y \end{pmatrix}$.

Example

Find the image of the square with vertices (0, 0), (1, 0), (1, 1) and (0, 1) under the transformation represented by the matrix $\begin{pmatrix} 1 & 2 \\ 0 & 1 \end{pmatrix}$.

Solution

$$\begin{pmatrix} 1 & 2 \\ 0 & 1 \end{pmatrix}\begin{pmatrix} 0 & 1 & 1 & 0 \\ 0 & 0 & 1 & 1 \end{pmatrix}=\begin{pmatrix} 0 & 1 & 3 & 2 \\ 0 & 0 & 1 & 1 \end{pmatrix}$$

Describing Transformation

To describe a transformation represented by a transformation matrix, use the transformation matrix to transform the unit square to a new image then draw both on the same axes and compare. Sometimes this is clear just by transforming the points (1, 0) and (0, 1).

Example

Describe the transformation represented by the matrix $\begin{pmatrix} -1 & 0 \\ 0 & 1 \end{pmatrix}$.

Solution

$$\begin{pmatrix} 1 \\ 0 \end{pmatrix} \mapsto \begin{pmatrix} -1 \\ 0 \end{pmatrix} \text{ and } \begin{pmatrix} 1 \\ 0 \end{pmatrix} \mapsto \begin{pmatrix} 1 \\ 0 \end{pmatrix}$$

The transformation is reflection in the line $x = 0$ (i.e. the y-axis).

Example

Using the rectangle with vertices $O(0,0)$, $A(3,0)$, $B(3,2)$ and $C(0,2)$ show that the matrix $\begin{pmatrix} 1 & 0 \\ 0 & -1 \end{pmatrix}$ is a matrix of reflection in the line $y = 0$ (i.e. the x-axis).

Solution

$$\begin{pmatrix} 1 & 0 \\ 0 & -1 \end{pmatrix}\begin{pmatrix} 0 & 3 & 3 & 0 \\ 0 & 0 & 2 & 2 \end{pmatrix}=\begin{pmatrix} 0 & 3 & 3 & 0 \\ 0 & 0 & -2 & -2 \end{pmatrix}$$

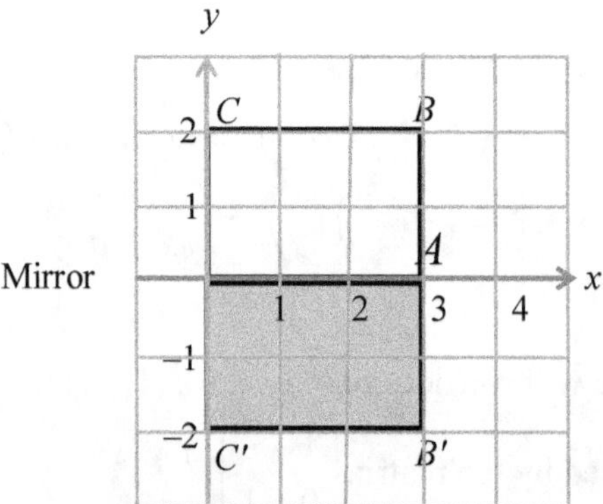

This is a matrix of reflection in the line $y = 0$ or the x-axis.

Example

The triangle with vertices $A\ (1, 1)$, $B\ (3, 1)$, $C\ (3, 2)$ is transformed by the matrix $\begin{pmatrix} 0 & 1 \\ -1 & 0 \end{pmatrix}$. Find the image of the transformation and describe the transformation completely.

Solution

$$\begin{pmatrix} 0 & 1 \\ -1 & 0 \end{pmatrix}\begin{pmatrix} 1 & 3 & 3 \\ 1 & 1 & 2 \end{pmatrix} = \begin{pmatrix} 1 & 1 & 2 \\ -1 & -3 & -3 \end{pmatrix}$$

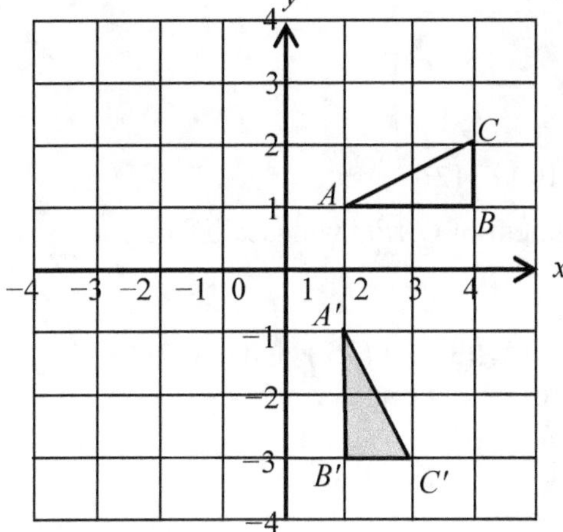

This is a matrix of rotation through 270° about (0,0) as center.

Remarks!

1. Rotations in the anti-clockwise sense are regarded as positive, while rotations in the clockwise sense are regarded as negative.
2. A rotation through 270° is the same as a rotation through –90°.
3. A rotation through 180° is the same as a rotation through –180°
4. A rotation through 360° is an invariant transformation so the transformation matrix is $\mathbf{I} = \begin{pmatrix} 1 & 0 \\ 0 & 1 \end{pmatrix}$.

Example

Find the image of the rectangle with vertices $O(0,0)$, $A(3,0)$, $B(3,2)$ and $C(0,2)$ under the transformation with matrix $\begin{pmatrix} 4 & 0 \\ 0 & 4 \end{pmatrix}$. Hence describe the transformation completely.

Solution

$$\begin{pmatrix} 4 & 0 \\ 0 & 4 \end{pmatrix}\begin{pmatrix} 0 & 3 & 3 & 0 \\ 0 & 0 & 2 & 2 \end{pmatrix} = \begin{pmatrix} 0 & 12 & 12 & 0 \\ 0 & 0 & 8 & 8 \end{pmatrix}$$

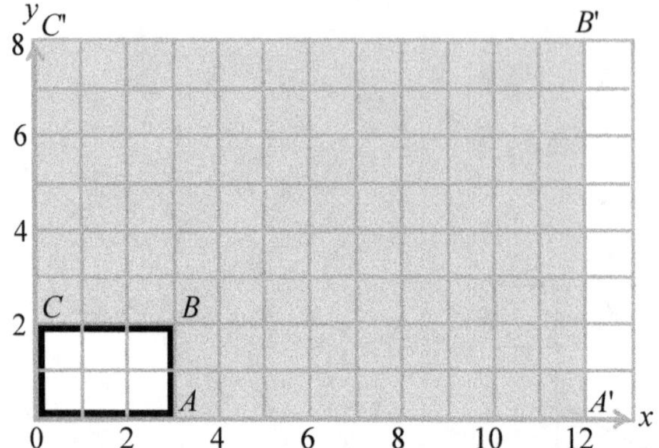

The image $OA'B'C'$ is an enlargement of $OABC$ with scale factor 4.

Example

Use the rectangle with vertices $O(0,0)$, $A(3,0)$, $B(3,2)$ and $C(0,2)$ to determine and describe each of the following transformations:

(i) $\begin{pmatrix} 1 & 2 \\ 0 & 1 \end{pmatrix}$ (ii) $\begin{pmatrix} 2 & 0 \\ 0 & 1 \end{pmatrix}$

Solution

(i) $$\begin{pmatrix} 1 & 2 \\ 0 & 1 \end{pmatrix}\begin{pmatrix} 0 & 3 & 3 & 0 \\ 0 & 0 & 2 & 2 \end{pmatrix} = \begin{pmatrix} 0 & 3 & 7 & 4 \\ 0 & 0 & 2 & 2 \end{pmatrix}$$

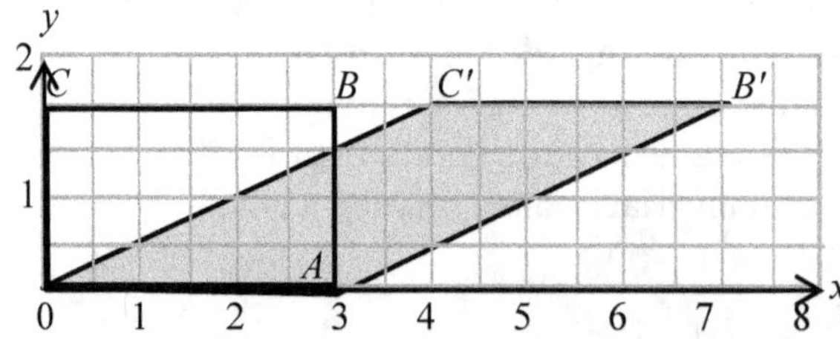

The transformation represents a shear with scale factor 2, parallel to the positive x- direction.

(ii) $$\begin{pmatrix} 2 & 0 \\ 0 & 1 \end{pmatrix}\begin{pmatrix} 0 & 3 & 3 & 0 \\ 0 & 0 & 2 & 2 \end{pmatrix} = \begin{pmatrix} 0 & 6 & 6 & 0 \\ 0 & 0 & 2 & 2 \end{pmatrix}$$

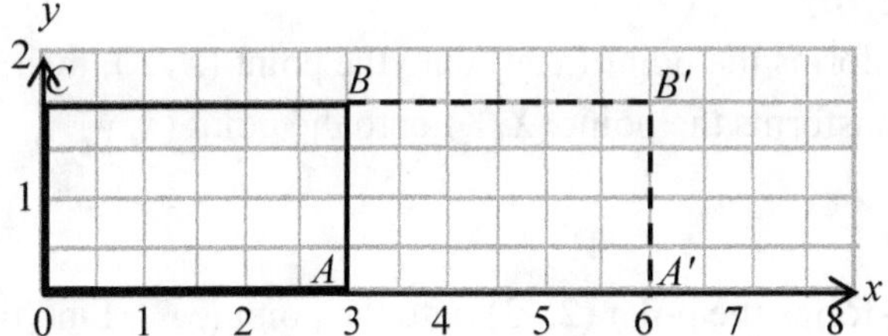

The transformation represents a stretch with scale factor 2, parallel to the positive x- direction.

Finding Transformation Matrices

Example

Under a certain transformation, $\begin{pmatrix}3\\-2\end{pmatrix} \mapsto \begin{pmatrix}9\\8\end{pmatrix}$and $\begin{pmatrix}5\\1\end{pmatrix} = \begin{pmatrix}-3\\4\end{pmatrix}$. Find the transformation matrix.

Solution

Let the transformation matrix be$\begin{pmatrix}a & b\\c & d\end{pmatrix}$.

Then, $\begin{pmatrix}a & b\\c & d\end{pmatrix}\begin{pmatrix}3\\-2\end{pmatrix} = \begin{pmatrix}9\\8\end{pmatrix}$.

$$\Rightarrow 3a - 2b = 9 \ldots\ldots\ldots\ldots\ldots\ldots\ldots\ldots \text{①}$$

$$3c - 2d = 8 \ldots\ldots\ldots\ldots\ldots\ldots\ldots\ldots \text{②}$$

$$\begin{pmatrix}a & b\\c & d\end{pmatrix}\begin{pmatrix}5\\1\end{pmatrix} = \begin{pmatrix}-3\\4\end{pmatrix}$$

$$\Rightarrow 5a + b = -3 \ldots\ldots\ldots\ldots\ldots\ldots\ldots \text{③}$$

$$5c + d = 4 \ldots\ldots\ldots\ldots\ldots\ldots\ldots\ldots \text{④}$$

$$2\times\text{③}+\text{①}:\ 13a = 3 \Rightarrow a = \frac{3}{13}$$

Substitute in ①: $3\left(\frac{3}{13}\right) - 2b = 8 \Rightarrow b = -\frac{54}{13}$

$$2\times\text{④}+\text{②}:\ 13c = 16 \Rightarrow c = \frac{16}{13}$$

Substitute in ②: $3\left(\frac{16}{13}\right) - 2d = 8 \Rightarrow d = -\frac{28}{13}$

Therefore, the transformation matrix is $\begin{pmatrix}\frac{3}{13} & -\frac{54}{13}\\ \frac{16}{13} & -\frac{28}{13}\end{pmatrix}$.

Inverse Transformations

If a matrix operator **A** transforms the point (x, y) onto the point (X, Y), then the matrix operator $\mathbf{A}^{-1}$ transforms the point (X, Y) onto the point (x, y).

Example

The matrix $\begin{pmatrix} 0 & -1 \\ 1 & 1 \end{pmatrix}$ transforms the point (2,–3) onto the point (3,2). Find the matrix which transforms the point (3,2) onto the point (2,–3).

Solution

Let $\begin{pmatrix} 0 & -1 \\ 1 & 1 \end{pmatrix} = \mathbf{A}$. Then the required matrix is $\mathbf{A}^{-1}$

$$\mathbf{A}^{-1} = \frac{1}{\det\mathbf{A}}(\mathrm{Adj}\mathbf{A}) = \frac{1}{1}\begin{pmatrix} 1 & 1 \\ -1 & 0 \end{pmatrix} = \begin{pmatrix} 1 & 1 \\ -1 & 0 \end{pmatrix}.$$

Example

The matrix $\mathbf{M} = \begin{pmatrix} 1 & 0 \\ -2 & 1 \end{pmatrix}$ maps a point (x, y) to (6,–9). Find the values of x and y.

Solution

$$\begin{pmatrix} 1 & 0 \\ -2 & 1 \end{pmatrix}\begin{pmatrix} x \\ y \end{pmatrix} = \begin{pmatrix} 6 \\ -9 \end{pmatrix}$$

$$\begin{pmatrix} 1 & 0 \\ -2 & 1 \end{pmatrix}^{-1} = \frac{1}{1}\begin{pmatrix} 1 & 0 \\ 2 & 1 \end{pmatrix} = \begin{pmatrix} 1 & 0 \\ 2 & 1 \end{pmatrix}$$

$$\Rightarrow \begin{pmatrix} x \\ y \end{pmatrix} = \begin{pmatrix} 1 & 0 \\ 2 & 1 \end{pmatrix}\begin{pmatrix} 6 \\ -9 \end{pmatrix} = \begin{pmatrix} 6 \\ 3 \end{pmatrix}$$

Therefore, the values of x and y are $x = 6$ and $y = 3$.

Transformations by Singular Matrices

Consider the transformation $\begin{pmatrix} 4 & 2 \\ 2 & 1 \end{pmatrix}\begin{pmatrix} 0 & 3 & 3 & 0 \\ 0 & 0 & 2 & 2 \end{pmatrix} = \begin{pmatrix} 0 & 12 & 16 & 4 \\ 0 & 6 & 8 & 2 \end{pmatrix}$.

The transformation is shown on the graph below. Notice that the matrix $\begin{pmatrix} 4 & 2 \\ 2 & 1 \end{pmatrix}$ is a singular matrix and that the square $\begin{pmatrix} 0 & 3 & 3 & 0 \\ 0 & 0 & 2 & 2 \end{pmatrix}$ to the straight line $\begin{pmatrix} 0 & 12 & 16 & 4 \\ 0 & 6 & 8 & 2 \end{pmatrix}$.

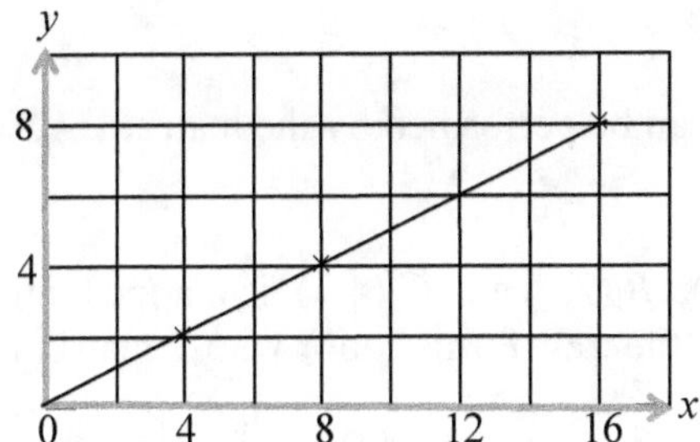

A **dilation** is a transformation which maps geometrical shapes to a straight line. All dilation matrices are singular matrices.

Transformations involving Change of Area

$$\text{image area} = |\det \mathbf{A}| \times \text{object area}$$

Example

A rectangle with vertices (4, 4), (4,–1), (10,–1), (10, 4) is transformed by the matrix $\begin{pmatrix} 3 & 3 \\ 2 & 3 \end{pmatrix}$. Find the area of its image.

Solution

Let $\mathbf{A} = \begin{pmatrix} 3 & 3 \\ 2 & 3 \end{pmatrix} \Rightarrow \det \mathbf{A} = 9 - 6 = 3.$

If we sketch this rectangle, we see that the length and width are 5 units and 6 units . $\Rightarrow$ Object area $= 5 \times 6 = 30 \text{ un}^2$

$\Rightarrow$ Image area $= |\det \mathbf{A}| \times$ object area $= 30 \times 3 = 90 \text{ un}^2$

Example

A transformation is defined by the matrix $\begin{pmatrix} 4 & 2 \\ 1 & 2 \end{pmatrix}$. Find the ratio of the image area to the object area.

Solution

Let $\mathbf{A} = \begin{pmatrix} 4 & 2 \\ 1 & 2 \end{pmatrix} \Rightarrow$ Det $\mathbf{A} = 8 - 2 = 6$

Image area $=$ $|\det \mathbf{A}| \times$ object area

$\Rightarrow$ Image area: object area $= |\det \boldsymbol{A}|: 1 = 6: 1$

Composite Transformations

A transformation **A** followed by **B** can be performed by the matrix **BA**.

Example

The triangle T_1 with vertices $A_1(1,1)$, $B_1(2,3)$ and $C_1(4,3)$, is mapped into triangle T_2 with vertices $A_2B_2C_2$ by the transformation P whose matrix is $\mathbf{P}=\begin{pmatrix}0 & 1\\ -1 & 0\end{pmatrix}$. The transformation Q whose matrix is $\mathbf{Q}=\begin{pmatrix}1 & 2\\ 0 & 1\end{pmatrix}$ further maps T_2 to T_3 whose vertices are $A_3B_3C_3$. Find the image of T_3 without finding the image of T_2.

Solution

$$\mathbf{QP}=\begin{pmatrix}1 & 2\\ 0 & 1\end{pmatrix}\begin{pmatrix}0 & 1\\ -1 & 0\end{pmatrix}=\begin{pmatrix}-2 & 1\\ -1 & 0\end{pmatrix}$$

$$\Rightarrow T_3=\begin{pmatrix}-2 & 1\\ -1 & 0\end{pmatrix}\begin{pmatrix}1 & 2 & 4\\ 1 & 3 & 3\end{pmatrix}=\begin{pmatrix}-1 & -1 & -5\\ -1 & -2 & -4\end{pmatrix}$$

Translations

A translation is a transformation in which every object point moves the same distance in the same direction. We represent translations by translation vectors.

The translation vector $\mathbf{t}=\begin{pmatrix}p\\ q\end{pmatrix}$ transforms any point A (a, b) by moving it p units in the Ox direction and q units in the Oy direction.

$$\mathbf{OA}'=\mathbf{OA}+\mathbf{t}=\begin{pmatrix}a\\ b\end{pmatrix}+\begin{pmatrix}p\\ q\end{pmatrix}=\begin{pmatrix}a+p\\ b+q\end{pmatrix}$$

Example

Determine the image of the triangle ABC whose vertices are A (1, 1), B (−2,1) and C (3, −4) under the transformation whose translation vector is $\begin{pmatrix}4\\ -7\end{pmatrix}$.

Solution

$\begin{pmatrix}1\\ 1\end{pmatrix}\to\begin{pmatrix}1\\ 1\end{pmatrix}+\begin{pmatrix}4\\ -7\end{pmatrix}=\begin{pmatrix}5\\ -6\end{pmatrix}$, $\begin{pmatrix}-2\\ 1\end{pmatrix}\to\begin{pmatrix}-2\\ 1\end{pmatrix}+\begin{pmatrix}4\\ -7\end{pmatrix}=\begin{pmatrix}2\\ -6\end{pmatrix}$ and

$$\begin{pmatrix} 3 \\ -4 \end{pmatrix} \rightarrow \begin{pmatrix} 3 \\ -4 \end{pmatrix} + \begin{pmatrix} 4 \\ -7 \end{pmatrix} = \begin{pmatrix} 7 \\ -11 \end{pmatrix}$$

Therefore, the image of the triangle ABC is
$A'(5,-6), B'(2,-6)$ and $C'(7,-11)$

SUMMARY

The following transformation matrices are very common at this level and note should be taken of them.

	Operator	Geometrical Effect
1	$\begin{pmatrix} 1 & 0 \\ 0 & -1 \end{pmatrix}$	A reflection in the x-axis
2	$\begin{pmatrix} -1 & 0 \\ 0 & 1 \end{pmatrix}$	A reflection in the y-axis
3	$\begin{pmatrix} 0 & 1 \\ 1 & 0 \end{pmatrix}$	A reflection in the line $y = x$
4	$\begin{pmatrix} 0 & -1 \\ -1 & 0 \end{pmatrix}$	A reflection in line $y = -x$
5	$\begin{pmatrix} 0 & -1 \\ 1 & 0 \end{pmatrix}$	An anticlockwise rotation through 90° about the origin (0,0)
6	$\begin{pmatrix} -1 & 0 \\ 0 & -1 \end{pmatrix}$	An anticlockwise rotation through 180° about the origin (0,0)
7	$\begin{pmatrix} 0 & 1 \\ -1 & 0 \end{pmatrix}$	An anticlockwise rotation through 270° about the origin (0,0)
8	$\begin{pmatrix} 1 & 0 \\ 0 & 1 \end{pmatrix}$	The identity transformation leaves all points invariant
9	$\begin{pmatrix} k & 0 \\ 0 & 1 \end{pmatrix}$	A stretch in the direction Ox, stretch factor k
10	$\begin{pmatrix} 1 & 0 \\ 0 & k \end{pmatrix}$	A stretch in the direction Oy, stretch factor k

	Operator	Geometrical Effect
11	$\begin{pmatrix} k & 0 \\ 0 & k \end{pmatrix}$	An enlargement centre (0,0), enlargement factor k
12	$\begin{pmatrix} 1 & k \\ 0 & 1 \end{pmatrix}$	A shear in the direction Ox, shear factor k
13	$\begin{pmatrix} 1 & 0 \\ k & 1 \end{pmatrix}$	A shear in the direction Oy, shear factor k

MULTIPLE CHOICE EXERCISE 9:3

1. A non-zero singular matrix:
   [A] has no inverse.
   [B] does not map any plane figure to a straight line.
   [C] does not have a determinant equal to zero.
   [D] can be a unit matrix.
2. A non-zero matrix whose determinant is zero:
   [A] has an inverse.
   [B] maps any plane figure to a straight line.
   [C] is not a singular matrix.
   [D] is an identity matrix.

3. The transformation matrix $\begin{pmatrix} 6 & 3 \\ 6 & 4 \end{pmatrix}$ maps the point $(x,1)$ onto the point (3, 4). The value of x is:
   [A] –3 [B] 4 [C] 0 [D] –6
4. The transformation matrix $\begin{pmatrix} a & 2a \\ b & 4b \end{pmatrix}$ maps the point (1, 2) to (5, 18). The values of a and $b$are:
   [A] $a = 2$ and $b = 1$ [B] $a = 1$ and $b = 2$
   [C] $a = 4$ and $b = 8$ [D] $a = 8$ and $b = 4$
5. The matrix which represents a reflection in the line $y = 0$ is:
   [A] $\begin{pmatrix} -1 & 0 \\ 0 & 1 \end{pmatrix}$ [B] $\begin{pmatrix} 0 & 1 \\ -1 & 0 \end{pmatrix}$ [C] $\begin{pmatrix} 0 & -1 \\ 1 & 0 \end{pmatrix}$ [D] $\begin{pmatrix} 1 & 0 \\ 0 & -1 \end{pmatrix}$
6. The transformation with matrix $\begin{pmatrix} -1 & 0 \\ 0 & 1 \end{pmatrix}$ is:
   [A] A reflection in the line $x = 0$. [B] A reflection in the line $y = 0$.

[C] A rotation through 180° center (0, 0) [D] A shear with shear factor -1

7. The transformation represented by the matrix $\begin{pmatrix} 1 & 0 \\ 2 & 1 \end{pmatrix}$ can completely be described as:
[A] A stretch with stretch factor 2 and points on the x-axis invariant.
[B] A stretch with stretch factor 2 and points on the y-axis invariant.
[C] A shear with shear factor 2 and points on the x-axis invariant.
[D] A shear with shear factor 2 and points on the y-axis invariant.

8. The transformation matrix $\begin{pmatrix} 2 & 4 \\ 0 & 2 \end{pmatrix}$ maps the point P onto P'. Given that P' is the point (9, 4), then P must be the point:
[A] $\left(\frac{1}{2}, 0\right)$ [B] (0, 2) [C] $\left(2, \frac{1}{2}\right)$ [D] $\left(\frac{1}{2}, 2\right)$

9. A rotation through 90° center (0, 0) can be represented by the matrix:
[A] $\begin{pmatrix} -1 & 0 \\ 0 & 1 \end{pmatrix}$ [B] $\begin{pmatrix} 0 & 1 \\ -1 & 0 \end{pmatrix}$ [C] $\begin{pmatrix} 0 & -1 \\ 1 & 0 \end{pmatrix}$ [D] $\begin{pmatrix} 1 & 0 \\ 0 & -1 \end{pmatrix}$

10. The transformation matrix $\begin{pmatrix} 2 & 4 \\ 0 & 2 \end{pmatrix}$ maps the point A onto B Given that A is the point (3, 2), then B is the point:
[A] (14, 4) [B] (4,14) [C] (6, 8) [D] (0, 4)

11. A linear transformation $\boldsymbol{M}: \mathbb{R}^2 \rightarrow \mathbb{R}^2$ is defined by $\mathbf{P} = \mathbf{Mq}$, where $\mathbf{M}$ is a 2×2 matrix and $\mathbf{p}$, $\mathbf{q}$ are 2×1 column vectors. Given that $\mathbf{p} = \begin{pmatrix} 3 \\ 7 \end{pmatrix}$ when $\mathbf{q} = \begin{pmatrix} 1 \\ 0 \end{pmatrix}$ and $\mathbf{p} = \begin{pmatrix} 6 \\ -1 \end{pmatrix}$ when $\mathbf{q} = \begin{pmatrix} 2 \\ -3 \end{pmatrix}$, $\mathbf{M} =$:
[A] $\begin{pmatrix} 3 & 0 \\ 7 & 5 \end{pmatrix}$ [B] $\begin{pmatrix} 5 & 7 \\ 0 & 3 \end{pmatrix}$ [C] $\begin{pmatrix} 0 & 5 \\ 3 & 7 \end{pmatrix}$ [D] $\begin{pmatrix} 7 & 3 \\ 5 & 0 \end{pmatrix}$

12. The triangle with vertices A (–1, –3), B (2, 1) and C(–2,2) is transformed by the matrix $\begin{pmatrix} x & 0 \\ 0 & y \end{pmatrix}$ into the triangle with vertices $A'(-2, -3)$, $B'(4,1)$, $C'(-4, 2)$. The values of x and y are respectively:
[A] 2 and 0 [B] 0 and 1 [C] 1 and 2 [D] 2 and 1

13. A certain transformation T is defined by $T: (x, y) \mapsto (2x + y, -x + y)$. The 2×2 matrix representing T is:
[A] $\begin{pmatrix} 2 & 1 \\ -1 & 1 \end{pmatrix}$ [B] $\begin{pmatrix} 1 & 1 \\ -1 & 2 \end{pmatrix}$ [C] $\begin{pmatrix} 1 & -1 \\ 1 & 2 \end{pmatrix}$ [D] $\begin{pmatrix} 1 & 1 \\ 2 & -1 \end{pmatrix}$

14. The quadrilateral $A(2,0)$, $B(2,2)$, $C(-2,-2)$, $D(-2,0)$ is transformed by the matrix $\mathbf{P} = \begin{pmatrix} 2 & -1 \\ 1 & 0 \end{pmatrix}$. The image $A'B'C'D'$ of $ABCD$ under $\mathbf{P}$ is:

[A] $A'(2,2), B'(6,4), C'(10,-6), D'(6,-4)$
[B] $A'(4,2), B'(-6,-2), C'(2,2), D'(-4,-2)$
[C] $A'(4,2), B'(2,2), C'(-2,-2), D'(-4,-2)$
[D] $A'(4,2), B'(2,2), C'(-4,-2), D'(-6,-2)$

15. A transformation is defined by $T\colon (x,y) \mapsto (2x, 2y)$. The 2×2 matrix representing T is:

[A] $\begin{pmatrix} 2 & 2 \\ 1 & 1 \end{pmatrix}$ [B] $\begin{pmatrix} 0 & 2 \\ 2 & 0 \end{pmatrix}$ [C] $\begin{pmatrix} 2 & 2 \\ 2 & 2 \end{pmatrix}$ [D] $\begin{pmatrix} 2 & 0 \\ 0 & 2 \end{pmatrix}$

16. The points $A(2,0)$, $B(2,2)$, $C(-2,-2)$ and $D(-2,0)$ are transformed by the matrix $\mathbf{P} = \begin{pmatrix} 2 & -1 \\ 1 & 0 \end{pmatrix}$. The invariant line is:

[A] the line $x = 0$ [B] the line $y = 0$ [C] the line $y = x$ [D] the line $y = 1$

17. A transformation is defined by $T\colon (x,y) \mapsto (2x, 2y)$. In words, the transformation T can be described as:

[A] A translation 2 units along Ox. [B] A shear scale factor 2
[C] An enlargement scale factor 2 [D] A stretch scale factor 2.

CHAPTER 10

VECTORS

Vector and Scalar Quantities

A **vector quantity** is a quantity which has magnitude, direction and sense. Examples of vector quantities are displacement, velocity, momentum, force, acceleration.
A **scalar quantity** or **scalar** is a quantity which has only magnitude no direction and no sense. Examples of scalar quantities are mass, temperature, distance, speed, area, length, volume, time, age, marks, price.

Representation of Vectors

A vector can be represented using a directed line segment.

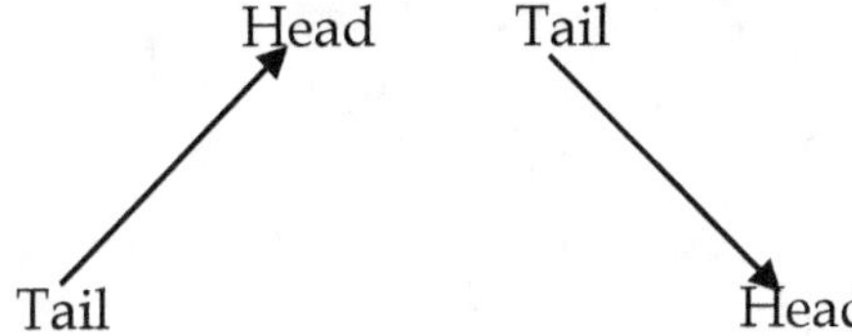

The orientation of a directed line segment, gives the direction while the arrow gives the sense of the vector.

Direction and Sense of a Vector

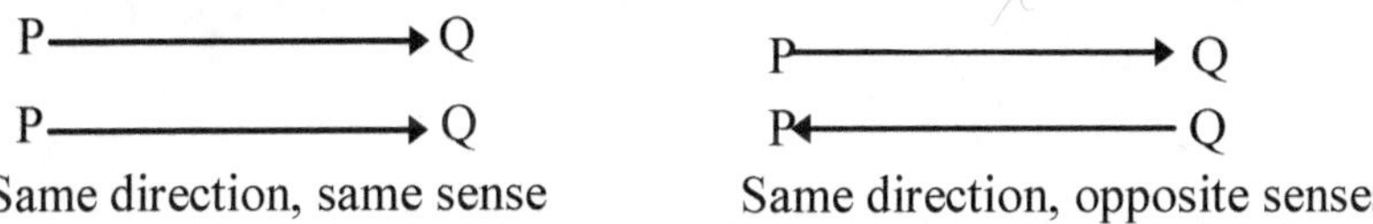

If two vectors lie on the same straight line, they are said to be **collinear**. Otherwise they are **non collinear**.

Vector Notation

The vector represented by the directed line segment AB below, is denoted by $\underline{a}$, $\vec{a}$ or $\overrightarrow{AB}$. In textbooks, bold print **AB** or **a** is used.

$$A \longrightarrow B$$

The magnitude of the vector **AB** or **a,** above is denoted by $|\underline{a}|$, $|\vec{a}|$ or $|\overrightarrow{AB}|$ or AB and is defined as the length of the line segment from A to B.

Column Vectors and Component Form Vectors

Base Vectors

Base vectors are vectors used to express other vectors in the same plane. In the *x-y* plane the base vectors are **i** and **j** called **unit base vectors** because they have unit length. A vector **AB** which is a units in the Ox direction and b units in the Oy direction is written as a column vector $\begin{pmatrix} a \\ b \end{pmatrix}$ or as a component form vector $a\mathbf{i} + b\mathbf{j}$.

Example

Write the vectors on the following Cartesian plane as
(a) Column vectors (b) component form vectors.

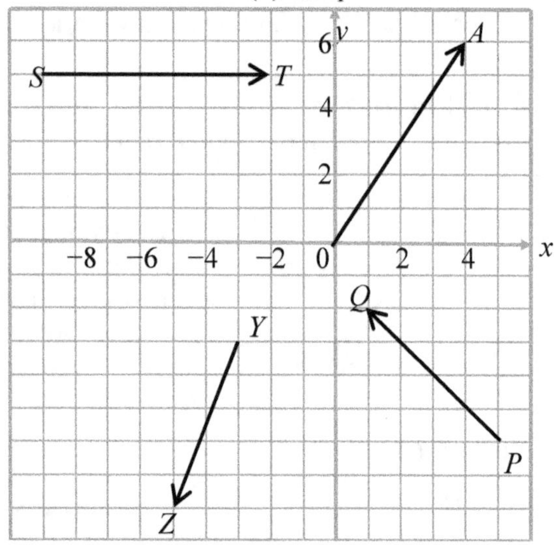

Solution

(a) $\mathbf{OA} = \begin{pmatrix} 4 \\ 6 \end{pmatrix}, \mathbf{ST} = \begin{pmatrix} 7 \\ 0 \end{pmatrix}, \mathbf{YZ} = \begin{pmatrix} -2 \\ -5 \end{pmatrix}, \mathbf{PQ} = \begin{pmatrix} -4 \\ 4 \end{pmatrix}$.

(b) $\mathbf{OA} = 4\mathbf{i} + 6\mathbf{j}, \mathbf{ST} = 7\mathbf{i},\ \mathbf{YZ} = -2\mathbf{i} - 5\mathbf{j}, \mathbf{PQ} = -4\mathbf{i} + 4\mathbf{j},$

The Magnitude of a Vector

The magnitude or modulus of $\mathbf{AB} = \begin{pmatrix} x \\ y \end{pmatrix} = x\mathbf{i} + y\mathbf{j}$ is given by

$|\mathbf{AB}| = \sqrt{x^2 + y^2}$.

The modulus of the vector **AB** with end points $A(x_1, y_1)$ and $B(x_2, y_2)$ is the distance between the points A and B given by

$$AB = \sqrt{(x_2 - x_1)^2 + (y_2 - y_1)^2}.$$

Example

Find the modulus of the following vectors:

(i) $PQ = \begin{pmatrix} -4 \\ 3 \end{pmatrix}$ (ii) $\mathbf{XY} = 5\mathbf{i} - 12\mathbf{j}$

Solution

(i) $\mathbf{PQ} = \sqrt{(-4)^2 + 3^2} = \sqrt{25} = 5$ units

(ii) $\mathbf{XY} = \sqrt{5^2 + (-12)^2} = \sqrt{169} = 13$ units

Example

Calculate the magnitude of the vector **YZ** where *Y* and *Z* are the points (–2, 2) and (4,–6) respectively.

Solution

$\mathbf{YZ} = \sqrt{(-6-2)^2 + (4-(-2))^2} = \sqrt{100} = 10$ units

VECTOR ALGEBRA

Equality of Vectors

Two vectors **a** and **b** are equal if and only if

(i) They have the same magnitude

(ii) They are in the same direction and sense (parallel or collinear)

Example

Which of the vectors in the following Cartesian plane are equal?

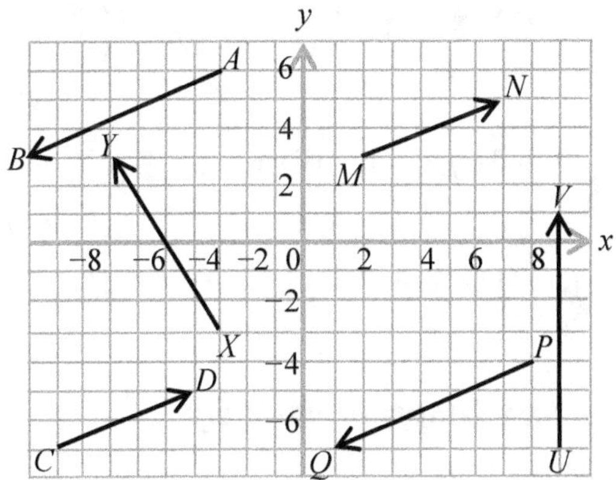

Solution

$$\mathbf{AB} = \mathbf{PQ} = \begin{pmatrix} -7 \\ -3 \end{pmatrix} \qquad \mathbf{CD} = \mathbf{MN} = \begin{pmatrix} 5 \\ 2 \end{pmatrix}$$

Fixed Vectors, Free Vectors and position Vectors

A **free vector** is a vector which can be situated anywhere in space.
A **fixed vector** is a vector which has a particular point of action.
A **position vector** is a fixed vector tied to the origin. The position vector of the point $P(x, y)$ is denoted by **r** and is written as $\boldsymbol{r} = \begin{pmatrix} x \\ y \end{pmatrix}$ or $\boldsymbol{r} = x\mathbf{i} + y\mathbf{j}$

Addition and Subtraction of Vectors

$$\begin{pmatrix} x_1 \\ y_1 \end{pmatrix} + \begin{pmatrix} x_2 \\ y_2 \end{pmatrix} = \begin{pmatrix} x_1 + x_2 \\ y_1 + y_2 \end{pmatrix} \text{ or}$$

$$(x_1\boldsymbol{i} + y_1\boldsymbol{j}) + (x_1\boldsymbol{i} + y_1\boldsymbol{j}) = (x_1 + x_2)\boldsymbol{i} + (y_1 + y_2)\boldsymbol{j})$$

$$\begin{pmatrix} x_1 \\ y_1 \end{pmatrix} - \begin{pmatrix} x_2 \\ y_2 \end{pmatrix} = \begin{pmatrix} x_1 - x_2 \\ y_1 - y_2 \end{pmatrix} \text{ or}$$

$$(x_1\boldsymbol{i} + y_1\boldsymbol{j}) - (x_1\boldsymbol{i} + y_1\boldsymbol{j}) = (x_1 - x_2)\boldsymbol{i} + (y_1 - y_2)\boldsymbol{j})$$

$\mathbf{a} + \mathbf{b} = \mathbf{b} + \mathbf{a}$. This means that vector addition is commutative.

Example

Given that $\mathbf{x} = \begin{pmatrix} 3 \\ 4 \end{pmatrix}$ and $\mathbf{y} = \begin{pmatrix} 1 \\ -2 \end{pmatrix}$, find (a) $\mathbf{x} + \mathbf{y}$ (b) $\mathbf{x} - \mathbf{y}$

Solution

(a) $\mathbf{x} + \mathbf{y} = \begin{pmatrix} 3 \\ 4 \end{pmatrix} + \begin{pmatrix} 1 \\ -2 \end{pmatrix} = \begin{pmatrix} 4 \\ 2 \end{pmatrix}$ (b) $\mathbf{x} - \mathbf{y} = \begin{pmatrix} 3 \\ 4 \end{pmatrix} - \begin{pmatrix} 1 \\ -2 \end{pmatrix} = \begin{pmatrix} 2 \\ 6 \end{pmatrix}$

Example

Given that $\mathbf{a} = 3\mathbf{i} + 4\mathbf{j}$ and $\mathbf{b} = \mathbf{i} + 2\mathbf{j}$, find (i) $\mathbf{a} + \mathbf{b}$ (ii) $\mathbf{a} - \mathbf{b}$

Solution

(i) $\mathbf{a} + \mathbf{b} = (3\mathbf{i} + 4\mathbf{j}) + (\mathbf{i} + 2\mathbf{j}) = 4\mathbf{i} + 6\mathbf{j}$
(ii) $\mathbf{a} - \mathbf{b} = (3\mathbf{i} + 4\mathbf{j}) - (\mathbf{i} + 2\mathbf{j}) = 2\mathbf{i} + 2\mathbf{j}$

Addition and Subtracting Vectors Diagrammatically

Diagrammatically vectors are added head to tail.
Subtracting a vector is the same as adding its inverse. i.e. $\mathbf{a} - \mathbf{b} = \mathbf{a} + (-\mathbf{b})$.

Example

The following is the representation of the vectors **a** and **b**. Draw a diagram to show the vectors (i) **a + b** (ii) **a – b**.

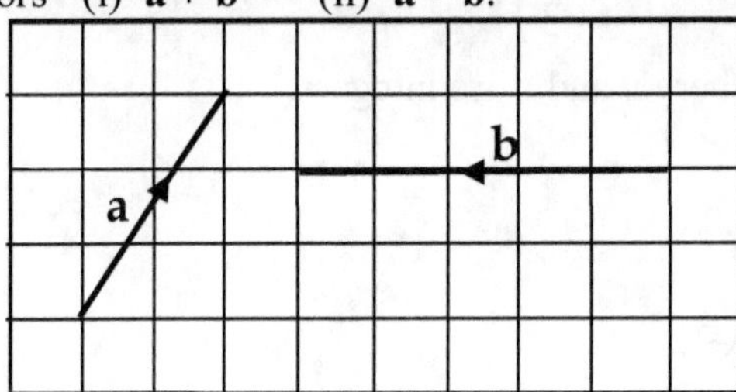

Solution

(i) **a + b**

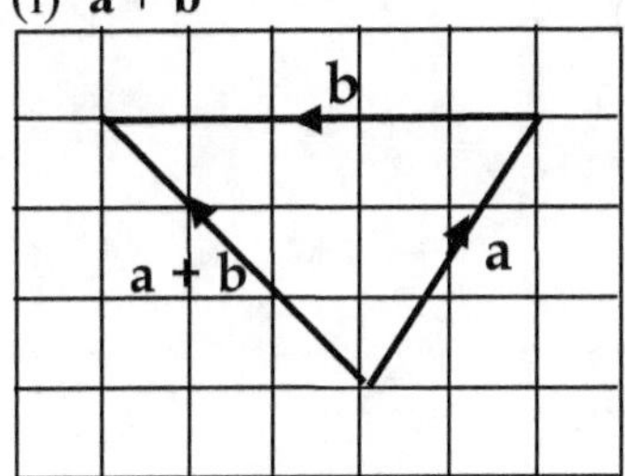

or

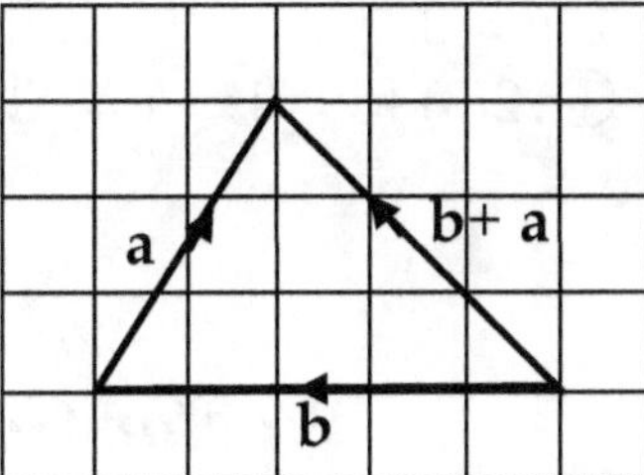

(ii) **a – b**

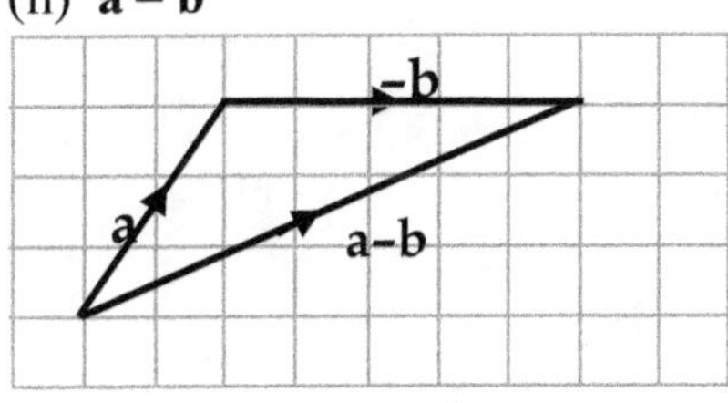

or

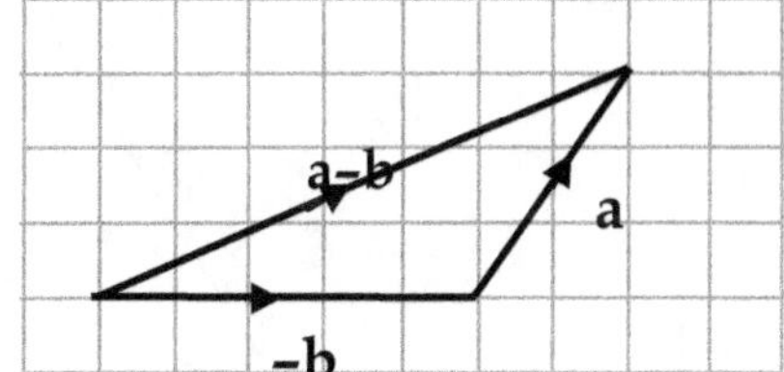

The Zero or Null Vector

A **zero** or **null vector 0** is a vector whose magnitude is zero. $\mathbf{0} = \begin{pmatrix} 0 \\ 0 \end{pmatrix}$.

Multiplication of Vectors by Scalars

$$\mu\begin{pmatrix} x \\ y \end{pmatrix} = \begin{pmatrix} \mu x \\ \mu y \end{pmatrix} \text{ or } \mu(x\mathbf{i} + y\mathbf{j}) = (\mu x\mathbf{i} + \mu y\mathbf{j}).$$

If **a** and **b** are parallel vectors then $\mathbf{a} = n\mathbf{b}, n \in \mathbb{R}$.

Example

If $\mathbf{a}=\begin{pmatrix}2\\5\end{pmatrix}$, $\mathbf{b}=\begin{pmatrix}4\\9\end{pmatrix}$ and $\mathbf{c}=\begin{pmatrix}1\\3\end{pmatrix}$, find a relationship between **a, b** and **c** in the form $\mathbf{b} = u\mathbf{a} + v\mathbf{c}$ where u and v are integers.

Solution

$$\begin{pmatrix}4\\9\end{pmatrix}=u\begin{pmatrix}2\\5\end{pmatrix}+v\begin{pmatrix}1\\3\end{pmatrix}$$

$\Rightarrow 2u+v=4$......................①

$5u+3v=9$......................②

① × 3 − ②: $u = 3$

Substitute in ①: $2(3)+v=4 \Rightarrow v=-2$

$\therefore \mathbf{b}=3\mathbf{a}-2\mathbf{c}$

Vector Geometry

The Section Theorem

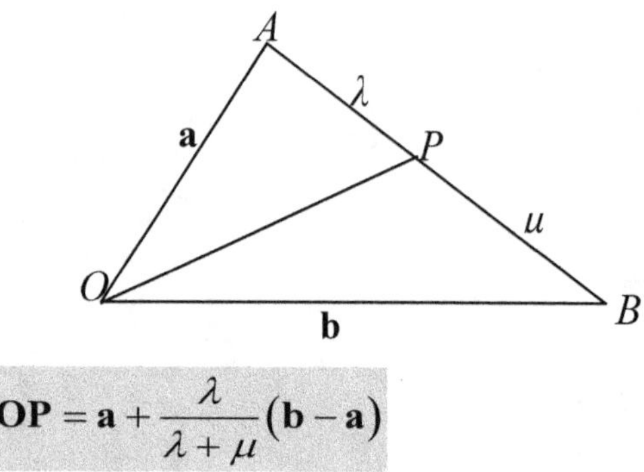

$$\mathbf{OP} = \mathbf{a} + \frac{\lambda}{\lambda+\mu}(\mathbf{b}-\mathbf{a})$$

The Position Vector of the Midpoint

The position vector of the midpoint M of a line segment is given by

$$\mathbf{OM} = \frac{1}{2}(\mathbf{a}+\mathbf{b})$$

Example

Given that **OP** = **p**, **OQ** = **q**, **OR** = **r** and that R divides PQ in the ratio 2:3. Express **r** in terms of **p** and **q**.

Solution

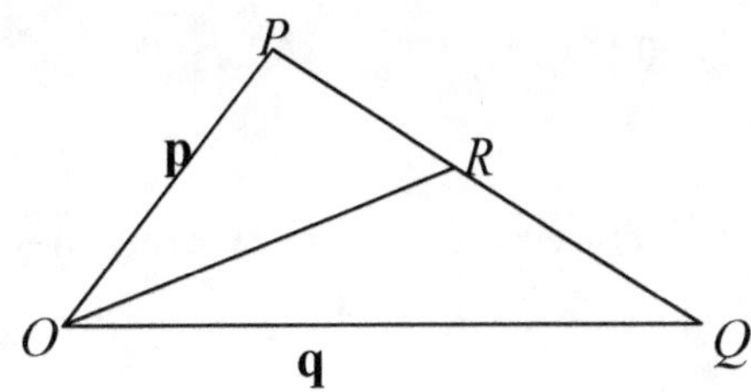

By section theorem,

$$\mathbf{r} = \mathbf{p} + \frac{2}{2+3}(\mathbf{q} - \mathbf{p})$$

$$\Rightarrow \mathbf{r} = \frac{3}{5}\mathbf{p} + \frac{2}{5}\mathbf{q}$$

Example

In the parallelogram *OACB*, the point *X* divides **OC** in the ratio 2:1. Given that $\mathbf{OA} = \mathbf{a}$ and $\mathbf{OB} = \mathbf{b}$, express the vector **XB** in terms of **a** and **b**.

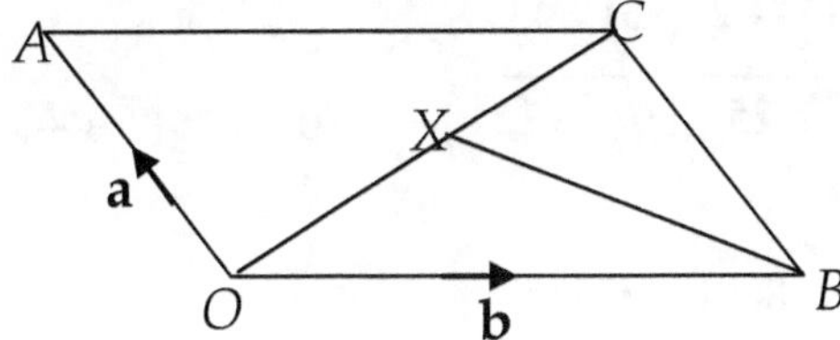

Solution

By the section theorem, $\mathbf{BX} = \mathbf{BC} + \frac{1}{2+1}(-\mathbf{OB} - \mathbf{BC})$

$\mathbf{OA} = \mathbf{BC} = \mathbf{a}$ [opposite sides of a parallelogram]

$$\Rightarrow \mathbf{BX} = \mathbf{a} + \frac{1}{2+1}(-\mathbf{b} - \mathbf{a}) = \frac{2}{3}\mathbf{a} - \frac{1}{3}\mathbf{b}$$

$$\mathbf{XB} = -\mathbf{BX} \Rightarrow \mathbf{XB} = -\frac{2}{3}\mathbf{a} + \frac{1}{3}\mathbf{b}$$

The Scalar or Dot Product

$\mathbf{a.b} = |\mathbf{a}||\mathbf{b}|\cos\theta$, where $0° \leq \theta \leq 180°$ is the angle between the two vectors **a** and **b.**

If $\boldsymbol{a}$ is perpendicular to **b,** $\mathbf{a \cdot b} = \mathbf{0}$ and if $\boldsymbol{a}$ is parallel to **b,** $\mathbf{a \cdot b} = |\mathbf{a}||\mathbf{b}|$.

$$(x_1\boldsymbol{i} + y_1\boldsymbol{j}) \cdot (x_2\boldsymbol{i} + y_2\boldsymbol{j}) = x_1x_2 + y_1y_2$$

Example

The magnitudes of two vectors **a** and **b** are 5 and 8 and the angle between them is 60°. Find $\mathbf{a} \cdot \mathbf{b}$.

Solution

$$\mathbf{a}.\mathbf{b} = |\mathbf{a}||\mathbf{b}|\cos\theta = (5)(8)\cos 60^\circ = (5)(8)\left(\frac{1}{2}\right) = 20$$

Example

Find the scalar product of the following pairs of vectors and say whether they are parallel or perpendicular.

1. $2\mathbf{i}+3\mathbf{j}$ and $3\mathbf{i}-2\mathbf{j}$
2. $10\mathbf{i}+25\mathbf{j}$ and $6\mathbf{i}+15\mathbf{j}$

Solution

1. $(2\mathbf{i}+3\mathbf{j})\cdot(3\mathbf{i}-2\mathbf{j}) = 2(3)+3(-2) = 0 \therefore (2\mathbf{i}+3\mathbf{j}) \perp (3\mathbf{i}-2\mathbf{j})$
2. $(10\mathbf{i}+25\mathbf{j})\cdot(6\mathbf{i}+15\mathbf{j}) = 10(6)+25(15) = 435$

$$|10\mathbf{i}+25\mathbf{j}||6\mathbf{i}+15\mathbf{j}| = \left(\sqrt{10^2+25^2}\right)\left(\sqrt{6^2+15^2}\right) \Rightarrow |10\mathbf{i}+25\mathbf{j}||6\mathbf{i}+15\mathbf{j}| = 435$$

Therefore, $10\mathbf{i}+25\mathbf{j}$ is parallel to $6\mathbf{i}+15\mathbf{j}$.

We can use the formula $\mathbf{a}.\mathbf{b} = |\mathbf{a}||\mathbf{b}|\cos\theta$ to find the angle between two vectors **a** and **b**.

Example

Find the angle between the vectors $6\mathbf{i} + 8\mathbf{j}$ and $4\mathbf{i} - 2\mathbf{j}$.

Solution

$(6\mathbf{i} + 8\mathbf{j}).(4\mathbf{i} - 2\mathbf{j}) = |6\mathbf{i} + 8\mathbf{j}||4\mathbf{i} - 2\mathbf{j}|\cos\theta.$

$$\Rightarrow 8 = 10\sqrt{20}\cos\theta \Rightarrow \theta = \cos^{-1}\left(\frac{8}{10\sqrt{20}}\right) = 79.7^\circ$$

MULTIPLE CHOICE EXERCISE 10:1

1. The modulus of $6\mathbf{i} + 8\mathbf{j}$ is:
   [A] $2\sqrt{7}$ [B] 8 [C] 10 [D] 6
2. If $\mathbf{a} = 3\mathbf{i} - \mathbf{j}$ and $\mathbf{b} = \mathbf{i} + 2\mathbf{j}$, then $\mathbf{a}\,.\,\mathbf{b}$ is equal to:
   [A] 1 [B] 5 [C] –5 [D] –1
3. Given that A, B and C are collinear and $\mathbf{OA} = \mathbf{i} + \mathbf{j}$, $\mathbf{OB} = 2\mathbf{i} - \mathbf{j}$ and $\mathbf{OC} = 3\mathbf{i} + a\mathbf{j}$. The value of a is:
   [A] 1 [B] –1 [C] –3 [D] –2

4. The vector $\begin{pmatrix}-4\\1\end{pmatrix}$ in **i** and **j** component form is:

[A] $4\mathbf{i}-\mathbf{j}$ [B] $\mathbf{i}-4\mathbf{j}$ [C] $-4\mathbf{i}+\mathbf{j}$ [D] $-\mathbf{i}-4\mathbf{j}$

5. The vector $2\mathbf{i}-3\mathbf{j}$ in column vector form is:

[A] $\begin{pmatrix}-3\\-2\end{pmatrix}$ [B] $\begin{pmatrix}3\\-2\end{pmatrix}$ [C] $\begin{pmatrix}-2\\3\end{pmatrix}$ [D] $\begin{pmatrix}2\\-3\end{pmatrix}$

6. The additive inverse of $3\mathrm{i}-5\mathrm{j}$ is:

[A] $3\mathbf{i}+5\mathbf{j}$ [B] $-3\mathbf{i}+5\mathbf{j}$ [C] $-3\mathbf{i}-5\mathbf{j}$ [D] $3\mathbf{i}-5\mathbf{j}$

7. A unit vector among the following is:

[A] $\frac{1}{\sqrt{2}}\mathbf{i}-\frac{1}{\sqrt{2}}\mathbf{j}$ [B] $\frac{1}{\sqrt{2}}\mathbf{i}+\frac{1}{\sqrt{2}}\mathbf{j}$ [C] $\mathbf{i}+\mathbf{j}$ [D] $\mathbf{i}-\mathbf{j}$

8. The additive inverse of $\begin{pmatrix}-3\\2\end{pmatrix}$ is:

[A] $\begin{pmatrix}-3\\-2\end{pmatrix}$ [B] $\begin{pmatrix}3\\-2\end{pmatrix}$ [C] $\begin{pmatrix}-2\\3\end{pmatrix}$ [D] $\begin{pmatrix}2\\-3\end{pmatrix}$

9. As a column vector, the vector represented by the directed line segment in the figure below is:

[A] $\begin{pmatrix}-3\\8\end{pmatrix}$ [B] $\begin{pmatrix}3\\-8\end{pmatrix}$ [C] $\begin{pmatrix}-8\\3\end{pmatrix}$ [D] $\begin{pmatrix}8\\-3\end{pmatrix}$

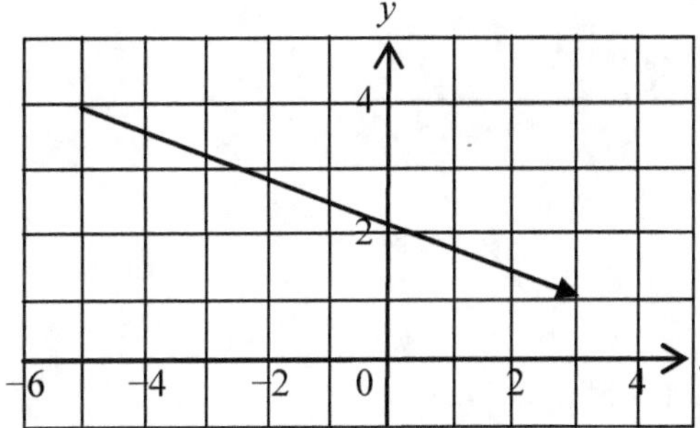

10. On the Cartesian plane, the free vector $-5\mathbf{i}+3\mathbf{j}$ could be represented as:

[A]

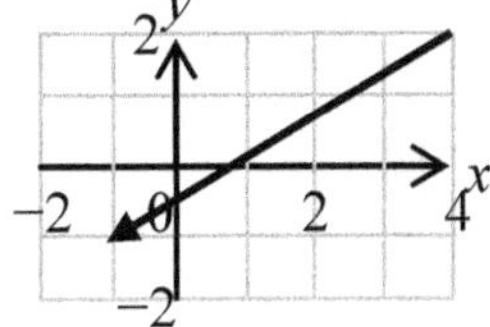

[B]

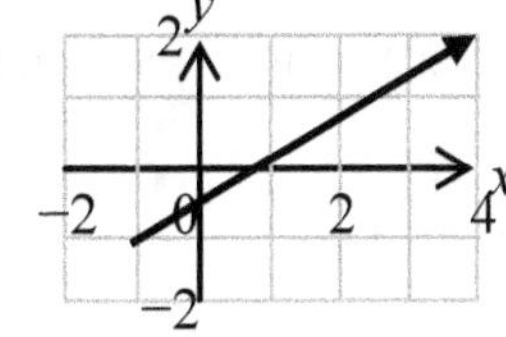

[C] 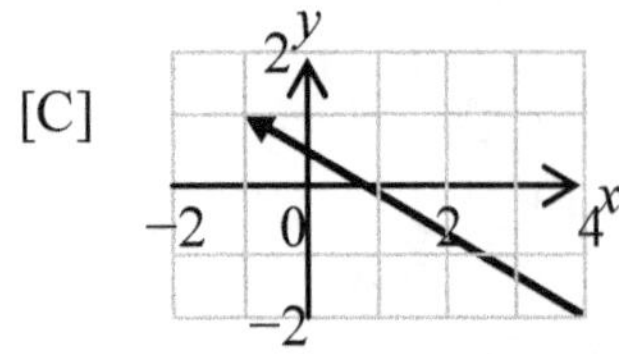

[D] 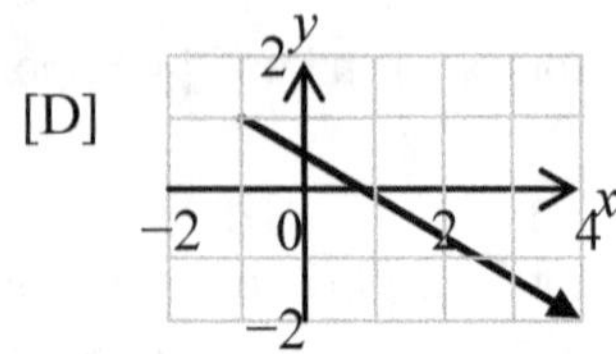

11. Given that $\mathbf{u} = \binom{2}{3}$ and $\mathbf{V} = \binom{0}{1}$, the numbers a and b such that

$a\mathbf{u} + b\mathbf{v} = \binom{4}{5}$

[A] $a = -1, b = 2$ [B] $a = -2, b = -1$ [C] $a = 2, b = -1$ [D] $a = 1, b = 2$

12. In the quadrilateral *OABC*, *D* is the mid-point of *BC*. Given that, $\mathbf{OA} = \mathbf{a}$, $\mathbf{OB} = \mathbf{b}$ and $\mathbf{OC} = \mathbf{c}$. In terms of **a**, **b** and **c** $\mathbf{OD}$ is:

[A] $\frac{1}{2}(\mathbf{b}-\mathbf{c})$ [B] $\frac{1}{2}(\mathbf{b}+\mathbf{c})$ [C] $\frac{1}{2}(-\mathbf{b}+\mathbf{c})$ [D] $-\frac{1}{2}(\mathbf{b}-\mathbf{c})$

13. In the quadrilateral *OABC*, *D* is the mid-point of *BC* and *G* is the point on *AD* such that $AG:GD = 2:1$. Given that, $\mathbf{OA} = \mathbf{a}, \mathbf{OB} = \mathbf{b}$ and $\mathbf{OC} = \mathbf{c}$. $\mathbf{OG}$ in terms of **a**, **b** and **c** is:

[A] $\frac{1}{3}(\boldsymbol{a}+\boldsymbol{b}-\boldsymbol{c})$ [B] $\frac{1}{3}(\boldsymbol{a}-\boldsymbol{b}+\boldsymbol{c})$ [C] $\frac{1}{3}(\boldsymbol{a}-\boldsymbol{b}-\boldsymbol{c})$ [D] $\frac{1}{3}(\boldsymbol{a}+\boldsymbol{b}+\boldsymbol{c})$

14-15 In the figure below, *OPQR* is a parallelogram $TR = 2OT$ and $RM = MQ$. Given that $\overline{OP} = \mathbf{p}$ and $\overline{OR} = \mathbf{r}$.

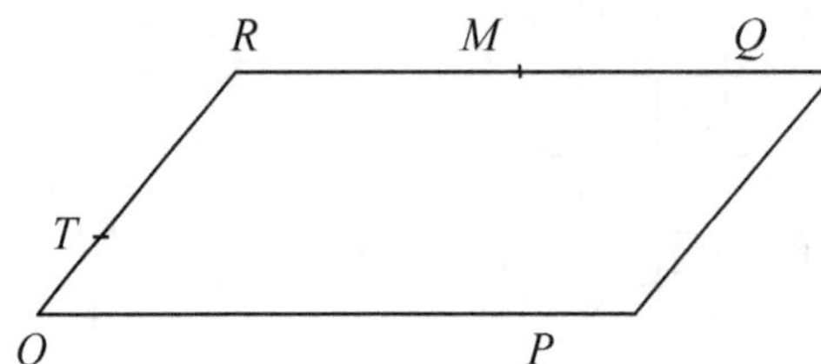

14. The vector **TM** in terms of **p** and **r** is:

[A] $\frac{2}{3}\mathbf{r}-\frac{1}{2}\mathbf{p}$ [B] $-\frac{2}{3}\mathbf{r}+\frac{1}{2}\mathbf{p}$ [C] $-\frac{2}{3}\mathbf{r}-\frac{1}{2}\mathbf{p}$ [D] $\frac{2}{3}\mathbf{r}+\frac{1}{2}\mathbf{p}$

15. The vector **PM** in terms of **p** and **r** is:

[A] $\frac{1}{2}\mathbf{p}+\mathbf{r}$ [B] $\mathbf{r}-\frac{1}{2}\mathbf{p}$ [C] $\frac{1}{2}\mathbf{r}-\mathbf{p}$ [D] $\mathbf{p}-\frac{1}{2}\mathbf{r}$

16. Let $\mathbf{a} = \begin{pmatrix} 2 \\ 5 \end{pmatrix}, \mathbf{b} = \begin{pmatrix} 4 \\ 4 \end{pmatrix}$ and $\mathbf{c} = \begin{pmatrix} 1 \\ 3 \end{pmatrix}$. The relationship between **a**, **b**, and **c**, in the form $\mathbf{b} = u\mathbf{a} + v\mathbf{c}$, where u and v are integers is:

[A] $\mathbf{b} = 3\mathbf{a} - 2\mathbf{c}$ [B] $\mathbf{b} = 3\mathbf{a} + 2\mathbf{c}$

[C] $\mathbf{b} = -2\mathbf{a} + 3\mathbf{c}$ [D] $\mathbf{b} = -3\mathbf{a} + 2\mathbf{c}$

17. The magnitude of $\mathbf{u} = 2\mathbf{i} + 3\mathbf{j}$ is:

[A] 5 [B] 13 [C] $\sqrt{13}$ [D] $\sqrt{5}$

18. The angle between the vectors $\mathbf{u} = 2\mathbf{i} + 3\mathbf{j}$ and $\mathbf{v} = 3\mathbf{i} - 2\mathbf{j}$ is:

[A] 30° [B] 45° [C] 60° [D] 90°

19-20 In the figure below, *OABC* is a trapezium with $\mathbf{OA} = \mathbf{a}$ and $\mathbf{OB} = \mathbf{b}$. The point *P* is the midpoint of *BC*, the point *X* is the midpoint of *OP* and the point *Y* is the midpoint of *AC*.

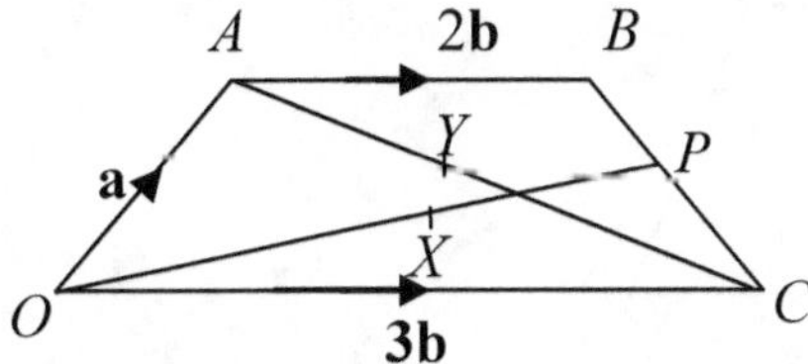

19. In terms of $\mathbf{a}$ and $\mathbf{b}$, $\mathbf{OX}$ is equal to:

[A] $\frac{1}{2}\mathbf{a} + \frac{5}{2}\mathbf{b}$ [B] $\frac{1}{4}(\mathbf{a} + 5\mathbf{b})$ [C] $\frac{1}{2}\mathbf{a} - \frac{5}{2}\mathbf{b}$ [D] $\frac{1}{4}(\mathbf{a} - 5\mathbf{b})$

20. In terms of $\mathbf{a}$ and $\mathbf{b}$, $\mathbf{OY}$ is equal to:

[A] $\frac{1}{4}(\mathbf{a} - 3\mathbf{b})$ [B] $\frac{1}{4}(\mathbf{a} + 3\mathbf{b})$ [C] $\frac{1}{2}\mathbf{a} - \frac{3}{2}\mathbf{b}$ [D] $\frac{1}{2}(\mathbf{a} + 3\mathbf{b})$

21. Given that $\mathbf{a} = 2\mathbf{i} + 5\mathbf{j}$, $\mathbf{b} = 4\mathbf{i} + 9\mathbf{j}$, the vector $\mathbf{c}$ such that $\mathbf{b} = 3\mathbf{a} - 2\mathbf{c}$ is:

[A] $\mathbf{i} - 3\mathbf{j}$ [B] $-\mathbf{i} + 3\mathbf{j}$ [C] $\mathbf{i} + 3\mathbf{j}$ [D] $2\mathbf{i} + 6\mathbf{j}$

22. In the figure below, $\mathbf{AC}$ in terms of $\mathbf{a}$ and $\mathbf{d}$ only is:

[A] $-\mathbf{a} + \mathbf{d}$ [B] $-\mathbf{a} - \mathbf{d}$ [C] $\mathbf{a} - \mathbf{d}$ [D] $\mathbf{a} + \mathbf{d}$

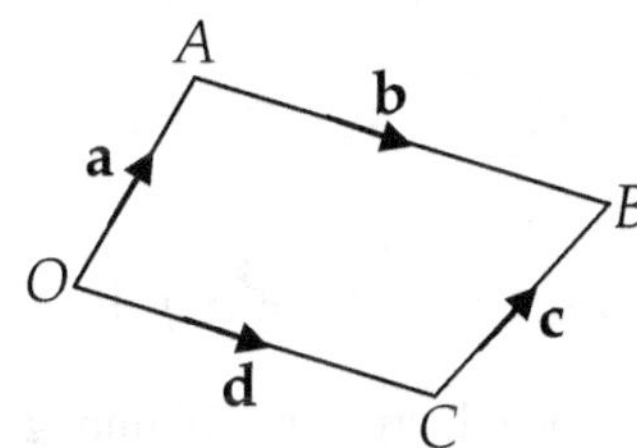

23. In the figure above, the right expression of $\mathbf{OB}$ in terms of $\mathbf{a}$ and $\mathbf{b}$ is:

[A] $-\mathbf{a} + \mathbf{b}$ [B] $-\mathbf{a} - \mathbf{b}$ [C] $\mathbf{a} - \mathbf{b}$ [D] $\mathbf{a} + \mathbf{b}$

24. In the figure above, the right expression of $\mathbf{OB}$ in terms of $\mathbf{c}$ and $\mathbf{d}$ is:

[A] $\mathbf{c} + \mathbf{d}$ [B] $-\mathbf{c} - \mathbf{d}$ [C] $-\mathbf{c} + \mathbf{d}$ [D] $\mathbf{c} - \mathbf{d}$

25. In the figure above, given that $\mathbf{b} = \frac{3}{2}\mathbf{d}$. In terms of $\mathbf{a}$ and $\mathbf{d}$, $\mathbf{c}$ is:

[A] $-\mathbf{a} + \frac{1}{2}\mathbf{d}$ [B] $-\mathbf{a} - \frac{1}{2}\mathbf{d}$ [C] $\mathbf{a} + \frac{1}{2}\mathbf{d}$ [D] $\mathbf{a} - \frac{1}{2}\mathbf{d}$

26. Given that the position vectors of *A*, *B*, *C* and *D* are respectively **a**, **b**, **c** and **d** where, $2\mathbf{c} = \mathbf{a}$ and $4\mathbf{d} = \mathbf{a} + \mathbf{b}$. **CD** can be written in terms of **AB** as:
    [A] $4\mathbf{CD} = \mathbf{AB}$ [B] $4\mathbf{CD} = -\mathbf{AB}$ [C] $\mathbf{CD} = 4\mathbf{AB}$ [D] $\mathbf{CD} = -4\mathbf{AB}$

27-28 In the figure below, *OPQR* is a trapezium with *OP* and *RQ* parallel.
Given that $\mathbf{RQ} = 2\mathbf{a}$, $\mathbf{OR} = \mathbf{b}$, $\mathbf{RQ} = \frac{2}{3}\mathbf{OP}$ and that *M* is the midpoint of *RQ*.

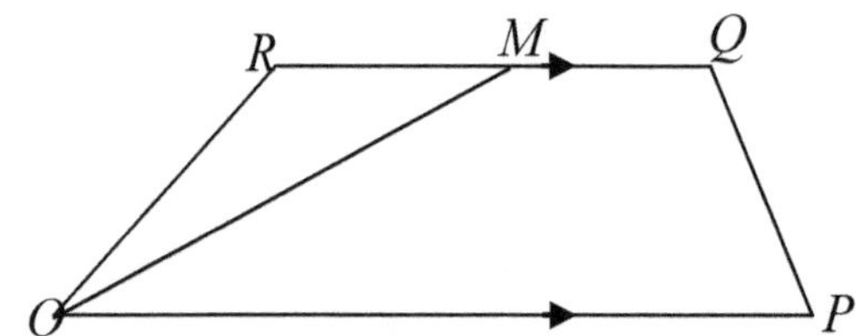

27. The vector **OM** in terms of **a** and **b** is:
    [A] $-\mathbf{a}+\mathbf{b}$ [B] $-\mathbf{a}-\mathbf{b}$ [C] $\mathbf{a}-\mathbf{b}$ [D] $\mathbf{a}+\mathbf{b}$
28. The vector **QP** in terms of **a** and **b** is:
    [A] $-\mathbf{a}+\mathbf{b}$ [B] $-\mathbf{a}-\mathbf{b}$ [C] $\mathbf{a}-\mathbf{b}$ [D] $\mathbf{a}+\mathbf{b}$
29. In the following triangle $\mathbf{OA} = \mathbf{a}$, $\mathbf{OB} = \mathbf{b}$ and $AM : MB = 2:1$. In terms of **a** and **b**, **AB** is equal to:
    [A] $-\mathbf{a}+\mathbf{b}$ [B] $-\mathbf{a}-\mathbf{b}$ [C] $\mathbf{a}-\mathbf{b}$ [D] $\mathbf{a}+\mathbf{b}$

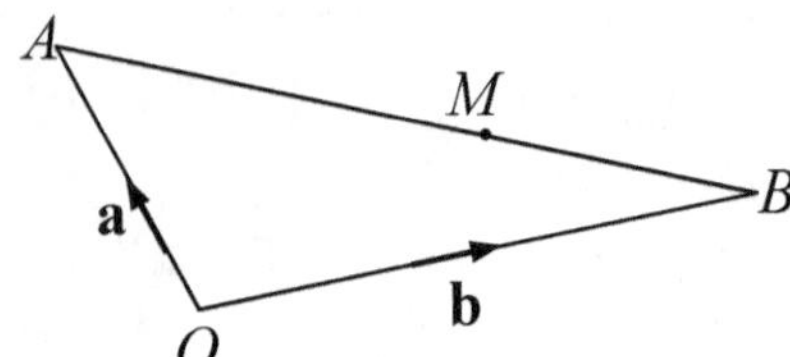

30. In the above triangle $\mathbf{OA} = \mathbf{a}$, $\mathbf{OB} = \mathbf{b}$ and $AM : MB = 2:1$. In terms of **a** and **b**, **OM** is equal to:
    [A] $-\frac{1}{3}(\mathbf{a}+2\mathbf{b})$ [B] $\frac{1}{3}(\mathbf{a}+2\mathbf{b})$ [C] $\frac{1}{3}(\mathbf{a}-2\mathbf{b})$ [D] $\frac{1}{3}(-\mathbf{a}+2\mathbf{b})$
31. The pair of vectors which are parallel among the following are:
    [A] **AB** and **NM** [B] **XY** and **MN** [C] **XY** and **PQ** [D] **BA** and **MN**

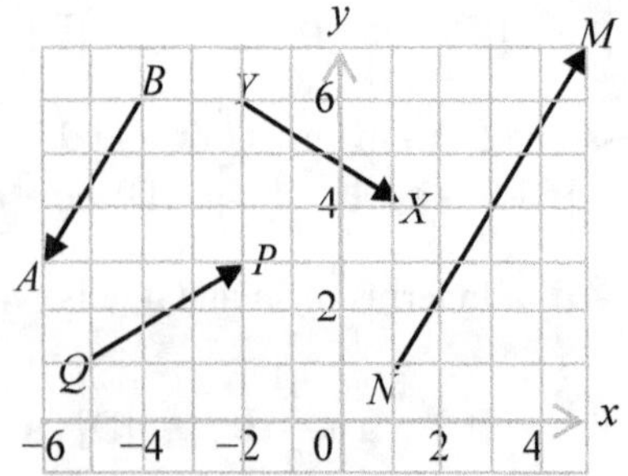

32. Given that, $\mathbf{a} = -5\mathbf{i}+12\mathbf{j}$, $\mathbf{b} = 12\mathbf{i}-5\mathbf{j}$, $\mathbf{c} = 5\mathbf{i} - 12\mathbf{j}$ and $\mathbf{d} = \begin{pmatrix} -5 \\ 12 \end{pmatrix}$. The vectors which are equal are:
[A] **a** and **b** [B] **a** and **c** [C] **a** and **d** [D] **c** and **d**

33. Given that, $\mathbf{a} = \begin{pmatrix} -9 \\ 12 \end{pmatrix}$, $\mathbf{b} = \begin{pmatrix} -12 \\ 9 \end{pmatrix}$, $\mathbf{c} = -9\mathbf{i}+12\mathbf{j}$ and $\mathbf{d} = 12\mathbf{i}-9\mathbf{j}$. The vectors which are equal are:
[A] **a** and **b** [B] **a** and **c** [C] **a** and **d** [D] **b** and **d**

34. The directed line segments [*AB*], [*QP*] and [*XY*] are parallel and equal in magnitude. It is true to say that:
[A] **AB** = **XY** [B] **PQ** = **XY** [C] **AB** = **PQ** [D] **AB** = **YX**

35. Two vectors are equal if and only if they have the same:
[A] Magnitude and direction. [B] Magnitude and are parallel.
[C] Magnitude and sense. [D] Sense and are parallel $\frac{1}{3}(\mathbf{a}+2\mathbf{b})$

CHAPTER 11

STATISTICS AND PROBABILITY

11.1 STATISTICS

Representation of Data-Statistical Graphs

Pie Charts or Circular Diagrams

A pie chart is a way of displaying statistical data using a circle divided into sectors such that the sizes of the angles of the sectors are proportional to the frequencies of the quantities they represent.

Example

A fruit dealer bought 200 pineapples, 400 bananas, 500 watermelons and 900 oranges. Represent this data on a pie chart.

Solution

Fruit	Number	Calculation	Angle
Pineapple(P)	200	$\frac{200}{2000} \times 360$	36°
Bananas (B)	400	$\frac{400}{2000} \times 360$	72°
Watermelons (W)	500	$\frac{500}{2000} \times 360$	90°
Oranges (O)	900	$\frac{900}{2000} \times 360$	162°
TOTAL	2000		360°

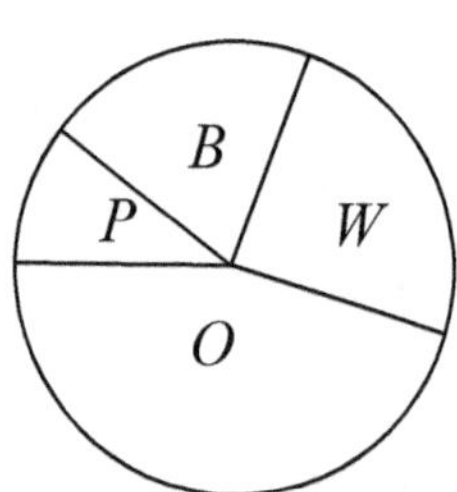

Bar Charts

A bar chart is a way of displaying statistical data using horizontal or vertical bars such that the heights or lengths of the bars are proportional to the quantities they represent.

Example

The table below shows the number of people who owned televisions in a certain town from 1985 to 1993. Draw a bar chart to represent this data.

Year	Number of people
1985	600
1986	800
1987	900
1988	1100
1989	1300
1990	1400
1991	1200
1992	1500
1993	1600

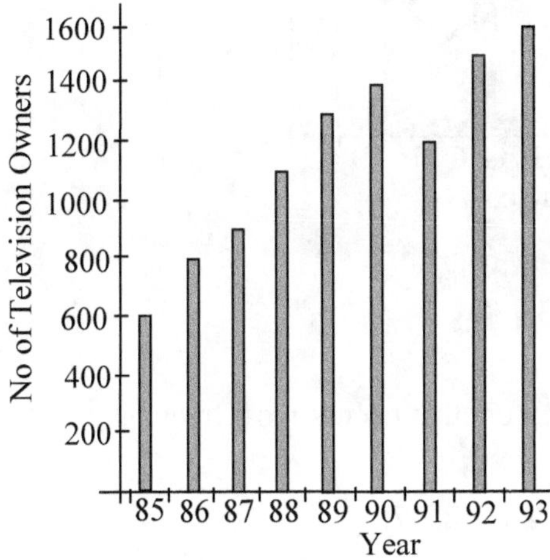

Histograms

A histogram is a statistical graph in which blocks with no spaces between them are drawn such that their areas (rather than their height, as in a bar chart) are proportional to the frequencies within a class or across several class boundaries.

Example

The following table shows the number of form 2 students in a certain school in the year 2002 who had the required textbooks for the subjects Mathematics (M), English (E), French (F), History (H), Geography (G), Chemistry (C), Physics (P), Biology (B) and Literature (L). Draw a histogram to represent this information.

Textbook	M	E	F	H	G	C	P	B	L
Number of students	14	15	11	3	9	5	7	1	13

Solution

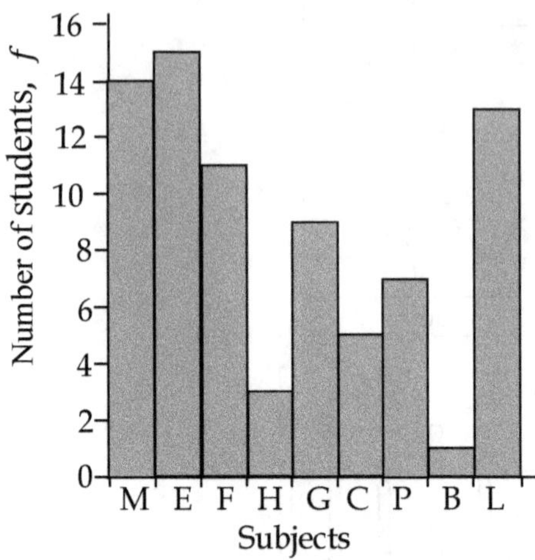

Measures of Central Tendencies

Mode

The mode of any given data is the item that occurs most frequent.

Example

Find the mode of the data 1,2,4,6,2,7,7,2,2,7.

Solution

x	1	2	6	7
f	1	4	1	3

The mode is 2.

Arithmetic Mean (Average or Mean)

The mean, $\bar{x} = \dfrac{\text{Sum of data}}{\text{Total frequency}}$ or $\bar{x} = \dfrac{\sum x}{\sum f}$

Example

Find the mean of 11, 9, 15, 12 and 13.

Solution

$$\bar{x} = \frac{\sum x}{\sum f} = \frac{11 + 9 + 15 + 12 + 13}{5} = \frac{60}{5} = 12$$

For repeated data the mean $\bar{x} = \dfrac{\sum fx}{\sum f}$

Example

The following shows the marks obtained by 30 students during a test. Calculate the average mark.

55	60	65	40	60	60
65	50	40	60	50	60
60	50	60	30	40	60
60	50	60	50	60	50
60	50	60	60	50	60

Solution

To ease the working we first make a frequency distribution table.

Mark, x	Frequency, f	fx
30	1	30
40	3	120
50	8	400
55	1	55
60	15	900
65	2	130
	$\sum f = 30$	$\sum fx = 1635$

$$\bar{x} = \frac{\sum fx}{\sum f} = \frac{1635}{30} = 54.5$$

Median

The median is the middle value in a set of statistical values that are arranged in ascending or descending order.

To obtain the median, first rank the data. For an odd number of numbers, the median is the middle number ranked $\frac{n+1}{2}$ and for an even number of numbers the median is the average of the two middle numbers ranked $\frac{n}{2}$ and $\frac{n}{2} + 1$.

Example

Find the median of (a) 12, 2, 7, 13, 6 (b) 12, 2, 7, 13, 6, 8

Solution

(a) Ranking we have 2, 6, 7, 12, 13 $\Rightarrow$median = 7

(b) Ranking we have 2, 6, 7, 8, 12, 13 $\Rightarrow$median = $\frac{7+8}{2} = 7.5$

GROUPED DATA

To analyze very large masses of raw data, distribute the data into **classes, class intervals** or **groups**.

Class Limits and Class Boundaries

For a class 30-39, the smaller number 30 is called the **lower class limit** and the larger number 39 is called the **upper class limit.** If the data were rounded up to the nearest whole number the **lower** and **upper class boundaries** will be 29.5 and 39.5 respectively.

$$\underbrace{30}_{\text{lower class limit}} - \underbrace{39}_{\text{upper class limit}}, \quad \underbrace{29.5}_{\text{lower class boundary}} - \underbrace{39.5}_{\text{upper class boundary}}$$

The class Size (class width or **class length** or **class interval)** c is given by

c = upper class boundary − lower class boundary.

Mid-interval value (class mark or **mid-point)** is given by

$$\text{class mark} = \frac{\text{upper class limit} + \text{lower class limit}}{2}.$$

When drawing histograms for grouped data, it is preferable to use the class boundaries and the class mark rather than the class limits.

The **modal class** is the class with the highest frequency represented by the tallest bar in a histogram. We can obtain the mode from a histogram by extrapolation.
If the class intervals are of equal sizes, we can also obtain the mode using the formula

$$\text{Mode} = L_1 + \left(\frac{\Delta_1}{\Delta_1 + \Delta_2}\right)c$$

where
L_1 = lower class boundary of the modal class
C = class width
Δ_1= modal class frequency – next lower class frequency
Δ_2= next upper class frequency – modal class frequency

Frequency Distribution Curve (Or Frequency Polygon)

We can draw another type of graph called a **frequency distribution curve** or **frequency polygon** after drawing a histogram, by joining the tips of the rectangles of the histogram.

Example

The following scores out of 100 were obtained by 36 students in a test.

25	49	76	12	51	56
81	50	45	92	58	67
55	52	43	31	48	84
66	56	44	39	45	22
56	74	98	67	34	41
34	68	69	70	85	51

(a) Starting with 0-9, arrange the marks in a grouped frequency table with class intervals of size 10.

(b) State the modal class.

(c) Draw a histogram to represent this data. Hence, obtain the mode of the distribution.

(d) By calculation obtain the mode of the distribution.

(e) Draw a frequency polygon of the distribution.

Solution

(a)

Marks, x	Class mark	Frequency, f
0-9	4.5	0
10-19	14.5	1
20-29	24.5	2
30-39	34.5	4
40-49	44.5	7
50-59	54.5	9
60-69	64.5	5
70-79	74.5	3
80-89	84.5	3
90-99	94.5	2

(b) The modal class is 50-59

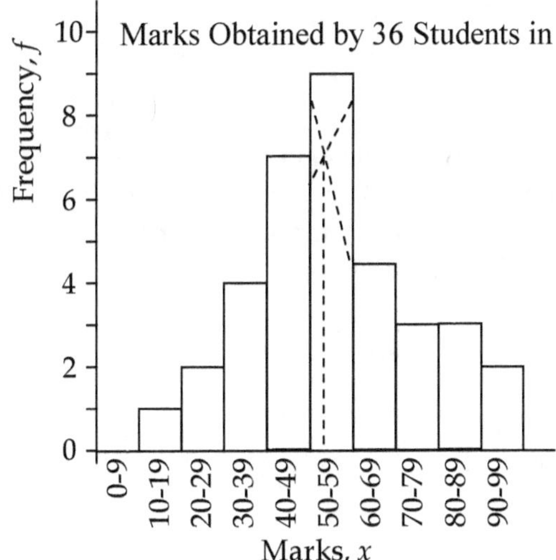

(c) From the histogram by extrapulation, the mode is 53.5.

(d) Mode $= L_1 + \left(\frac{\Delta_1}{\Delta_1+\Delta_2}\right)c = 49.5 + \left(\frac{2}{2+4}\right)10 = 52.8$

(e) The frequency polygon is as shown below.

Marks Obtained by 36 Students in a test

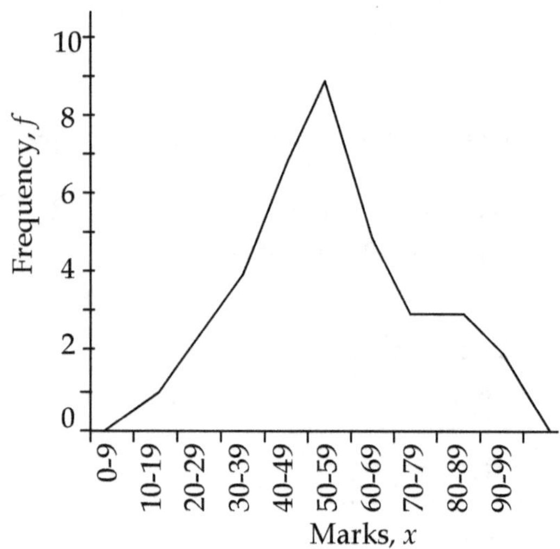

Cumulative Frequency Curves

Cumulative frequency curves (or **ogives**) are graphs of cumulative frequency against upper class boundary of each class.

Quantiles

When data is ranked; the median M divides the data into two equal parts, the **quartiles** Q_1, Q_2 and Q_3 divide the data into 4 equal parts, the **deciles** D_1, D_2, D_3, D_4, $D_5 \ldots D_9$ divide the data into 10 equal parts and the **percentiles** P_1, P_2, P_3, $P_4 \ldots P_{99}$ divide the data into 100 equal parts.

Median $M = P_{50} = D_5 = Q_2$, $P_{25} = Q_1$ and $P_{75} = Q_3$.

Example

The marks obtained by 80 students in an examination are arranged in a frequency distribution in the following table.

Marks, x	Frequency, f
1-10	3
11-20	4
21-30	6
31-40	8
41-50	12
51-60	15
61-70	13
71-80	9
81-90	6
91-100	4

(a) Make a cumulative frequency table for this distribution.

(b) Taking 1cm to represent 10 units on each axis, draw a graph of this distribution.

(c) Use your graph to obtain the (i) Median, M (ii) Lower quartile, Q_1 (iii) Upper quartile, Q_3 (iv) 90th percentile, P_{90} (v) 10th percentile, P_1

Solution

(a)

x	f	C.F.
≤ 10	3	3
≤ 20	4	7
≤ 30	6	13
≤ 40	8	21
≤ 50	12	33
≤ 60	15	48
≤ 70	13	61
≤ 80	9	70
≤ 90	6	76
≤ 100	4	80

(b) See graph below.

(c) From the graph,

(i) The cumulative frequency corresponding to median is $\frac{1}{2}$ of $80 = 40$

$\Rightarrow M = 55.$

(ii) The cumulative frequency corresponding to the lower quartile is 25% of $80 = 20 \Rightarrow Q_1 = 39$

(iii) The cumulative frequency corresponding to the upper quartile is 75% of $80 = 60 \Rightarrow Q_3 = 68$

(iv) The cumulative frequency corresponding to 90th percentile is 90% of 80 = 72 $\Rightarrow P_{90} = 82$.

(v) The cumulative frequency corresponding to 10th percentile is 10% of 80 = 8 $\Rightarrow P_{10} = 20$

(b) Marks out of 100 Obtained by 80 Students

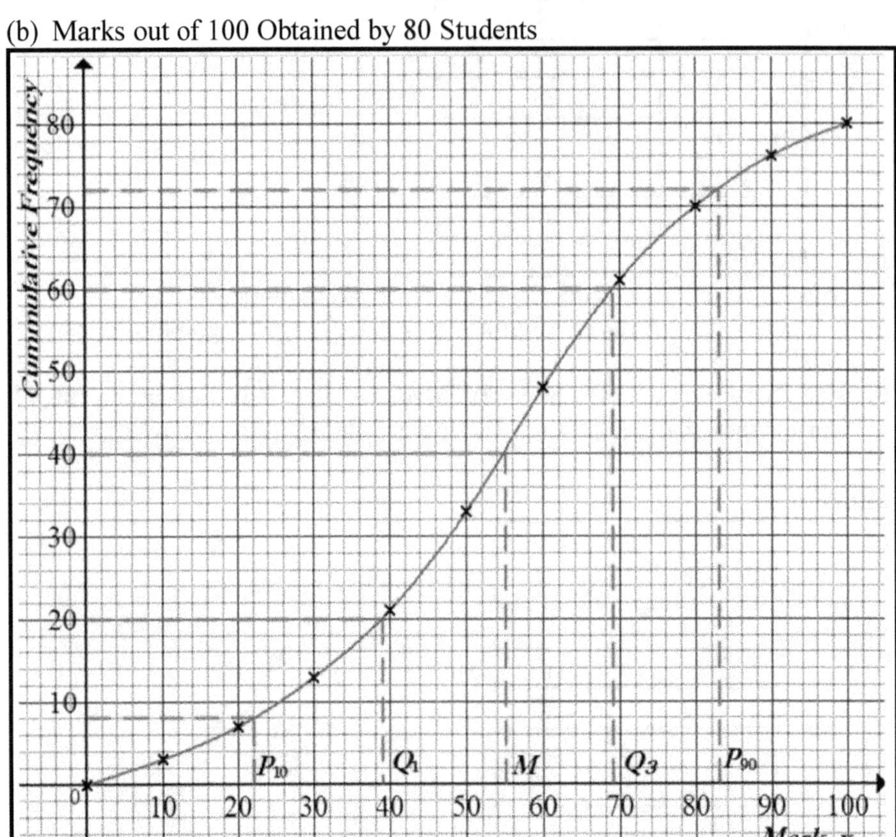

Median of Grouped Data by Calculation

$$\text{Median} = L_1 + \left(\frac{\frac{\sum f}{2} - \left(\sum f\right)_1}{f_{median}} \right) c$$

Where,

L_1 = lower class boundary of the median class.

$\sum f$ = total frequency

$\left(\sum f\right)_1$ = sum of frequency of all classes below the median class

f_{median} = median class frequency

c = median class size

Example
Calculate the median of the data in the previous example.

Solution

$$\text{Median} = L_1 + \left(\frac{\frac{\Sigma f}{2} - (\Sigma f)_1}{f_{median}}\right)c = 51.5 + \left(\frac{\frac{80}{2} - 33}{15}\right)10 = 56.2$$

Measures of Dispersion or Variation (*Spread or Scatter of Data*)

The measures of dispersion are the degrees to which numerical data turns to spread about an average. The most common methods used to measure the spread of numerical data are the **range**, the **mean deviation**, the **inter quartile range** the **10- 90 percentile range**, and the **standard deviation.**

The Range

The range of a set of data is the difference between the smallest and the largest statistic in the set.

Example
Find the range of the following data. 2, 7, 3, 7, 8, 21, 17, 35, 4, 39.

Solution

Range = largest statistic-smallest statistic = 39−2 = 37

Inter Quartile Range

Inter quartile range = $Q_3 - Q_1$

Semi-inter Quartile Range

$$\text{Semi-interquartile range} = \frac{Q_3 - Q_1}{2}$$

The 10 - 90 percentile ranges

10-90 percentile range = $P_{90} - P_{10}$

Example
Using the cumulative frequency curve above find:
(a) The inter quartile range and hence calculate the semi-inter quartile range of the data.
(b) The 10-90 percentile range of the data.

Solution

(a) Inter quartile range = $Q_3 - Q_1 = 68 - 38 = 30$

$$\Rightarrow \text{Semi-interquartile range} = \frac{Q_3 - Q_1}{2} = \frac{30}{2} = 15$$

(b) 10-90 percentile range = $P_{90} - P_{10} = 82 - 20 = 62$

Mean of Grouped Data

To find the mean of grouped data, the class mark is taken to represent the mark for the whole class.

Mean Deviation from the Mean

The mean deviation from the mean denoted by *M.D.* is given by

$$M.D. = \frac{\sum_{i=1}^{n} |x_i - \bar{x}|}{n} \quad \text{or} \quad M.D. = \frac{\sum_{i=1}^{n} f_i |x_i - \bar{x}|}{\sum_{i-1}^{n} f_i}$$

Example
Calculate the mean deviation from the mean of the set of numbers 2, 3, 6, 8, 11.

Solution

$$\bar{x} = \frac{2+3+6+8+11}{5} = \frac{30}{5} = 6$$

$$\Rightarrow M.D. = \frac{|2-6| + |3-6| + |6-6| + |8-6| + |11-6|}{5}$$

$$= \frac{4+3+0+2+5}{5} = \frac{14}{5} = 2.8$$

Example

Find the mean and hence the mean deviation from the mean of the following data.

x	1-10	11-20	21-30	31-40	41-50	51-60	61-70	71-80	81-90	91-100
y	3	5	5	9	11	15	14	8	6	4

Solution

Marks, x	f	fx	$\lvert x_i - \bar{x}\rvert$	$f\lvert x_i - \bar{x}\rvert$
5.5	3	16.5	48.125	144.375
15.5	5	77.5	38.125	190.625
25.5	5	127.5	28.125	140.625
35.5	9	319.5	18.125	163.125
45.5	11	500.5	8.125	89.375
55.5	15	832.5	1.875	28.125
65.5	14	917	11.875	166.250
75.5	8	604	21.875	175.000
85.5	6	513	31.875	191.250
95.5	4	382	41.875	167.500
	$\Sigma f = 80$	$\Sigma fx = 4290$		1456.25

$$\bar{x} = \frac{\Sigma fx}{\Sigma f} = 53.625$$

$$M.D. = \frac{\sum_{i=1}^{5} f\lvert x_i - \bar{x}\rvert}{\sum_{i=1}^{5} f} = \frac{1456.25}{80} = 18.20 \text{ to 2 d.p.s}$$

Standard Deviation and Variance

The variance, σ^2 of a set of n numbers $x_1, x_2, x_3, ..., x_n$ with mean $\bar{x}$ is given by

$$\sigma^2 = \frac{\sum_{i=1}^{n} (x_i - \bar{x})^2}{n}$$

The standard deviation, σ of a set of n numbers $x_1, x_2, x_3, \ldots, x_n$ with mean $\overline{x}$

is given by $\sigma = \sqrt{\dfrac{\sum_{i=1}^{n}(x_i - \overline{x})^2}{n}}$ or $\sigma = \sqrt{\dfrac{\sum_{i=1}^{n} f_i (x_i - \overline{x})^2}{\sum_{i=1}^{n} f_i}}$.

Example

Find the variance and hence the standard deviation of the set of data 9, 3, 8, 8, 9, 8, 9, 18.

Solution

$$\sigma^2 = \frac{\sum_{i=1}^{n} f_i (x_i - \overline{x})^2}{\sum_{i=1}^{n} f_i}, \quad \overline{x} = \frac{\sum_{i=1}^{n} f_i x_i}{\sum_{i=1}^{n} f_i}.$$

x_i	f_i	$x_i f_i$	$x_i - \overline{x}$	$f_i (x_i - \overline{x})^2$
3	1	3	-6	36
8	3	24	-1	3
9	3	27	0	0
18	1	18	9	81
$\sum$	8	72		120

$$\Rightarrow \overline{x} = \frac{72}{8} = 9$$

$$\sigma^2 = \frac{120}{8} = 15$$

$$\Rightarrow \sigma = \sqrt{15} = 3.87$$

MULTIPLE CHOICE EXERCISE 11:1

1. A survey of people and/or property is called:
   [A] Census [B] Data [C] Population [D] Sample
2. The tally marks 卌 卌 卌 卌 卌 ||| represent the number:
   [A] 18 [B] 23 [C] 28 [D] 33
3. The correct tally representation of 17 students is:
   [A] 卌 卌 卌 [B] 卌 卌 ||||||| [C] ||||| |||| |||| || [D] 卌 卌 卌 ||
4. In recording data, the tally marks 卌 卌 卌 ||| will be recorded as:
   [A] 18 [B] 15 [C] 13 [D] 20

5. In a school examination 480 out of a total of 720 candidates were awarded a *D* Grade. On a pie-chart showing all the grades the angle at the centre for the *D* grade is:
   [A] 270° [B] 240° [C] 210° [D] 180°
6. The measure of central tendency a shoe company will be most interested in is:
   [A] mean [B] mode [C] median [D] mean and median
7. The average of 0, 1, 6, 7, 9 and 19 is:
   [A] 9 [B] 6 [C] 7 [D] 10
8. The average of 1, 2, 5, 7, and 15 is:
   [A] 6 [B] 30 [C] 7 [D] 15
9. A group of four people found that their heights were 1.38 m, 1.71 m, 1.23 m and 1.40 m. Their average height (in metres) is:
   [A] 1.145 [B] 1.18 [C] 1.39 [D] 1.43
10. The average wage bill in FCFA of 40 men who collectively earn 3,540,000 FCFA is:
    [A] 87,000 [B] 29,500 [C] 88,500 [D] 31,700
11. The mean of 9,13,16,17,19,23,24 correct to two decimal places is:
    [A] 23.00 [B] 17.29 [C] 16.50 [D] 16.33
12. The average of the first four prime numbers greater than 10 is:
    [A] 20 [B] 19 [C] 17 [D] 15
13. The mean of 20 observations is 4. The observed largest value 23 is removed. the mean of the remaining observations is:
    [A] 4 [B] 3 [C] 2.85 [D] 2.60
14. The mean heights of the three groups of students consisting respectively of 20, 16 and 14 students are 1.67 m, 1.50 m and 1.40 m respectively. The mean height of all the students is:
    [A] 1.52 m [B] 1.53 m [C] 1.54 m [D] 1.55 m
15. The mean of 30 observations is 5. The observed largest value of 34 is deleted. The mean of the remaining observations is:
    [A] 4 [B] 3.8 [C] 3.4 [D] 5
16. The following table shows the scores of some students in a test. The average score is 3.5. The value of x is:

Scores	1	2	3	4	5	6
No. of students	1	4	5	6	x	2

    [A] 1 [B] 2 [C] 3 [D] 4
17. The mean of 9, x and 13 is 11. The value of x is:
    [A] 7 [B] 8 [C] 11 [D] 13
18. The value of x which qualifies 4 as the mean of the data 4, $3x$, 0 and 3 is:
    [A] 1 [B] 2 [C] 3 [D] 4
19. The mean of five observations is 15. Four of them are 11, 12, 19 and 20. The fifth is:
    [A] 10 [B] 25 [C] 20 [D] 13

20. A pie chart is drawn to represent the percentages 20%, 50%, 25% and 5%. The angle which represents 5% is:
[A] 5° [B] 18° [C] 25° [D] 126°
21. Given the scores –3, 4, 0, 4,–2,–5, 1, 7,10,5 the median of the scores is:
[A] 2.5 [B] 2 [C] 4 [D] 3.5
22. From the following table, the mean number of male children per family is:
[A] 5 [B] 4 [C] 3 [D] 2

No. of male children	0	1	2	3	4
No. of families	4	8	6	2	7

23. The mean of four numbers a, b, c and d is 6. The mean of 5 numbers a, b, c, d and e is 10. The value of e is:
[A] 24 [B] 25 [C] 26 [D] 27
24. The average age of five boys is 11 years. A sixth boy whose age is 17 years is added, the mean age in years will now be:
[A] 14 [B] 12 [C] 13 [D] 11
25. The median of 8, 10, 9, 6, 7, 10, 12, 8, 9, 8 is:
[A] 7.5 [B] 8 [C] 8.5 [D] 8.7
26. The median of the set of scores 65, 75, 55, 48, 78 is:
[A] 55 [B] 60 [C] 72 [D] 65
27. The median of the set of numbers 2.64, 2.50, 2.72, 2.91, 2.35 is:
[A] 2.72 [B] 2.64 [C] 2.50 [D] 2.35
28. Given the set of numbers 12,15,13,14,12 and 12. The median is:
[A] 12.5 [B] 12 [C] 13 [D] 13.5
29. The following table shows the age distribution of a group of children. Their median age is:
[A] 4 years [B] 7 years [C] 8 years [D] 9 years

Age (in years)	4	5	6	7	8	9	10
Frequency	2	1	2	4	3	6	2

30. The following table gives the marks scored by a group of students in a test. The median mark is:
[A] 4 [B] 3 [C] 2 [D] 1

Mark	0	1	2	3	4	5
Frequency	1	2	7	5	4	3

31. The mode of the numbers 8, 10, 9, 9, 10, 8, 11, 8, 10, 9, 8 and 14 is:
[A] 8 [B] 9 [C] 10 [D] 11
32. A group of students measured a certain angle (to the nearest degree) and obtained the following results.

75° 76° 72° 73° 74° 79° 72°
72° 77° 72° 71° 70° 78° 73°

The mode of their measurements is:

[A] 78° [B] 74° [C] 73° [D] 72°

33. The measure which is not a measure of dispersion is:
[A] Mode [B] mean deviation
[C] Inter-quartile range [D] standard deviation
34. It is true to say that the measure which is not measure of dispersion is:
[A] Range [B] Variance [C] Mode [D] Percentile range
35. The Variance of a given distribution is 25. The standard deviation is:
[A] 625 [B] 75 [C] 25 [D] 5
36. The standard deviation of the marks 2, 3, 6, 2, 5, 0, 4, 2 is:
[A] 1.5 [B] 1.7 [C] 1.8 [D] 1.9
37. The standard deviation of the numbers 2, 5, 6, 4 and 8 is:
[A] 2 [B] 4 [C] 6 [D] 7
38. The mode of the distribution in table below is:
[A] 2 [B] 3 [C] 4 [D] 5

Score	0	1	2	3	4	5
Frequency	2	3	4	2	7	2

39. The mean score of the distribution in table above is:
[A] 1.75 [B] 2 [C] 2.5 [D] 2.75
40. The median score of the distribution in table above is:
[A] 0 [B] 2.5 [C] 3 [D] 5
41. For a class of 30 students, the scores in a test out of 10 marks were as in the data below. The mode is:
[A] 3 [B] 5 [C] 6 [D] 7

4	5	7	2	3	6	5	5	8	9
5	4	2	3	7	9	8	7	7	7
3	4	5	5	2	3	6	7	7	2

42. For a class of 30 students, the scores in a test out of 10 marks were as in the data above. The median score is:
[A] 3 [B] 5 [C] 6 [D] 7
43. For a class of 30 students, the scores in a test out of 10 marks were as in the data above. The range of the distribution is:
[A] 7 [B] 2 [C] 8 [D] 9
44. The following table shows the tithes in thousand FCFA, collected in a church. The mode is:
[A] 3 [B] 6 [C] 9 [D] 12

Amount (thousand FCFA)	3	6	9	12	15	18
No. of Christians	3	9	6	15	3	12

45. The table above shows the tithes in thousand FCFA, collected in a church. The median of the distribution is:
[A] 3 [B] 9 [C] 12 [D] 15
46. The following table shows the frequency distribution of a number of chairs in each rooms of a hotel. The mean of the distribution is:

[A] 3.5 [B] 4.0 [C] 4.4 [D] 5.0

Number of chairs	1	2	3	4	5	6	7
Frequency	2	7	5	4	9	7	6

47. The table above shows the frequency distribution of a number of chairs in each rooms of a hotel. The mode of the distribution is:
[A] 2 [B] 5 [C] 7 [D] 9
48. The table above shows the frequency distribution of a number of chairs in each rooms of a hotel. The median of the distribution is:
[A] 4 [B] 4.5 [C] 5 [D] 5.5
49. The following table shows the frequency distribution of marks scored by a group of students in a test. The number of students who took the test is:
[A] 14 [B] 15 [C] 18 [D] 20

Marks	2	3	4	5	6
Frequency	2	4	5	3	1

50. The table above shows the frequency distribution of marks scored by a group of students in a test. The modal score is:
[A] 2 [B] 3 [C] 4 [D] 5
51. The table above shows the frequency distribution of marks scored by a group of students in a test. The mean mark is:
[A] 1.3 [B] 2 [C] 3 [D] 3.8
52. The following table shows the scores of 15 students in a physics test. The number of students who scored at least 5 is:
[A] 6 [B] 8 [C] 9 [D] 7

Marks	1	2	3	4	5	6	7	8	9	10
No. of students	1	3	2	0	1	6	1	0	1	0

53. The table above shows the scores of 15 students in a Physics test. The median score is:
[A] 5 [B] 6 [C] 7 [D] 8
54. The following table shows the scores of a group of 40 students in a Biology test. If the mode is m and the median is n then (m, n) as an ordered pair is:
[A] (5,5) [B] (5,6) [C] (6,5) [D] (9,4)

Score	1	2	3	4	5	6	7	8	9
Frequency	2	3	6	7	9	6	2	2	3

55. The table above shows the scores of a group of 40 students in a physics test. The mean of the distribution is:
[A] 4.5 [B] 4.8 [C] 5.0 [D] 5.2
56. The number of goals scored by a football team in 20 matches is shown in the table below. The mean number of goals scored is:
[A] 1.75 [B] 1.9 [C] 2 [D] 2.15

Number of goals	0	1	2	3	4	5
Number of matches	3	5	7	4	1	0

57. The number of goals scored by a football team in 20 matches is shown in the table above. The modal number of goals scored is:
[A] 1 [B] 2 [C] 5 [D] 7

58. The distribution by Region of 840 students in the faculty of science of the University of Buea in a certain session is as follows:

Adamawa Region	45
North West Region	410
Littoral Region	105
Western Region	126
South West Region	154

In a pie chart drawn to represent this distribution, the angle subtended by Western Region is:
[A] 42° [B] 45° [C] 48° [D] 54°

59. The pie chart below, illustrates the amount of private time a student spends in a week studying various subjects. The value of k is:
[A] 15° [B] 30° [C] 60° [D] 90°

60. The pie chart below, illustrates the amount of private time a student spends in a week studying various subjects. Given that he spends 2 and a half hours on science, the total number of hours he studies in a week is:
[A] $3\frac{1}{2}$ [B] 5 [C] 8 [D] 12

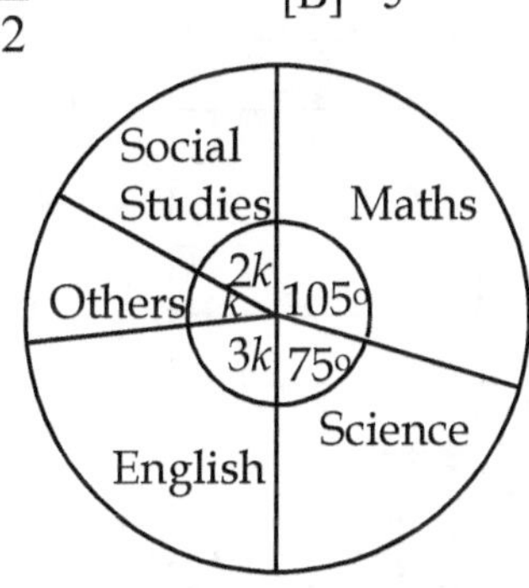

61. The pie chart below represents the number of fruits on display in a grocery shop. Given that there are 60 oranges in display, the number of apples is:

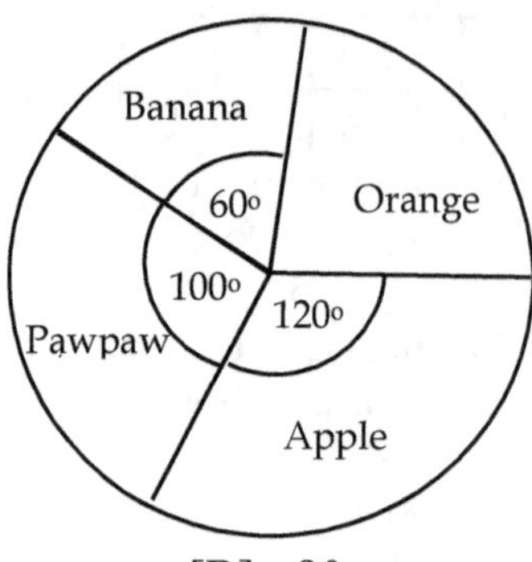

[A] 40 [B] 80 [C] 90 [D] 120

62. The marks obtained by pupils of a certain class are grouped as shown below; 0-4, 5-9, 10-14, 15-19. It is true to say that:
I: The mid values of the grouped marks are 2,7,12, and 17.
II: The class interval is 4.
III: The class boundaries are 0.5, 4.5, 9.5, 14.5 and 19.5.
[A] I only [B] II only [C] III only [D] I and II

63. The histogram below shows the number of candidates, in thousands who obtained given ranges of marks in an entrance examination. The total number of candidates who sat for the examination is:
[A] 120,000 [B] 250,000 [C] 260,000 [D] 270,000

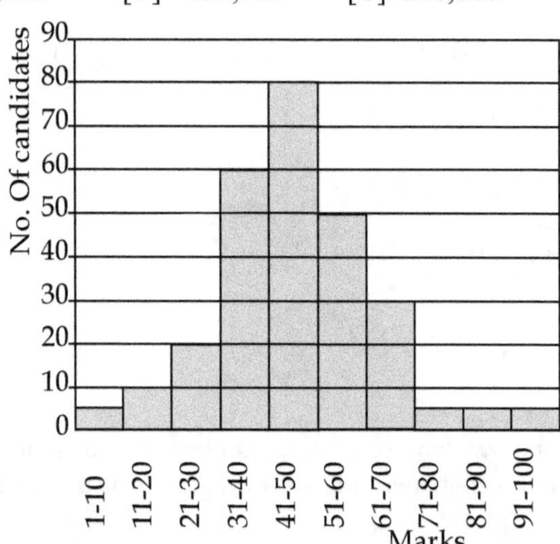

64. The histogram above shows the number of candidates, in thousands who obtained given ranges of marks in an entrance examination. The number of candidates who scored at most 30% is:
[A] 20,000 [B] 25,000 [C] 35,000 [D] 60,000

65. The histogram below shows the distribution of a group of students according to their ages. The range of their ages is:

[A] 14 years [B] 20 years [C] 30.5 years [D] 31 years

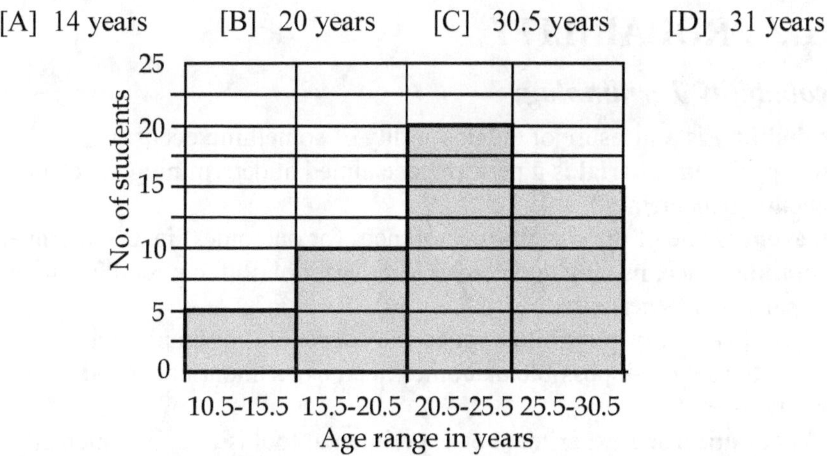

66. The histogram above shows the distribution of a group of students according to their ages. The mode of their ages is:
    [A] 22.5 years [B] 23.0 years [C] 24.0 years [D] 24.5 years
67. For six sequences, Ngange scored 76, 57, 97, 86, 86, 70 in Mathematics. If Ngange wants to convince his parents of his strength in Mathematics the measure he should use should be:
    [A] Mean [B] median [C] mode [D] range
68. Six employees earn 800 FCFA, 850 FCFA, 900 FCFA, 950 FCFA, 1000 FCFA, and 2350 per hour. The manager claims that the median of the hourly wages is 925 FCFA. The manager is:
    [A] wrong because 925 FCFA is the mode.
    [B] wrong because he seems to ignore the amount 2350 FCFA.
    [C] wrong because 925 FCFA is the mean.
    [D] right because 925 FCFA is the correct median.
69. The president of a certain Credit Union used the data in the following table to find the mean monthly salary of the Credit Union staff.

Monthly Salary	No. of workers
26,000 FCFA	7
30,000 FCFA	8
240,000 FCFA	1
260,000 FCFA	1
300,000 FCFA	3

In a report he stated, "The typical salary at the Credit Union is about 92,000 FCFA." His statement is:
[A] misleading because salaries of five staff are far above those of the other fifteen.
[B] misleading because the mean of the data is not 92,000 FCFA.
[C] misleading because the president ignored the highest salary.
[D] not misleading.

11.2 PROBABILITY

Probability Terminology

Probability is a measure of the possibility of something occurring.
An **experiment** or **trial** is a performance aimed at determining the chance of something occurring.
An **event** is one of the possible occurrences (or outcomes) in an experiment.
A **sample space**, usually denoted by S is the set of all the possible outcomes in a particular experiment.
An **event subset** or **possibility space** is a subset of the sample space which defines the set of all possible outcomes in an experiment under specified conditions.
A **fair** or **unbiased** experiment or experimental tool is one for which all events have equal chances of occurrence.
An **unfair** or **biased** experiment or experimental tool is one for which the events have unequal chances of occurrence.
Equiprobable or **equally likely** events are events which have equal chances of occurrence.
At most means "less than or equal to a given number".
At least means "greater than or equal to a given number".

Probability as a Number

$0 \leq P(E) \leq 1$

$P(A) = 1 \Leftrightarrow$ event A must occur.

$P(B) = 0 \Leftrightarrow$ event B cannot occur.

If there are n equally likely events, the probability $P(E)$ of one of the events occurring is given by $P(E) = \frac{1}{n}$.

Given a sample space S consisting of a finite number of equiprobable outcomes, then the probability of an event E occurring is defined as:

$$\text{Probability of } E = \frac{\text{Number of outcomes in the event subset } E}{\text{Total number of outcomes}}$$

i.e. $P(E) = \dfrac{n(E)}{n(S)}$

Example

A card is picked at random from a well shuffled pack of 52 playing cards. What is the probability that it is (i) an Ace of heart (ii) a king

Solution

$$n(S) = 52$$

(i) $n(\text{Ace of hearts}) = 1 \Rightarrow P(\text{Ace of hearts}) = \frac{n(\text{Ace of hearts})}{n(\text{S})} = \frac{1}{52}$

(ii) $n(\text{Kings}) = 4 \Rightarrow P(\text{Kings}) = \frac{n(\text{Kings})}{n(\text{S})} = \frac{4}{52} = \frac{1}{13}$

Complementary Events

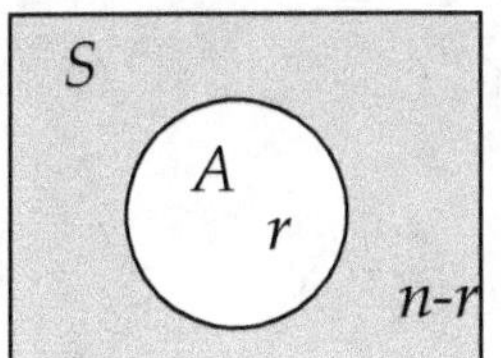

$n(A') = n(s) - n(A)$

$n(A') = n - r$

$P(A) + P(A') = 1 \Leftrightarrow P(A') = 1 - P(A)$

Example

A fair die is tossed find the probability that a 2 will not be obtained.

Solution

Let the event of obtaining a two be T, and the event of not obtaining a two be T'.

$P(T) = \frac{1}{6}$ and $P(T') = 1 - P(T) \Rightarrow P(T') = 1 - \frac{1}{6} = \frac{5}{6}$

Addition Laws of Probability (*P* (*or*))

Intersecting Events

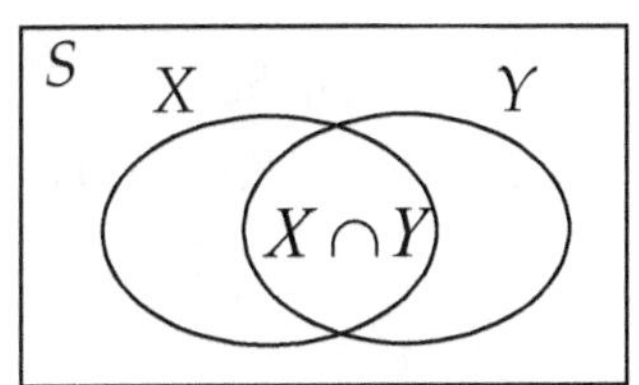

$P(X \text{ or } Y) = P(X \cup Y) = P(X) + P(Y) - P(X \cap Y)$

Example

If X and Y are two events such that $P(X \cap Y) = \frac{1}{9}$, $P(X) = \frac{1}{3}$ and

$P(X \cup Y) = \frac{4}{9}$. Find $P(Y)$.

Solution

$$P(X \cup Y) = P(X) + P(Y) - P(X \cap Y)$$
$$\Rightarrow P(Y) = P(X \cup Y) + P(X \cap Y) - P(X) = \frac{4}{9} + \frac{1}{9} - \frac{3}{9} = \frac{2}{9}$$

Mutually Exclusive Events

For two mutually exclusive events $P(X \text{ or } Y) = P(X \cup Y) = P(X) + P(Y)$

In general for n mutually exclusive events $E_1, E_2, \ldots, E_n$,

$$P(E_1 \cup E_2 \cup \cdots \cup E_n) = P(E_1) + P(E_2) + \cdots P(E_n)$$

Example

A coin is tossed twice. Find the probability that the result is either two tails or a tail and a head respectively.

Solution

$$P(H) = \frac{1}{2}, P(T) = \frac{1}{2}$$
$$P(TT) = \frac{1}{2} \times \frac{1}{2} = \frac{1}{4}, \quad P(HT) = \frac{1}{2} \times \frac{1}{2} = \frac{1}{4}, \quad P(TH) = \frac{1}{2} \times \frac{1}{2} = \frac{1}{4}$$
$$P(\text{two tails or a tail and a head}) = P(TT) + P(TH) + P(HT)$$
$$= \frac{1}{4} + \frac{1}{4} + \frac{1}{4} = \frac{3}{4}$$

Dependent Events

If two events X and Y are such that $P(X) \neq 0$ and $P(Y) \neq 0$, then the probability of X given that Y has already occurred is denoted by $P(X/Y)$ and is given by

$$P(X/Y) = \frac{n(X \cap Y)}{n(Y)} = \frac{P(X \cap Y)}{P(Y)} \Rightarrow P(X \cap Y) = P(Y).P(X/Y)$$

This is known as the multiplication law for dependent events.

Example
Given that a heart card is picked at random from a pack of 52 playing cards, find the probability that the next card chosen is a picture card.

Solution
$P(H) = \frac{13}{52}, P(P \cap H) = \frac{1}{52}$

$$P(P/H) = \frac{P(P \cap H)}{P(H)} = \frac{1}{52} \div \frac{13}{52} = \frac{1}{52} \times \frac{52}{13} = \frac{1}{13}$$

Independent Events

Two events X and Y are independent $\Rightarrow$ $P(X \cap Y) = P(X).P(Y)$

In general for n independent events

$$P(E_1 \cap E_2 \cap \cdots \cap E_n) = P(E_1).P(E_2).\cdots.P(E_n)$$

Example
A die and a coin are thrown in succession. Find the probability that a head is shown on the die and a 2 is shown on the coin.

Solution

$$P(2 \cap H) = P(2) \cdot P(H) = \frac{1}{6}.\frac{1}{2} = \frac{1}{12}$$

Conditional Probability

Drawing With and Without Replacement

Example
A bag contains 10 balls, 7 of which are red and 3 of which are white.
(a) What is the probability that a ball chosen at random from the bag is
(i) Red? (ii) White?

Suppose that the ball chosen is red and is not replaced.

(b) What is the probability that the second ball chosen at random from the bag is (i) Red? (ii) White?

Suppose that the secnd ball chosen is red again and is not replaced but 3 white balls are added.

(c) What is the probability that the third ball chosen at random from the bag is
(i) Red? (ii) White?

Solution
Let the events "choosing a red ball" and "choosing a white ball" be R and W respectively. Then,

(a) (i) $P(R) = \frac{7}{10}$ (ii) $P(W) = \frac{3}{10}$

(b) (i) $P(R) = \frac{6}{9} = \frac{2}{3}$ (ii) $P(W) = \frac{3}{9} = \frac{1}{3}$

(c) (i) $P(R) = \frac{5}{11}$ (ii) $P(W) = \frac{6}{11}$

Repeated Trials

When an experiment is repeated and the conditions are unaltered the events can be considered independent.

Example
A biased coin is tossed five times. If $P(H) = 2P(T)$, find the probability of having a head, a head, a tail, a head, and a tail in that order.

Solution

$P(H) + P(T) = 1$ and $P(H) = 2P(T)$

$\Rightarrow 2P(T) + P(T) = 1 \Rightarrow P(T) = \dfrac{1}{3}$

$$\begin{aligned} P(HHTHT) &= P(H).P(H).P(T).P(H).P(T) \\ &= 2P(T).2P(T).P(T).2P(T).P(T) \\ &= 8P(T)^5 = 8\left(\tfrac{1}{3}\right)^5 = \tfrac{8}{243} \end{aligned}$$

Tree Diagrams

Example
A coin is tossed 4 times. Find the probability of obtaining 3 heads and one tail.

Solution
By drawing the tree diagram as shown on the next page, we can obtain the probability of 3 heads and one tail as follows.

Probability of 3 heads and one tail

$$\begin{aligned} &= P(HHHT) + P(HTHH) + P(HHTH) + P(HHHT) \\ &= \tfrac{1}{2}\left(\tfrac{1}{2}\right)\left(\tfrac{1}{2}\right)\left(\tfrac{1}{2}\right) + \tfrac{1}{2}\left(\tfrac{1}{2}\right)\left(\tfrac{1}{2}\right)\left(\tfrac{1}{2}\right) + \tfrac{1}{2}\left(\tfrac{1}{2}\right)\left(\tfrac{1}{2}\right)\left(\tfrac{1}{2}\right) + \tfrac{1}{2}\left(\tfrac{1}{2}\right)\left(\tfrac{1}{2}\right)\left(\tfrac{1}{2}\right) \\ &= \frac{1}{16} + \frac{1}{16} + \frac{1}{16} + \frac{1}{16} \end{aligned}$$

$\therefore$ Probability of 3 heads and 1 tail $= 4\left(\frac{1}{16}\right) = \frac{1}{4}$.

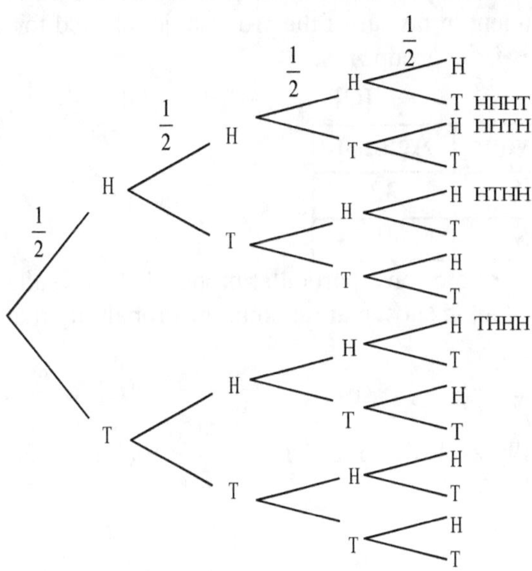

MULTIPLE CHOICE EXERCISE 11:2

1. There are m boys and 12 girls in class. The probability of selecting at random a girl from the class is:
   [A] $\frac{m}{12}$ [B] $\frac{12}{m}$ [C] $\frac{12}{m+12}$ [D] $\frac{m}{m+12}$
2. The following table gives the marks scored by a group of students in a test. The probability of selecting a student from the group that scored 2 or 3 is:
   [A] $\frac{1}{11}$ [B] $\frac{5}{25}$ [C] $\frac{7}{22}$ [D] $\frac{6}{11}$

Mark	0	1	2	3	4	5
Frequency	1	2	7	5	4	3

3. The probability of having an odd number in a single toss of a fair die is:
   [A] $\frac{2}{3}$ [B] $\frac{1}{6}$ [C] $\frac{1}{3}$ [D] $\frac{1}{2}$
4. The following table gives the scores of a group of students in an English Language test. A student is chosen at random from the group. The probability that he scored at least 6 marks is:
   [A] $\frac{3}{4}$ [B] $\frac{1}{5}$ [C] $\frac{1}{4}$ [D] $\frac{3}{10}$

Score	2	3	4	5	6	7
Number of students	2	4	7	2	3	2

5. Two groups of students cast their votes on a particular proposal. The results are as in the table below. A student in favour of the proposal is selected for a post, the probability that he is from group A is:

   [A] $\frac{8}{9}$ [B] $\frac{16}{35}$ [C] $\frac{4}{5}$ [D] $\frac{4}{7}$

	In favour	Against
Group A	128	32
Group B	96	48

6. Two groups of students cast their votes on a particular proposal. The results are as in the table above. A student is chosen at random, the probability that he is against the proposal is:

   [A] $\frac{3}{19}$ [B] $\frac{4}{19}$ [C] $\frac{5}{19}$ [D] $\frac{9}{19}$

7. The events X and Y are mutually exclusive and $P(X) = \frac{1}{3}$, $P(Y) = \frac{2}{5}$.

   $P(X \cap Y)$ is:

   [A] 0 [B] $\frac{2}{15}$ [C] $\frac{4}{15}$ [D] $\frac{11}{15}$

8. The events X and Y are mutually exclusive and $P(X) = \frac{1}{3}$, $P(Y) = \frac{2}{5}$.

   $P(X \cup Y)$ is:

   [A] 0 [B] $\frac{2}{15}$ [C] $\frac{4}{15}$ [D] $\frac{11}{15}$

9. A box contains 2 white and 3 blue identical marbles. Two marbles are picked at random one after the other without replacement. The probability of picking two marbles of different colours is:

   [A] $\frac{2}{3}$ [B] $\frac{3}{5}$ [C] $\frac{2}{5}$ [D] $\frac{3}{10}$

10. Mrs. Ngala is expecting a baby. The probability of a boy is $\frac{1}{2}$ and the probability that the baby will have blue eyes is $\frac{1}{4}$. The probability that she will have a blue-eyed boy is:

    [A] $\frac{1}{8}$ [B] $\frac{1}{4}$ [C] $\frac{3}{8}$ [D] $\frac{3}{4}$

11. A number is chosen at random from the set $\{1, 2, 3, ..., 9, 10\}$. The probability that the number is greater than or equal to 7 is:

    [A] $\frac{1}{10}$ [B] $\frac{3}{10}$ [C] $\frac{2}{5}$ [D] $\frac{3}{5}$

12. The probability of throwing a number greater than 2 with a single fair die is:

[A] $\frac{1}{6}$ [B] $\frac{1}{3}$ [C] $\frac{1}{2}$ [D] $\frac{2}{3}$

13. A fair die is rolled once. The probability of obtaining 4 or 6 is:

[A] $\frac{2}{3}$ [B] $\frac{1}{6}$ [C] $\frac{1}{3}$ [D] $\frac{1}{2}$

14. Three balls are drawn one after the other with replacement, from a bag containing 5 red, 9 white and 4 blue identical balls. The probability that they are one red, one white and one blue is:

[A] $\frac{5}{81}$ [B] $\frac{5}{27}$ [C] $\frac{5}{162}$ [D] $\frac{5}{243}$

15. The probability that an integer selected from the set of integers {20,21,...,30} is a prime number is:

[A] $\frac{2}{11}$ [B] $\frac{5}{11}$ [C] $\frac{6}{11}$ [D] $\frac{9}{11}$

16. A fair die is rolled once. The probability of obtaining a number less than 3 is:

[A] $\frac{1}{6}$ [B] $\frac{1}{3}$ [C] $\frac{1}{2}$ [D] $\frac{2}{3}$

17-18 The data below shows the number of workers employed in the various sections of a construction company in Yaounde. Use the information to answer questions 17 to 18. 24 Carpenters, 27 Labourers, 12 Plumbers, 15 Plasterers, 9 Painters, 3 Messengers and 18 Bricklayers.

17. One of the workers is absent on a day. The probability that he is a bricklayer is:

[A] $\frac{1}{9}$ [B] $\frac{2}{9}$ [C] $\frac{1}{6}$ [D] $\frac{1}{4}$

18. A worker is retrenched. The probability that he is a plumber or a plasterer is:

[A] $\frac{3}{4}$ [B] $\frac{1}{9}$ [C] $\frac{5}{36}$ [D] $\frac{1}{4}$

19. The probability that a total sum of seven would appear with two tosses of a fair die is:

[A] $\frac{5}{36}$ [B] $\frac{1}{6}$ [C] $\frac{1}{36}$ [D] $\frac{5}{6}$

20. A die is rolled 200 times. The outcomes obtained are shown in the following table. The probability of obtaining a 2 is:

[A] 0.002 [B] 0.015 [C] 0.15 [D] 0.16

21. A die is rolled 200 times. The outcomes obtained are shown in the following table. The probability of obtaining a number less than 3 is:

[A] 0.125 [B] 0.150 [C] 0.275 [D] 0.500

Number	1	2	3	4	5	6
Number of times	25	30	45	28	40	32

22. Two cards are drawn one after the other with replacement from a well shuffled ordinary deck of 52 cards containing four aces. The probability that they are both aces is:

[A] $\frac{1}{13}$ [B] $\frac{1}{169}$ [C] $\frac{1}{52}$ [D] $\frac{1}{26}$

23. The probability that a number selected from the numbers 30 to 50 inclusive is a prime is:

[A] $\frac{1}{4}$ [B] $\frac{5}{21}$ [C] $\frac{3}{7}$ [D] $\frac{1}{3}$

24. Two fair dice are tossed together once. The probability that the sum of the outcome is at least ten is:

[A] $\frac{1}{12}$ [B] $\frac{5}{36}$ [C] $\frac{1}{6}$ [D] $\frac{5}{18}$

25. From a box containing 2 red, 6 white and 5 blackballs; a ball is randomly selected. The probability that the ball selected is black is:

[A] $\frac{2}{13}$ [B] $\frac{5}{13}$ [C] $\frac{5}{11}$ [D] $\frac{11}{13}$

26. A bag contains 3 red, 4 black and 5 green identical balls. Two balls are picked at random one after the other without replacement. The probability that one is red and the other is green is:

[A] $\frac{5}{48}$ [B] $\frac{5}{11}$ [C] $\frac{5}{22}$ [D] $\frac{5}{44}$

27. The following table gives the distribution of outcomes obtained when a die was roll 100 times. The experimental probability that it shows at most 4 when rolled again is:

[A] $\frac{8}{25}$ [B] $\frac{12}{25}$ [C] $\frac{13}{25}$ [D] $\frac{17}{25}$

Number of die	1	2	3	4	5	6
Frequency	18	14	20	16	15	17

28. A bag contains red, black, and green identical balls. A ball is picked and replaced. The following table shows the result of 100 trials. The experimental probability of picking a green ball is:

Colour of ball	red	black	green
Number of occurrences	54	30	16

[A] $\frac{4}{25}$ [B] $\frac{21}{25}$ [C] $\frac{1}{3}$ [D] $\frac{4}{21}$

29. A box contains 2 white and 3 blue identical balls. 2 balls are picked at random one after the other, without replacement. The probability of picking 2 balls of different colours is:

[A] $\frac{6}{25}$ [B] $\frac{7}{20}$ [C] $\frac{3}{5}$ [D] $\frac{2}{3}$

30. A group of eleven people can speak either English or French or both. Seven can speak English and six can speak French. The probability that a person chosen at random can speak both English and French is:

[A] $\frac{2}{11}$ [B] $\frac{4}{11}$ [C] $\frac{5}{11}$ [D] $\frac{11}{13}$

31. The probabilities that Awah and Suh pass an examination are $\frac{3}{4}$ and $\frac{3}{5}$ respectively. The probability of both boys failing the examination is:

[A] $\frac{2}{3}$ [B] $\frac{3}{10}$ [C] $\frac{9}{10}$ [D] $\frac{1}{10}$

32. A box contains 5 red, 3 green and 4 blue balls. A boy is allowed to take away two balls at random from the box. The probability that the two balls are red is:

[A] $\frac{5}{33}$ [B] $\frac{5}{36}$ [C] $\frac{103}{132}$ [D] $\frac{31}{36}$

33. A box contains 5 red, 3 green and 4 blue balls. A boy is allowed to take away two balls at random from the box. The probability that one is green and the other is blue is:

[A] $\frac{2}{11}$ [B] $\frac{5}{12}$ [C] $\frac{8}{12}$ [D] $\frac{7}{11}$

34. The probability that an event will occur is p and the probability that it will not occur is q. The true assertion is:

[A] $p-q=1$ [B] $q-p=0$ [C] $p+q=1$ [D] $p+q=0$

35. A number is selected at random from the set $Y=\{18, 19, 20,\ldots,28, 29\}$. The probability that the number is a prime number is:

[A] $\frac{1}{4}$ [B] $\frac{3}{11}$ [C] $\frac{1}{2}$ [D] $\frac{3}{4}$

36. The numbers of goals scored by a school team in 10 netball matches are: 3,5,7,7,8,8,8,11,11,12. The probability that in a match, the school team will score at most 8 goals is:

[A] $\frac{1}{5}$ [B] $\frac{2}{5}$ [C] $\frac{3}{5}$ [D] $\frac{7}{10}$

37. The probability that a number chosen at random from $\{2, 3, 4 \ldots 9, 10\}$ is either a prime number or a multiple of 3 is:

[A] $\frac{5}{9}$ [B] $\frac{2}{3}$ [C] $\frac{6}{7}$ [D] $\frac{5}{7}$

38. The probabilities that Ade and his dog will be alive in 10 years time are 0.8 and 0.3 respectively. The probability that they will both be alive in 10 years time is:

[A] 1.00 [B] 0.50 [C] 0.24 [D] 0.06

39. The probabilities of Fru and Nsang passing an examination are $\frac{3}{4}$ and $\frac{5}{8}$ respectively. The probability that the two boys fail the examination is:

[A] $\frac{3}{32}$ [B] $\frac{3}{8}$ [C] $\frac{15}{32}$ [D] $\frac{5}{8}$

40. Two beads are drawn at random, one after the other with replacement, from a box containing 5 red and 7 white identical beads. The probability that the beads are the same colour is:

[A] $\frac{119}{144}$ [B] $\frac{95}{144}$ [C] $\frac{37}{72}$ [D] $\frac{48}{144}$

41. The probability that John passes the GCE is $\frac{2}{3}$ and the probability that Paul fails the same exam is $\frac{1}{4}$. The probability that both John and Paul pass is:

[A] $\frac{1}{6}$ [B] $\frac{1}{2}$ [C] $\frac{3}{4}$ [D] 1

42. Given that a coin is tossed twice, then the probability of obtaining exactly one head is:

[A] $\frac{1}{4}$ [B] $\frac{1}{2}$ [C] $\frac{3}{4}$ [D] 1